THE BIOLOGICAL CHEMISTRY OF THE ELEMENTS

The Biological Chemistry of the Elements

The Inorganic Chemistry of Life

J. J. R. FRAÚSTO da SILVA

Professor of Analytical Chemistry, Instituto Superior Técnico,
Universidade Técnica de Lisboa

and

R. J. P. WILLIAMS

Royal Society Napier Research Professor,
University of Oxford

CLARENDON PRESS · OXFORD
1991

Oxford University Press, Walton Street, Oxford OX2 6DP

Oxford New York Toronto
Delhi Bombay Calcutta Madras Karachi
Petaling Jaya Singapore Hong Kong Tokyo
Nairobi Dar es Salaam Cape Town
Melbourne Auckland

and associated companies in
Berlin Ibadan

Oxford is a trade mark of Oxford University Press

Published in the United States
by Oxford University Press, New York

A catalogue record for this book is available from the British Library

Library of Congress Cataloging in Publication Data
da Silva, J. J. R. Fraústo.
The biological chemistry of the elements: the inorganic chemistry
of life/J. J. R. Fraústo da Silva and R. J. P. Williams.
1. Bioinorganic chemistry. I. Williams, R. J. P. [Robert Joseph Paton]
II. Title.
QP531.S54 1991 574.19'214—dc20 91-11585 CIP
ISBN 0-19-855598-9

Typeset by Cotswold Typesetting Ltd, Gloucester
Printed in Great Britain by
Butler & Tanner Ltd, Frome, Somerset

Preface

The majority of present-day living organisms require rather strict environmental conditions: temperature from 0 to about 40°C, pressure of the order of 1 atm., salinity up to about 4 per cent, pH in the range 4 to 9, redox potential in the region of stability of water (-0.4 to $+0.8$ V at pH$=7$ versus the standard H^+/H_2 electrode), water activity between 0.7 and 1.0 mole fraction. In a few cases these limits can be greatly exceeded and it is known that, in the earth's past, conditions existed which were quite different from the present ones but did not hinder the development of life, although they did impose constraints upon the *forms* of life that could exist. Indeed, this is an essential feature of life itself: the continuous adaptation of living organisms to changing environmental conditions through the development of defence devices and mechanisms against those changes, while sometimes turning new (adverse) factors into favourable ones. The result was, and still is, the survival and predominant proliferation of the fittest, and the history of evolution, which started as development under constraints of fixed conditions, became the history of the continuous adaptation of living organisms to changing conditions. In fact, and as we shall see, organisms themselves have caused much of the chemical environmental change.

Although ideas concerning the evolution of species and forms are well accepted, they are less recognized when transferred to the small molecule chemistry scale, even though it has become customary to call 'chemical evolution' the series of events which is thought to have taken place between the beginnings of prebiotic synthesis of large organic molecules and the appearance of the first vesicles ('protocells') to which one might assign some 'vital' properties. Throughout the (unfinished) biological evolution there has been, in addition, a continual selection toward the best chemical components for the necessary functions and a continuous improvement in the conditions of their utilization, leading to the formation of separate compartments in cells, to the functional separation of cells, to the specialization of different multicellular organisms and to their diversification and development. As the species became more complex, so did their chemistry, as evidenced by the increased complexity of the materials used and the manipulation of dynamic factors. The development of materials is again well recognized in the evolution of biological polymers but it can only be thoroughly appreciated, we believe, in terms of the structures and patterns of the flow of elements and chemicals, of ionic and electronic charge, and of energy directed in biological space. The strict environmental conditions that are required today by living organisms also point to the necessity of strict internal control and regulation of these processes, and this again means time and space organization of the biological substances, both large and small, and of their reactions. Much of this control is included in the now well-known classical treatment of the organic chemistry of

life but, diverging from the classical tradition, we shall regard it in this book as the interplay of not just the great number of small and very large organic molecules but also of a group of about 25 chemical elements, free as ions, combined as complexes, or as precipitates, organized in biological space and time. Most of these elements are classified as 'inorganic'. We shall go on to insist that a proper appreciation of living systems and their evolution cannot be obtained without reference to these elements.

Central to this group of elements is a series of metals, each one of them used in many ways within a given biological system. The position of each metal element in space and the variation of this position in time, as well as the exact chemical partners with which it interacts, have been selected and perfected by the demands of evolution seemingly in such a way that the element's function has become virtually optimal and its role separate and distinct. Hence, several such elements are essential for life. In order to fully appreciate the subtleties of the function that Nature makes of these $ca.$ 15 metal elements, it is necessary to understand how they are selected, in terms of uptake, and how they are combined, transported, stored, and tuned to their specific functions. Evolution refined function by acting on the interaction of the metals with their partners (usually proteins) or by abandoning some in favour of others more appropriate to conditions. Much of this chemistry involves ionic interactions.

Of course, in the description of the way in which the chemical elements are involved in life we cannot concentrate on metal ions alone. Some non-metals actually function in ways not too different from those of the metals, but as anions not cations, e.g. Cl^-. Others, the major elements of life, C, N, O, and S, to which we refer to rather a small degree since their biological chemistry is well described elsewhere, behave very differently, much as their chemistry is also very different, being loosely called covalent rather than ionic. Between the extremes lies the chemistry of phosphorus, as phosphate in biology, and hydrogen, which have both ionic and covalent chemistry. Our point will be that all the elements we find in biology are used in some optimal manner developed from their chemical possibilities. Biological chemistry is concerned, therefore, with this intricate, interwoven, 'designed', and sometimes symbiotic use of organic (non-metal) molecules and inorganic (metal) chemical elements in cells and tissues and cannot be restricted or related to so-called organic chemistry rather than to so-called inorganic chemistry. This book is then about biological chemistry seen from a particular inorganic perspective where a certain inorganic *bias* is inevitable to compensate for the traditional approach which leans heavily on organic chemistry. Accordingly, we have named it *The biological chemistry of the elements*.

Before entering the main body of the text, the reader is warned that every biological device is unlikely to be the best of all possible devices. Not only has biology general chemical limitations, but it is also evolving in changing conditions, many of which are created by life itself. A consequence of this slow progression is that now and then a less than perfect feature will survive, which has evolved from a committed route earlier in evolution and then found a new semi-effective use. It also follows that, should environmental conditions, the air, the sea, or the land, change very rapidly, complex life forms are at risk. There is a very general warning here, as man's activities saturate the ability of the environment to buffer the chemical changes he makes.

The book is organized in the following manner. In Chapter 1 we analyse the occurrence of the chemical elements in biology showing that of the 92 elements in the periodic table some 25–30 are especially important in biology. Much of the reason behind this selection lies in the limitations imposed by the abundance of the elements (even in the universe) and their availability on earth. Availability is further constrained by the concentration in waters and in the air and depends on local conditions. In the absence of organic chemicals most elements are bound in simple oxygen- or sulphur-containing compounds, as hydrates or precipitates. On entering biology the elements undergo special combinations with organic molecules. In the next five chapters we shall be considering this chemical speciation in a great variety of compounds in biology. We shall describe in the first instance in Chapter 2 how the biological uptake of the elements from natural sources can be controlled by selective (energized) chemistry of the elements and leads on to selective chemical combinations, e.g. of metals, M, with organic ligands, L. In Chapter 3 we extend the discussion to physical transfer and separation of M, L, or ML across membranes into compartments. Only at this stage will it be possible to start to explore function. Function is dependent upon the ability of elements to undergo chemical change (within compounds). The kinetics of activity will therefore be analysed in Chapter 4. Chapter 5 will illustrate how energy is put into compounds within biology. This is an obvious necessity both in synthesis and in transfer. We shall then turn in Chapter 6 to one major concern of this book—the functionally necessary requirement for life of a variety of elements. Chapter 7 will be a rather lengthy but necessary interjection concerning the biopolymers of life. The second part of the book, Chapters 8 to 19, will consider the essential functions of individual inorganic elements in life's processes in detail and especially in relation to the principles we outline in the first part, before we attempt to pull all the (inorganic) chemistry of life together, in the last part of the book.

We begin the difficult final analysis with an enquiry into the nature of minerals in biology (Chapter 20), since this allows us to look at a peculiar biological problem, the relationship between chemical activity and morphology. Shape is so clearly an essential feature of living things. Having established some simple principles, we move in Chapter 21 to a consideration of the simultaneous homeostasis of shape and chemistry. It is here that the interactive feedback relationships between the functions of elements become so intriguing. It almost goes without saying at this stage that the balance of elements in life is related to the elements available and it follows that changing the availability must affect life. We look at problems of man's interference with the natural supply of elements in Chapter 22.

Each chapter has only a limited number of references and most of them are to reviews. We realize that this approach may offend some but we found it impossible to be sure that we could give proper priority amongst the vast number of primary sources we have used.

Lisbon and Oxford JJRFdS
January 1991 RJPW

Acknowledgements

Our special thanks go to Mrs S. Compton and Mrs J. Edwards who typed the manuscript and to Oxford University Press for their assistance in its final preparation.

The work reviewed here reflects in large part the efforts of a now very distinguished group of research scientists from all over the world, who have been with us. We wish to thank them for all that they brought to the research. We trust that the integrated view in this book will be seen as the result of an international effort to examine a further part of the wonders of chemistry within biology.

J. J. R. Fraústo da Silva acknowledges a travel grant for the preparation of this book from the Junta Nacional de Investigação Cientifica e Tecnologica (JNICT)—Portugal. Acknowledgements are also due to the Instituto Nacional de Investigação Cientifica (INIC)—Portugal for general support of research activities, to Mrs Teresa Maria Carreiras da Silva and Mrs Maria Idalisa Figueiredo dos Santos for secretarial help and to Mrs Ilda Proença for typing parts of the preliminary versions of the manuscript.

R. J. P. Williams acknowledges the generous support of Wadham College, Oxford, of Oxford University, of two British Research Councils (the Medical Research Council and the Science and Engineering Research Council), and of The Royal Society over some forty years.

Contents

PART I · The chemical and physical factors controlling the elements of life

1 The chemical elements in biology 3
 1.1 The natural selection of the elements 3
 1.1.1 Abundance in living systems 3
 1.1.2 Chemical speciation: limitations of life chemistry and physics 6
 1.2 An aside: some biological chemistry of hydrogen 7
 1.3 The economical use of resources: abundance and availability 8
 1.4 Biological environment and element availability 11
 1.4.1 pH and redox potential considerations 11
 1.4.2 Element concentrations in aerobic environments and in some organisms 14
 1.4.3 Solubility products 18
 1.5 Sulphide chemistry in water (anaerobes) 18
 1.6 A note on homeostasis 20
 1.7 Conclusion 22

2 The principles of the uptake and chemical speciation of the elements in biology 23
 2.1 General aspects 24
 2.2 The separations that biology can achieve 26
 2.3 The selective uptake of metal ions 28
 2.4 Effective stability constants 31
 2.5 Element uptake at 'equilibrium'—the selectivity of the uptake 33
 2.5.1 Selection by charge type 36
 2.5.2 Selection by ion size 37
 2.5.3 Combination of radius and charge effect 40
 2.5.4 Selection by liganding atom 40
 2.5.5 Selection by preferential co-ordination geometry 41
 2.5.6 Selection by spin-pairing stabilization 43
 2.5.7 Selection by binding in clusters 44
 2.6 Hydrolysis 45
 2.7 Selective control of oxidation states of metals 47
 2.8 Selection by control of concentration by buffering 48
 2.9 Selection by transfer coefficients from water to proteins 50
 2.10 Soft and hard acids, solvents, and extraction 52

2.11 The selectivity of channels 53
2.12 Kinetic effects and control 54
 2.12.1 The distribution coefficient D for ML_1 54
 2.12.2 Kinetics of transport 54
 2.12.3 Insertion in pre-formed holes or chelating rings 55
 2.12.4 Energy coupling to the selective movement of M (gates and pumps) 56
 2.12.5 A summary of the kinetics of uptake of metal ions 57
2.13 Rejection of metal ions 58
2.14 General aspects of the uptake of non-metals 58
2.15 Mechanisms of selection of anions based on thermodynamic properties 60
 2.15.1 Selection by charge and size; the hydration of anions and their binding to proteins 60
 2.15.2 Selection of anions by differences in binding affinity for different types of cationic centres 62
 2.15.3 Kinetic binding traps for anions with or without accompanying redox reactions 63
2.16 Redox incorporation of anions 64
2.17 Coenzymes 65
2.18 Concluding remarks: element handling in biological systems 67
2.19 Precipitates in compartments 68

3 Physical separations of elements: compartments and zones in biology 71
3.1 General aspects 71
3.2 The nature of compartments 73
3.3 The chemical solutions and physical states of compartments 73
3.4 The role and nature of membranes 76
3.5 Different types of membrane 77
 3.5.1 Internal and external particles and their relationship to membranes 78
 3.5.2 Lateral membrane organization and in-membrane flow 79
 3.5.3 Transmembrane organization and flow 82
 3.5.4 Flow of vesicles in cells 83
3.6 Special solution conditions in vesicles 86
3.7 Co-operativity of separations and localizations 87
3.8 Summary of metal ion positioning 87
3.9 Symbiosis and multicellular systems 88
3.10 Spatial distribution of non-metals 89
3.11 The spatial transfer of H, C, N, S: mobile and fixed coenzymes 90
 3.11.1 H (and O) transport 90
 3.11.2 Phosphate transport 91
 3.11.3 Carbon transport 92
 3.11.4 Nitrogen transport 92
 3.11.5 Sulphur transport 93
3.12 Summary of non-metal transport 93

4 Kinetic considerations of chemical reactions, catalysis, and control ... 95

4.1 Introduction ... 95
4.2 Chemical transformations ... 96
4.3 The nature of acid–base reactions: hydrolysis and condensation ... 97
4.4 The hydrolysis of proteins, RNA, DNA, and other polymers ... 99
4.5 The nature of ion flow ... 100
 4.5.1 Signals and controls ... 100
 4.5.2 Ion migration rates: electrolytics ... 101
4.6 On/off reactions: control systems ... 103
4.7 Electron transfer reactions: electronics ... 105
4.8 Redox potentials of complexes ... 108
4.9 Atom transfer reactions ... 109
4.10 Group transfer ... 111
4.11 The nature of transition-metal centres in catalysis ... 112
4.12 Free radical reactions ... 114
4.13 Sizes of atoms, stereochemistry, and reaction paths ... 115
4.14 The creation of local small spaces: mechanical devices for transfer reactions ... 116
4.15 Summary ... 118

5 Energy in biological systems and hydrogen biochemistry ... 120

5.1 Introduction ... 120
5.2 Biological systems and light ... 122
5.3 Oxidative energy: mitochondria ... 124
5.4 Proton migration coupled to redox reaction ... 126
5.5 The coupling of gradients to ATP formation ... 126
5.6 Anaerobes in the dark ... 128
5.7 Compartments, energy, and metabolism ... 131
5.8 Organization and tension ... 132
5.9 Motion of organisms ... 133
5.10 Local storage of elements ... 134
5.11 The evolution of the compartments for energy generation ... 135
5.12 Early sources of energy ... 135
5.13 Sulphur compounds ... 136
5.14 Feedback ... 136
5.15 Flow of material ... 136
5.16 The role of the proton in homeostasis ... 137
5.17 The proton and redox potentials ... 138

6 The functional value of the chemical elements in biological systems ... 140

6.1 Introduction ... 140
6.2 Major chemical properties of elements in aqueous solutions ... 141

6.3	Biochemical functions of the chemical elements	143
6.4	The living process	146
6.5	The chemical flow in biology	149
	6.5.1 The synthesis and degradation of the polymers of life	149
	6.5.2 Synthesis and degradation of monomers	151
	6.5.3 The uptake and loss of elements	154
	6.5.4 The flow of energy in compounds	154
	6.5.5 Self-organization and flow patterns	155
6.6	Organization and flow	156
	6.6.1 Patterns of electron flow: electrical potentials	156
	6.6.2 Patterns of proton gradients and flow electrolytic potentials	157
	6.6.3 Patterns of ion gradients and flow: communication	158
	6.6.4 Tension patterns and flow	159
6.7	The integration of activity	160
6.8	Conclusion	160
	6.8.1 The biological selection of elements	160
	6.8.2 The evolution of the chemistry of life	161

7	**The role of biological macromolecules and polymers**	**163**
A	Proteins and nucleic acids	163
7.1	Introduction	163
7.2	Protein composition and basic structure	164
7.3	The protein fold and the internal motions in solution	171
7.4	The amino-acid composition of specific proteins	173
7.5	Structural proteins and mechanical devices	178
7.6	Matching of proteins and organic and inorganic ions	178
7.7	Enzymes	180
7.8	States of metal ions in proteins	182
	7.8.1 General aspects	182
	7.8.2 The copper blue centres	182
	7.8.3 Protein/protein assemblies and metal ions	184
7.9	Summary of proteins	185
7.10	Nucleic-acid composition and outline structure	185
7.11	Metal-ion binding to polynucleotides	188
	7.11.1 Cavities and metal-ion binding in DNA	188
	7.11.2 DNA sequences and mutation	190
7.12	Nucleic acids and proteins	190
7.13	Polymer synthesis and degradation	191
B	Polysaccharides and lipids	193
7.14	Introduction to polysaccharides	193
7.15	Backbones of biological polymers	195
	7.15.1 Proteins	195
	7.15.2 Nucleic acids	195
	7.15.3 Polysaccharides	195
7.16	Sidechains: physical properties	197
	7.16.1 Proteins	197

	7.16.2 Nucleic acids	197
	7.16.3 Polysaccharides	197
7.17	The information content of biopolymers	198
7.18	Space-filling	199
	7.18.1 Proteins	199
	7.18.2 Nucleic acids	199
	7.18.3 Polysaccharides	200
7.19	Chain branching in polysaccharides	201
7.20	Polysaccharide sidechains: chemical nature	201
7.21	Functions of biological polymers and oligomers	202
	7.21.1 General notes	202
	7.21.2 Roles of polysaccharides: relatively rigid structures	202
	7.21.3 Recognizability of short mobile polysaccharide structures	203
7.22	Larger random polysaccharides	204
	7.22.1 Non-interactive polysaccharides	204
	7.22.2 Random extensions of glycoproteins: cell surface packing	205
	7.22.3 Polysaccharides showing chain–chain repulsion or attraction	206
	7.22.4 Anionic linear polymers	206
7.23	The cation interactions	210
7.24	Conclusions about polysaccharides	210
7.25	Properties of lipids	210
	7.25.1 Lipids in membranes as macromolecular assemblies	210
	7.25.2 Composition and general functions of lipids	211
	7.25.3 Membrane functions and their lateral anisotropic organization	214
	7.25.4 Ion association with lipid surfaces	215
	7.25.5 The influences of selected inorganic elements	216
	7.25.6 Potentials on membrane surfaces	216
7.26	The connection between lipids, hormones, and coenzymes	216
	7.26.1 Cholesterol	216
	7.26.2 Essential fatty acids	217
7.27	Summary of biopolymers	217

PART II · The roles of individual elements in biology

8	**Sodium, potassium, and chlorine: osmotic control, electrolytic equilibria, and currents**	223
	8.1 Introduction	223
	8.2 Passive diffusion	225
	8.3 Gated channels	230
	8.4 Channel selectivity and possible constructions	230
	8.5 Active transport: pumps	232
	8.6 The nature of selective pumps	236
	8.7 Building electrolytic circuits	236

8.7.1	The positions of pumps and channels	236
8.7.2	The ATPase pumps: some general comments	236
8.7.3	Exchangers	237
8.7.4	Organic anions and cations as current carriers	238
8.7.5	Currents of ions and morphogenic patterns: growth	239
8.8	Simple salts and the conditions of polyelectrolytes	239
8.9	Enzymes requiring potassium	240
8.10	Ion energies: summary	240
8.11	Conclusions	241

9 The biological chemistry of magnesium: phosphate metabolism **243**

9.1	Introduction	243
9.2	The spatial distribution of magnesium	244
9.3	Magnesium chemistry	245
9.4	Magnesium pumping in biology	248
9.5	Very strong binding of magnesium	249
9.6	Magnesium in walls and membranes	250
9.7	Magnesium enzymes: magnesium and phosphates	250
9.8	Magnesium and muscle cells	255
9.8.1	Activation of tension	255
9.8.2	Magnesium and muscle relaxation: calcium buffering	255
9.9	Magnesium and polynucleotides	256
9.9.1	Structures of DNA and RNA particles	256
9.9.2	Ribozymes	257
9.10	Competition with polyamines	257
9.11	Polymeric equilibria: tubulins, DNA, and the cell cycle	260
9.12	Sol/gel equilibria and magnesium	261
9.13	Magnesium and lipids	261
9.14	Magnesium and chlorophyll	262
9.14.1	Chlorophyll binding in proteins	262
9.14.2	Magnesium insertion in chlorophyll	263
9.15	The use of manganese as a magnesium probe	265
9.16	Pollution and Mg^{2+} biochemistry	266
9.17	Lithium and magnesium	266
9.18	Conclusion	266

10 Calcium: controls and triggers **268**

10.1	Introduction	268
10.2	Free calcium ion levels	270
10.3	The calcium ion	271
10.4	Protein ligands for calcium	272
10.5	Magnesium/calcium competition	275
10.6	The resting state of calcium in cells	276
10.7	Calcium triggering: calmodulins	276
10.8	The calcium trigger proteins	278
10.9	S-100 proteins	280

10.10 Other triggering modes 281
 10.10.1 Vesicular contents and their release: exocytosis 281
 10.10.2 Calcium, filaments, and cell shape 282
10.11 Calcium and protein phosphorylation 283
10.12 Calcium buffering and calcium transport in cells 283
10.13 Calcium currents: movement through membranes, channels, gates, and pumps 284
10.14 Internal calcium-induced proteases 286
10.15 General remarks concerning control systems in cells 287
10.16 Extracytoplasmic calcium 288
 10.16.1 Vesicular calcium: exchangers 288
 10.16.2 Organelle calcium 289
 10.16.3 Calcium in circulating fluids 289
10.17 Extracellular enzymes and calcium: digestion and blood clotting 289
10.18 Calcium proteins of biominerals 292
10.19 Calcium biominerals 293
10.20 Intra- and extracytoplasmic calcium balances 294
 10.20.1 Multiple sites for calcium 294
 10.20.2 Cell shape 295
 10.20.3 Cell–cell interaction 295
 10.20.4 Cell morphogenesis 295
10.21 Summary 296

11 **Zinc: Lewis acid catalysis and regulation** 299
11.1 Introduction to Lewis acids 299
11.2 Zinc in biological space 302
11.3 Availability and concentration of free Zn^{2+} ions 303
11.4 Types of protein associated with zinc 303
11.5 Zinc exchange rates 307
11.6 The number and selectivity of ligands to zinc 307
11.7 Zinc as a catalytic group in enzymes 308
 11.7.1 Zinc as a Lewis acid 308
 11.7.2 Zinc and redox reactions 311
 11.7.3 The entatic state and probes 311
11.8 The use of zinc in biological space and time 312
11.9 Regulatory roles of zinc 312
11.10 Zinc and peptide hormones 314
11.11 Summary of intracellular zinc: is zinc a master hormone? 314
11.12 Extracellular zinc 315
 11.12.1 The export of zinc enzymes: digestion 315
 11.12.2 Cross-linking and hardening of extracellular matrices 316
 11.12.3 Zinc in solid-state devices 317
11.13 Conclusion 317
 Note. Probes of metal sites by isomorphous substitution 318

12 Non-haem iron: redox reactions and controls 319

12.1 General introduction to transition metals in biology 319
12.2 Introduction to iron biological chemistry 322
12.3 Iron uptake 323
12.4 The non-haem iron proteins 326
12.5 The iron/sulphur proteins 326
12.6 Fe/S centres active as enzymes 329
12.7 Location of Fe/S proteins 330
12.8 Why are there Fe_nS_n clusters in biology? 331
12.9 The organization and synthesis of ferredoxins 332
12.10 The Fe–O–Fe cluster 333
12.11 Mononuclear non-haem/non-Fe/S iron and oxidative enzymes 335
12.12 Iron and secondary metabolism 337
12.13 Binding ligands in non-haem/non-Fe/S proteins 338
12.14 Iron as an acid catalyst 338
12.15 Summary of non-haem iron functions 339
12.16 Iron regulation and homeostasis 339
12.17 Iron buffering 340
Note. The discovery of Fe/S centres 341

13 Haem iron: coupled redox reactions 343

13.1 Iron in porphyrins 343
13.2 Properties of isolated haem units 344
13.3 Classification of haem proteins by iron properties 347
13.4 Classification of haem proteins by secondary structure 348
13.5 Where are haem proteins in cells? 352
13.6 Haem-protein functions I: electron transfer 353
13.6.1 Simple electron transfer proteins 353
13.6.2 Multihaem electron transfer proteins 354
13.6.3 Electron transfer/proton transfer coupling: models 354
13.6.4 The membrane electron transfer proteins: protein coupling 356
13.7 The surfaces of haem proteins 357
13.8 Haem-protein functions II: storage and transport 358
13.8.1 Dioxygen: storage, transport, and signalling 358
13.8.2 Nitric oxide and haem receptors 359
13.8.3 Cytochrome *c* and cyanide 360
13.9 Haem-protein functions III: oxidases and dioxygenases 361
13.9.1 Cytochrome oxidase 362
13.9.2 Cytochrome P-450 363
13.9.3 Peroxidases and catalases 364
13.10 Substrates of haem enzymes and secondary metabolism 365
13.11 Haem in controls 366
13.12 The synthesis of haem 367
13.13 Summary of haem-iron functions 367
13.14 Control of cell activity and iron 368

14 Manganese: dioxygen evolution and glycosylation 370
 14.1 Introduction 370
 14.2 Manganese chemistry 372
 14.2.1 Oxidation states 372
 14.2.2 Structures of Mn(II) complexes with organic ligands 372
 14.2.3 Manganese(II) equilibria and biochemistry 373
 14.2.4 The kinetics of Mn(II) complexes 375
 14.3 Monomeric Mn(III) and Mn(IV) chemistry 376
 14.3.1 Structures 376
 14.3.2 Thermodynamics 376
 14.3.3 Kinetics of Mn(III) and Mn(IV) reactions 377
 14.3.4 Manganese cluster chemistry 377
 14.4 The biological chemistry of manganese 379
 14.4.1 The occurrence and availability to biology of manganese 379
 14.4.2 The uptake of manganese 379
 14.4.3 Pumping of manganese gradients 379
 14.4.4 The distribution of Mn in biological systems 380
 14.5 Manganese control systems 380
 14.6 The production of dioxygen 381
 14.7 Manganese and dioxygen metabolism 382
 14.8 Manganese precipitates 383
 14.9 The evolution of manganese functions 385
 14.10 Summary 386

15 Copper: extracytoplasmic oxidases and matrix formation 388
 15.1 Introduction 388
 15.2 Copper and electron transfer 389
 15.3 Copper and dioxygen 391
 15.3.1 Haemocyanin 391
 15.3.2 A comment on copper of cytochrome oxidase 391
 15.3.3 Copper and extracellular oxidases 392
 15.3.4 Substrates of copper oxidases 393
 15.3.5 Copper proteins and coenzyme PQQ 394
 15.4 Copper enzymes and nitrogen oxides 394
 15.5 Superoxide dismutase 395
 15.5.1 The location 395
 15.5.2 The reactions of superoxide dismutases 395
 15.6 The transport and homeostasis of copper 396
 15.6.1 The binding proteins 396
 15.6.2 Copper exchange rates 396
 15.6.3 Regulatory copper proteins 397
 15.7 The overall functions of copper 397

16 Nickel and cobalt: remnants of early life? 400
 16.1 Introduction 400
 16.2 The chemistry of cobalt and nickel 402
 16.3 Hydrogenases 405

16.4	The reactions of vitamin B_{12}	406
16.5	The organometallic chemistry of life	408
16.6	Urease: nickel in vacuoles	409
16.7	Nickel uptake	410
16.8	Conclusion	410

17		Molybdenum, vanadium, and tungsten	411
	17.1	Molybdenum: introduction	411
	17.2	The availability of molybdenum	412
	17.3	The molybdenum enzymes: a first overview	413
	17.4	Physical methods for enzyme and model studies	415
	17.5	Relevant model chemistry of molybdenum	416
		17.5.1 Structures of molybdenum complexes including clusters	416
		17.5.2 The elementary aqueous chemistry of molybdenum	416
		17.5.3 The stability of model complexes	417
		17.5.4 Condensation and complexation	418
		17.5.5 The competition with iron(III) and copper(II) for biological ligands	419
		17.5.6 The special ligand for MoO_2^{2+} in biology	420
	17.6	Rates of reaction	422
	17.7	Biological chemistry of molybdenum: general considerations	422
		17.7.1 Uptake of molybdate	422
		17.7.2 Oxidation states and redox potentials in enzymes	423
		17.7.3 Oxygen-atom transfer reactions	424
		17.7.4 Nitrogenase and its reaction centre: FeMoco	425
		17.7.5 Molybdenum exchange and possible control functions	426
	17.8	Molybdenum: conclusions	427
	17.9	Vanadium chemistry and biochemistry	427
		17.9.1 The biological forms and functions of vanadium	428
		17.9.2 Vanadium and iron	430
	17.10	Tungsten biological chemistry	431
	17.11	Molybdenum, vanadium, and tungsten in evolution	432

18		Phosphate, silica, and chloride: acid–base non-metals	436
	18.1	Introduction to the non-metals	436
	18.2	Phosphate biochemistry	437
		18.2.1 Introduction to the functions of phosphate in biology	437
		18.2.2 The chemistry of phosphate	437
	18.3	The forms and energies of bound phosphate	438
		18.3.1 Phosphate in large molecules or assemblies in biology	438
		18.3.2 The energy of pyrophosphates and the resting state of cells	439
		18.3.3 Phosphate controls	440
		18.3.4 Phosphate messengers	441
		18.3.5 Phosphate switches: protein phosphorylation and time hierarchy	443

	18.3.6	Phosphates and gene transcription	444
	18.3.7	Phosphate minerals	445
18.4		Summary of phosphates	445
18.5		Anion balance: chloride, phosphate, sulphate, and carboxylates	445
18.6		Chloride channels and exchangers	445
18.7		Silicon biochemistry	446
	18.7.1	The chemistry of silica in biology	449
18.8		Boron in biology	451
18.9		Non-metal trace elements	451

19 Sulphur, selenium, and halogens: redox non-metals 453

19.1		Introduction	453
19.2		Sulphur biochemistry	454
	19.2.1	Oxidation states and redox reactions	454
	19.2.2	Acid–base reactions	456
	19.2.3	Group transfer reactions: coupling	457
	19.2.4	The sulphur cycle	458
19.3		Summary of sulphur biochemistry	459
19.4		The biochemistry of selenium	460
19.5		Evolution and selenium	461
19.6		Notes on halogen redox reactions	462

PART III · Biological shape and the integration of element functions

20 Biological minerals and biological shapes 467

20.1		Introduction	467
20.2		Biominerals viewed as structures	469
	20.2.1	Physicochemical character	469
	20.2.2	Vesicles and polymer matrices	471
20.3		Formation of inorganic mineral structures	473
	20.3.1	Nucleation	473
	20.3.2	An explanation of Lussac's law	474
	20.3.3	Growth of inorganic crystals	474
	20.3.4	Ripening (ageing)	475
20.4		Biological single crystals in relation to single-cell morphology	476
20.5		Relationships between biomineral formation and morphological structure: examples	477
	20.5.1	*Acantharia*	477
	20.5.2	Biominerals formed outside cells	481
	20.5.3	Coccolithiderae	482
	20.5.4	Radiolarians	483
	20.5.5	Sensors: barium and strontium sulphates in other organisms	484
	20.5.6	Magnetotactic bacteria	484

20.6	Unicellular eukaryotes: summary	485
20.7	Development and growth in multicellular animals	487
20.8	Extracellular biominerals in multicellular organisms	489
20.8.1	Shells: their chemistry and physics	489
20.8.2	Bones	490
20.9	Summary of skeletal crystals in multicellular organisms	491
20.10	The organic extracellular matrix for mineralization	491
20.11	Sensors	493
20.11.1	Calcium carbonate in sensors	494

21	An integrated view: the elements in homeostasis, morphogenesis, and evolution	495
21.1	Introduction	495
21.1.1	Homeostasis and morphogenesis: definitions	496
21.1.2	Shape in biology and its evolution	498
21.2	The cytoplasmic homeostasis of basic elements	499
21.2.1	Phosphate	499
21.2.2	Phosphates, magnesium, and protons	501
21.2.3	Sodium, potassium, chloride, and calcium	502
21.2.4	Hydrogen, sulphur, and iron in primitive prokaryotes	502
21.2.5	Regulation	503
21.3	A detailed case: the homeostasis of iron in prokaryotes today	503
21.4	A summary of cytoplasmic homeostasis early in evolution	506
21.5	The environment of early cells	507
21.5.1	Prokaryote morphology	507
21.5.2	The supply of organic chemicals: the role of zinc	508
21.5.3	Slow and fast controls	508
21.5.4	Evolution of ion currents	509
21.6	Eukaryotes	510
21.6.1	The value of shape: internal filaments and vesicles	510
21.6.2	Calcium, shape, and the environment	510
21.7	The evolutionary switch in homeostasis	512
21.7.1	The advent of dioxygen	512
21.7.2	Cell/cell combination	514
21.7.3	Copper and zinc: the extracellular connective matrix	515
21.8	Controls over calcium external flow and mineral formation	516
21.8.1	The circulating fluids of multicellular organisms	516
21.8.2	Eukaryotic DNA and zinc: hormonal connections	517
21.9	A summary of zinc controls	519
21.10	Organic signals	520
21.10.1	Iron, dioxygen, poisons, and hormones	520
21.10.2	Evolution of secondary metabolism	521
21.10.3	Overall schemes of metal linkages in controls	523
21.11	Changes in functions of other elements	525

21.11.1 New roles for molybdenum and vanadium 525
21.11.2 Nickel and cobalt in eukaryotes 525
21.12 Morphogenesis 526
21.13 Concluding remarks 527

22 Man's use of the chemical elements in biological environments 531
22.1 Introduction 531
22.2 Man's agriculture 534
22.3 Industry and pollution 535
22.4 Toxicity 536
 22.4.1 The 'a'-class (hard) metals 536
 22.4.2 The intermediate metals 537
 22.4.3 The 'b'-class (soft) metals 538
 22.4.4 Non-metal elements 539
22.5 Diseases associated with abnormalities in concentrations of required elements 540
 22.5.1 General examples 540
 22.5.2 Toxicity and the immune system 542
22.6 Poisons and evolution 543
22.7 Elements in medicine 544
 22.7.1 Hard, 'a'-class metals 545
 22.7.2 Soft, 'b'-class metals 546
 22.7.3 Intermediate elements 547
 22.7.4 Removal of excess of metal elements 548
 22.7.5 Non-metal elements in medicines 549
 22.7.6 Solid-state inorganic compounds in medicine 549
22.8 Conclusion 550

Index 553

PART I

The chemical and physical factors controlling the elements of life

1

The chemical elements in biology

1.1	The natural selection of the elements	3
1.2	An aside: some biological chemistry of hydrogen	7
1.3	The economical use of resources: abundance and availability	8
1.4	Biological environment and element availability	11
1.5	Sulphide chemistry in water (anaerobes)	18
1.6	A note on homeostasis	20
1.7	Conclusion	22

1.1 The natural selection of the elements

1.1.1 Abundance in living systems

Analyses of a few hundred plant species, among the existing 0.5 million, and of about two hundred among the probable 1.1 million animal species, as well as of organs, tissues, and other substances of biological origin, have enabled us to establish the number and identity of the chemical elements present in biological systems and to recognize those that are 'essential' for plant and animal life. It became clear in these and related studies that, amongst the chemical elements, biological systems concentrate certain elements while rejecting others and that some of the processes involved require energy. One could, therefore, speak of a natural selection of the chemical elements by biological systems, which involves a readjustment of the element distribution on the earth's local scale by utilizing energy ultimately provided by the sun.

Analysis has shown that only *eleven* elements appear to be approximately constant and predominant in all biological systems. In the human body these constitute 99.9 per cent of the total number of atoms present but just *four* of them—carbon, oxygen, hydrogen, and nitrogen—correspond to 99 per cent of that total (see Table 1.1). The very large percentages of oxygen and hydrogen arise from the high water content of all living systems. Carbon and nitrogen, next in importance in Table 1.1, together with oxygen and hydrogen, are the basic elements of the organic structures and metabolites of living systems.

The other *seven* elements, which together represent about 0.9 per cent of the total number of atoms in the human body, are sodium, potassium, calcium, magnesium, phosphorus, sulphur, and chlorine. Besides these 11 elements,

Table 1.1 Percentage of atoms in the human body

Element	Atom percentage (wet wt)
Hydrogen	62.8
Oxygen	25.4
Carbon	9.4
Nitrogen	1.4
Others	1.0

which are absolutely essential, some *ten* others—metals and non-metals italicized in the list that follows—are required by most biological systems, but not necessarily by *all* biological systems. There are still a further seven elements, some of which may be generally required by plants and animals, whereas others may be required by just plants or just animals or sometimes by a relatively few species of plants or animals. These last 17 elements are: vanadium, chromium, *manganese, iron, cobalt, nickel, copper, zinc, molybdenum, boron, silicon, selenium,* fluorine, iodine, arsenic(?), bromine(?), and tin(?). The levels of requirement of these 17 elements are generally very different and we note particularly the relatively large amounts of iron and zinc present in most if not all species. Finally, we note that elements such as tungsten, strontium, and barium are known to be important in the chemistry of one or two particular species.

The limited number of species examined, the difficulties of the analytical work (particularly for elements at the trace level) and the lack of detailed knowledge concerning the role of each element make it necessary to check their hypothetical 'essentiality' by delicate tests, usually by following the developmental growth of species, animal or plant, while giving them carefully prepared 'diets' deficient in the particular element considered.

In some cases these studies have indeed enabled not only the classification of a certain element as 'essential' or 'not essential' for a particular species but also the evaluation of the effects of its deficiency. In other cases the results are ambiguous and no definite conclusion has been reached. At present, elements such as arsenic, bromine, and tin are considered as 'possibly essential' while others, like aluminium (see Table 1.2), are considered to have no definite general biological role, perhaps because the effects of their deficiency have not

Table 1.2 Unusual accumulation of elements by plants or animals

Element	Plant	Animal
Al	Club moss, hydrangea, tea plant	
As	Brown algae	Coelenterates
B	Brown algae, *Plumbaginaceae*	Sponges
Br	Brown algae	
Ba	Rhizopods, Brazil nuts, *Loxedes*	Coelenterates, molluscs
Cl	Numerous marine and salt desert plants	Soft coelenterates
Cu	*Caryophyllaceae*	Annelids, molluscs, etc.
F	*Dichapetalum cymosum*	Vertebrates
I	Brown algae, diatoms	Marine annelids
Li	*Thalictrum* sp., *Cirsium* sp.	
Mn	Ferns, *Digitalis purpurea*	Marine crustacea
Mo	*Papilonaceae*	
Ni	*Alyssium* sp., *Hybanthus floribundus*	
Se	*Micrococcus selenifera, Cruciferae, Astragalus racemosus*	
Si	Horse tails, *Radiolaria*	Protozoa, sponges
Sr	Brown algae	Alcyonaria, octocorallia Acantharia
V	Some fungi, brown algae	Ascidians
Zn	*Thlaspi calaminare*	Coelenterates

Sources: Bowen, H. J. M. (1966). *Trace elements in biochemistry.* Academic Press, London.
Hewit, E. J. and Smith, T. A. (1975). *Plant mineral nutrition.* The English Universities Press, London.

been sufficiently tested or because the requirements by the species used in the tests are so low that the needs are satisfied with the extremely low amounts present even in carefully purified diets.

Uptake of a certain element by a biological system, taken as isolated evidence, is no proof or guarantee of essentiality and even less is the detection of a certain (trace) element in a particular biological system. The analytical difficulties, the possibility of contaminations at the trace and ultratrace level, and the likelihood of competition of different elements for the same chemical sites leave room for considerable doubts as to the conclusions to be extracted from these experiments. Moreover, some biological systems accumulate elements even though they may not need them (Table 1.2). It follows that the classification of a certain element as essential or not essential implies a prescriptive definition of 'essentiality'. In general terms we will consider as essential an element consistently present in a certain biological species such that its deficiency in the nutritive sources of that species leads to disease, metabolic anomalies, or perturbations in its development. More precise criteria have been put forward and are commonly adopted by nutritionists.

The results obtained in the studies of essentiality are summarized in Fig. 1.1—the periodic table of the elements—where the elements that are today considered to be essential for life are conveniently identified. A first conclusion to be derived from this figure is that the 'chemistry of life' is, predominantly, the chemistry of the *lighter elements*, metals and non-metals of atomic number less than 35. Of required elements only molybdenum and iodine have higher atomic numbers and neither of these seems to be generally required by *all* living species. On the other hand, the biological elements seem to have been selected from practically all groups and subgroups of the periodic table (the only exceptions are groups IIIA and IVA, besides that of the inert gases) and this means that practically all kinds of chemical properties are associated with life processes within the limits imposed by environmental constraints. A little thought will show that such a selection is to be expected from an evolving system.

Fig. 1.1 The distribution of elements essential for life in the periodic table.

1.1.2 Chemical speciation: limitations of life chemistry and physics

To describe the limitations which the elements (in compounds) face in life one must recall that oxygen and hydrogen are, to a large extent, combined with each other in the form of water and that most life chemistry takes place in aqueous media. This fact implies that, generally, only processes compatible with the presence of this solvent are possible, restricting the utilizable range of redox potentials, pH, and temperature. Bulk metals or alloys as such, for instance, cannot be used by biological systems since they are not preparable in water and usually are unstable in it. Processes depending on metallic electronic conduction which might be valuable, e.g. to send messages, must be carried out by other chemical agents or systems. The same can be said for temperature constraints. Thus, today's industry uses operations requiring high energy input to produce products that biological systems also need, e.g. NH_3 obtained from N_2; in biological systems these operations must be carried out at normal physiological temperatures which is only possible by using extremely sophisticated catalytic processes developed over millions of years.

To explore these aspects in somewhat more detail we note that the elements in the periodic table which concern us are not just those that constitute the majority of the organic biomolecules, but also include some closed shell metals, especially of groups I and II, most of the first series transition metals, some B-subgroup metals, and also a more extensive group of non-metals, of groups IVB to VIIB, including (B), F, Si, P, S, Cl, Se, Br, and I, besides molybdenum, which can be considered, in certain respects, as a non-metal. Given that water is the solvent, these later non-metal elements tend to be found as anions, in contrast with the elements Na, K, Mg, Ca, and several transition metals which are found as cations.

Although the distinctive properties *in aqueous solution* of each of these conventionally separated four classes of elements, i.e. groups IA to IIIA, transition metals, B-subgroup metals, and non-metals, do not always provide sharp divisions, it is useful to distinguish some general chemical properties associated with these classes. Hence, groups IA and IIA elements are associated with equilibrium ionic-model chemistry. The transition metals are also generally associated with aqueous ionic chemistry, apart from organo-metallic chemistry which is unusual in biology, and additionally with one-electron redox reactions but, across the series, exhibit increasing covalent chemistry and increasing Lewis acid strength not necessarily at equilibrium with their surroundings. This means that in biology they become largely bound to organic biological molecules in which equilibrium is harder to achieve. Group IIB metal ions, e.g. zinc, display a compromise in their chemistry having strong Lewis acid properties while maintaining faster ionic equilibration but with limited or no redox activity. They too usually become bound in biological systems. There is next a class of non-metals (H, C, N, O, S, P) in which covalence introduces gross kinetic control in aqueous solution. It is this that allows the development of the biopolymers as well as kinetic control over two- and (a few) one-electron oxidation state changes and over many acid–base reactions, with Lewis base properties dominating over Lewis acid properties. The extreme non-metals appear overwhelmingly as simple anions (we specially note the presence of chlorine as chloride), when their chemistry is

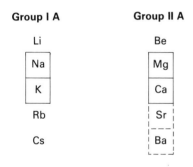

Group I A

Li

Na

K

Rb

Cs

Group II A

Be

Mg

Ca

Sr

Ba

Non-redox non-metals

B

Si P

Group II B

Zn

Cd

Hg

Redox non-metals

H

C N O

S

Se

Group VII

F

Cl

Br

I

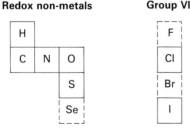

Redox metals

Cr Mn Fe Co Ni Cu

Mo

classed with the group IA and IIA elements. The aqueous conditions which allow the classification are maintained in biological systems.

Biological systems took immediate advantage of these differences of chemical properties, but we shall see that the distinctions between the various individual elements have been heightened during evolution in such a way that each of them has been selected to perform one or more particular functions with high specificity, much more so than in ordinary bench chemistry. This is not just a matter of selecting partners but will involve an energized state of the element brought about by separation across membranes, for example. These general points are so important that it is worth digressing to give a simple example.

1.2 An aside: some biological chemistry of hydrogen

Fig. 1.2 Some chemical functional values of the element hydrogen which have been realized in biology. Hydrogen atoms are initially incorporated into carbon compounds by the action of light. They can be transferred in different redox states e.g. H^+, $H\cdot$, or H^-. If the transfer operations are energized and diffusion-limited, say, by a membrane, while electrons are moved separately as shown, then H^+/OH^- gradients can be built. Hydrogen is the only element which can be used in this manner to generate electrolytic cell energy from light. The H^+ gradient is subsequently used to aid the transfer of other ions and molecules across membranes or, as shown, to make adenosine triphosphate, ATP, the first form of chemically stored energy (see Chapter 5 for a more detailed explanation). Notice that a redox potential difference is established across the membrane.

There is one very special element in chemistry and biology—hydrogen. It is special in that it has three very distinct states, H^+ (a cation), H– (a covalently bound state), and H^- (an anion). As H^+, the proton, it is an extremely strong acid and extremely mobile in protic media such as water and the water/polymer interfaces of biology. In other words, as an acid it is unique. As H– it is a special element in that it forms very stable non-metal bonds in the form of C–H and N–H which are terminal blocking units limiting reactivity at carbon and nitrogen. These bonds are kinetically stable even in the presence of dioxygen and provide great stability to organic molecules in biology. Third, under the influence of catalysts hydrogen can be transferred from a non-metal such as carbon not just as H^+ or H– but as H^-. Which of these forms is transferred depends upon the group which the hydrogen is leaving. Thus, hydrogen atoms can be involved in one- or two-electron processes. Great use of this redox chemistry is made in biological redox reactions. Even more importantly, the internal redox energy of organic molecules, C–H bonds, versus dioxygen, both derived from light (Fig. 1.2), can be converted within membrane systems, using oxidation/reduction, to different forms of more usable energy. Here hydrogen is transferred first as H^-, then effectively as $H\cdot$, and finally as H^+, to produce eventually proton gradients across membranes in water (Fig. 1.2). The reactions require the separate movement of electrons, of course. The strong acid, H^+, which is freely mobile, is seen to be readily connected to covalent, kinetically-controlled, redox chemistry as well as to ionic acid/base chemistry. The last step in Fig. 1.2 is to make ATP, a form of pyrophosphate useful in synthesis, from the proton gradient. *No other element has these different capacities to the same degree* in water and, as a consequence, hydrogen is at the heart of energy transduction mechanisms in biology (see Chapter 5).

We give this example to stress the way in which life's chemistry is built around elements which have essential functional potential, here for energy capture, but in the case of other elements for the transmission of information (Na^+, K^+, Ca^{2+}), for catalysis (Fe, Cu, Zn, Mo), for the transfer of electrons (Fe), and for the building of solid structures (Ca, P, Si). All of these features are essential in the same sense as it is true to say that C, N, H, and O are essential for the synthesis of biological macromolecules (Fig. 1.3).

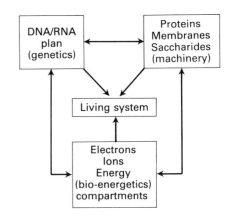

Fig. 1.3 The scheme shows the interrelationship between the three interlocking features of a biological system. The DNA/RNA unit is that of inheritable information. The protein unit covers structural aspects, including the lipid membranes, the proteins, and the polysaccharides, within which there is the mechanical and catalytic machinery of the cell. The third unit is a source of energy which is necessary for the activity of cells including all syntheses and the uptake of chemicals. The role of inorganic elements is dominant in the last unit, is generally very important within the machinery, and has a place in the DNA/RNA unit.

Before delving further into these matters, it is appropriate to give some justification for the various trends observed in the distribution in biology of the elements of the periodic table.

1.3 The economical use of resources: abundance and availability

In general, an element that is to be used by biological systems must have a useful function but it must also be abundant in the environment and must be in an easily extractable form, i.e. it must be 'biologically available' from the environment. A cursory examination of Fig. 1.1 immediately shows that the elements essential for life are the lighter elements which are also the most abundant in the universe (Fig. 1.4). They are also the elements which are most abundant on earth. It is true that the abundance in the planets, particularly in the earth, does not exactly correspond to that of the solar system and deeper considerations of this aspect must go down to the level of the detailed composition of the earth's crust, oceans, atmosphere, etc. Even then, however, and considering the amounts of the relevant elements locally, it is not necessary to modify the discussion greatly: generally, biological elements are still those which are the most universally abundant.

Some elements, e.g. lithium, caesium, beryllium, and strontium, are either of low abundance or are less abundant than others which have analogous properties and can play a similar biological role. Presumably one reason why the bulk biological metals, sodium, potassium, magnesium, and calcium, were selected in preference to lithium, rubidium, caesium, beryllium, strontium, and barium is that they are more abundant and can perform in large measure the tasks that the other elements of the same subgroups of the periodic table can perform. The periodic table (Fig. 1.1) is, to some degree, a table of redundancy. This, like many other generalizations, has interesting exceptions, which will be considered later in this section.

Besides the problem of abundance there is that of availability in water solutions. Availability is concerned with chemical combination and solubility of compounds as well as abundance. All the compounds of the elements of groups I, II, and VII as well as of O and S in group VI are soluble and found in the sea in quantity (see Fig. 1.5). The parallel between the elemental

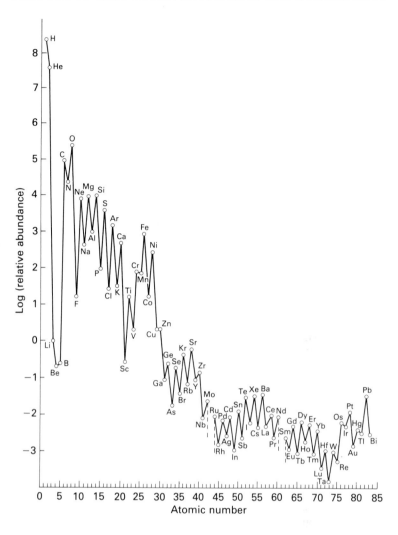

Fig. 1.4 Abundance of the elements in the universe relative to $Si = 10^4$ as a function of the atomic number.

composition of animals or plants and that of sea water in some respects is quite striking, but, since life has modified the composition of the sea itself, no simple deductions are possible. Group IIIA and IVA elements are of relatively low availability, i.e., although they form amphoteric oxides of high abundance, they are insoluble and it requires great energy to obtain measurable amounts of such elements at pH = 7.0 (Fig. 1.6). In fact, these elements are notable for their absence from biology.

If we extend this analysis to other groups, one or two elements turn out to be common in biology but not really available. A striking problem, revealed in Fig. 1.6, is the difficulty biological systems face today with the lack of availability of iron. The special lengths to which biology has been forced in order to obtain this precious element will be seen later to have a controlling influence on the whole of living systems, yet biology created this difficulty for itself by generating dioxygen (see Chapter 21). Inspection of Figs 1.5 and 1.6 and the data given earlier for the element content of biological systems indicates how absolute is the demand for some specific elements. Why?

One may therefore say, in a general way, that in its selection of chemical

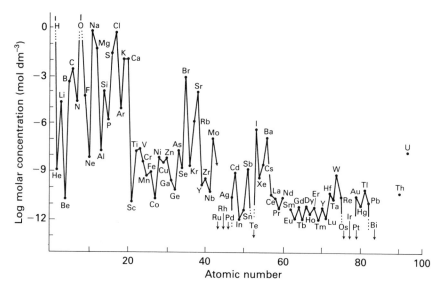

Fig. 1.5 The concentrations of elements in the sea. (After P. A. Cox, see 'Further Reading'.)

elements Nature responded to abundance and availability following a principle of 'the economical utilization of resources', i.e. choosing those elements less costly in terms of energy required for uptake, *given the function for which they are required*.

It is important not to consider this principle as other than dynamic, since biological systems are adaptable and can to some extent manipulate their chemistry and their position in the environment. While it would appear that the solvent for life has always been water, other conditions, such as those of temperature and the dominant chemical species in the atmosphere, have changed with time. Life, while itself producing many of these changes, has evolved with the changes. This will always involve compromises, such as the greater ease of access to one element at the cost of lesser access to another. Again, geological conditions today provide and always have provided a variety of niches, which are indeed special environments into which species may 'guide' themselves (see Chapter 20). It is also the case that different elements are accumulated in biological systems to take advantage of special conditions. A simple example is the accumulation of large quantities of dense $SrSO_4$ and $BaSO_4$ in some protozoa and algae (Table 1.2), which allows these organisms to remain in relatively deep (anaerobic) waters of relatively high density. We must note too that the changes in the atmosphere and the consequent changes in the sea and in the earth's surface over millenia have had a considerable effect especially on rare element distribution and uptake. Some of the elements found in biology were selected a long time ago in very different circumstances; iron is a striking example. Even so the remarks made above concerning the economy of uptake have a general quality across many species.

Against the background of these figures we turn to the detailed reasons today for the observed availability of the elements and their speciation in water at pH = 7, given exposure to molecular dioxygen of the air. At this stage of discussion we are not concerned with organic chemicals at all but only with the elements and their simple oxides and hydrides which will be used later to form all compounds *in vivo*.

Fig. 1.6 The figure shows the ease of extraction of elements from the earth to the sea and is a guide to the ease with which biological systems will have access to elements, no matter what the absolute abundance. The top of the figure indicates the speciation.

1.4 Biological environment and element availability

1.4.1 pH and redox potential considerations

If we are to appreciate the skill displayed by the living systems in the uptake and utilization of the chemical elements, both those they need as well as those they reject because of their potential toxicity, we must study the chemical forms, the speciation, in which all these elements occur in the natural surroundings of life (Fig. 1.6). While doing this we must also take into account the limitations imposed not only by the solvent water and its pH but also the prevailing (present-day) redox conditions.

Recalling that at pH 7 the redox potentials in water are largely restricted to the range -0.4 to $+0.8$ V relative to the standard hydrogen electrode, which corresponds to the systems H^+/H_2 and O_2/OH^-, i.e. to the values below which and above which water is reduced to H_2 or oxidized to O_2, respectively, then a transition metal as an ion has only a small range of stable oxidation states, e.g. iron $(+2, +3)$, copper $(+1, +2)$, etc., but a non-metal may have a very wide range, e.g. sulphur $(-2$ to $+6)$, carbon $(-4$ to $+4)$—see the oxidation state diagrams in Figs 1.7 and 1.8. This difference arises from the different electronic energies involved in the two cases. In the first transition series it is the 3d electron core that is ionized and this process, which is of rapidly increasing energy with increasing positive charge, cannot be easily compensated by chemical binding to water, but in the non-metals the ionization is of sp electrons and compensation by chemical bond formation is largely possible. Note that we shall use both arabic and roman numerals alternatively for oxidation state, n: M^{n+} stresses free ion character while $M(III)$ stresses bound character, e.g. compare Fe^{3+} and $Fe(III)$ (see Phillips and Williams 1966 in the 'Further reading' section).

Fig. 1.7 Oxidation state diagram for some metals at pH = 7 ($n\,E$ versus n) taking into account the formation of hydroxides, where n is the oxidation state and E the electrode potential from the element to that state.

The only element of biological interest for which these generalizations fail is molybdenum, which is a second-row transition *metal* element but behaves more like a non-metal. It has a wide variety of oxidation states in water within the referred restricted range of potentials (Fig. 1.7). Note that here it is the 4d core which is ionized. The case of vanadium is also interesting in that it has an unusual set of available high-valence states. Although its biological role is not so extensive, it will be seen later to afford another example of unexpected solutions provided by Nature to face specific necessities in particular cases (see Chapter 17). Molybdenum and vanadium can perform similar functions, e.g. N_2- fixation.

Another conclusion that can be derived from the oxidation state diagrams is that there are metals and non-metals with only one very stable state. This is the case of the alkali and alkaline-earth metals restricted, respectively, to the +1 and +2 oxidation states; of zinc and cadmium, restricted to the +2 oxidation state; and of several non-metals, notably boron, silicon, and phosphorus, which appear only in the highest oxidation states of +3, +4, and +5,

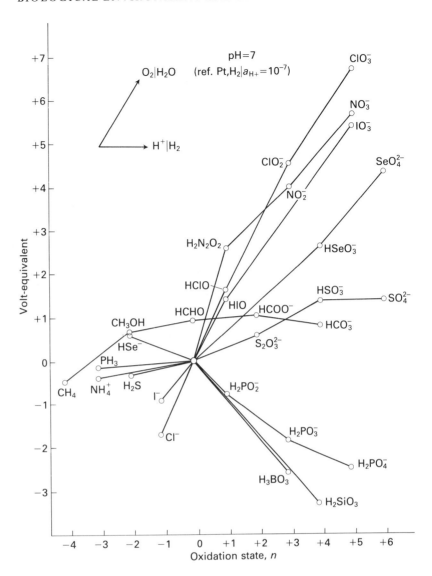

Fig. 1.8 Oxidation state diagram for non-metals at pH = 7 (nE versus n) taking into account protonation reactions.

respectively, as borate, silica, and phosphate. The predominant form of chlorine is as a chloride ion and the few rare compounds of biological origin containing covalently combined chlorine are often antibiotics. (This observation should be carefully examined in its implications for the rather careless use that man still makes of organo-chlorine compounds, insecticides, and other industrial products, which are being banned by progressively more restrictive legislation due to their detrimental biological effects.) As examples of the opposite behaviour, a few other non-metals are unstable at the extremes of their available redox states, for example, nitrogen as nitrate, which tends to be run down to NH compounds, and selenate which is incorporated as covalently combined seleno-thiol and analogous species.

This shows that only a very limited number of non-metals can be used by life as *reversible* redox switches. Hydrogen (see Section 1.2) and carbon occupy overwhelmingly dominant positions, with available oxidation states from −1

to $+1$ and -4 to $+4$, respectively, and all are interconvertible at redox potentials close to zero. Sulphur comes close to carbon and selenium comes next, but the number of sulphur and selenium states in biologically relevant chemical species is much more limited than those of carbon. Against the background of this non-metal chemistry it is interesting to note again the restricted valence state range of the transition metals (Fig. 1.9) as is also illustrated by the steepness of the curves in Fig. 1.7 as opposed to those in the centre of Fig. 1.8.

Fig. 1.9 Oxidation state patterns for the elements from potassium to zinc, compiled from a survey of aqueous monatomic ions and aqueous oxyions, stoichiometric oxides, halides, and other simple salts.

1.4.2 Element concentrations in aerobic environments and in some organisms

From what has been said, it can be concluded that the biological elements are either available from the atmosphere, i.e. as gases, in the cases of C (CO_2), of O (O_2), of N (N_2), and of H (H_2O), or from water solutions in which they are likely to be present in simple or not very complex species as shown in Tables 1.3 and 1.4 due to the oxidizing conditions of present-day waters.

Some of the peculiarities of the influences of redox potentials of the elements at pH $= 7$ appear in these tables but, in addition, it is the hydrolytic reactions of the elements with water in the presence of dioxygen which have determined the availability of elements today. Thus, in the transition series, iron is only present as a minute trace in water as the redox potential restricts this element to the state Fe^{3+}, which precipitates as insoluble $Fe(OH)_3$, so that it cannot be obtained as a soluble cation, e.g. Fe^{2+}, or a soluble anion, such as FeO_4^{2-}. The above environmental availability restricts to some extent the possibilities of development of life, but, as we have seen, Nature has found ways of circumventing this type of limitation when there is a strong requirement for

Table 1.3 Forms in which main biological elements occur in aerated soil, water, rivers, lakes, and sea (or in blood plasma)—simple species

Cations	Anions	Neutral species
NH_4^+, H_3O^+	HCO_3^-, CO_3^{2-}, NO_3^-	H_2O, $B(OH)_3$
Na^+, K^+	$H_2PO_4^-$, HPO_4^{2-}	CO_2, $SiO_2 \cdot nH_2O$
Mg^{2+}, Ca^{2+}	OH^-, F^-, Cl^-, Br^-, I^-, SO_4^{2-}	N_2, NH_3, O_2

Table 1.4 Inorganic speciation of some trace metals (essential, not essential, and toxic) in marine systems at 25°C, 1 atm. pressure, 3.5 per cent salinity, and pH = 8 (computer model calculation from unknown total concentrations and available stability constant data)

Metal	Main species
Al^{3+}	$Al(OH)_3$ 100%
Cr^{3+}	$Cr(OH)_3$ 100%
Mn^{2+}	Mn^{2+} 58%, $MnCl^+$ 37%, $MnSO_4$ 4%, $MnCO_3$ 1%
Fe^{2+}	Fe^{2+} 69%, $FeCl^+$ 20%, $FeCO_3$ 5%, $FeSO_4$ 4%, $Fe(OH)^+$ 2%
Fe^{3+}	$Fe(OH)_3$ 100%
Co^{2+}	Co^{2+} 58%, $CoCl^+$ 30%, $CoCO_3$ 6%, $CoSO_4$ 5%, $Co(OH)^+$ 1%
Ni^{2+}	Ni^{2+} 47%, $NiCl^+$ 34%, $NiCO_3$ 14%, $NiSO_4$ 4%, $Ni(OH)^+$ 1%
Cu^{2+}	Cu^{2+} 9%, $CuCO_3$ 79%, $Cu(OH)^+$ 8%, $CuCl^+$ 3%, $CuSO_4$ 1%
Zn^{2+}	Zn^{2+} 46%, $ZnCl^+$ 35%, $Zn(OH)^+$ 12%, $ZnSO_4$ 4%, $ZnCO_3$ 3%
Mo(VI)	MoO_4^{2-} (100%)
Cd^{2+}	Cd^{2+} 3%, $CdCl^+$ 97%
Pb^{2+}	Pb^{2+} 3%, $PbCl^+$ 47%, $PbCO_3$ 41%, $Pb(OH)^+$ 9%, $Pb(SO_4)$ 1%
Si(IV)	H_4SiO_4 92%, $H_3SiO_4^-$ 5%

Data from: Turner, D., Whitfield, M., and Dickson, A. G. (1981). *Geochimica et Cosmochimica Acta*, **45**, 855.

some function as a result of evolutionary pressure and can extract the elements it needs, sometimes to a quite remarkable extent, as can be appreciated from Table 1.5, and Fig. 1.10.

It is worth stressing that Table 1.5 includes elements that are not considered as 'essential', e.g. Ti, Nb, and Tl. Their extraction may be determined by competition with the 'naturally' required element, by substitutive capabilities (as is the case for Nb, which replaces V in some marine species of *Ascidiae*) or because the element extracted is indeed essential for the species concerned but essentiality has not yet been established. This same reasoning applies to many other elements which occur irregularly in living tissues and can be accumulated by certain species of plants and animals (cf. Table 1.2) but do not meet the requirements for essentiality as far as such studies have gone.

Table 1.5 Abundance of chemical elements in sea water and in some marine organisms (dry weight)

Chemical element	Total concentration in sea water (*) (p.p.b.)	Maximum concentration in marine organisms(†) (p.p.b.)
Ti	<1	10 000
V	1.5	280 000
Cr	0.3	1 400
Mn	0.03	4 100
Fe	0.05	86 000
Cu	0.3	7 500
Mo	10	6 400
Nb	0.01	98 000
Tl	0.01	700

p.p.b., parts per billion.
* Bruland, K. W. (1983). In *Chemical oceanography*, Vol. 8 (ed. J. P. Riley and R. Chester), pp. 157–220. Academic Press, London.
† Goldberg, E. D. (1957). *Memoirs of the Geological Society of America*, **67**, 345.

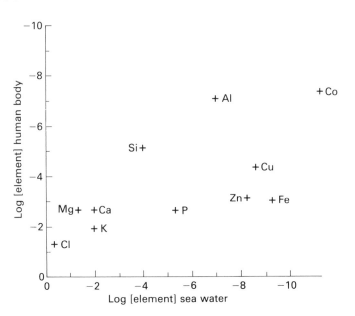

Fig. 1.10 A comparison between the molar concentrations of elements in the human body and in sea water. Against a general trend there are strong variations indicative of some of the problems faced in maintaining an element balance in man.

Elements such as aluminium, barium, bismuth, germanium, lithium, lead, rubidium, and titanium are likely candidates, but the prevalent view is still that these or other (non-essential) elements (about 30 of them) may have been acquired through contact with the environment, either by physical processes (adsorption, ingestion of particles, contamination of skin, etc.) or by chemical processes in competition with the 'normal' essential elements for their binding sites in biological systems. In this case, it is possible, if the competition becomes permanent, that some of these biological systems will adapt to the 'abnormal' conditions and change the (in principle) adverse effect of a non-essential element into a beneficial one. They will then have started to 'use' the new element; this is actually one of the general principles of evolution whereby, by adapting to new conditions, living species become more fit and increase their possibilities of survival. We shall look carefully into the history of the use of copper in the distant past (1×10^9 years ago) in Chapters 15 and 21 and at what man is doing today with the distribution of this and other elements (Chapter 22). A particularly interesting example is that of mercury.

Once again we stress too that the above description of hydrolytic equilibria applies to the majority of environments of life today. It does not apply to strictly anaerobic environments, to high-temperature, reducing waters, such as are found in deep ocean vents, nor did it apply in the early history of life. Table 1.6 shows some changes which occur on changing from anaerobic (early history of earth) to aerobic (recent history of earth) conditions. In fact life underwent a dramatic transition with the coming of dioxygen (Fig. 1.11). We return to this problem in Section 1.5.

Figures 1.4 and 1.5 point to another problem for living systems. The availability of the elements is a bio-geological problem on an immense time scale. In the past it has adjusted slowly allowing adaptation of species and development, but man is now having an effect on element availability over a short period of time. He has learned to apply enormous energy sources in an intense fashion to mining and agriculture (note the effect on the atmosphere). The benefits to human kind today are very substantial, but a glance at the figures in this chapter gives cause for concern. Biological systems are acting in

Table 1.6 Forms of elements and (availability)

Element	Reducing environment	Oxidizing environment
Iron	Fe(II) (high)	Fe(III) (low)
Copper	As sulphide (low)	Cu(II) (moderate)
Sulphur	HS^- (high)	SO_4^{2-} (high)
Molybdenum	$[MoO_nS_{4-n}]^{2-}$, MoS_2 (low)	MoO_4^{2-} (moderate)
Vanadium	V^{3+}, V(IV) sulphides	VO_4^{3-} (moderate)

Note the switch in availability of iron relative to copper that may have had a profound effect on evolution as we will show later. Vanadium sulphides are quite soluble.

adverse circumstances keeping out some common elements and taking in others which are not very available. What will happen if the availability of any element, even a rare one, changes by a factor of, say, 10 within, say, a period of 500 years? One particular concern today is with the acidity of water and release of aluminium. While concentrating upon this problem, constant vigilance will be needed in the immediate future to avoid the adjustment of availability of many other elements, e.g. Sn, Pb, Hg, Ni, and Be. Added to these problems are those of radioactive nuclides which are not of concern in this book. A further problem is the changing amount of O_3, CO_2, and so on in the atmosphere (see Chapter 22), but these last are problems of speciation (compounds) not of elements.

Fig. 1.11 The involvement of the elements in life has to be seen against the progression of the development of organization over historical time. We suppose earth has existed for over 4×10^9 years. We guess that organic molecules were formed haphazardly for a long period both through the action of light and on catalytic surfaces. Organization of chemistry in space and time evolved and is called living once it could reproduce. For a long period only simple cells, prokaryotes, developed. They had no internal boundaries. At some stage from about 2.7×10^9 years an evolution of dioxygen in the atmosphere developed due to prokaryote metabolism. A vast change took place by some 570×10^6 years ago bringing internal cellular structures of great diversity and multi-cellular organisms. (This figure is after Bernal (1967) and will be used in a somewhat different form in Chapter 21.)

Fig. 1.12 The solubility products of hydroxides (– – – –) and sulphides (———) which have been of extreme importance in evolution. The horizontal line indicates that solubility product which gives a precipitate at pH = 7.0 when [M] = 10^{-4} M.

1.4.3 Solubility products

In the oxidation state diagram for metals at pH = 7.0 (Fig. 1.7) certain of the elements are in simple hydrated ionic states or are present as simple anions. As we have stated, it is this that makes them highly available. Other elements are present as insoluble precipitates, e.g. TiO_2, and in such states these elements are virtually not available. The magnitude of solubility products of the hydroxides (oxides) of monovalent and divalent ions (Fig. 1.12) does not cause serious precipitation of any of the related elements of major biological interest, although there is an increase in insolubility following the Irving–Williams series. The situation for the trivalent ions is very different and all trivalent hydroxides are very insoluble at pH = 7; for example, $[Fe^{3+}]$ cannot exceed 10^{-18} M. It is this that removes elements such as Sc, Ti, Cr, Mn, Fe, and Ga from aqueous oxygenated waters. (Note that Mn^{2+}, Co^{2+}, Ni^{2+}, and Cu^{2+} do not give precipitates from 10^{-4} M solutions while Ti (TiO_2), Cr^{3+}, Fe^{3+}, Co^{3+}, and Ga^{3+} ($M(OH)_3$) all do). While these precipitations had and have a profound effect on life in aerobic conditions, much as does the insolubility of calcium, strontium, and barium (large cations) with oxyanions (such as phosphate and sulphate), there must have been an even more important influence on early life through the insolubility of sulphides (Fig. 1.12).

1.5 Sulphide chemistry in water (anaerobes)

On good grounds it is believed that the early atmosphere and the early sea did not have a redox potential range between the H_2 and the O_2 potentials at pH = 7.0. These are the conditions we have described for aqueous solution chemistry so far. It is much more probable that the early reaction systems in water including life were limited by the H_2S/S_n and H_2/H^+ potentials which cover the range 0.0 to -0.4 V at pH = 7. We should therefore reconsider the oxidation state diagrams, Figs 1.7 and 1.8, so that we include the reactions of H_2S/S_n as being more important than H_2O/O_2 and ignore the possibility of the establishment of redox potentials outside the anaerobic range. Since the redox potential of $(CHOH)_n$ type organic compounds is also around 0.0 V both in oxidizing and reducing directions this allows the full range of organic compounds from CO_2 to CH_4. Nitrogen may have been reduced to NH_3 or HCN when it would have been available. However, these early redox conditions had a dramatic effect on the availability of other elements, especially metals, and their reactions (Table 1.6). We attempt a semi-quantitative analysis, and to do so a value has to be put on the free H_2S concentration. We shall assume that it was around 10^{-3} M in water.

We shall start the discussion of metal partition between oxides and sulphides with one of the most important reactions for biology or indeed mineralogy under these conditions which can be written schematically as a competition between sulphide and oxide for hydrogen and metals

$$MO + H_2S \rightleftharpoons MS + H_2O.$$

Here we include under MO and MS the variety of hydrolysed forms of oxides

and sulphides from OH^- and SH^- compounds to the oxy- and thiolate anions such as MO_4^{2-} and MS_4^{2-}. Early in the periodic table in both of the first two rows, i.e. from hydrogen to chlorine, and in the first four A subgroups, sulphide chemistry is of little consequence. However, as soon as we consider transition metal or B subgroup chemistry we find remarkably stable sulphur compounds in water given even this relatively low H_2S presence. The easiest way in which to see this chemistry quickly is to look at the old-fashioned (unfortunately) qualitative analysis tables for students. The separation procedures used in those tables depend on the above equilibria and allow first quantitative separation in acid solution of the most insoluble sulphides, i.e. of the elements Mo, Cu, Ag, Cd, Sn, Pb, As, Sb, Bi, and many later second and third transition series metals which are never seriously involved in biological systems except as drugs. Notice immediately that, while copper is precipitated as Cu(I), many of the other elements are precipitated in high oxidation states, e.g. Mo, Sb, Sn, and As. Of these metals, Mo, Sn, As, and Sb sulphides dissolve in an alkali H_2S solution to give thio-anions. V and W sulphides are soluble in both acid and alkali, forming soluble anions, but they precipitate in neutral solution too. Sulphide removed Cu, Ag, Cd, Hg, and Pb very effectively from participation in early biology but it left V, Mo, W, Sn, As, and Sb as available elements at pH = 7.

Notice too that the chemical switch from oxide to sulphide chemistry going from P to As (and Sb), from Ge to Sn, from Cr to Mo is very dramatic and allows strict separation of these pairs of elements to be made easily. The distinction occurs in both the higher and the lower valence states of these elements and, whatever the theoretical explanation given, it was critical for the development of the geochemistry of the earth and its biological systems as we shall see in detail later.

Now in alkaline solution the qualitative analysis tables show that a further group of elements precipitate as sulphides including Fe, Co, Ni, Zn, and, less markedly, Mn (see Fig. 1.12). In the presence of 10^{-3} M H_2S, i.e. the presumed condition of early life, the concentration of all these metals was limited, but we must remember that at a pH close to the solubility product of any precipitate there will always exist small clusters in solution. We may well believe that biology took advantage of these sulphide clusters in the cases of Fe, Co, and Ni while Cu was virtually excluded since it is so insoluble (Fig. 1.12). There is little or no copper in archaebacteria to this day.

Returning to the equilibrium,

$$MO + H_2S \rightleftharpoons MS + H_2O,$$

Fig. 1.12 shows that for two elements, Mn and Fe, the equation is balanced in a very intriguing and different manner. The situation is complicated by the variety of oxidation states and the geometry demands (electronic states) of the two elements. However, the following points can be established: firstly, in the presence of H_2S at pH = 7 we need only consider oxidation states of these elements from II to IV, since higher oxidation states are very powerful oxidizing agents which are not stable with either oxide or sulphide as ligands in water but generate dioxygen and sulphur respectively. Considering the lower oxidation states we find that the tendency to form sulphide rather than oxide links is

$$Fe(II) > Fe(III) \gtrsim Mn(II) > Mn(III) \gtrsim Mn(IV) \gtrsim Fe(IV).$$

(It is probable that this order is a consequence of the exclusion of sulphide chemistry by redox reactions of Fe(IV) and Mn(III) and by the geometric demands by Mn(IV).) The consequences of these stabilities allowed Fe/S cluster chemistry, Fe(II) and Fe(III), to dominate early biology since iron must have been the most abundant metal and the Fe(II) sulphide is quite soluble. The link between availability of metal sulphide complexes and the evolution of biological systems will be stressed later in the cases of nickel, molybdenum, vanadium, and tungsten, as these elements with iron still dominate the simple chemistry of N_2, H_2, O_2, CO, CH_4, etc. in biology and within sulphide compounds.

Turning to manganese we see that its chemistry is more closely connected to the A subgroups in that the binding to sulphide is never dominant over binding to oxide (Fig. 1.12). The fact that it also has, apart from Mn(II), two just accessible oxidation states Mn(III) and Mn(IV) (see Fig. 1.7) made it possible for the fluctuating fortunes of evolution to uncover the photochemical generation of dioxygen unimpeded by sulphide reactions on manganese centres. The whole of advanced life depended and depends on the fact that the chemistry of manganese is dominated by oxide and not by sulphide under biological conditions, while the reverse is true for Fe, Ni, Zn, Mo, and Cu.

Amongst the non-metals we have already mentioned that oxide chemistry dominates them all except for hydrogen itself (H_2S) and sulphur (and selenium). In the reducing conditions *inside* cells to this day H_2S has a limited presence while sulphate is not stable. Outside cells in aerobic conditions sulphide is not stable whilst sulphate is.

It can be seen from this section that the forms of life are dependent on the state (and availability) of chemical elements, while the state of the chemical elements can slowly become dependent on the nature of life (Fig. 1.11). The chemistry of the surface of any planet dictates living forms and, as growth intensifies, this surface chemistry is adjusted by life and the system, planet and life, evolves interactively until they could become almost one. Lovelock and others claim that earth has reached this last condition—the so-called Gaia hypothesis—while others view the idea with deep doubt.

Against this background we introduce a note about the care with which biological systems maintain the element and small compound concentrations within themselves, so-called homeostasis, which in a sense is opposed to evolution. There is a very strong conservative character in biology which makes its forms very slowly adaptable. The importance of homeostasis will become apparent in Chapter 21 and here we give but a brief note.

1.6 A note on homeostasis

The famous French physiologist Claude Bernard called attention to the fact that all living species have two different 'environments': one, the external surroundings, the 'habitat', and the other, the internal 'milieu', made up of the biological fluids, e.g. the blood plasma. The external environment in its physical condition or chemical composition can change to a large extent and even abruptly, e.g. in tidal waters, but the internal medium must be kept approximately constant or else the metabolic reactions cannot be co-

ordinated, the mechanisms of adaptation cannot function, and the species will die. (Bernard was not to know that the environment is largely biologically generated as far as, especially, its inorganic chemistry is concerned.)

To see this clearly it is enough to observe some simple physical facts as to what happens in our everyday life: the body temperature may not be below 35° nor above 42°C and, when the external conditions are such that there is a risk of these limits being exceeded, several corrective mechanisms start operating to restore the normal conditions, e.g. muscle shivering to increase the temperature and sweating to lower it. For other animals there are similar mechanisms and protective devices, such as insulating fat or fur in cold environments, increased surface to ease losses by convection (e.g. an elephant's ears) or to promote evaporation (e.g. a dog's tongue) in the cases of excess heat. The same can be said for the chemical composition of biological fluids, e.g. even small deviations from the normal salt (NaCl) composition of the blood reflect a state of disease and prolonged abnormal situations will endanger life. This means that only those species with adequate regulating internal mechanisms can survive in external environments subject to changes. Many plant and animal species do not have very well regulated mechanisms and they must remain in biological niches where conditions are stable. This is the case, for example, for some primitive organisms, e.g. archaebacteria, which, lacking the adequate adaptative possibilities, are forced to live in anoxic media. Similarly, we can only live in aerobic media.

The same argument has been invoked to support the sea or, perhaps better, some sheltered closed ponds, as the birthplace or at least the major development environment for life. The earth's crust is too much exposed to sudden changes of temperature, pressure, or humidity and to natural phenomena, e.g. winds, volcanoes, freezing, etc., to be suitable for the staging of the first phases of the processes leading to life. The large water masses afford far better conditions since the high specific heat and viscosity of water ensure fewer thermic or mechanical disturbances and aquatic media are clearly suitable for chemical reactions, enable easier transport of nutrients, and increase their availability. As mentioned in Section 1.5, even CO_2, H_2S, and O_2 are reasonably soluble in water. Of course, there are other possibilities here aiding the origin of life, amongst which is the interface between water and a solid surface which will be discussed later.

This conservation of conditions in a living system, homeostasis, does not correspond to a state of equilibrium in thermodynamic terms (which would be equivalent to death) but to a series of related and controlled steady states in a dynamic process of continuous material, charge, and energy flows through the cells, with forced and fixed directions. Inside the cells there will then be a vast array of chemical transformations, leading to hundreds of compounds, development, and reproduction, and the overall process requires energy to effect reactions (Fig. 1.3). It is necessary to pump in reagents, to pump out the products, and to maintain the physical stress needed for the structural organization of these cells, as well as the osmotic balance and material or electric potential gradients. Due to these gradients the cell can respond to the outside environment and its internal metabolic reactions can be compensated for changes of external conditions, given that the system as a whole has feedback regulation. We shall have to examine these points very carefully. We shall see that they are very dependent on inorganic chemicals.

We can therefore say that the steady homeostatic states are definable in terms of energy supply and available concentrations of simple compounds *plus* energy utilization and concentration gradients. This addition to the more traditional concept of homeostasis is so important that it may be at the origin of the first manifestation of 'life' properties. We come again to this subject in Chapter 21.

1.7 Conclusion

In this chapter we have outlined the abundance and availability of the elements and their overall distribution in living systems. The only chemical combinations of elements with other elements which have been considered are those between all the elements and hydrogen, oxygen, and sulphur. These are the redox–hydrolytic equilibria of elements in water. They dominate availability. The nature of living systems is different. Through the use of energy they make a great variety of C, N, O, H, etc., compounds to which all other elements can gain controlled access and with which they combine. In turn, this access helps to create new organic/inorganic chemistry. In the next chapters we shall be considering this chemical speciation in a great variety of compounds in biology but we must remember that the elements which are ultimately found in biology are there for functional value.

Further reading

Bernal, J. D. (1967). *The origins of life*. Weidenfeld and Nicholson, London.

Bowen, H. J. M. (1966). *Trace elements in biochemistry*. Academic Press, London.

Cox, P. A. (1989). *The elements*. Oxford University Press, Oxford.

Goldberg, E. D. (1957). The biogeochemistry of trace elements. *Memoirs of the Geological Society of America*, **67**, 345.

Hewitt, E. J. and Smith, T. A. (1985). *Plant mineral nutrition*. The English Universities Press, London.

Lovelock, J. (1979). *Gaia—a new look at life on earth*. Oxford University Press, Oxford.

Lovelock, J. (1988). *The ages of Gaia*. Oxford University Press, Oxford.

Margulis, L. and Sagan, D. (1986). *Microcosmos*. Simon and Schuster, New York.

Phillips, C. S. G. and Williams, R. J. P. (1966). *Inorganic chemistry*. Oxford University Press, Oxford.

Turner, D., Whitfield, M., and Dickson, A. G. (1981). Inorganic speciation of trace metals. *Geochimica Cosmochimica Acta*, **45**, 855.

Williams, R. J. P. (1961). Functions of chains of catalysts. *Journal of Theoretical Biology*, **1**, 1–13.

Woese, C. R. (1987). Bacterial evolution. *Microbiological Reviews*, **51**, 221–71.

2

The principles of the uptake and chemical speciation of the elements in biology

2.1	General aspects	24
2.2	The separations that biology can achieve	26
2.3	The selective uptake of metal ions	28
2.4	Effective stability constants	31
2.5	Element uptake at equilibrium: the selectivity of the uptake	33
2.6	Hydrolysis	45
2.7	Selective control of oxidation states of metals	47
2.8	Selection by control of concentration by buffering	48
2.9	Selection by transfer coefficients from water to proteins	50
2.10	Soft and hard acids, solvents, and extraction	52
2.11	The selectivity of channels	53
2.12	Kinetic effects and control	54
2.13	Rejection of metal ions	58
2.14	General aspects of the uptake of non-metals	58
2.15	Mechanisms of selection of anions based on thermodynamic properties	60
2.16	Redox incorporation of anions	64
2.17	Coenzymes	65
2.18	Concluding remarks: element handling in biological systems	67
2.19	Precipitates in compartments	68

2.1 General aspects

We shall now examine how biological systems can take up and distribute differentially the elements they need for their functions. This will require us to discuss in detail the general mechanism used to achieve physicochemical movement and capture and the basis of the competition of the elements for biological organic molecular or solid structures. Since the plants are the primary sources of chemical elements for animal life they will be considered first, but even plants depend on micro-organisms for several elements they need. We will have to be concerned with the mechanisms operating at the cell level in different types of cells. General principles are discernible, but not all biological systems behave alike with regard to all elements.

Let us briefly examine how a plant appropriates elements present in its natural surroundings, e.g. the soil. The uptake of a certain element, M, involves a series of reactions which are interrelated as

The reaction scheme can be further generalized as in Fig. 2.1.

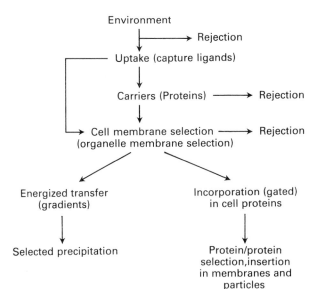

Fig. 2.1 Some pathways for the uptake and incorporation of elements.

Many of these reactions will not come to rapid equilibrium and we shall consider both thermodynamic and kinetic controls in the overall uptake. Step 1 is determined, essentially, by the occurrence and then the solubility in water of the mineral containing the element M in the soil (see Chapter 1). In some cases step 1' is operative, when the plant (or micro-organism) sends out a complexing agent L which solubilizes and captures the necessary element. Note that some plants acidify the soil around their roots to aid these steps. Step 2 is a diffusional process and depends on the nature of the chemical species containing M, on the distance to the plant roots, and on the nature (physical and chemical) of the soils. Plants break up soils to aid diffusion. Steps 1', 3, 4, and 5 are complexing reactions under thermodynamic or kinetic biological control; in particular, step 5 can be a kinetic device to ensure irreversible uptake of an element which may take the form of a simple barrier to free diffusion or an irreversible reaction (Fig. 2.1). Note that energy is used in two ways: (1) in synthesis of ligands; and (2) in pumping of M, possibly as ML, across boundaries. As one might expect, in some of these steps oxidation–reduction reactions may occur, as is the case for the metals Fe^{3+} and MoO_4^{2-} and the non-metals NO_3^-, SO_4^{2-}, O_2, and N_2, and many of these redox steps are also effectively irreversible.

Many of the uptake mechanisms are not truly specific, at least in the primary steps, and, given the possibility of competition, initial selection is based on the concentration and binding of the various elements near the plant root (or the absorbing cells). Since the affinity of the elements for common surface binding centres (organic carboxylates, phosphates, sulphates, etc.) does not vary much, selectivity is poor there (Table 2.1). For instance, the affinity of the alkaline-earth and first-series transition-metal ions for such ligands, represented by the stability constants of their complexes, shows little selectivity (see Table 2.2). Hence, many plants and especially lichens can

Table 2.1 Affinity of metal ions for some plant surface tissues

Ion	Species	Organ or tissue	log K
K^+	*Hordeum vulgare* (barley)	Excised roots	1.68
K^+	*Zea mays* (maize)	Leaf tissue	1.42
K^+	*Elodea densa* (waterweed)	Leaf (whole)	1.00
Rb^+	*Hordeum vulgare* (barley)	Excised roots	1.77
Rb^+	*Zea mays* (maize)	Leaf tissue	1.82
Sr^{2+}	*C. alpestris* (lichen)	Powdered plant	1.95
Ni^{2+}	*U. muhlenbergii*	Powdered plant ($pK_a = 5.40$)	3.30
Ni^{2+}	*U. muhlenbergii*	Powdered plant ($pK_a = 2.85$)	3.05
Cu^{2+}	Sugar cane	Leaf	1.87
Zn^{2+}	Sugar cane	Leaf	1.95
Mn^{2+}	Sugar cane	Leaf	1.79

K (for acid centres) $= [MA^+]/[M^{2+}] \cdot [A^-]$ where A is a surface group. The pK_a, where given, shows the proton affinity for A.

N.B. There is evidence that lichens have special binding enolates for higher valent metals which makes lichens peculiarly useful in studies of heavy metal pollution. See (for lichens): Tuominen, Y. (1964). *Annales Botanici Fennici*, **4**, 1. Nieboer, E., Puckett, K. J., and Grace, B. (1976). *Canadian Journal of Botany*, **54**, 724. Data for other plants adapted from: Clarkson, D. (1974). *Ion transport and cell structure in plants*, pp. 282–3. McGraw Hill, Maidenhead.

Table 2.2 Stability constants (log K_{ML}) of alkaline-earth and first-series transition metal ion 1:1 complexes with some carboxylate, phosphate, and sulphate ligands ($T = 25°C$)

	log K_{ML}									
	Mg^{2+}	Ca^{2+}	Sr^{2+}	Ba^{2+}	Mn^{2+}	Fe^{2+}	Co^{2+}	Ni^{2+}	Cu^{2+}	Zn^{2+}
Acetate	1.25	1.24	1.19	1.15	1.40	—	1.46	1.43	2.23	1.57
Malonate	1.95	1.85	(1.25)	1.34	(2.3)	(2.5)	2.98	3.30	5.55	2.97
Phosphate	1.60	1.33	1.0	—	2.58	—	2.18	2.08	3.2	2.4
Pyrophosphate	5.4	4.9	(4.7)	(4.6)	—	—	6.1	6.98	7.3	(5.1)
Methylphosphate	1.52	1.49	—	—	2.19	—	2.00	1.91	—	2.16
Sulphate	2.20	2.31	2.3	2.3	2.0	2.3	2.47	2.40	2.4	2.3
Thiosulphate	1.84	1.98	2.04	2.21	1.95	2.17	2.05	2.06	—	2.30

Source: Chemical Society (1964). *Stability constants,* The Chemical Society Special Publications, Nos 17 and 25. The Chemical Society, London.

concentrate elements on their surfaces, some of which are not considered as 'biological' (cf. Table 1.2 and Fig. 1.2) and, in some cases, can be toxic. *All* elements, even the essential ones, can be toxic if the uptake is excessive. We will return to this subject in Chapter 22.

The overall process described so far readily explains why, when non-essential elements are relatively abundant and become available to plants (micro-organisms or animals) for natural reasons or because of accidental contamination or environmental pollution, they may compete successfully with the essential elements and may even be preferentially captured. Over the long term this change in availability is part of evolution.

2.2 The separations that biology can achieve

A simple (prokaryotic) cell has two main features on top of the above external surface as traps for chemical elements: an external membrane system which does not allow diffusion of ions or large molecules; and an internal (cytoplasmic) concentration of trapped synthesized organic molecules, especially polymers, that can bind to many elements. Some such cells have two membranes within the outer wall and the space between them is called the periplasmic space. More evolved cells also have internal compartments (organelles and vesicles) to which the same considerations apply. A cell can, therefore, utilize five types of aqueous traps for elements (see Fig. 2.2).

(1) The cell membrane can be used as a physical barrier to diffusion of a free or bound ion.

(2) An element can be stored, free or bound, within an internal vesicle or organelle after passing a second membrane.

(3) Internal polymers can bind to an element at equilibrium.

(4) Internal polymers can incorporate an element forming thermodynamically unstable but kinetically stable bonds, e.g. covalent bonds in organic compounds (insertion reactions may be required).

(5) Two trapped elements can combine to form a precipitate at equilibrium.

(Note in passing that many of these steps can be used to reject elements.)

Fig. 2.2 A scheme for steps in the transfer of elements, M, through membranes into cells using ligands, L. Note the energy-dependent control of pH, $E°$, and synthesis.

Traps 1, 2, and 4 are always kinetically controlled; 1, 2, and 5 hinder free diffusion, and 4 eliminates the possibility of exchange, though the polymer may or may not diffuse. The non-metals H, C, N, O, P, S, and Se are very largely trapped according to trap 4, i.e. by forming kinetically stable but not thermodynamically stable molecules and polymers, such as proteins and polynucleotides. In a few cases this is also true for the halogens Cl, Br, and I, but their incorporation requires unusual oxidizing conditions and organic molecules with covalently bound chlorine and bromine rarely occur in higher plants or animals.

This incorporation of the more common non-metals is usually carried out in three steps: reduction of simple molecules or ions, e.g. CO_2, N_2, O_2, SO_4^{2-}, and SeO_4^{2-}, and then condensation of the products into monomeric small molecules and finally into polymers. The processes require energy and, although the subsequent degradation is usually a downhill reaction, it has a high activation energy; hence the incorporation of these elements is associated with strong kinetic barriers. (Notice that there is always a steady state concentration of small-molecule non-metal ionic forms too, e.g. HCO_3^-, HPO_4^{2-}, OH^-, H^+, amino acids, and a great variety of small 'substrates'.)

The metals cannot be readily trapped in this way. Sodium and potassium, particularly, form relatively weak complexes; hence they exchange and diffuse rapidly as simple ions. The major possible ways, steps 1 and 2, in which to retain them are by physical barriers to ionic diffusion, i.e. by membranes enclosing aqueous domains into which these ions must be driven using the energy of 'pumps'. This is not to deny that even these ions can be retained in special cyclic chelates, but this is an unusual mode of retention. The small anions containing non-metals can be retained in the same way.

For magnesium and calcium, as well as for the transition and group IB and IIB metal ions, it becomes increasingly possible to retain them by thermodynamic traps (step 3) in polymers (proteins, polysaccharides) but their final site often is required to be in a state from which exchange is very slow, i.e. for which the thermodynamic binding strength and/or the kinetic activation energy for escape is made very large. It is immediately obvious that carrier and transport complexes (L_1) in Fig. 2.2 must differ chemically from final sites (L_2) since a required property for the first is easy release of the metal.

In some systems a cation and an anion are pumped into a cell or one of its compartments until the solubility product of the salt ML is reached, i.e. mode 5. The solid is sometimes expelled out of the cells. Frequently, it grows within an organic matrix to produce the common biominerals: $SiO_2(H_2O)_n$, $CaCO_3$, $Ca_2(OH)PO_4$, $Fe(OH)_3$, or $Ca(C_2O_4)$. Once again these minerals must be correctly placed in space and the control of shape in the process of formation of the solid phases is of utmost importance. The minerals must remain in some steady state relationship with their free ions.

The way in which a certain element is trapped immediately imposes constraints upon its functional role. Very mobile elements, like Na^+, K^+, or Cl^-, are of little use as reaction centres, for instance in enzymes, but they can transfer ionic charge, a function that fixed ions cannot accomplish. But before we turn to the uses of the chemical elements in their selected sites we must see in detail how many of them can be selectively captured by more specific chemistry and sometimes even forced into particular sites, and here we shall consider passage across a single membrane only. This will be the purpose of

the next sections. Note that we shall be interested in the principles of *chemical* selectivity. Chapter 3 describes the several *physically* separated compartments into which different chemicals are distributed. When we put the two together we shall see how selectively elements are handled in biology (Chapters 8 to 19).

2.3 The selective uptake of metal ions

To a first approximation we can consider that when any organic surface is immersed in an aqueous media, e.g. roots in soil or plants in the sea, the uptake is based upon equilibrium considerations, i.e.

Amount bound varies as $K_{ML} \cdot$ [concentration of free ion]

where K_{ML} is the (effective) stability (affinity) constant (see Section 2.4) of the 1:1 complex of each metal ion with the ligand(s) at the surface. Since the concentration of the free metal ions in natural aqueous media falls dramatically (see Chapter 1), say, roughly as

K^+, Na^+	Mg^{2+}, Ca^{2+}	Zn^{2+}	Cu^{2+}	Fe^{3+}
10^{-1} M	10^{-3} M	$< 10^{-9}$ M	$< 10^{-12}$ M	10^{-17} M,

it is clearly difficult to incorporate large quantities of any other metal ions except K^+, Na^+, Mg^{2+}, and Ca^{2+}. Some special chemicals are clearly necessary to make K_{ML} values large and different from metal ion to metal ion, e.g. selective ligands based on different donor atoms (Table 2.3), to retain most metals. We know that for the alkali metals these are neutral oxygen atom, O, donors, e.g. crown ethers (see Fig. 2.6), but for the transition metals they are N/S donors (see Table 2.4). This means that deeper behind cell membranes there must be much stronger binding sites than those at the surface described in Tables 2.1 and 2.2, and, indeed, several lines of evidence show that these metal binding sites exist, undergo slower exchange, may become saturated, and may be coupled to energized steps which transfer the element inward. In order that the various metal ions can be selectively captured, different types of selective ligands must therefore be used, obeying conditions such as those shown in Table 2.4 where some simple ligands are suggested merely as examples.

The separation of elements into compounds by group, such as the three types in Table 2.4, may therefore be easily achieved by chemical synthesis of reagents. A parallel description at this level will be given in Sections 2.14 and 2.15 for the selective uptake of classes of anions. It is still not clear at this stage how the individual ions in each of the three groups in the table can be separated during their uptake into cells.

This complicated problem can be approached by separating all the individual possible steps as in Fig. 2.2 (neglecting, for the moment, competition between different elements, including the proton, and different ligands) and considering the following:

Table 2.3 Selective ligands based on different donor atoms

O-donor	$^-O_2C - CH_2 - CO - CH_2 - CO_2^-$
N,O-donor	$^-O_2C - CH_2 - NH - CH_2 - CO_2^-$
N-donor	$H_2N - CH_2 - CH_2 - NH - CH_2 - CH_2 - NH_2$
O-donor (phenolate)	benzene ring with two O^- groups (catecholate)
N-donor (aromatic)	2,2'-bipyridine
O,S-donor	$^-O_2C - CH_2 - S^-$
N,S-donor	$H_2N - CH_2 - CH_2 - S^-$
S,S-donor	$^-S - CH_2 - CH_2 - S^-$
O-donor (hydroxamate)	$R - CH_2 - N(O^-) - C(=O)R$
EDTA	$(^-O_2C - CH_2)_2 N - CH_2 - CH_2 - N(CH_2 - CO_2^-)_2$

Table 2.4 Differentiating ligands for some cations: stability constant (K_{ML}) and uptake factor ($K_{ML} \cdot [M]$)

	Na^+, K^+	Mg^{2+}, Ca^{2+}	Zn^{2+}, Cu^{2+}	Class of ligand
Type I				
K_{ML}	>10	$<10^2$	$<10^6$	O-Macrocycles (large hole)
Uptake factor	>1	<0.1	<0.1	(Fig. 2.6)
Type II				
K_{ML}	1.0	$>10^3$	$<10^6$	Di-or-tri carboxylates
Uptake factor	<0.1	>1.0	<0.1	(O-donors)
Type III				
K_{ML}	0.1	$<10^2$	$>10^6$	Nitrogen (or sulphur)
Uptake factor	<0.1	<0.1	>1.0	ligands

The concentration of M, [M], is given in Section 2.3.

(1) the ability of different ligands L_1 to compete with an external ligand, L_3, for M. This gives the thermodynamic ability to capture metals, M, and is in part a problem of relative stability constants, but ML_1 must be soluble in the organic membrane phase, hence we must consider, besides the stability constants K_{ML} in aqueous media, as above, the distribution coefficient between water and the liquid membrane, D, for ML_1 (which may be governed by a receptor). The ability of M to enter a membrane is then proportional to the product $K_{ML_1} \cdot D$;

(2) the rates of formation and of dissociation of the complexes ML_3 and ML_1 and any rate steps involved in crossing the interfaces;

(3) the further formation of ML_2 when similar considerations apply;

(4) the input of synthetic and degradative energy in the binding of M to L_1 and L_2 in the cell, taking into account the possibility of variations of pH and any redox changes of M due to the metabolic activity of the cell, as well as the stabilization of L_2 by M in ML_2;

(5) an additional energy-coupled transport of ML_1 through the membrane for which the necessary energy is provided by special energy-giving chemical processes coupled to the receptor in Fig. 2.2;

(6) the movement of ions M through selective channels, energized or not, that bypass the need for the formation of ML_1 (see the energy pump in Fig. 2.2);

(7) finally, the synthesis of scavenging and carrier ligands can be coupled to the concentration levels of the metal so that there is a feedback control of uptake even back to gene expression (see Chapters 12 and 21).

The possibilities considered under (5) and (in some cases) (6) are what we call active or energized transport. The processes (1) to (4) can be treated as equilibration within a steady state of the cell. Over relatively short periods this second assumption is valid since any further passage of M from ML_2 to, say, ML_4 can be considered as a slow irreversible step. A full rigorous treatment of all these processes is obviously impossible for lack of the necessary information, but some simplifying assumptions can be introduced and consequences derived on a semiquantitative basis.

We will illustrate an application of the principles for the case of a single cell with an outer bathing solution, a porous wall (outside the cell), an ion-impermeable membrane, an inner cytoplasm (inside the cell), and no other divisions or compartments. This is very like a prokaryote cell, e.g. a bacterium. We shall assume that the wall contains, overwhelmingly, (charged) O-donor ligands, but that the membrane and the cytoplasm have a range of N/S-donors as well as O-donors (mostly proteins). Taking the outside solution to be high in Na, K, Mg, and Ca, such that these elements are in excess over the ligands and relatively low in other metals (see beginning of this section), then the outer face of the cell will probably be a kind of Ca–Mg loaded ion-exchange resin surrounded by an aqueous phase rich in these free cations with which the ions in the wall can exchange. (This is in fact the case but note peculiarities in some cells, e.g. of lichens; see Table 2.1.) Other ions and ML_3 species will pass through this wall probably by simple diffusion. These trace elements can be complexed outside the cells by special ligands L_1 or L_3, e.g. iron by the siderophores, or exchanged from the wall ligands to other ligands L_1 at the

membrane. Generally, the alkali (and alkaline earth) metals do not form sufficiently stable species to be dealt with by these mechanisms, and their uptake (or their rejection) is usually by selective gated channels and pumping devices which pass them through membranes. In fact it is rather their loss from cells which is mediated by ionophore complexing drugs which allow transport of them out through the membrane.

Inside cells (and this includes the membranes), besides other O-donor ligands, there are a variety of N/S-donor sites designed to capture particular *trace* elements and to protect against damage from unwanted elements (Table 2.4). The unwanted elements may well be rejected.

From this general outline of possible events we can go to the finer details of the uptake and separation of the individual metal ions, but two important aspects must be stressed: first, there is a fundamental difference between the conditions outside the cells and inside the cells (or in the membrane), including differences in ion concentration and redox potential as well as pH. Outside cells the concentration of ions can be much higher than that of the (weak) ligands at the surface, whereas inside cells and in the membrane the opposite is true, i.e. the concentration of stronger ligands is under metabolic control and probably is (just) slightly higher than that of the essential trace ions transported. This suggests immediately that the supply of ligands and control of transport is under feedback control through their genetic apparatus for the production of L. This point will be shown using iron as an example in Chapter 12 and in more detail in Chap. 21. In the second place, the stability constants that should be used to describe the binding of metal ions to the ligands are not usual stoichiometric stability constants K_{ML} determined in aqueous solutions for simple metal–ligand equilibria, but the so-called *conditional* or *effective* binding constants K_{ML}^{eff}, which take into account the competition of the proton and of other metal ions for the sites of M and also the competition of other ligands present for M. This competition has very important effects on selectivity.

2.4 Effective stability constants

The effective stability constants are related to the conventional stability constants by the expressions (considering competitive complex formation for only two metals M_1 and M_2 and two ligands L_1 and L_2 and when only M_1L_1 and M_2L_1 complexes are formed)

$$\log K_{M_1L_1}^{eff} = \log K_{M_1L_1} - \log \alpha_L - \log \alpha_M$$

where

$$\alpha_M = 1 + \sum_{i=1}^{i=N} \beta_i \cdot [L_2]^i$$

for $M_1(L_2)_i$ complexes (and the simplest case is when there is only one of these complexes, M_1L_2) and

$$\alpha_L = 1 + \sum_{i=1}^{i=N} \beta_i^H [H^+]^i + K_{M_2L_1}[M_2].$$

The β_i^H are the overall formation constants of the species L_1H_i, $K_{M_2L_1}$ is the stability constant of the complex formed between M_2 and L_1, and the β_i are the overall stability constants of any $M_1(L_2)_i$ complexes formed. In effect, α_L is the reduction in L_1 available to M_1 due to the presence of H^+ and M_2, while α_M is the reduction of M_1 available to L_1 due to the removal of M_1 by binding to L_2. The values of $K_{M_1L_1}^{eff}$ may be much smaller than $K_{M_1L_1}$. The various equilibria locked in K_{ML}^{eff} generally are frequently so complex that only with the aid of computers can we assess the likelihood of formation of any ML in a determined set of conditions (see Da Silva and Williams (1976) in the 'Further reading' section and the diagram in the margin).

To illustrate the considerations affecting effective constants one may recall some simple examples, e.g. that of nickel ions that bind to ammonia molecules with a 1:1 binding constant for $[Ni(NH_3)(H_2O)_5]^{2+}$ of $\log K = 3.0$. However, protons bind to NH_3 with a $\log K = 9.0$; hence, at $pH < 6$, nickel ions do not bind NH_3 appreciably, i.e. the ability of ammonia to bind to nickel is limited. For plants growing in nickel-rich soils, the acidity of the storage vacuoles, for example, does not allow simple amines to bind to nickel and storage is associated with ligands of low acid dissociation constants (pK_a), e.g. organic carboxylates. In this case, the *effective* binding constant to nickel ions is indeed higher for an oxygen-donor ligand than for a nitrogen-donor ligand, which is the reverse order to that of the overall stability constants.

A similar type of limitation arises as the result of competitive binding by different elements M to a ligand L. The stability constants of their respective complexes may differ by a factor that largely overcomes differences in concentration and even small excesses of certain foreign elements over ligands may deprive some biological elements of their biological carriers or retaining ligands. In this respect, copper and mercury are particularly dangerous, since they usually form complexes of high stability with a wide range of ligands, and metal ions of elements such as Pb, Cd, Ni, Co, and Zn are also very competitive, especially for centres that bind through sulphur and/or nitrogen groups. Clearly, free metal ions of this kind must not be present except at very low levels, just as the proton concentration must be low for it also binds to these centres. Those aspects that concern toxicity will be dealt with in greater detail in Chapter 22.

Finally, it must be noted that most of the stability constants available are for simple small-molecule ligands, which may or may not be good models for the sites of the metal ions in biological molecules and were not measured under identical conditions. As will be discussed later, there are several differences which will necessarily be reflected in the values of the stability constants for the real complexes with proteins, for example. To a first approximation we shall take the experimentally determined model-ligand stability constants of their respective complexes multiplied by a factor which converts them into the values valid for a protein which employs exactly the same set of donor atoms. This aspect must be borne in mind in later discussion and we shall include a special section on the influence of protein fold stability on the binding of metal ions (Chapter 7).

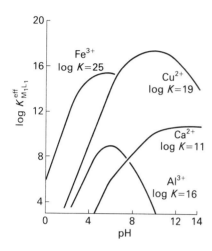

The effective stability constants at different pH values for EDTA complexes of four metal ions

2.5 Element uptake at equilibrium: the selectivity of the uptake

Note : no membrane processes involved

Fig. 2.3 A simplified scheme taken from Fig. 2.2 to illustrate uptake of M at equilibrium.

We have seen that the uptake of metal ions M is almost invariably energized: consequently, it is not a true equilibrium process. However, if we assume a steady state in certain steps, we can treat the uptake problem as one in which the element M distributes between only two very different phases containing, say, L_3 and L_2 separately and between which only M itself can equilibrate. We call this 'equilibrium uptake' (Fig. 2.3) since the process of transport in Fig. 2.2 is presumed not to interfere. Selected uptake can then be dealt with through analysis of the binding constants of M in two different compartments: the outer aqueous phase and the inner aqueous phase. The uptake of M will depend only on the stability of ML_2 relatively to ML_3 and no kinetic considerations or irreversible energized steps will be invoked in this approximation. (Energy is in fact put into the system just because L_2 and L_3 are synthesized in a controlled way and kept separate.)

On the other hand, since the values of K_{ML_3} in the outer phase do not vary much from ion to ion (see Table 2.2), the discussion of selectivity can be centred just on the values of K_{ML_2} (or of K_{ML_1} for the uptake into a membrane assuming $D=1$, see Fig. 2.2) and thus becomes a simple problem of co-ordination chemistry.

Now, general co-ordination chemistry teaches us that the thermodynamic binding of cations to organic chelating agents in mononuclear complexes follows certain trends and that the different metal ions have preferences for particular donor atoms and for particular coordination geometries. Some general characteristics and the differential properties induced are shown in Table 2.5 to indicate directions in which we must be thinking. The ligand types and geometry preferences are summarized in Tables 2.6 and 2.7 and in Fig. 2.4.

Fortunately, the stability of the complexes formed by the different types of ions follow sequences which are, to a large degree, independent of the ligand when this ligand is built from an internally mobile organic ligand, i.e. where steric selectivity is minimal. For monovalent ions the binding constants K_{ML}

Table 2.5 Some general characteristics of metal ion complexes present in biological systems

Characteristics	Na$^+$, K$^+$	Mg^{2+}, Ca^{2+}	Zn^{2+}, Ni^{2+}	Fe, Cu, Co, Mo, Mn
Predominant oxidation state	+1	+2	+2	More than one state
Stability of the complexes	Very low	Low or Medium	High	High, but medium in the cases of Mn^{2+} and Fe^{2+}
Preferred donor atoms	Oxygen	Oxygen	Sulphur or nitrogen	Nitrogen or sulphur (and especially oxygen in the cases of some Fe and Mn complexes)
Mobility in biological media	Very mobile	Semi-mobile	Static, Ni; usually static, Zn	Static (semi-mobile in the cases of Mn^{2+} and Fe^{2+})

(a)

Tetrahedral

Octahedral

Tetragonal

(b)

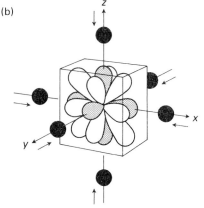

Fig. 2.4 (a) Common co-ordination geometries around a central metal ion. (b) d-orbitals in an octahedral field, where e_g (shaded) $= \sigma$-bonding and t_{2g} (unshaded) $= \pi$-bonding.

Table 2.6 Ligands preferred by different metal ions in simple co-ordination compounds

Metal ion	Ligands
Na^+, K^+	Oxygen-donor ligands, neutral or of low charge (-1)
Mg^{2+}	Carboxylate, phosphate, and polyphosphate (total charge > -2). N-donation (special)
Ca^{2+}	Carboxylate, phosphate (less than Mg^{2+}), some neutral O-donors
Mn^{2+}, Fe^{2+}	Carboxylate, phosphate, and nitrogen donors combined, (thiolate)
Fe^{2+} (special)	Unsaturated amines (particularly porphyrins)
Cr^{3+}, Mn^{3+}	Phenolate (e.g. tyrosine), hydroxamate, hydroxide
Fe^{3+}, Co^{3+}	Carboxylate, N-donors, polypyrroles, (thiolate)
Ni^{2+}	Thiolate (e.g. cysteine), unsaturated amines, polypyrroles
Cu^{2+}, Cu^+	Amines, ionized peptide $> N^-$, thiolate
Zn^{2+}	Amines, thiolate, carboxylate

The case of Fe^{2+} is especially complicated by spin-state changes and polymerization of mixed oxidation states.

Table 2.7 Preferred geometries in simple co-ordination compounds

Metal ion	Preferred geometries
Cu^{2+}, Mn^{3+}	Tetragonal $>$ 5-co-ordination $>$ tetrahedral
Ni^{2+}	Octahedral $>$ others
Co^{2+}	Octahedral $>$ tetrahedral $>$ others
Zn^{2+}	Tetrahedral $>$ octahedral $>$ 5-co-ordination
Mn^{2+}	Octahedral $>$ others
Fe^{3+}, Co^{3+}, Cr^{3+}	Octahedral $>$ others

* NB. In this table all metal ions are in high-spin states and ligands atoms are small O, N donors. S-donors favour lower co-ordination numbers. Ligand-field theory, that is polarization of and binding by the core electrons and orbitals of the metal ion (Fig. 2.4(b)), can explain the above observations; see inorganic chemistry textbooks in the 'Further reading'.

are $Cu^+ > Na^+$, K^+ and, for divalent ions, one has the Irving–Williams stability order $(Mg^{2+}, Ca^{2+}) < Mn^{2+} < Fe^{2+} < Co^{2+} < Ni^{2+} < Cu^{2+} > Zn^{2+}$, again largely independent of the ligand. Few stability constants are available for trivalent ions, but, in general, we expect and find the order Co^{3+}(low-spin) $> Mn^{3+} > Fe^{3+} > Cr^{3+} > Al^{3+}$.

These trends can be explained to a first approximation in terms of the ionic potential z/r (electrostatic binding) where z is the ionic charge and r the ionic radius of an ion relative to I_n the ionization potential of M to M^{n+}, i.e. the electron acceptor power of M^{n+}. The plot of z/r against I_2 for divalent ions (Fig. 2.5) shows a clear separation of the so-called 'hard' (a-class) and 'soft' (b-class) ions (see also Table 2.8) and also generates the Irving–Williams' stability order (see Fig. 2.8). (Similar plots are readily obtained for other oxidation states.) While this plot helps to get a quick overall view of stability constant trends such as those in Table 2.5, a more detailed description is required if a deep understanding of the selectivity factors which operate in both chemistry and biology are to be appreciated. We must introduce for

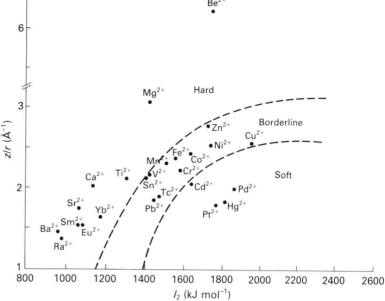

Fig. 2.5 A plot of z/r against I_2, ionization energy from M to M^{2+}, for some elements (see text).

Table 2.8 Hard and soft acids and bases

Hard (class (a))	Borderline	Soft (class (b))
Acids		
H^+, Li^+, Na^+, K^+, Be^{2+}, Mg^{2+}, Ca^{2+}, BF_3, BCl_3, $B(OR)_3$, Al^{3+}, $AlCl_3$, $Al(CH_3)_3$, Sc^{3+}, Ti^{4+}, VO^{2+}, Cr^{3+}, Fe^{3+}, Co^{3+}	Fe^{2+}, Co^{2+}, Ni^{2+}, Cu^{2+}, Zn^{2+}, Rh^{3+}, $B(CH_3)_3$, R_3C^+, Pb^{2+}, Sn^{2+}	Cu^+, Ag^+, Au^+, Cd^{2+}, Hg^{2+}, Pt^{2+}, Pt^{4+}, MoO_2^{2+}, Pd^{2+}
Bases		
NH_3, RNH_2, N_2H_4, H_2O, OH^-, O^{2-}, ROH, RO^-, CO_3^{2-}, SO_4^{2-}, ClO_4^-, F^-	$C_6H_5NH_2$, N_3^-, N_2, Br^-, Cl^-	H^-, R^-, C_2H_4, C_6H_6, CN^-, CO, SCN^-, R_3P, R_2S, RSH, RS^-, I^-

NB. Hardness is defined, see textbooks in the references, as $\eta_M = \frac{1}{2}(I - A_e)$ where I is the ionization potential of the state of M concerned and A_e is its electron affinity. I dominates for positive ions, whence the usefulness of figures such as Fig. 2.5, but in aqueous solution must be considered relative to z/r.

Ahrland, S., Chatt, J., and Davies, N. R. (1958). The relative affinities of ligand atoms for acceptor molecules and ions, *Quarterly Review*, **12**, 265–76.

Pearson, R. G. (1963). Hard and soft acids and bases, *Journal of the American Chemical Society*, **85**, 3533–9.

example stereochemical properties of ligands and preferred symmetry of sites, ligand–field effects, etc. for different metal ions.

The different ways of achieving selective thermodynamic binding, based on size, donor type, and shape of the ligand that we shall discuss, are, therefore:

(1) selection by charge;

(2) selection by ion size;

(3) selection by liganding donor atom;

(4) selection by preferential co-ordination geometry;

(5) selection by spin-pairing stabilization;

(6) selection by binding in clusters.

After discussing these thermodynamic factors we look at biological factors:

(7) selection by control of concentration of M and L;

(8) selection by transfer coefficients from water to proteins (or membranes).

Naturally, several of these mechanisms operate simultaneously and their separation in the discussion only aims at better identification of each possible basis of selection. While most of the discussion is based on model complex ion chemistry we shall refer time and again to biological molecules which illustrate the principles. Except where stated we refer to aqueous solutions.

2.5.1 Selection by charge type

In aerated aqueous media the elements of the first transition series are normally found in oxidation states $+3$ from scandium to iron and $+2$ from cobalt to zinc (Fig. 1.9); hence these two groups can be differentiated by those ligands preferring ions of higher charge. It is found that singly charged ligands of small size, hard ligands, prefer the highly charged, small, class 'a' (hard) cations and this means that appropriate biological ligands for the tri-positive ions are small first-row anions such as R_2N^-, RO^-, and F^-. Since F^- is uncommon in biology this ion cannot easily be used for capture devices. On the other hand, the concentration of R_2N^- will normally be small at pH ~ 7 since the pK_a of R_2NH is high. This leaves RO^-, which occurs in hydroxamates, enolates, phenolates, etc., as the adequate ligand to take up trivalent ions, e.g. Fe^{3+}, V^{3+}, and Mn^{3+}, preferentially. Chromium should also be captured by such anions, but there is yet no firm evidence for the occurrence of the corresponding species in biological systems. Aluminium is bound but more weakly and we find that, interestingly, aluminium is not taken up to any great degree. One of the interesting features here is the competition between organic ligands, e.g. phenolates, and inorganic hydroxide. Where the difference is small, hydrolysis to hydroxide precipitates prevents uptake, e.g. of aluminium, but this competition is affected by acidity, e.g. in acid rain conditions. At the beginning of the first row of the transition metals, scandium and titanium also are not taken up (Chapter 1) probably due to hydrolysis. Elements which would form even more highly charged cations are hydrolysed to anions. Vanadium, and molybdenum in the second row, are probably handled initially as oxoanions, VO_4^{3-} and MoO_4^{2-}. We return to a more detailed analysis of this problem in Chapter 17.

In conclusion highly negatively charged ligands having RO^- donors are useful in the selective uptake of a very few *trivalent* transition metal ions, e.g. Fe^{3+} and Mn^{3+}, while reduction to low oxidation states, e.g. Co^{2+}, Ni^{2+}, or greater relative hydrolysis of ions such as Al^{3+} removes these elements from competition for this type of ligand. Inside cells RS^- ligands assume greater importance and, surprisingly, are even found with trivalent metals, e.g. Fe^{3+}, see Section 2.10. Finally, we have to take into account competition from the

doubly charged anions O^{2-} and S^{2-} which can lead on to precipitation, see Section 2.19.

2.5.2 Selection by ion size

The order of stability of different metal complexes would be that of the ionic potential of the metal ions z/r if covalence, steric hindrance, or other specific factors did not operate. Hence, for the ionic complexes of the alkali and alkaline-earth metals separately one would expect the following orders of stability to be obeyed from their ionic radii

$$Li^+ > Na^+ > K^+ > Rb^+ > Cs^+,$$
$$Be^{2+} > Mg^{2+} > Ca^{2+} > Sr^{2+} > Ba^{2+}.$$

Li^+	Be^{2+}
0.6	0.4
Na^+	Mg^{2+}
0.95	0.65
K^+	Ca^{2+}
1.33	1.00
Rb^+	Sr^{2+}
1.48	1.13
Cs^+	Ba^{2+}
1.69	1.35

Ion sizes (Å)

However, these orders are seldom found except with the smallest anions, such as F^- and OH^-, and many inversions of order occur, depending on the type of ligand used. The inversion is the more noticeable the larger the anion, e.g. SO_4^{2-} against OH^-. (NB $BaSO_4$ is insoluble but $MgSO_4$ is not.)

The simplest reason for the inversion of an order is of a stereochemical nature, which is readily understood with polydentate ligands which cannot fold around the smallest metal ions, e.g. Li^+, Be^{2+}, and Mg^{2+}, due to the crowding strain caused in the ligand on co-ordinating these ions. The ligands therefore do not displace water of hydration easily (water is a small ligand). We can analyse the problem on a semi-quantitative basis starting from the solubility of salts MX. The equilibrium governing selectivity is

$$M^{n+} \text{ (hydrate)} + X^{n-} \text{ (hydrate)} \rightleftharpoons MX \text{ (hydrate)} \downarrow + m\, H_2O.$$

Apart from the free energy gain of the water on the right, it is clear that the difference in electrostatic energy between the product and the reactant (ions of charge z^{n+} or z^{n-} and radius r) is related to a difference between individual ion hydration energies and lattice energies of MX salts,

$$\Delta G \propto C \left\{ \frac{z^{n+}}{r_M} + \frac{z^{n-}}{r_X} \right\} - B \left\{ \frac{z^{n+} z^{n-}}{d_{MX}} \right\} \tag{2.1}$$

where d_{MX} of the lattice $= r_M + r_X$ to a first approximation and C and B are constants. By using C and B as fitting constants we are taking repulsions into account approximately. Simple examination of this function for a fixed r_X shows that at either very large or very small values of r_M in the first term, the hydration energy, dominates. In a somewhat obvious way the lattice is most stable somewhere near the equality $r_M = r_X$. Detailed analysis allowing for $d_{MX} \neq r_M + r_X$ only reduces the importance of the lattice term. This means that ions of approximately equal size pack together best. The solubility of salts can be discussed in detail using eqn (2.1) and it can be demonstrated that even the residual structural hydration of ions in lattices in the most stable form of $MX(\text{hydrate})_n$ is decided by the radius ratio of anion and cation. As the anion gets larger relative to the cation there is an increasing probability that a hydrated cation, not the bare cation, will come out of solution with a given anion and that the hydrated salts are then the most insoluble, e.g. $BaSO_4$ and $SrSO_4$, but $CaSO_4 \cdot H_2O$ and $MgSO_4 \cdot 7\,H_2O$, which is to be contrasted with the series with a small anion, $Ba(OH)_2 \cdot 2H_2O$, but $Sr(OH)_2$, $Ca(OH)_2$,

$Mg(OH)_2$, and $Be_nO(OH)_m$. In biology we find $BaSO_4$, $SrSO_4$, and $CaCO_3$. The data obviously reflect the drive toward matching of sizes of ions.

Next we shall consider the effect of radius ratio on the formation of ion pairs in solution. There is a close parallel between the above analyses of solubility and that of the reaction for complex ion or ion pair formation

$$M^{n+}(\text{hydrate}) + X^{n-}(\text{hydrate}) \rightleftharpoons$$
$$[M^{n+} \cdot X^{n-}](H_2O)_y \text{ or } [M^{n+}(H_2O)_y \cdot X^{n-}] \quad + \quad m\,H_2O.$$

The equation for the formation energy of the complex is the same as that of eqn (2.1) but the constant B is different. In solution we expect therefore that size factors, of ions and water, will be important so that small anions and cations on one hand and large cations and anions on the other tend to associate. Additionally, in some complexes it will be the hydrated ion which dominates to give the species $M(H_2O)_yX$, whereas in others the bare ion will dominate to give $MX(H_2O)_y$. The difficulty in proving the point in solution lies in recognizing the species. We must go to spectroscopic studies in order to recognize them, but the central point is that ion-pairs always have structure-dependent energies as well as metal ion radii dependent energies. We have analysed the situation for small inorganic anions and cations but the analysis is equally applicable to all sizes of charged molecules—substrates, proteins, and DNA, for example.

Selection by complexation based on size is greatly used by biological systems. The ligands synthesized are often cyclic and create a central cavity which fits most closely a particular cation size, e.g. either binding is $K^+ > Na^+$ or $Na^+ > K^+$. It has been possible recently to synthesize similar compounds which are so tailored that they strongly favour the complexation of one particular ion, namely the so-called 'crown-ethers' and other macrocyclic ligands (Fig. 2.6(a), (b)). There are well documented parallel examples in bacterial antibiotics, e.g. nonactin and valinomycin (Fig. 2.6(c)), where the bactericidal action is due to the leakiness of membranes to K^+ or Na^+ through the presence of these ligands. (Note the ability to enter membranes (Fig. 2.2) as well as good binding.) Magnesium is taken up very effectively in a porphyrin-type cyclic ligand of fixed geometry and cavity size, chlorophyll, in clear contrast with calcium, which is excluded due to its large size, but it must be pointed out that the insertion of magnesium into this nitrogen heterocycle requires special conditions (see Section 3.12).

A second general method to generate size selectivity based on covalent ligand design is by protein folding. Proteins are large molecules and they can generate by folding different types of sites particularly suited for binding certain metals (and indeed many non-metals, cofactors, and coenzymes (Fig. 2.7). Here the fold strength is such that it provides a site (hole) which either: (1) can collapse around all ions and therefore is a better site for small ions; or (2) can only collapse to a particular large hole favouring large ions or even clusters. Typical examples are magnesium in phosphomutases (Chapter 9) or the calcium sites in many calcium-binding proteins, e.g. calmodulin, parvalbumin, thermolysin, etc. (Chapter 10). In most cases the proteins are folded in such a way as to bind preferentially one of these two metal ions virtually exclusively given the biochemical concentration conditions.

Fig. 2.6 Illustrations of biological ring chelates made from neutral O-donors for selective binding to ions such as K^+. (a) The cryptate (2,2,2)-diamine; (b) stability constants for complexes of the above cryptate with the group IA cations; (c) structure of the rubidium cryptate-222 cation; (d) structures of two naturally occurring macrocyclic antibiotics, nonactin and valinomycin; (e) structures of the potassium complexes of naturally occurring macrocyclics enniatin B (D-hydroxy-isovaleric acid-*N*-methyl-L-valine) and nonactin.

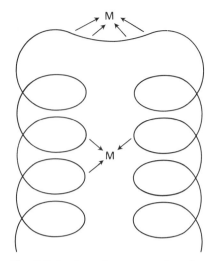

Fig. 2.7 A schematic representation of two ways in which a protein fold can generate sites of restricted size and geometry for any element M.

Fig. 2.8 A plot of binding constants, log K, for a series of single ligands with the divalent ions from Mn^{2+} to Zn^{2+}. The ligands are cystein (\bigcirc); ethylenediamine (\blacktriangle); glycine (\square); and oxalate (\bullet).

In many cases, however, size is not the only determinant factor and is complemented, even in the case of simple alkali metal ions, by the preference for a well-defined stereochemistry of the ligands (octahedral in the case of magnesium), the nature of the liganding atoms, and other effects. For instance, in phosphomutases co-ordination of magnesium appears to be by two carboxylates plus one imidazole group in an octahedral geometry where the metal ion retains three water molecules. Calcium binds relatively poorly to such a site. It prefers binding to carboxylate groups, together with carbonyl groups, in irregular geometry. This difference is seen in inorganic chemistry too. Discrimination between the two metals is not achieved only by the difference in ionic size, but clearly depends on stereochemistry and ligand–donor nature. We return to these problems in Chapters 8 to 10.

Finally, size selectivity is sensitive and can affect ions of quite similar size, e.g. compare the stability data with EDTA-like ligands for Mn^{2+} (0.75 Å radius) and Zn^{2+} (0.60 Å radius) (see Chapter 14, Fig. 14.2). This is also important in considerations of redox potentials where ion sizes change with charge.

In conclusion, ligands can be synthesized such that the constraints on the hole size which they generate or can generate are a powerful selectivity factor.

2.5.3 Combination of radius and charge effect

Clearly, a description of selectivity based on electrostatics for ions of both different charge and size needs a quantitative analysis of both the attractive (z/r) and repulsive energies. Such a treatment cannot be given here without interrupting the general consideration of factors involved in selective ion uptake. We shall therefore take up the whole matter of the electrostatic selective interactions in Chapters 8–10 where we analyse the biological chemistry of group IA and IIA cations more thoroughly. In particular, it will be important to notice that the problem of understanding in biology is made much more complex by the occurrence of organic cations, e.g. RNH_3^+, which compete for the anionic centres (Chapters 8 and 9).

2.5.4 Selection by liganding atom

This mechanism is based on the affinity of the a-group metal (hard) ions for oxygen (O)-donor ligands, whereas the b-group metal (soft) ions prefer to co-ordinate to sulphur (S) or nitrogen (N) donors. This topic is well dealt with in standard textbooks and reviews (see 'Further reading' section) and we will not discuss it in detail. The differences of stability between the complexes of the transition metals with ligands containing different donor atoms become greater the more polarizable the donors, i.e. they increase in the order O ligands < N ligands < S ligands (see Fig. 2.8). This figure, which again illustrates the Irving–Williams' order of stability, shows that the divalent transition metals form complexes that may differ in their thermodynamic stability by several orders of magnitude if the liganding atoms are appropriately chosen.

Now we have pointed out that outside the cell most donors are of the O-donor variety and RSH ligands are oxidized away. This means that uptake, inside/outside selectivity, can be based on soft/hard discrimination and this

would apply to all the metals and ligands equally if there were no control over ligand synthesis and metal uptake. Copper ions bind most strongly and, if copper were in excess over the total free N/S ligand concentrations, all the other divalent transition metals would be excluded from N/S centres. However, through biological control mechanisms (Section 4.4), particular ligands inside or outside the cells are likely to be present in only slight excess relative to particular metals, which are limited in availability. Both ligands and metal ions may also be restricted by selective pumping across membranes. Competition can now be such that each metal occupies its proper site and does not interfere with the distribution of those that form less stable complexes. The *effective* stability constants K_M^{eff} are operative in determining an adequate distribution. For this reason copper is bound by the best donor sites, e.g. 4(N) (as in superoxide dismutase) inside cells and 4(N/S) (as in plastocyanin-like proteins outside cells or in vesicles), overcoming competition with zinc which is more available. Once almost all copper is bound and there is no excess of its ligands, then for any site of lower donor power K_{Zn}^{eff} [Zn] will exceed K_{Cu}^{eff} [Cu]; hence the remaining 4N or N/S sites (or centres of two or three N/S donors, perhaps with one carboxylate ligand) bind zinc, e.g. carboxypeptidase or carbonic anhydrase. Nickel and cobalt are relatively rare elements especially in advanced biological cells so K_M^{eff} [M] for these metals is generally low and very special capture devices (vitamin B_{12} and factor F-430) are needed for them. (There are exceptional anaerobic environments and organisms for which this is not true of nickel, see below.) *Divalent* Mn and Fe, although abundant, do not bind strongly to any of these donors in sites of low co-ordination number and do not interfere in the distribution—they must be dealt with by biological systems in a different manner. They often bind to a variety of centres from four histidines (N) to a two-carboxylate and one-nitrogen centre but always in higher, often *octahedral*, co-ordination and they are then in fast exchange. Their selective geometric demands are fully utilized (Table 2.7). (We return in Section 2.7 to the ability to form mixed oxidation state clusters.) It will now be seen that combination of metal and ligands is dependent on the donor atom but also on the input (or exit) rates of metal ion content in the cell and on the rates of synthesis and degradation of ligands. This is most important; it is a reminder that biological systems are kinetic states of matter.

The selectivity of organic O-donors relative to N- and S-donors is part of the 'a'/'b' or hard/soft classification of metal ions generally and does not just apply to mono and divalent ions. Amongst isolated trivalent ions, M^{3+}, only occasional affinity for organic S-donors is apparent within the metals of biological interest, e.g. Fe^{3+}, but note that this is not true for Tl^{3+}, Bi^{3+}, and the M^{3+} ions of later elements in the second and third transition metal series. Such competing ions are likely to be very poisonous since they can displace required divalent ions.

In conclusion, selectivity by donor atom type is very powerful if the concentrations of ligand and metal are controlled.

2.5.5 Selection by preferential co-ordination geometry

In some cases, use can be made of the preference of certain metal ions for co-ordinating sites of particular geometry. The cases of zinc(II) and copper(II) are

given first as an example. Both cations are only available in trace amounts, though zinc may be 100 times more readily available in soils and water than is copper. To counterbalance this, copper binds to most ligand centres at least 100 times more strongly than does zinc (see Section 2.5.4). Thus, log K_M^{eff} [M] is similar for both metals with many conformationally mobile ligands and we must ask how can selective uptake be achieved with over 99 per cent efficiency as is required in both copper and zinc enzymes. We believe that selectivity here is based upon their different polarizing power, the use of different donors, and their different stereochemical demands. (Here we ignore the influence of selection for copper(I), which is critical in some cases, and the enforced movement of copper and zinc into selected biological spaces (Chapter 3).)

In the first place, copper(II) shows a remarkable ability not shared with zinc to form a bond at pH = 7 to peptide nitrogen in its ionized state. In this case the selectivity factor copper/zinc is perhaps as great as 10^4. Positive selection for copper is further enhanced by utilizing the preference of this metal for a tetragonal hole, (A), which arises from the ligand-field Jahn–Teller effect. An example of one of these effects is found in the copper-carrying plasma protein, an albumin. One or two of the metal–nitrogen bonds it forms are ionized amides while the geometry of the metal site must be that of a roughly square planar, four-nitrogen ligand, chelate. The site geometry is a property of the ligand conformation. (Here is an example of selectivity *outside* a cell which has few RS$^-$ ligands.)

Positive selectivity for zinc can be based on its different demand now for tetrahedral geometry, (B), with nitrogen or sulphur ligands, and many examples of this are known in important biological zinc sites, e.g. in carbonic anhydrase and alcohol dehydrogenase.

A fine example of simultaneous zinc and copper selectivity appears in the blue-green superoxide dismutase: the zinc site is close to tetrahedral and the copper site is a distorted tetragonal hole (Fig. 15.6). Both sites have nitrogen bases as ligands and one of these (for copper) is an ionized nitrogen of imidazole.

There is a second pair of ions in the first-row transition series where strong advantage can be taken of different preferences for their co-ordination geometry: Mn^{3+} and Fe^{3+}. The first ion prefers tetragonal sites, like Cu^{2+}, whereas Fe^{3+} prefers tetrahedral or octahedral co-ordinating geometries. Due to the higher charge of the ions the preferred ligands in these cases are the phenolates of tyrosines and OH$^-$. Examples of the two situations are found in a second type of superoxide dismutase (SOD), the wine-red Mn^{3+}-SOD and the yellow Fe^{3+}-SOD. The dioxygen-release manganese centres of chloroplasts may select Mn against Fe in this way too but we must be careful— biology can select by preferential pumping.

There is a most peculiar feature of the association of particular stereochemistries, ligand field stabilization, and functional use in biology. The two elements from the first transition metal series that in hydrates and oxide rocks favour most strongly an octahedral geometry are Cr^{3+} and Ni^{2+}, (C). We know that both are relatively available in the sea (Fig. 17.1), but neither has an important role in advanced, aerobic, organisms in marked contrast with Mn, Fe, Co, Cu, and Zn. It is as if the strong preference for a particular octahedral geometry and the possession of virtually no redox chemistry led to the exclusion of these elements. However, in primitive anaerobic organisms nickel is used in special ways in part associated with quite different geometries and

(A)

Tetragonal Cu^{2+} bis ethylenediamine

(B)

Tetrahedral Zn^{2+} bis ethylenediamine

(C)

**Octahedral Ni^{2+} bis
ethylenediamine dihydrate**

oxidation states. We shall return to these problems of choice of metal ion in biology later since they may well concern the kinetics of reactions rather than thermodynamic selectivity (Chapter 4).

While the metal ions have preferred sites, it is the organic ligands which generate holes of particular sizes and shapes in biology. An example from model chemistry is the formation from three molecules of orthophenanthroline (D) of an octahedral arrangement around low-spin Fe^{2+}, see margin, but the sterically restricted formation from two molecules of 2,9 dimethyl orthophenanthroline (E) to give a tetrahedron around Cu^+ (Section 2.7). These two reagents are greatly used to analyse for Fe^{2+} and Cu^+ selectively and they illustrate the parallel selectivity in biology by haem proteins for Fe^{2+} octahedral, and plastocyanin for Cu^+, tetrahedral.

We have preferred to illustrate the ligand-field polarization of the metal ions rather than give a more detailed theoretical outline since this is available in many books (see the 'Further reading' section). The theory, well supported by experiment, shows that factors of up to 10^3 in stability can arise from a matching of ligand structures with energetically preferred metal ion polarizations (see Table 2.7).

Combining stereochemical preferences and types of ligand we can build a table of selectivity in which we list the preference of four of the ions considered for different sites (Table 2.9). Note that Mn^{2+} prefers octahedral co-ordination of N, O mixed donors (Table 2.9) to a greater degree than Zn^{2+} in a competition in which the supply of ligands is limited.

In conclusion, metal ion polarizability (ligand-field stereochemical effects) generates considerable selectivity on top of charge, size, and ligand polarizability differences.

(D)

(E)

Table 2.9 Table of selectivity for Cu^{2+}, Zn^{2+}, Mn^{3+}, and Fe^{3+} (selection by adequate combination of donor atom and geometry of site)

Geometry of site	Donor atom			
	Phenolate RO⁻	Ionized peptide >N⁻	Thiolate RS⁻	Nitrogen >N or >N, O
Tetrahedral	Fe^{3+}	—	Zn^{2+}, Fe^{3+}, Cu^+	Zn^{2+}
Tetragonal	Mn^{3+}	Cu^{2+}	Cu^{2+}	Cu^{2+}
Octahedral	Fe^{3+}	—	(Mo(V))	Mn^{2+}, Fe^{2+}

2.5.6 Selection by spin-pairing stabilization

For transition metal ions there is a further possibility to increase selectivity, based on a spin-state change of a metal on binding to some ligands. The spin states, i.e. the way electrons are packed in 3d orbitals in the presence of different ligands, will generally be found to be as in Tables 2.10 and 2.11. Notice in Fig. 2.9 that changes of spin state affect ion size, and this is put to considerable use in biology, e.g. in haemoglobin (Chapter 13).

The selection of iron in preference to copper in porphyrins could perhaps be based on the low-spin stabilization of Fe^{2+} and the same may be said for the selection of cobalt in vitamin B_{12} in the presence of other divalent metals. However, later we shall show that kinetic effects are much more likely to dominate these incorporations. An important point again is that the

Table 2.10 Usual spin states for some metal ions

Metal	M^{2+}	M^{3+}
Mn	High-spin (d^5)	High-spin (d^4)
Fe	Low-spin, high-spin (d^6)	High-spin (d^5)
Co	High-spin (d^7)	Low-spin (d^6)
Ni	High-spin (d^8)	Low-spin (d^7)

Table 2.11 Spin-state switches for some metal complexes

Cation	d-electrons	High-spin	Low-spin
Mn^{2+}	5	The vast majority	CN^-
Fe^{2+}	6	NH_3, H_2O, Cl^-	CN^-, NO_2^-, phen, cytochromes
Co^{2+}	7	The vast majority	CN^-, vitamin B_{12}
Ni^{2+}	8	The vast majority	CN^-, F-430
Cr^{3+}	3	All	None
Mn^{3+}	4	The vast majority	CN^-, phen
Fe^{3+}	5	H_2O, F^-, Cl^-	CN^-, phen, cytochromes
Co^{3+}	6	F^-	All the others
Ni^{3+}	7	O-donors	CN^-, many

phen is 1,10 orthophenanthroline. Note the change to low-spin, paired d electron, states gains σ-acceptor and π-donor strength offsetting spin-pairing energy.

Fig. 2.9 Radii of the divalent ions Ca^{2+} to Zn^{2+} (left) and trivalent ions from Sc^{3+} to Ga^{3+} (right) as a function of the number of 3d electrons. Low-spin ions are shown as solid circles. (From Shannon, R. D. and Prewitt, C. T. (1970). *Acta Crystallographica*, **B26**, 1076.) 1 pm = 100 Å.

selectivity factor will only operate if some metal ion concentrations, e.g. Cu^{2+}, are lowered by their removal in a complex with a controlled supply of a more powerful S,N ligand donor.

It is then expected that strong N/S-donor ligands will bind low-spin $Co^{3+} > Mn^{3+} > Fe^{3+} (> Cr^{3+})$ and this may also apply with RO^- ligands, while the order for low-spin divalent ions will be $Fe^{2+} > Co^{2+} > Mn^{2+}$. Model ligands such as 1,10 orthophenanthroline can be used to illustrate spin-state changes (Table 2.11).

2.5.7 Selection by binding in clusters

The binding of metallic elements to proteins developed in a further way. In their higher oxidation states the hydrolysis of ions at pH ~ 7 is quite extensive and some low molecular mass species of hydrolysed units occur. These states will be present in all surface waters decreasing the availability of the simple ions. The most likely species to be present are Fe_2O^{4+}, $Cu_2(OH)_2^{2+}$, and Mn_2O^{4+}, highly charged cations with small, hard, anions. All three dimeric units can form complexes and, in fact, these dimer centres are present in some proteins where, in the course of evolution, they have undergone subtle changes so as to become of functional value both outside and inside cells. Interesting examples are those of haemerythrin, containing a Fe_2 dimer linked by one

O^{2-} and two carboxylate bridges (Fig. 12.9), the copper dimer in copper oxidases (Fig. 15.13), and the manganese units in the dioxygen-producing enzyme of photosynthesis (Chapter 14).

Inside cells (and also outside cells during the early history of life), sulphide ions are also present, and can react exactly as the hydroxide ions do, to give polysulphide units, e.g. $Fe_4S_4^{4+}$, based on linked tetrahedra of iron in the Fe^{3+} or both the Fe^{3+} and Fe^{2+} states, but not Fe^{2+} alone. These units can also bind to small molecules or proteins and widely known examples are those of the Fe_2, Fe_3, and Fe_4 units of ferredoxins, where S^{2-} acts as a bridge and cysteinate (RS^-) as the protein ligand. The selectivity between oxide and sulphide in dimers is in favour of sulphide in the order $(Mo) > Fe^{3+} \geqslant (Cu)^{2+} > Mn$ (see Chapter 1) as a consequence of the changing ratio of $I/(z/r)$ (see Fig. 2.5). Mn–S units have not been found in proteins. There is also the possibility of formation of mixed high-valence clusters and this may be the basis for the finding of Mo–Fe–S and V–Fe–S species in nitrogenase and for the formation of some nickel centres. Generally, clusters are stabilized by highly charged cations since they involve highly charged anions, O^{2-} and S^{2-} and in biology are only observed in proteins (see Section 2.6).

Note that this binding of metal ions in pairs, or even more at a time, is naturally opposed by entropic considerations, but the unfavourable term in the association is greatly reduced if the binding organic anionic centres are already built into a preformer, e.g. a protein, such that the protein is largely folded into the correct conformation for the uptake of the clusters. This is a gross size selection much as we have seen to occur on a smaller scale for simple metal ions. This type of co-operativity effect will be discussed later in a wider context. The selection here also depends on the highly charged states of the cations (see Section 2.5.1) and clearly operates in favour of strong binding to organic anions; in this sense it assists $Fe^{3+} = Mn^{3+} > Cu^{2+}$. Clusters are usually found bound to anion protein sidechains. Note that the trace element zinc is excluded from this type of site since it only forms weak clusters.

Oxo-hydroxo cluster

Fe_4S_4 cluster

2.6 Hydrolysis

Cluster formation is a type of hydrolysis (sulpholysis). As stated in Section 2.5.1, in the case of trivalent and more highly charged cations there is competition between complex ion formation with organic ligands and hydrolysis. Furthermore, the hydrolysis is of two kinds leading to hydroxy-species and, ultimately, hydroxide polymers and precipitates, and oxo(oxene) species which can also take part in polymerization as oxides (see Table 2.12). These reactions allow the appearance in biology of a large variety of shapes and sizes of oxygen-bound metal ions which must be held in controlled amounts. The reaction scheme is

$$M + H_2O \rightleftharpoons MOH(+H^+) \rightleftharpoons MO(+H^+)$$
$$[M(OH)]_n \qquad\qquad [MO]_n$$

As the pH of a solution of Fe^{3+} is raised, the amount of Fe_2O^{4+} rises rapidly as does the amount of a variety of oxy-/hydroxy-polymers such as $Fe(OH)_3$,

Table 2.12 Hydrolysis and sulpholysis of ions

General formulae	M^{2+} (M(II)) Simple ions	M^{3+} (M(III))	M^{4+} (M (IV))	M^{5+} (M(V))	M^{6+} (M(VI))
		MO^+	MO^{2+}	MO^{3+}	MO_2^{2+}
		$M–O–M$	$M–O–M$	MO_2^+	MO_4^{2-}
		$M–S–M$		MO_4^-	MS_4^{2-}
Specific examples	Mg^{2+}, Zn^{2+}	Fe^{3+}	MnO^{2+}	MoO^{3+}	MoO_2^{2+}
		$Fe–O–Fe$	FeO^{2+}	MoO_2^+	MoS_4^{2-}
		$Mn–OH$	VO^{2+}	$MoOS^+$	$MoOS^{2+}$

There are also many mixed states in polymers where there are internal O or S bridges and terminal groups. Note that S replaces O more readily down any group of the periodic table and increasingly toward the end of the transition metal series and in B-subgroups.

ferrihydrite, $FeO(OH)$, e.g. goethite, and Fe_2O_3. There is interconversion of the solids with ageing. We shall find these and other hydroxy/oxide precipitates in biology in Chapter 20.

A contrasting behaviour is that of V(IV) which has a variety of forms in biological media including VO^{2+} and V^{4+} in complexes. (For comparison we have the well-known inorganic species UO_2^{2+}, which is related to U^{6+}.) The formation of oxene species can also occur in well separated stages so that we may write for Mo(VI), for example,

$$MoO^{4+} \rightleftharpoons MoO_2^{2+} \rightleftharpoons MoO_3\downarrow \rightleftharpoons MoO_4^{2-}$$

with hydroxy-variations.

Whereas with the uranyl ion, and to some extent in vanadyl, VO^{2+}, chemistry, most reactions favour the maintenance of oxene units, in the case of molybdenum the different oxo groups can be replaced by organic chelating agents so that Mo(VI) chemistry in biology extends from units in which there are no oxene units to all the other possible chemistries of MoO^{4+}, MoO_2^{2+} and MoO_4^{2-}. The units can all be selectively recognized by proteins as we shall see in Chapter 17. We shall find that in biology the oxo group of VO^{2+} may also be replaced by ligands in some situations.

As already stated, oxygen is only one element which can engage in this type of chemistry; outstandingly in the inner compartments of biological systems today (and in early history in the environment) another available element is sulphur which can give the states M(SH) or MS. There is then competition between the deprotonation of H_2O and that of H_2S, which is also biased by redox reactions in the cellular compartments. Several factors are important in this case: (1) the dissociation constants of H_2S are much lower, making SH^- and S^{2-} much more available; (2) however, the concentration of H_2O exceeds that of the maximum H_2S concentration by some 10^5 and in conventional aerobic biological media this figure approaches perhaps 10^{10}, when there is very little H_2S; (3) the relative affinities of metal ions for sulphur- and oxygen-donor ligands (H_2S and H_2O) are metal-selective. In the presence of reasonable amounts of H_2S we know these affinities from the old student group analysis tables (sulphide affinities in precipitates for divalent ions)

$$Hg, Cu, Cd, Pb > Ni, Co, Zn > Mn, Fe > Mg, Ca.$$

(This is, in part, the Irving–Williams stability series.) For higher valent ions the sulphide preference is

Mo, As, Sb, Bi, Sn > V, Fe > Mn > Al.

Although this background to the oxide and sulphide chemistry is useful, it will not give more than general guidance for an understanding of the complexity of biological reactions but it allows us to note with no great sense of surprise that the oxo- and sulpho-complexes we meet in biology are only of the kinds

FeO^{2+}, MoO^{4+}, MnO^{2+}, VO_2^+, VO^{2+}, and possibly CuO^+ (mononuclear),

$[Fe_2O]^{4+}$, $[Mn_2O]^{4+}$, and possibly $[Cu_2O]^{2+}$, MoS^{4+}, Fe_nS_n,

$(Fe, Mo)S_n$, $(Fe, V)S_n$

where only Mo or V carry terminal S and n is variable perhaps from 2 up to 6–8.

The difficulty of understanding the selective appearance of the different types of hydrolysis product and non-hydrolysed ions does not lie only in the inorganic chemistry but in the *controlled evolved* selection of proteins to incorporate the different shapes and sizes of these single sites and clusters (see Chapter 12). One general rule appears to be that oxide species appear with O- and N-donors from proteins and have high co-ordination numbers while sulphide species appear with S-donors and have low co-ordination numbers.

Finally, we observe that it is relatively rare to observe sulphide precipitates in biology, but there are anaerobic species which contain them.

At this stage we hope to have introduced all the possible ways in which bonding aids selectivity of combination. Very much has been learned from the studies on small model complexes.

2.7 Selective control of oxidation states of metals

One of the major tools of the analytical chemist is to use the redox potential of the medium to control differentially the redox states of metal ions under analysis. Since this control involves compartments in biology (see Chapter 3) only a few illustrative examples are given here.

The easiest way in which to incorporate copper selectively into organic ligands is to reduce this metal ion to Cu(I). In analytical chemistry the use of reagents such as 2,9 dimethyl orthophenanthroline in reducing conditions allows selective estimation of copper(I). It is the combined use of reducing conditions, the strong binding of Cu(I) to nitrogen donors, plus the demand for tetrahedral geometry by the Cu(I) ion, see margin, which makes for selection. We shall see that biological systems have taken full advantage of this combination of factors in copper proteins.

A second example which we gave earlier was the use of anionic chelating agents to scavenge for iron(III) outside cells. The redox conditions and the selective reaction based on high charge effectively excludes divalent ions of metals such as Cu, Co, Ni, and Zn from phenolate complexes.

These manipulations of elements that depend on recognition of particular redox states require us to note carefully the redox conditions in compartments and to have knowledge of the redox potentials in different complexes. The potentials are not analysed separately in this book since they follow from the

redox potentials in the oxidation state diagrams of Chapter 1 and the stability constants of individual states using the equation

$$E = E^0 + \frac{RT}{n\mathscr{F}} \ln K_{\text{red}}/K_{\text{ox}} \tag{2.2}$$

where K_{red} and K_{ox} are the stability constants of the two states concerned.

2.8 Selection by control of concentration by buffering

In the following sections situations in which model chemistry is less helpful are analysed. We have seen that selectivity can be greatly enhanced by control over [M] in the expression $K_{\text{ML}}^{\text{eff}}$[M]. There may be conditions in which it will be advantageous if sites that are not truly specific will have to capture certain elements from among others which are more available so as to generate some special biological (chemical) function. A possible mechanism, certainly used by biological systems, is that of 'sequestering' the more common elements preferentially by using an even more powerful second ligand. The concentration of the 'interfering' elements will be lowered and their product K^{eff}[M] for the given site will be made smaller than that for the element to be captured. Let us look a little more closely at the nature of the discussion in Section 2.4.

Disregarding acidity effects for the sake of simplicity, i.e. assuming that the proton does not complete for the site, the desired situation can be represented as

$$M_1 + M_2 + L_1 + L_2 \rightarrow M_1L_1 + M_2L_2,$$

M_1L_2—excluded by lack of M_1, removed by L_1,

M_2L_1—excluded by lack of L_1, removed by M_1.

The generation of this circumstance can be approached in terms of effective stability constants

$$K_{M_1L_1}^{\text{eff}} = \frac{[M_1L_1]}{[M_1]^*[L_1]^*}$$

where $[M_1]^*$ is the concentration of M_1 *not bound* to L_1 and $[L_1]^*$ is the concentration of L_1 *not bound* to M_1, i.e. in each case it is a sum which one can easily derive as

$$[M_1]^* = [M_1]\{1 + K_{M_1L_2}[L_2]\},$$

$$[L_1]^* = [L_1]\{1 + K_{M_2L_1}[M_2]\}.$$

If binding of M_2 to L_1 is to be avoided, we have to decrease $[M_2]$ to such an extent that $K_{M_2L_1} \cdot [M_2]$ becomes smaller than say 0.01 (to be neglected in the expression for $[L_1]^*$). This can be done by preferential binding of M_2 to L_2. In this case, since

$$[M_2] = \frac{[M_2L_2]}{[L_2] \cdot K_{M_2L_2}},$$

one obtains

$$\frac{K_{M_2L_1} \cdot [M_2L_2]}{K_{M_2L_2} \cdot [L_2]} < 10^{-2}.$$

Now, if L_2 is in excess relatively to M_2, e.g. twice as abundant as M_2, it is enough that $K_{M_2L_2} > 10^2 K_{M_2L_1}$ to decrease $[M_2]$ to a level such that its competition with M_1 for L_1 will be less than 1 per cent. Larger ratios in the constants will allow the same effect with just a small excess of L_2 over M_2.

To avoid binding of M_1 by L_2 the same reasoning can be applied and one obtains

$$\frac{K_{M_1L_2} \cdot [M_1L_1]}{K_{M_1L_1} \cdot [L_1]} < 10^{-2},$$

and the problem is now related to the degree of saturation of L_1 by M_1 or to the differences in the values of the constants.

In general, for approximately equal concentrations of M_1 and M_2 and controlled small excesses of L_1 and L_2, selection giving only M_1L_1 and M_2L_2 will be achieved up to 99 per cent accuracy by sites differing by 10^2 in the ratio of their stability constants, i.e.

$$\frac{K_{M_2L_2}}{M_{M_2L_1}} > 10^2 \quad \text{and} \quad \frac{K_{M_1L_2}}{K_{M_1L_1}} < 10^2.$$

This requires that the order of ligand preference by different metals is different. In fact, this is the case for ions such as Mg^{2+} and Ca^{2+} compared with Cu^{2+}, as can be seen from Table 2.4 where Mg^{2+} and Ca^{2+} prefer O-donors but Cu^{2+} prefers N or S donors. Now, however, Cu^{2+} and Ni^{2+} both prefer N-donors over O-donors and here it is necessary that the stability constants are such that there is a higher selectivity in favour of Cu^{2+} with any one ligand in order to bind copper almost exclusively, see margin. Figure 2.8 shows that this condition will easily be achieved in biological systems using N/S ligands. Therefore, selection is dependent on the restrictions on concentrations of metal ions relative to ligands as well as on stability constants of the respective metal complexes.

Finally, M_2 can be H^+ when the approach given earlier in the chapter on effective stability constants reveals whether M_1 or H^+ will bind to any given L. The acidity of a compartment is critical.

This amounts to the statement that, provided biology limits free ligand and free metal ion concentrations (by controlled feedback mechanisms acting on synthesis and hydrolysis of the ligand and on the uptake and rejection of the metal ion, respectively), the selectivity of metal ion/ligand combination can be

An example of a selective anionic nitrogen ligand for Cu^{2+} in a tetragonal field

Table 2.13 Carriers in blood plasma

Metal ion	Free M	Carrier	Basis of selectivity
Iron (and manganese)	10^{-16}	Transferrin	Phenolates bind M^{3+} octahedral complex
Copper	10^{-14}	An albumin	Ionized peptide $>N^-$ binds Cu^{2+} in tetragonal complex
Zinc	10^{-8}	An albumin	N(S) donors in tetrahedral geometry bind Zn^{2+}
Cobalt (as vitamin B_{12})	$\ll 10^{-10}$	Carrier protein	The B_{12} sidechains are recognized
Calcium	10^{-3}	Phosphoproteins (when needed)	Weak binding for most metals
Magnesium, sodium potassium, (calcium)	10^{-3}–10^{-1}	None	Freely circulating as ions

well-nigh complete given known stability constants. This proviso will prove of the greatest possible interest to us.

An example of control of concentration to achieve selectivity is that of the blood plasma, where the metal ions are removed one by one by powerful ligands so as to avoid mutual interferences (see Table 2.13). Thus, copper is removed first by an albumin containing an ionized peptide bond (after complexation). Iron is oxidized and the blood-protein transferrin picks-up the Fe^{3+} ion using phenolate binding groups. Zinc is also removed by an albumin which may have three or four neutral N-donors in a tetrahedral geometry. Manganese, Mn^{3+}, may be also transported by transferrin (same site as Fe^{3+}). Cobalt as Co^{3+} is carried as vitamin B_{12} (insertion control; Section 2.12.3); sodium, potassium, magnesium, and, to some extent, calcium are left free in the blood solution. If extra Ca^{2+} is required, it cannot travel as free Ca^{2+} in the blood since an increase in concentration would exceed the solubility products of its phosphate or carbonate. There is then a special calcium carrier protein with phosphorylated sidechains, which is generated under hormonal control. If Cu, Fe, Mn, and Zn had not been removed, these ions would bind strongly to the calcium sites and the process would be impossible. We see here that it must be the combination of kinetic limitations on ligand synthesis and on the transfer of ligands and ions in and out of compartments together with inorganic chemical selectivity which dominate observed combinations. In all cases it is K^{eff} which we must consider. The consequences for biology are very considerable. Many states of disease relate to failures in these control processes. In turn, this means that we must analyse compartmentalization (Chapter 3), kinetics (Chapter 4), and biological synthesis of ligands (Chapter 7). The fact that pumps can be used to transfer any of the components becomes a dominant factor over selectivity since they control, with synthesis, the concentration terms.

2.9 Selection by transfer coefficients from water to proteins

So far, the discussions on selectivity rest, for the most part, on the assumption that a protein behaves like a model complexing agent made from small monodentate ligands even in a membrane. In fact, a protein adjusts its fold around a metal ion and the changes in protein fold energy are likely to be commensurate with the metal–ligand bond energy. A compromise in co-ordination numbers, bond lengths, and bond angles is forced upon the metal and the fold of the protein imposing a strain on both. The result can be a distorted energized situation to generate an activity which has been called an 'entatic' (under tension) state of the metal ion.

But it is not just a matter of energetic effects in the first co-ordination sphere of the metal, since the fold generates an environment for the whole complex ML, where L is the equivalent set of metal ligands in a model complex. The energies of transfer of a simple complex ML from an aqueous medium (aq) to the inside of a protein (in a membrane or not) of lower dielectric constant can be very large and also selective, e.g. for M^{2+} against M^{3+}. The overall stability of a metal–protein complex is then related to a binding constant log $K_{ML(aq)}$ for the free unlinked amino acids that compose the binding groups by a transfer

Iron porphyrin. Particular porphyrins have multiple ring substitutions. The protein-binding groups are attached above and below the Fe.

Table 2.14 Complexes of iron porphyrin IX with proteins

Protein	Binding groups	$E°$ (mV)
Myoglobin	His or His, H_2O	$> +100$
Haemoglobin	His or Hir, H_2O	$> +100$
Cytochrome b_5	His, His	0
Cytochrome b_{562}	His, Met	$< +100$
Peroxidase	His or His, H_2O	-200
Catalase	Tyr or Tyr, H_2O	-500

coefficient p of ML from water into a real protein. This involves local strain in ML and a considerable interaction of ML with second- and third-sphere neighbours, which can be favourable or unfavourable, and we now show that p can generate selectivity just as much as log $K_{ML(aq)}$ through long-range energetics.

This can be illustrated by reference to the uptake of haem (iron), i.e. a special ML, into haem proteins, see (A) and (B). It is known that the haem can be bound in different ways (see Table 2.14) and these generate changes of the normal redox potential over a range of more than 600 mV. Hence, different proteins can selectively distinguish complexes of Fe^{3+} from Fe^{2+}, and therefore, in principle, of any M^{3+} from any M^{2+} by a factor of more than 10^{10} simply by changes of *one* or at most *two* ligands, keeping the other four fixed. Even without change of ligands, however, Table 2.14 shows that protein folds can distinguish Fe^{3+} from Fe^{2+} by a factor of 10^5, e.g. in peroxidase and myoglobin; this results just from a change of strain or 'solvation' within the protein, i.e. the redox potential of the iron couple is in part controlled by the degree of exposure of haem to solvent water and to protein hydrophobic sidechains. In effect this result implies that a selectivity of any M^{3+} from any M^{2+} can be obtained by a change of internal dielectric constant, i.e. that of the protein. Similar selectivity applies to the partition of ML complexes into organic phases, e.g. membranes.

We must not be surprised, therefore, if sometimes M^{3+} and $(M')^{2+}$, e.g. Fe^{3+} and Zn^{2+}, are selected by the same co-ordination partners in the same geometric environment but by different protein sequences. This is observed in rubredoxin (Fe) and alcohol dehydrogenase (Zn) where the metal ions have four thiolate ligands. Similar factors can select Cu^+ against Zn^{2+} and again we know that iron–sulphur proteins select between $|Fe_4S_4|^+$, $|Fe_4S_4|^{2+}$, or $|Fe_4S_4|^{3+}$ and also select between $|Fe_4S_4|$, $|Fe_3S_4|$, $|Fe_2S_2|$, and Fe all bound to the same tetrahedral group of four thiolate ligands. The protein fold decides the complex that is captured using many factors. Now if it were so that different metals favoured different types of sulphide complexes M_nS_m, then proteins could also select between them, e.g. in the uptake of Mo–Fe–S as opposed to Fe–S species. Even when $n = m$ the size and charge of M_nS_m will be sufficiently different that several metals can appear in the same general co-ordination sphere selected by different proteins on the basis of folding energy, i.e. 'p' can be very discriminating.

It also follows that the fold itself will be dependent on the exact metal or metal complex so that the surface of the protein can vary depending on the metal or metal complex and, of course, this dependence is variable being related to the strength of the protein fold. Here there is total co-operativity

(A)

(B)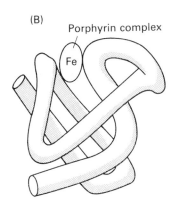

(A) The strained co-ordination geometry at iron in an iron porphyrin bound in a protein as at (B)

between metal ion geometry and protein fold. This may affect considerably the distribution coefficient D of M (in a protein) between an aqueous and an organic phase which is again a means of achieving selectivity. Examples are the folding of metallothionine (different for Cu, Cd, and Zn) and their passage through membranes. Of course, the metalloprotein may be recognized by other proteins as well as by membranes based on this protein fold difference. Here we begin to see how a particular ML may be positioned in a cell after uptake (Chapter 3).

Finally, we note that where the protein folding has open to it several conformations of somewhat equal free energy then different binding sites on a protein become interactive, co-operative, or antico-operative. The possibility of the adjustment of one site by another is called an allosteric or rack interaction, e.g. as is found in dioxygen uptake by haemoglobin. The different sites can be for a metal ion and, say, two proteins when metal/protein A and protein A/protein B build a communication network dependent on the binding and on the redox state of the metal ion.

Before discussing this aspect in greater detail, especially its kinetic implications, we must stress again that in the previous considerations we have often made the important assumption that a given ligand (protein) is in only slight excess over the essential trace metal ion that it binds. This will only arise in biology if the synthesis of the protein is stopped once this small excess is achieved and there must therefore be a requirement for feedback homeostasis which effectively buffers free [M] and free [L] which may only be overridden locally by hormonal effects. The feedback becomes specific to M if there are different folds for different metal ions. We will come later to this aspect since we believe it contains the kernel of why element chemistry has so dominant a role to play in biological activity.

The above discussion of selectivity principles can obviously be applied to effects upon redox potentials between any two oxidation states and it applies equally to anions (see Section 2.15).

2.10 Soft and hard acids, solvents, and extraction

The classification of 'soft' and 'hard' acids and bases given in Table 2.8 is for aqueous solutions as far as ions are concerned. It is therefore solvent-dependent. Free ions behave quite generally more as 'a'-class (apparently *hard*) the lower the dielectric constant is, i.e. in organic solvents. However, when we consider systems of two immiscible solvents and the partition of complexes between them

$$M^+ + X^- \rightleftharpoons MX \rightleftharpoons MX$$
$$\text{aquo} \qquad \text{aquo} \qquad \text{organic solvent}$$

it is usually the case that MX in the organic phase is relatively more strongly favoured the more 'b' class the anion and the cation. The electrostatic controls over these equilibria are obviously considerable. Now, the interior of a protein or of a membrane is more like an organic solvent than an aqueous solution and the anion, X, may be trapped there so that 'b' class behaviour of all metal elements may be more prominent in proteins than expected from aqueous

solution equilibria. There is a selectivity term here too in that complexes of one metal ion can have a different partition coefficient, D, from others. Thus metal ions can be phase-separated.

2.11 The selectivity of channels

Elevation section

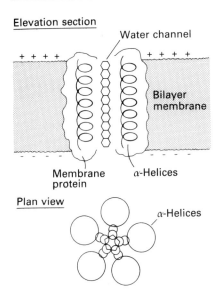

Fig. 2.10 A simplified view of a channel in a membrane. In real systems the channels can be in open or closed states and the state is gated in various ways, e.g. electrically or conformationally.

Fig. 2.11 The possible H-bond network for a proton channel.

The existence of selective ion channels in the membranes is the basis of a special case of transport that is under thermodynamic control. To be selective a channel does not have to bind an element M; it is enough that it has the same radius as that of the (hydrated) ion, since being somewhat mobile and flexible molecules, proteins can simulate 'holes' of variable size and perform a 'sieving' action, allowing the free flow of certain ions while preventing that of others (Fig. 2.10). The sieving does not require that the interior of the channel is not hydrated and evidence suggests that the channel diameter for some ions is large, say 5–10 Å diameter, yet it still sieves (see Chapters 5 and 8). Selectivity, in this case, is shown by the fact that there are highly discriminatory potassium channels, sodium channels, calcium channels, and even proton channels (as well as various anion channels). Typical highly selective channels are the potassium channels of nerves, which show a selectivity $Li < Na \ll K > Rb > Cs$, and the calcium channels which have selectivity $Mg \ll Ca > Sr > Ba$ (see Chapters 8 and 9). Since for these ions the energy at a site is a function of the inverse of the radius, it is clear that a channel hole can only match one ion closely for which there can be no real binding since it can pass unimpeded. This, in effect, is the same radius ratio effect that controls binding to chelating agents but here there is a further problem, that of mobility, which requires a different process ensuring free movement *in* and *out* of the cell. Selection is based in part on repulsion, not on attraction: smaller ions do not cross the channels since they are too *big* to enter—they carry more water molecules and cannot be dehydrated except at the considerable expense of energy; larger ions cannot enter due to their size. Selection is then very effective and for these reasons Na^+ and Ca^{2+} channels exclude K^+, Li^+, Mg^{2+}, i.e. larger and smaller ions independently from charge considerations by factors of, say, 10- to 100-fold. The lanthanides block the calcium channels, and barium blocks the potassium channels, showing that there is some interaction with channel surfaces since $La^{3+} = Ca^{2+}$ and $Ba^{2+} = K^+$ in size. In practice many channels are controlled by gates, kinetic devices allowing or preventing free flow. The 'gates' may be charge movements in the channels due to external electrostatic potentials.

The case of the proton is clearly different from that of other cations since it binds easily to several side-groups of proteins. A unique possibility is that of successive protonation–deprotonation of water molecules (or other OH groups) inside a protein channel (Fig. 2.11). The exact network of hydrogen is not confined to linked H_2O; it could be any H-bond system which is continuous and can perform the necessary motions. Such channels are extremely important in energy capture devices (Chapter 5). Usually we do not think in terms of selectivity for protons, but biological systems have an outstanding need to make such a selection. The acidity of different biological compartments is an example of the result of such membrane selection.

2.12 Kinetic effects and control

The next step in our analysis of the uptake of metal ions will focus on the rate of movement of M from the external to the internal aqueous phase (see Figs. 2.1 and 2.2). There are several possibilities which we consider in turn. These considerations could well oppose the thermodynamic assumptions made earlier (see Fig. 2.3), but we consider that it is more probable that they will act so as to enhance the thermodynamic selection within a steady state.

2.12.1 The distribution coefficient D for ML_1

The first limitation on this movement is the restriction upon the concentration of M in the membrane due to a generally low distribution coefficient D for the ML_2 complex. This reduces the rate at which ML_2 can form (Fig. 2.2). However, the effect can be made more selective if there is a special membrane protein, a receptor, that recognizes L_1 virtually only when it is bound to a particular metal ion M through stereochemical factors. The rate at which the metal crosses the membrane depends on the existence of such transport systems that may be refined to the level of protein–protein interactions and which may, through highly specific receptor recognition processes, act so as to restrict the distribution of different metals in different cells and cell compartments. An example of this is the special uptake of iron from transferrin into bone marrow cells upon which we will comment later. There is also a most intriguing system of transport proteins, called Ton B, for the movement of metal chelates, such as vitamin B_{12} molecules and siderophores, into *E. coli* (see Chapter 12).

A particular case arises when an adventitious ligand L_1 is sent out of the cell to aid in the capture or sequestration and removal of some elements. The availability of M to L_1 is then dependent on the controlled synthesis of L_1, which is also a kinetic device. In higher organisms, L_1 can be a trap for an element which removes it into one organ, e.g. the kidney, so that it is not available to another. We describe in Chapter 11 the particular case of iron. Once again feedback control can determine that, once a cell has a sufficiency of M, it no longer synthesizes the ligand for capturing this metal ion. The distribution rate now depends on the rate of synthesis and cycling of the ligand, L_1.

2.12.2 Kinetics of transport

Although the thermodynamic binding features of ML_1 and ML_2 (Fig. 2.2) are very similar, a characteristic of transport systems, as opposed to those of retaining centres, is that the former must bind and release M reasonably quickly since transport must be continuous. The release of M may be achieved by lowering the oxidation state of the metal (to decrease the binding constant), by changes of pH (to decrease the effective binding constant), by changes in the conformation of the protein (to change the co-ordination site of the metal), or by hydrolysis of the ligand.

2.12.3 Insertion in pre-formed holes or chelating rings

A particular method of selection by kinetic control may also be considered: that of insertion into a pre-formed hole or chelating ring, from which exchange of the metal ion concerned is virtually impossible. This reaction is not rapid, due to steric constraints, as mentioned before. Consider the case where, while the binding of M to L_1 is governed by thermodynamic factors, it could well be that the rates of transfer of M from ML_1 to L_2 are M-dependent and, in fact, L_1 could merely act to control those rates

$$M + L_1 \underset{k'_{ML}}{\overset{k_{ML}}{\rightleftharpoons}} ML_1 \xrightarrow[+L_2]{k} ML_2 + L_1 .$$

$$\qquad\qquad\quad \text{initial} \qquad \text{final}$$
$$\qquad\qquad\quad \text{uptake} \qquad \text{uptake}$$

An example makes clear the selectivity that can be introduced by this mechanism. The insertion of magnesium in the chlorin ring of chlorophyll cannot be made directly *in vitro* in the presence of other free metal ions, like copper, zinc, or even nickel and cobalt. It is hard to see how any reasonably devised intermediate step could prevent the latter ions from entering preferentially a strong porphyrin-type ligand like chlorin at *equilibrium* (see Fig. 2.8 for the relative binding strengths).

Now let us suppose that the step $ML_1 \rightarrow ML_2$ is irreversible and that the kinetic constants k are very different for different M. This could arise in part because the transition metals are complexed by other ligands (L_4) and the concentration of their free ions are low compared to that of magnesium, which is then complexed to a larger degree to a relatively poor ligand L_1, e.g. an O-donor protein site such as ATP, or because the MgL_1 complex is specifically recognized by L_2. It could even be that any ML_1 complex of the transition metals is not recognized due to conformational factors within a carefully modulated, even energized, protein transfer system. As a consequence of one or another of these factors, k would be much higher for the reaction $MgL_1 \rightarrow MgL_2$ than for any transition metal ion and, since the step is irreversible, this reaction would hinder the possibility of occurrence of all the other metal ions in ML_2. In different words, those weak-binding metal ions, which are in excess over ligands but react rapidly, can block the entry of ions that bind better to L_2 in an absolute sense, provided the kinetics of transfer are suitably arranged. It is a striking observation that, while *in vivo* the insertion is performed correctly giving magnesium chlorophyll, *in vitro* it is very difficult to avoid the production of zinc chlorin when this metal is present even in trace amounts.

In fact it seems likely that the correct insertion of each metal ion with less than 0.1 per cent error is achieved for magnesium in chlorin, iron in protoporphyrin, cobalt in corrin, nickel in F-430 (nirrin), and copper in uroporphyrin (see Table 2.15). This result could only be achieved by kinetic control based on specific recognition of ML_1 complexes (Fig. 2.12). Thus the design of these ligands is to prevent easy access and so control associations.

Once in its new site, the complex ML_2 is recognized specifically and bound by a protein in such a way that it is still harder to exchange the metal M (see Fig. 2.7). In effect, ML_2 is almost a new 'element' in that it allows Fe, Co, Ni, Cu, Mg to be used specifically in new ways. As proteins developed to bind

Table 2.15 Polypyrrole chelating agents and their selection of metal ions

Chelating agent	Metal	Binding site
Chlorin	Mg	Chlorophyll proteins
Protoporphyrin	Fe	Haemoglobin, cytochromes
Corrin	Co	Cobalamin (vitamin B_{12}) proteins
(Nirrin) F-430	Ni	Factor F-430 proteins
Uroporphyrin	Cu	Turacin (pigment), free(?)

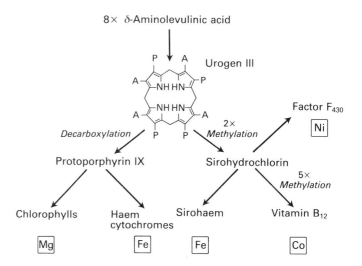

Fig. 2.12 The diverse ring chelates produced for different metal ions from one precursor.

these special complexes so the use of the elements expanded: it allowed mononuclear ring-chelate species of Fe^{2+} (and Fe^{3+}) in different spin states, of cobalt and nickel in at least two oxidation states and in low spin states, of magnesium chlorophyll in membrane proteins, just as it was possible to retain copper, zinc, and molybdenum directly bound to the side-groups of the proteins. Previous to this development intracellular iron was overwhelmingly bound as Fe_nS_n complexes, i.e. mixed valence clusters, cobalt might not be captured, and magnesium was associated only with rapidly equilibrating reactions of phosphate in water. Very large functional advantages accrue from the design of these small, relatively rigid molecule chelates as will be seen later.

The selectivity requirements described in this section require that the different chelating rings produced for the different metal ions are synthesized in controlled amounts. Once more there must be feedback from synthesized product to the genes.

2.12.4 Energy coupling to the selective movement of M (gates and pumps)

Energy can be applied to a membrane in many ways; in the first, energy can exert non-specific effects due to the fields, electrical potential differences, across membranes. These fields can be produced by light or metabolism and

they can be general over the whole of the membrane or localized. The membrane here is nothing else but a restriction on diffusion, while the energy input separates negative and positive charges across it. Either general neutralization of these charges or general exchange of like charges across the membrane can produce energized uptake and will give a concentration gradient of perhaps several orders of magnitude. A specific pathway in the membrane, a channel, makes the device specific to, say, the proton, the sodium ion, or the calcium ion, but the membrane here is passive even though it has structure and a transmembrane charge gradient.

An alternative is to involve the membrane components in the transport activity, i.e. to put energy into the membrane structure. For example, a membrane protein can change its state in a cyclic fashion under the impact of energy. The general case is as follows: a structure E in a membrane accepts n M (or n ML_1) from an outer aqueous space with a binding constant $D \cdot K_{aq} \cdot p$. Energy is now applied to change E such that E' is no longer open to the outer phase but is open to the inner phase and such that E' now binds n M (or n ML_1) with a different constant $D' \cdot K'_{aq} \cdot p'$ which is not large enough to retain these species. They are, therefore, released to the inner phase and E reverts to its initial state. Energy losses are involved in the cycle and the transport is irreversible in a thermodynamic sense (heat lost) but the pathway can be reversed by suitable energy input. We shall have to consider what types of protein could carry out this activity (Chapter 7). In general, proteins depend on the supply of energy from pyrophosphate of ATP to force their changes in conformation in membranes. Figure 2.13 gives an example.

Fig. 2.13 One proposed cycle of states for driving protons, or other ions across membranes. I, in; O, out.

2.12.5 A summary of the kinetics of uptake of metal ions

From what has been said we now have the following rates to consider:

(1) in-flow of M to the environment of the cell, which can be assisted by a certain ligand L_1;

(2) out-flow of M from the environment of the cell, which can be assisted by another ligand L';

(3) transfer in and out of cells involving properties of ML_1, K, p, D, and pores and gates all of which can be energized;

(4) (irreversible) binding of M by L_2 inside the cell;

(5) later we shall have to add the possibility that M reacts with a group L to give a precipitate which may or may not be in equilibrium with free M and free L.

The actual capture of M by a cell is then no longer a problem of equilibrium thermodynamics, although binding constants are relevant and partially determinant. Rather we see that element uptake and rejection is a matter of a steady state in which paths and rates of various processes are as important as uptake equilibria. Remember too that a higher cell has many internal membranes so that different metals are concentrated by transfers in different vesicles, organelles and so on. Table 2.16 gives some examples. This brings us again to the topic of homeostasis but we will deal with it in a separate chapter, Chapter 21.

Table 2.16 Pumped gradients of metals and their complexes

Metal	Inward to cytoplasm	Outward to	Organelle or vesicle
Fe^{3+}	Ferroxamines, etc.	(Ferritins?)	
	Transferrin	Citrate	Mitochondria
Co^{2+}	Vitamin B_{12}	?	
Ca^{2+}	(Free ion)	Free ion	Reticula, outside
Na^+	(Free ion)	Free ion	Outside
Mg^{2+}	Free ion	(Free ion)	(Outside)
H^+	Free ion and many ligands	Free ion and many ligands	Vesicles, outside
K^+	Free ion	(Free ion)	
Cd^{2+}	?	Metallothionein	Outside
HPO_4^{2-}	Free ion	Several organic phosphates	Mitochondria
Cl^-	Free ion	Free ion	Outside
Cu^{2+}	Albumin	Metallothionein	Outside
Zn^{2+}	Albumin	?	(Vesicles)

2.13 Rejection of metal ions

Before describing how anions are selected we must return to Fig. 2.1 to examine rejection. Rejection of elements M can occur in three ways:

(1) pumping of M through membranes to external fluids or to the air either in the form of ions or special molecules (reversal of Fig. 2.13);

(2) pumping into storage vesicles, vacuoles, again as ions or compounds;

(3) precipitation as solids in vesicles which may be rejected as pellets or retained.

Clearly, the principles behind rejection are very similar to those for uptake, but we do need to repeat that the operation is very selective. Note that Na^+, Ca^{2+}, and many M^{3+} (Al, Sc, lanthanides) ions are generally rejected.

We have now considered all the many ways in which metal ions can be handled and we shall see how these principles apply to particular metal ions in Chapters 8–19. Similar principles apply to non-metals and we give next an outline of their uptake.

2.14 General aspects of the uptake of non-metals

The trapping of non-metals by biological systems must also depend on the use of membrane restrictions, thermodynamic binding traps, and kinetically energized binding traps and steps. However, unlike ions, neutral molecules such as $N_2, O_2, H_2O, CO_2, NH_3, Si(OH)_4$, and $B(OH)_3$ carry N, O, H, C, Si, and B to all parts of a cell and cannot be prevented from free diffusion across membranes. Of these, compounds such as $B(OH)_3$ are readily trapped by binding to cis-diols (see margin), probably to specific polysaccharides, giving moderately labile condensation products.

Apparently similar reactions can be written for $Si(OH)_4$ or any weak dibasic acid and we find these structures for phosphorus in cyclic adenosine phosphate. We also note the occurrence of boron in the natural compound boromycin and it could well be that silicon in mucopolysaccharides is another example of this type of binding. In this context an element such as molybdenum, now seen as $Mo(OH)_6$, is similar to $Si(OH)_4$ (Chapter 17). The ring structures decrease in thermodynamic stability, in the order $B > Si > P$, so that for the most part, the later molecular hydroxides are retained by hydrogen bonding rather than covalent complexing, at least initially.

The other molecules, H_2, N_2, O_2, H_2O, CO_2, and NH_3, are less reactive and some need to be activated, reduced, or oxidized, before they can be bound. Their uptake is controlled by use of energy and kinetics of reactions and not by thermodynamics (see Fig. 1.8). After the initial capture, they usually pass into small molecule intermediates, such as charged amino acids and sugars, which remain trapped in cells by the membranes. These intermediates are *covalent* (as opposed to ionic) kinetic traps and, as small charged molecules, can be retained behind membranes before further synthesis leads to polymer formation.

Among these small non-metal molecules, dioxygen and dinitrogen are first absorbed at metal centres. Organic molecules can hardly activate them. For O_2 several types of centre are known, such as haemoglobin (Fe(II)), haemerythrin (Fe(II) ... Fe(II)), and haemocyanin (Cu(I) ... Cu(I)). In the last two cases oxygen is held as O_2^{2-} while in haemoglobin it binds reversibly as O_2, although there is partial charge transfer from iron to give a formal O_2^- state. If oxygen is to be incorporated, the final capture of the element involves the attack on a substrate by O_2 activated at a metallo-enzyme centre similar to the dioxygen carrier centres to give oxygen containing organic molecules. In some cases O_2 is reduced to superoxide and then to peroxide, which can also be utilized in various reactions or taken down to H_2O, the last stage of reduction. The uptake of N_2 is known to be by the iron–molybdenum enzyme nitrogenase which carries out the reaction

$$N_2 + 6H_2O + 6e \xrightarrow{\text{many ATP molecules}} 2NH_3 + 6OH^-.$$

The reduction is very energy-expensive in that it uses many ATP molecules. It is effectively irreversible through this use of energy, but also because the resulting ammonia goes on to be incorporated in organic molecules by condensation reactions. The question why the more active O_2 is not taken up by the N_2-absorbing sites is puzzling and still unsolved. Perhaps O_2 is removed first and its access to the site is prevented, or N_2 is indeed picked up by a special centre with little or no affinity at all for dioxygen, but this is not very common in chemistry, except for compounds involving metals that are not used in biology, e.g. ruthenium.

As to CO_2, it is rapidly converted to bicarbonate by the zinc-enzyme carbonic anhydrase or reacts directly with an activated organic molecule. The incorporation is again energy-driven by ATP. In this case there is no interference by O_2 or N_2, since CO_2 is a quite different molecule with a quadrupole moment that allows a different type of binding. After incorporation, the CO_2 is reduced to states equivalent to HCHO, $-CH_2-$, etc. in organic

$HCO_3^-, HPO_4^{2-},$
Cl^-, SO_4^{2-}, NO_3^-

Fig. 2.14 Some of the incorporation movements of anions as they pass membranes or are activated into kinetic traps by enzymes.

compounds. Incorporation of C from CH_4 involves typical organometallic chemistry (see Chapter 16).

Some of the above and all the other non-metals are also available to biological systems as simple anions, e.g. F^-, Cl^-, Br^-, I^-, or oxyanions such as CO_3^{2-}, HCO_3^-, $H_3SiO_4^-$, NO_3^-, $H_2PO_4^-$, HPO_4^{2-}, $B(OH)_4^-$, SO_4^{2-}, and SeO_4^{2-}. Molybdenum and vanadium can also be included in this group as MoO_4^{2-} and VO_4^{3-} and the principles of uptake for all these elements are similar. Naturally, the ions cannot cross membranes without carriers, and, once captured, they can be retained by such physical separation devices. The problem of capturing these non-metal ions by biological systems is therefore not very different from that of the capture of metals. We shall need to look for anion pumps and channels and their selectivity principles as well as at simple thermodynamic control over incorporation (see Fig. 2.14). Once inside a cell several of them can be reduced and differential non-metal incorporation is then greatly assisted.

2.15 Mechanisms of selection of anions based on thermodynamic properties

We will consider two major selection mechanisms based on 'equilibrium' binding properties in order to shorten discussion which should follow the full pattern of possibilities outlined for cations:

(1) selection by size and charge, taking into account the overwhelming influence of the hydration of anions on their binding to proteins;

(2) Selection by differences in binding affinity for different types of cationic centres, e.g. H^+, $-NH_3^+$, M^{n+} (compare $-OH$, RNH_2, RS^-, etc. as binding centres for cations).

2.15.1 Selection by charge and size; the hydration of anions and their binding to proteins

Table 2.17 Pauling's crystal radii of common anions

Ion	Radius (Å)
F^-	1.36
Cl^-	1.81
Br^-	1.95
I^-	2.16
O^{2-}	1.40
S^{2-}	1.84
Se^{2-}	1.98
Te^{2-}	2.21
OH^-	1.40
SH^-	1.95

Source: Pauling, L. (1960). *The nature of the chemical bond* (3rd edn). Cornell University Press, Ithaca, New York.

The various anions of group VII and oxyanions of groups IV to VI show considerable variation in size (see Tables 2.17 and 2.18). F^-, Cl^-, Br^-, and I^-, for example, should be readily separated by the correct design of absorbing holes or channels. There is good reason to suppose that the associated thermodynamic selectivity can outweigh the vast differences in abundance and availability of these elements. On the basis of similar sizes only the anions F^- and OH^- or Br^- and SH^- show competition.

There are, however, some differences as to the nature of the binding sites for anions relative to those used in separating cations. First, anions are negatively charged and, secondly, anions have a relatively lower charge density as they are large ($F^- \cong K^+$). This means that, since it is difficult to build a molecule with a *positively* charged dipole surface (compare cyclic polyethers), anions must bind to clustered, positively charged groups, such as the ammonium groups $-NH_3^+$, the guanidinium group, and metal ions, and/or through very

extensive hydrogen bonding. Rare cases are known in which there occurs binding at hydrophobic centres or regions containing $-OCH_3$, $-SCH_3$, $-CH_2-$, phenyl, etc., despite the fact they are of some polarity. Generally, these centres will only take up anions if they are surrounded by a 'buried' positive charge generated by the fold of the protein. This type of binding to buried charges is possible just because the charge density of anions is low. A deeper understanding of these observations comes from the balance of energy effects involved in the hydration of the anion and of the binding site in the interaction of these two entities, i.e. the energetics of the reaction (as for cations; Section 2.5.2),

$$X^{n-} \cdot (H_2O)_x + site^+ \cdot (H_2O)_y \rightarrow site^+ X^{n-} \cdot (H_2O)_z + (x+y-z)H_2O.$$

This balance can indeed generate orders which differ from those simply based on ionic sizes or potentials. The basic electrostatic factors that regulate it are those that regulate processes such as the preferential precipitation of large cations by large anions or small cations by small anions, e.g. Cs^+ by ClO_4^- but Li^+ by F^-.

The net free energy change in binding to a site from aqueous solution including terms other than electrostatic can be written as

$$\Delta G_{binding} = \Delta G_{elect} + (any \ \Delta G_{specific}) - \Delta G_{hyd}. \qquad (2.3)$$

Here $\Delta G_{binding}$ is the net free energy balance between the bound (hydrated) state and the free hydrated forms; $\Delta G_{elect.}$ is the free energy change corresponding to the electrostatic interactions; $\Delta G_{specific}$ is the contribution to the free energy change derived from other interactions, e.g. covalent binding, hydrogen-bonding, stereochemical or other effects; and ΔG_{hyd} is the algebraic sum of the different hydration energies of the free and bound species, symbolized by $(x+y-z)$ in eqn (2.3).

Although the different terms are difficult to evaluate, it can be seen from eqn (2.3) that different trends of $\Delta G_{binding}$ can be generated depending on the relative values of the separate terms. Generally speaking, hydration opposes binding if the binding sites are hydrophobic and the less hydrated anions will then be bound preferentially. The so-called 'lyotropic' or Hofmeister series is obtained, which is just the reverse of the hydration free energy, i.e. in binding strength

$$I^- > Br^- > Cl^- > F^-,$$
$$ClO_3^- > BrO_3^- > IO_3^- > CO_3^{2-} > HPO_4^{2-}.$$

The non-polar sites in proteins which give such series often contain (buried) charged histidine residues or, not so frequently, arginine and lysine residues. (Notice how these inverted orders compare with the inverted orders of radii for cation binding, Section 2.5.)

However, in other cases, the opposite order is found, i.e. the more hydrated anions, like SO_4^{2-} and $P_2O_7^{4-}$, are exactly those which are bound preferentially and they are often the substrates of corresponding enzymes. This occurs when the binding sites are polar and heavily hydrated. X-ray studies have shown that, quite often, these regions contain arginine and lysine amino groups (see Section 17.7).

Table 2.18 Thermochemical radii of common anions

Ion	Radius (Å)
CO_3^{2-}	1.85
HCO_3^-	1.63
SiO_4^{2-}	2.40
BO_3^{3-}	1.91
MoO_4^{2-}	2.54
NO_3^-	1.89
PO_4^{3-}	2.38
AsO_4^{3-}	2.48
SbO_4^{3-}	2.60
OH^-	1.40
SO_4^{2-}	2.30
SeO_4^{2-}	2.3
TeO_4^{2-}	2.54
ClO_4^-	2.36
IO_4^-	2.49

Source: Waddington, T. C. (1959). *Advances in inorganic chemistry and radiochemistry*, Vol. 1, p. 180. Academic Press, London.

2.15.2 Selection of anions by differences in binding affinity for different types of cationic centres

The affinity of anions for different types of cationic centres can be illustrated by data from the binding of free aqueous anions to free aqueous cations. Two different types of behaviour can be recognized: for class 'a' (hard) cations the order of binding by anions is $F^- > Cl^- > Br^- > I^-$ and $O^{2-} > S^{2-}$; for class 'b' (soft) cations the order is the reverse, i.e. they bind to the heavier donors rather than to the lighter ones (Table 2.8). As stated in Section 2.5, the order of binding of anions to class 'a' metal ions is overwhelmingly due to electrostatic terms, but binding to class 'b' metal ions is driven by stronger covalence.

Now the main cations concerned in biological systems are Mg^{2+}, Ca^{2+}, Cu^{2+}, Cu^+, Zn^{2+}, Co^{2+}, Co^{3+}, Fe^{2+}, Fe^{3+}, Mn^{2+}, Mn^{3+}, but only Cu^+ is a class 'b' cation. Na^+ and K^+ cannot be considered as probable anion binding sites since they form only very weak complexes and are highly mobile, but they may have a role in some specific cases (see the end of this section).

Thus, for free aqueous ions, Cu^+ will amplify the lyotropic effect, while all other hydrated cations should oppose it to different degrees

$$Mg^{2+} > Ca^{2+} \qquad \text{and} \qquad Mg^{2+} > Mn^{2+} > Zn^{2+} > Cu^{2+}.$$

When we consider the metal ions in proteins the situation is found to be very different. Consider:

(1) highly charged positive sites in water-free regions or regions of restricted access of water in a protein, e.g. carbonic anhydrase where Zn^{2+} is bound to neutral ligands only but the metal is deep in a narrow pocket;

(2) more lowly charged positive sites in highly hydrated sites, e.g. carboxy peptidase where Zn^{2+} is in an open pocket;

(3) lowly-charged sites, e.g. in alcohol dehydrogenase where the metal (zinc) is bound in a pocket to two R–S$^-$ groups or in the haem-proteins where Fe^{2+} is similarly neutralized by the two negative charges of the ligand;

(4) copper sites which utilize O_2 in oxidation reactions and go through an OH^- bound state, e.g. caeruloplasmin.

The third type of site will bind anions very poorly, as there is little electrostatic effect, and it is to be expected that binding agents for such sites will be neutral molecules which bind to metal ions, e.g. imidazole, or those anions that form covalent bounds, e.g. RS^-, but not any of the usual ions of the Hofmeister series. The substrate of alcohol dehydrogenase is C_2H_5OH.

The site (1), zinc in carbonic anhydrase, was referred to in Section 2.15.1. Being in a deep pocket, which can only be of low hydration, it gives the lyotropic series, e.g.

$$I^- > Br^- > Cl^- > F^-; \qquad CNO^- > SCN^- > N_3^- > ClO_4^- > NO_3^- > CH_3COO^-.$$

This is what we would expect for a class 'b' cation which zinc is *not*; hence the protein pocket has changed the balance of energy changes involved and it is site hydrophobicity that determines the order of binding rather than the nature of the aqueous metal ion. The separation of metal ions into 'a' and 'b' classes is solvent-dependent (protein is here the effective solvent). Carbonic

anhydrase can therefore act as a good Cl^- binding protein even in the presence of either SO_4^{2-} or HPO_4^{2-}. Sites (2) and (4) show the opposite properties. We now expect binding to follow the order of the ionic potential, dominated by overall favourable entropy effects. Actually, we observe frequently intermediate behaviour, depending on the degree of hydration of the metal in its site. Examples are the already mentioned zinc centre in carboxypeptidase and the single- and multi-copper centres of caeruloplasmin and laccase. The observed binding orders to these latter centres are

Carboxypeptidase: $F^- \cong Cl^- \cong Br^-$,
Laccase: $F^- > N_3^- > SCN^- > CNO^- \gg Cl^-$.

Note that in the case of laccase binding is to Cu(II), not to Cu(I) that does not bind the anion.

Typical of the zinc centres is that they bind simple carboxylate anions weakly, but this binding can be increased greatly by increasing the hydrophobicity of the sidechain, e.g. β-phenylpropionate \gg acetate.

There is still one case of the binding of anions to proteins which deserves special attention: that of simultaneous uptake of an anion and a cation. The cations concerned must be those in fast exchange between their protein binding sites and the free aqueous state, i.e. Mg^{2+}, Ca^{2+}, K^+, Na^+, and certain small organic cations. The most frequently met example is that of the binding of polyphosphates such as ATP and ADP together with Mg^{2+}. The binding appears to be to very hydrophylic protein regions containing additional positive charges due to basic sidechains of the proteins and can be very selective since the combined requirements of the anion and the cation have to be satisfied simultaneously.

2.15.3 Kinetic binding traps for anions, with or without accompanying redox reactions

We have already seen that some neutral molecules can be trapped by condensation reactions with various organic compounds forming kinetically stable covalent bonds, e.g. $B(OH)_3$ which condenses with cis-diols. Other neutral molecules must be activated, reduced, or oxidized before they can bind to other compounds through kinetically stable covalent bonds. These can be considered together with the non-metal elements that are captured as anions rather than neutral molecules.

The phosphate anions can only be retained weakly by ionic interaction with positively charged groups, e.g. $-NH_3^+$, but they can be retained in a *not too stable* (energized) kinetic trap by condensing with $-OH$ groups

$$HPO_4^{2-} + HOR \rightarrow R{-}O{-}PO_3^{2-} + H_2O.$$

In these forms phosphate is transported, e.g. as sugar phosphate, stored, e.g. as polyphosphate and ATP, and transferred to other molecules, simple or polymerized, to give a variety of substances, e.g. RNA and DNA. Among the other anions, only SO_4^{2-} can, to a lesser extent, give similar condensation products, but only as the simple ion, not polymerized.

In the above sections describing the behaviour of anions we have set aside the binding of sulphide. This anion was generally available in the early

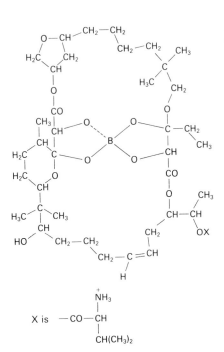

Boromycin

conditions of evolution and is still plentiful in anaerobic zones. It binds overwhelmingly and selectively in biology to metal centres, especially Fe and Mo, often in metal clusters. The nature of the competition between this anion and oxide has been described in Chapter 1 and section 2.5.7. The selectivity of its reactions is re-introduced here since we wish to point out that in aerobic systems the successive stages of uptake of sulphur must start from the chemistry of sulphate and go through many steps to sulphide incorporation. In the next chapter we shall see that the part of space in which the element is found decides its oxidation state and its incorporation mode.

2.16 Redox incorporation of anions

Inside cells, SO_4^{2-} is reduced and sulphur is retained in sulphide, S^{2-}, the thiol R–SH, or the disulphide –S–S– form. The same reactions occur with SeO_4^{2-}, which is bound as selenethiol, and with NO_3^-, which is incorporated as an amino group or other nitrogen-containing molecule. In the halide group, F^- and Cl^- are usually handled as such, but I^- and Br^-, once taken into a cell, can be selectively absorbed by enzymes and the fact that they can be much more easily oxidized than Cl^- permits the reactions $Br^- \rightarrow Br$ and $I^- \rightarrow I$ in the presence of Cl^-, e.g. by the action of many peroxidases. The reactive radicals so produced attack organic compounds readily, e.g. phenols, hence Br and I are inserted covalently linked into organic molecules. A good example is the thyroid hormone. In some cases the free halogen can be formed and liberated, e.g. I_2 in some seaweeds, and it is said that this (or perhaps the volatile halogenated hydrocarbons formed by insertion of bromine and iodine in organic molecules together with some $(CH_3)_2S$) accounts for the smell of the sea-shore (not ozone).

Finally, there are the cases in which reduction can change an anion into a cation. The uptake of molybdenum and of vanadium initially as MoO_4^{2-} and HVO_4^{2-}, respectively, i.e. in a non-metal form, is based on this principle. They are then reduced to oxocations or simple cations and bound into proteins or cofactors. Oxovanadium(IV), VO^{2+}, compares with nickel in its complexes with polyaminocarboxylic acids and binds strongly enough to N/O donors at $pH \sim 7$, but with certain strong complexing agents it can lose the oxo-group to give V(IV) complexes. The example known, the vanadium-containing toadstool, *Amanita muscaria*, contains this species, V(IV), bound to an N-hydroxy derivative of iminodiacetic acid forming a very stable octaco-ordinated complex in which the N-hydroxy group is ionized and binds the metal, see margin. The unusual accumulation of vanadium in tunicates as V(III) or VO^{2+}, from the low concentrations of HVO_4^{2-} found in the environment, is probably due to this change in oxidation state, so that there is no 'saturation' inside cells relatively to the species available outside even when no internal ligand appears to be present and the retention of the ion (V^{3+} or VO^{2+}) is ensured by storage in vesicles (see Chapter 17). The case of molybdenum is very different.

The reactions of molybdenum are quite idiosyncratic since as a metal element it is the closest to the non-metals. Thus its chemistry can be considered starting from MoO_4^{2-} (SO_4^{2-}). On entering cellular systems the hydroxide form of this ion $MoO_2(OH)_4^{2-}$ (formally written) can react not just with OH groups to form MoO_2^{2+} bound species (compare UO_2^{2+}) but, since it is a soft 'b'-class metal ion, it can give

$$MoO_2(OH)_4^{2-} + 2RSH \longrightarrow [MoO_2(OH)_2(RS)_2]^{2-} + 2H_2O.$$

Now the more stable forms of this anion are probably $[MoO_3(RS)_2]^{2-}$. Here there is a resemblance to non-metals in that these compounds can be quite stable kinetically if $(RS)_2$ is a chelate. In fact molybdenum appears to be transported in biology as the cofactor complexed metal with only *two* (covalent) bonds (Fig. 17.4). This is in contrast with common metal ions, which need three or more bonds, and with sulphate, which can be transported with but one covalent link, ADP–$[OSO_3]$. The intermediate nature of molybdenum chemistry is fully exploited in biology (see Chapter 17).

2.17 Coenzymes

Table 2.19 Some coenzymes which carry non-metals

Non-metal	Coenzyme or cofactor
H	NADH, NADPH, flavin Quinones
P	ATP (NTP), IP_3
C	CoA, vitamin B_{12}
S	PAPS

While we were describing metal ions we stated that in some cases there were kinetic traps which removed the element making it into an effective 'extra' element. Examples were haem, chlorin, and vitamin B_{12}. Some literature refers to these bound entities as cofactors. When we look at the corresponding systems for non-metals, often called coenzymes (Table 2.19 and Fig. 2.15), we observe two broad differences from metal containing cofactors: (1) the kinetically stable unit holds the element by but one bond leaving it very accessible for transfer by catalysts, enzymes, which recognize the whole coenzyme and not the element: for example, phosphate is bound in the five nucleotide triphosphates, NTP (adenine, uridine, cytidine, guanidine, and thymidine) and hence phosphate can enter five selective transport (of phosphate) systems; (2) the coenzymes may be freely mobile or at least of restricted mobility, e.g. the NTP molecules. In addition, notice that each element, H, C, S, etc. is carried by coenzymes which belong to it alone so that pathways of each are not open to blocking by the coenzymes of another element. Recognition is by the coenzyme tail and not by the element—contrast calcium and compare cobalt. The need for these properties in several different coenzymes is that the elements H, C, N, P, and S are parts of many synthetic substrate pathways as well as having group activities and these pathways must not meet haphazardly. The chemical form of the transported element is interesting: phosphorus always as phosphate, hydrogen as H^-, but carbon as $-COCH_3$, $-CH_3$, $-sugar$, etc. and nitrogen in amino-acids (glutamate). At the same time there are coenzymes which are immobile holders of non-metals, such as flavins, biotin, and pyridoxal phosphate. We shall return to the use of different coenzymes in different pathways in Chapters 3 and 18 since they, like metal ions, are differentially concentrated in space.

Pyridoxal phosphate

Adenosine triphosphate (ATP)

Quinone

Coenzyme A (CoA)

Flavin

Nicotinamide coenzymes

Naphthoquinone

S-Adenosyl methionine (SAM)

Vitamin K

Fig. 2.15 The formula of the organic moieties which act as the coenzymes for carrying amines, phosphate, acetyl, hydride, and methyl groups.

2.18 Concluding remarks: element handling in biological systems

Periplasm
Cu, Mo, Mn, Ni

Fig. 2.16 The distribution of some 16 elements in a present-day prokaryote cell with no inner compartments.

The above discussion on selectivity provides most of the chemical information needed to understand how the different biological elements are captured and which kind of device is used to retain them inside cells or to keep them outside cells. We have considered metals and non-metals, cationic, anionic, and neutral species, thermodynamic and kinetic effects, and we have also briefly discussed some problems related to binding by proteins and in membranes. It is, therefore, appropriate to check now some actual situations in biological systems against the predictions and we will do this first by examining the traffic of the biological elements in a generalized cell and its environment. A simplified model is that of Fig. 2.16 where only general trends are considered for the sake of simplicity.

Figure 2.16 shows that some chemical elements are captured by cells in which they are retained by thermodynamically stable binding or by kinetic devices (covalent bonding, membranes, precipitates, etc.). Some other elements are rejected by the cells and remain in the external medium as free ions, combined with transporting ligands or as solid phases.

In their various situations, these elements follow the principles extensively discussed, which concern the type of traps adopted to retain them, the preferential binding to some atoms, and the geometry of the binding sites. Tables 2.20 and 2.21 exemplify these situations. The combination of the different characteristics, affinities for certain atoms, preferences for determined geometries and all possible devices for retention and rejection available in biological systems, allows each element to be organized in the biological space in very specific patterns and thus acquire also very specific functional significance; role overlap occurs, but to a very limited extent, and closer examination usually reveals that either prevalent local conditions dictate duplications (e.g. the Fe^{3+}, Mn^{3+}, and Cu^{2+}/Zn^{2+} superoxide dismutases) or that they may be derived from the coexistence of different evolutionary stages (probably the case of V(V) and Fe-peroxidases and that of V- and Mo-nitrogenases). Examples of both situations will be discussed later in this book, Chapter 17.

Table 2.20 Traps for different biological elements

Element	Trap
Na^+, K^+, Mg^{2+}, Ca^{2+}, Cl^- ($H_2PO_4^-$, HPO_4^{2-}, SO_4^{2-})	Physical retention of ions by membranes (fast chemical exchange)
Cu^{2+}, Zn^{2+}, Ni^{2+}, Mo, Si, B, Fe^{3+}, Mn^{3+}	Thermodynamically and kinetically stable binding to polymers, proteins, and polysaccharides (slow exchange)
Fe^{2+}, Co^{2+}, Ni^{2+}, Mg^{2+}	Additional gated, thermodynamic and kinetic traps in small chelates, e.g. polypyrroles bound to proteins
H, C, N, O, P, S, Se, I, Br	Kinetic retention in thermodynamically unstable small or large molecules by covalent bonds
Si, Ca, P, C, S, F	Precipitates (biominerals)

Table 2.21 Biological elements: some binding situations in biological systems

Binding atom (water excluded)	Element	Examples (with nearest simple geometries of site, which is usually distorted)
S only	Fe	Ferredoxin (tetrahedral)
	Zn	Alcohol dehydrogenase (tetrahedral)
S, O	Mo	Molybdenum cofactor
N, S	Cu	Plastocyanin (tetrahedral)
	Fe	Cytochrome c (octahedral)
	Zn	Alcohol dehydrogenase (tetrahedral)
N only	Fe	Cytochromes (octahedral)
	Cu	Superoxide dismutase (tetragonal)
	Zn	Carbonic anhydrase (tetrahedral)
	Mg	Chlorophyll (square-planar)
N, O	Zn	Carboxypeptidase (tetrahedral)
	Mn	Concanavallin (octahedral)
	Fe	Haemerythrin (octahedral)
O only	Ca, Mg	Calmodulin (Ca, 7/8 co-ordinate ionic binding)
	Mg	ATP 6-co-ordinate (octahedral)
	P, Si, B, S	XO_4 groups condensed with ROH, e.g. in polysaccharides (tetrahedral)
N, C	Co	Coenzyme B_{12} (octahedral)
C only	S, Se, I	Cysteine(S), selenocysteine(Se), thyroxine(I): single covalent bonds

2.19 Precipitates in compartments

We have left the description of precipitation as an uptake mode until last since precipitation requires uptake of X^- (the anion) and M^+ (the cation) of the MX precipitate into the same (vesicle) compartment. The selection here is by solubility product (S.P.), which can be appreciated by reference to the data for sparingly soluble salts found in Table 2.22. The selectivity of anion for cation in biology is precisely as expected from considerations of inorganic reactions, e.g. $BaSO_4$, $CaCO_3$, $Fe(OH)_3$, $SiO(OH)_2$; however, biological selection of a solid state goes much deeper in that there is choice amongst the allotropic forms of one salt, e.g. calcite or aragonite, both $CaCO_3$, and there is also a choice of shape and size of crystallites. These considerations will be examined in Chapter 20.

The solubility products in Table 2.22 hold a message and a warning for us when we inspect biological systems. It is undoubtedly true that free Fe^{3+} and Ca^{2+} are very close to their solubility limits as hydroxides and carbonates/phosphates, respectively. These ions, whose roles are dominant in biology, may well control in part homeostasis of cell metabolism since their precipitation reactions link them to the pH of the solution as well as to the levels of critical anions such as phosphate and carbonate

Table 2.22 Solubility-product constants (S.P.)

Substance	Formula	Solubility product
Aluminium hydroxide	$Al(OH)_3$	2×10^{-32}
Barium carbonate	$BaCO_3$	5.1×10^{-9}
Barium oxalate	BaC_2O_4	2.3×10^{-8}
Barium sulphate	$BaSO_4$	1.3×10^{-10}
Cadmium hydroxide	$Cd(OH)_2$	5.9×10^{-15}
Cadmium oxalate	CdC_2O_4	9×10^{-8}
Cadmium sulphide	CdS	2×10^{-28}
Calcium carbonate	$CaCO_3$	4.8×10^{-9}
Calcium fluoride	CaF_2	4.9×10^{-11}
Calcium oxalate	CaC_2O_4	2.3×10^{-9}
Calcium sulphate	$CaSO_4$	2.6×10^{-5}
Copper(II) hydroxide	$Cu(OH)_2$	1.6×10^{-19}
Copper(II) sulphide	CuS	6×10^{-36}
Iron(II) hydroxide	$Fe(OH)_2$	8×10^{-16}
Iron (II) sulphide	FeS	6×10^{-18}
Iron(III) hydroxide	$Fe(OH)_3$	4×10^{-38}
Magnesium ammonium phosphate	$MgNH_4PO_4$	3×10^{-15}
Magnesium carbonate	$MgCO_3$	1×10^{-5}
Magnesium hydroxide	$Mg(OH)_2$	1.8×10^{-11}
Magnesium oxalate	MgC_2O_4	8.6×10^{-5}
Manganese(II) hydroxide	$Mn(OH)_2$	1.9×10^{-15}
Manganese(II) sulphide	MnS	3×10^{-15}
Strontium oxalate	SrC_2O_4	5.6×10^{-8}
Strontium sulphate	$SrSO_4$	3.2×10^{-7}
Zinc hydroxide	$Zn(OH)_2$	1.2×10^{-17}
Zinc oxalate	ZnC_2O_4	7.5×10^{-9}
Zinc sulphide	ZnS	4.5×10^{-24}

Data from *Stability Constants* (1964), Spec. Pub. No. 7. The Chemical Society, London. Note the preferred use of Roman labels of oxidation states.

$$Fe^{3+} + 3H_2O \rightleftharpoons Fe(OH)_3 + 3H^+,$$
$$2Ca^{2+} + H_2O + HPO_4^{2-} \rightleftharpoons Ca_2(OH)PO_4 + 2H^+,$$
$$Ca^{2+} + HCO_3^- \rightleftharpoons CaCO_3 + H^+.$$

The message is clear that we must not look upon biological minerals as just inert structural deposits: they must be made and sustained in communication with the levels of free ion concentrations.

The fact that biological systems run with concentration levels of several pairs of ions close to their solubility products has obvious dangers too. The precipitation of calcium salts is a major hazard in old age and deposits of iron oxides in haemosiderin are frequent in treatments of people with blood diseases. There is also some suggestion of a correlation between aluminium silicate deposits and Alzheimer's dementia. It will be necessary for us to see how in normal cellular systems the precipitation reactions are used and controlled so that diseased states are avoided for long periods of life.

Finally in an atmosphere of H_2S the removal of elements such as copper and the much reduced concentration of zinc gave easier access of iron(II) to thiolate ligands, Fig. 1.12.

Note

It is very difficult to transfer the complex language of inorganic chemistry to biological chemistry since authors of different specialist origins invent codes. In this book the classification of metal ions by very basic factors is to be preferred to ever more complex descriptions based on theory, we believe. Thus charge, size, and ionization energy are easily recognized, but 'hard' and 'soft' HOMO and LUMO have quantitative value only when handled with care. Much though these last terms may be useful in chemistry, we shall not need them here and will avoid their use for the most part. For those interested we refer to the books in the 'Further reading' section which follows. These books describe in depth the main parts of this chapter which is largely concerned with inorganic chemistry.

Further reading

Burger, K. (1990). *Biocoordination chemistry*. Ellis Horwood, New York.

Cotton, F. A. and Wilkinson, G. (1988). *Advanced inorganic chemistry* (5th edn). Wiley, New York.

Da Silva, J. J. R. F. and Williams, R. J. P. (1976). The uptake of elements by biological systems. *Structure and Bonding*, **29**, 67–122.

Greenwood, N. N. and Earnshaw, A. (1984). *Chemistry of the elements*. Pergamon Press, Oxford.

Jensen, W. B. (1980). *The Lewis acid–base concepts*. Wiley, New York.

Kolthoff, L. M. and Elving, P. J. (ed.). (1978). *Treatise on analytical chemistry*. Interscience, New York.

Phillips, C. S. G. and Williams, R. J. P. (1966). *Inorganic chemistry*. Oxford University Press, Oxford.

Schwarzenbach, G. and Flashka, H. (1969). *Complexometric titrations* (translation by H. M. N. H. Irving). Methuen, London.

Shriver, D. F., Atkins, P. W., and Langford, C. H. (1990). *Inorganic chemistry*. Oxford University Press, Oxford.

3

Physical separations of elements: compartments and zones in biology

3.1 General aspects 71

3.2 The nature of compartments 73

3.3 The chemical solutions and physical states of compartments 73

3.4 The role and nature of membranes 76

3.5 Different types of membrane 77

3.6 Special solution conditions in vesicles 86

3.7 Co-operativity of separations and localizations 87

3.8 Summary of metal ion positioning 87

3.9 Symbiosis and multicellular systems 88

3.10 Spatial distribution of non-metals 89

3.11 The spatial transfer of H, C, N, S: mobile and fixed coenzymes 90

3.12 Summary of non-metal transport 93

3.1 General aspects

In order to understand the observed binding of the chemical elements, metals and non-metals, to particular chemical sites in biological systems, we have discussed three factors: (1) the limitations on the concentration of chemical elements imposed by their abundance and availability in the environment; (2) the possibility of adjusting this availability in a cell by pumping an element or compound across a single membrane into or out of its inner aqueous phase; (3) the intrinsic stability, thermodynamic and kinetic, of the association of the chemical elements with particular 'ligand' frameworks which have 'designed' structures, either in solution or in a condensed phase.

Whereas in primitive cells there was apparently only one internal aqueous compartment, i.e. one surrounding cytoplasmic membrane, and, in some cases, a periplasmic (outer) controlled spatial compartment, evolutionary

pressures led to progressive modifications which included: (1) the differentiation of internal compartments by extra membranes in cells (Fig. 3.1(a) gives a real example and Fig. 3.1(b) is a simplified diagram for discussion purposes); (2) the functional specialization of cells together with the development of multicellular organisms. In the present chapter we will be concerned mainly with the first of these aspects, but we must notice immediately that even in cells without internal, membrane-limited, compartments there must be some kind of internal organization; otherwise it would be practically impossible to ensure that the thousands of reactions which take place there could proceed without confusion. 'Modern' cells are complex co-operative organizations, with various internal divisions, some of which give simple physical meaning to our idea of 'compartment' since they have essentially different aqueous phases separated by membranes, but there are others where the rates of diffusion and reaction define *zones* less precisely. The inside of a cell is a network of filaments and vesicles, not space open to free diffusion. Special proteins and elements are bound in particular locations and we must not forget that each membrane is a defined zone or series of zones (see Section 3.4) with its own complement of proteins.

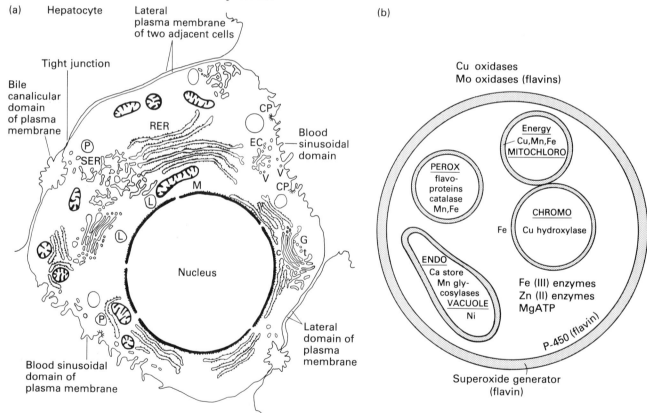

Fig. 3.1 (a) Section through an hepatocyte illustrating the various organelles and membrane systems. EC, endocytic compartment; G, Golgi apparatus (c = cis face; t = trans face); L, lysosome; M, mitochondrion; P, peroxisome; RER, rough endoplasmic reticulum; SER, smooth endoplasmic reticulum; V, vesicle; CP, coated pit. (b) A schematic indication of some of the different membrane separated compartments in an advanced cell. PEROX is a peroxisome; MITOCHLORO is either a mitochondrion or a chloroplast; CHROMO is a vesicle of, say, the chromaffin granule; ENDO is a reticulum, e.g. the endoplasmic reticulum. Other compartments are lysosomes, vacuoles, calcisomes, and so on. Localized metal concentrations are shown.

3.2 The nature of compartments

Fig. 3.2 Diagram of a mitochrondrion; see also Figs 3.1(a) and 3.11.

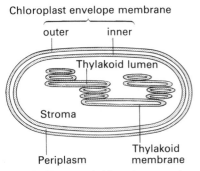

Fig. 3.3 Diagram of chloroplast; see also Fig. 3.10.

We start this description with the two major organelles. These are compartments which may be remnants of prokaryote cells trapped inside a eukaryote cell. (Note that a eukaryote cell has a nuclear membrane enclosing its DNA.) The two are the mitochondrion (Fig. 3.2) and the chloroplast (Fig.3.3). Each has its own DNA, not enclosed, and its own system of membranes. The mitochondrion has two membranes separated by a periplasmic space as in a prokaryote; the chloroplast has two outer membranes and an inner vesicle, the thylakoid. The inner membrane of the mitochondrion carries out oxidative phosphorylation; the thylakoid membrane is responsible for light capture and transduction. In mitochondria the metabolism is largely that of fats and the citric acid cycle and is oxidative; in chloroplasts the reactions generate free dioxygen and reductive incorporation of CO_2. Both organelles generate ATP; indeed they are the major energy-generating compartments.

Other membrane-isolated compartments within cells contain no DNA; they are often used for storage or delivery of chemicals. There are a multitude of such vesicles containing transmitters, hormones, digestive proteins, and so on, according to cell type. Vesicles such as lysosomes contain enzymes that break down polymers and supply the cell with basic chemicals, while peroxisomes generate H_2O_2 and generally drive certain oxidative reactions, but they may also release enzymes externally. Pervading many cells is a reticulum which acts as a communicating network and on stimulation releases, for example, calcium from its internal store; it is usually a very extensive tube-like zone. Finally, there is the Golgi apparatus in which many proteins are modified before export (note especially glycosylation and sulphation). The Golgi apparatus also directs vesicles to different parts of a cell (see Section 3.5.4).

3.3 The chemical solutions and physical states of compartments

Across the membranes, as across cells, there is traffic of ions, ligands (including proteins), and complexes. Energy can be applied to these processes or to a whole phase by adjusting, for example, the pH or the redox potential in a compartment. If we take a reference compartment to be at a standard state pH = 7 and at a redox potential equivalent to that of O_2/H_2O, i.e. +0.8 V relative to the standard hydrogen electrode, then we would expect that the oxidation states and, perhaps through evolutionary origins, even the concentrations of the different aquo ions in this compartment would be similar to those in sea water. The extracellular fluids of blood are not too far from this condition. The situation in any cell compartment will be changed by ion pumping, as described in Chapter 2, by pH control (that is by metabolism or by proton pumping) and redox control, by oxidative or reductive metabolism, or in other ways, which may even include mechanical pressure, as described in

Section 3.4, so each compartment will be 'energized' relative to the reference state in many ways.

Furthermore, since the ions considered will combine with different ligands L and their functional value depends on this combination in most cases, we must look into the factors that regulate the availability and concentration of L in the various compartments. Here we find that some of these factors are common to those for the 'energization' of the metals, namely pH and redox potential. For example, the state of a ligand carrying two thiol groups can be $(-SH)_2$, $(-S^-)_2$, or $-S-S-$, depending on the values of those two variables in the compartment into which the compound is pumped. In addition, every ligand, including any protein, is synthesized in variable amounts by cell and organelle metabolism and may be pumped into a given compartment. Finally, it is observed that some of the ligands L are in fixed positions, say, on a filamentous construction or a membrane, so that ML is part of a surface and not free in solution. The position of ML in a cell is connected to function in addition to the controlled chemical combination itself.

The amount of ML formed in each compartment and its standing concentration is, therefore, decided by four new factors to be added to those stressed in Chapter 2 and at the beginning of Section 3.1:

(1) the values of pH and redox potential of the phase in a particular compartment, affecting the states of both M and L;

(2) the limitation to the delivery of M and L into that compartment, including the possibility of its local synthesis (note that L can be a protein);

(3) the further possible movement of the complex ML from that compartment to other parts of the cellular organization (Fig. 3.4);

(4) binding to surfaces, when we may well be interested in bound ML than free ML.

Given the multiplicity of factors controlling the siting of a given metal in any biological compartment it is not possible, with the present state of knowledge, to deduce where it will be found or where its different complexes are to be placed. We may, however, be able to work backwards from the knowledge of

Fig. 3.4 The movement of elements M in and out of vesicular spaces with proteins P. As an example consider M_1 to be iron. It is picked up in a protein P_1, by a ligand L, and/or by a receptor (coated pit) P_2 and transported in a vesicle to a lysozome. Here it is released to be distributed in the cell to give a precipitate ML_2, or to a protein P_3 for export in the Golgi. It could also go into an organelle, which almost behaves as a cell within a cell, where it is processed into haem. Other elements M_2 enter the cell through pumping of ions. The whole arrangement may well have three-dimensional organization, directed flow, and time dependence. Within the cell cytoplasm copper and zinc circulate using metallothionein and iron circulates in ferritin. Both carriers, metallothionein and ferritin, are then homeostatic proteins.

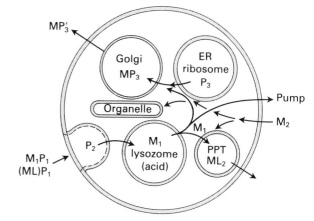

where the free ions or their complexes are found to the reasons for their respective distribution and then try to relate this distribution to the functions performed.

A first striking fact, which is becoming more and more apparent from recent data on the localization of the various elements in biological systems, is the asymmetry of their distribution. Figure 3.1(b) is a simplified description of some of the compartments. We have already observed that sodium and calcium are rejected from the cell cytoplasm whereas potassium and magnesium are concentrated in it, but similar observations can be made for the trace elements and more detailed information about compartments is now available for the oligoelements referred to above. For example, manganese is generally located in organelles and reticula, including the Golgi vesicles, but not in cytoplasm; in plant cells it is either in the vacuole or in the chloroplast organelle, but perhaps not so markedly in the mitochondria as in animal cells. Zinc, like calcium and manganese, is also sometimes strongly concentrated in special vesicular or reticular spaces, but not necessarily in the same compartments as manganese. Examples of these special spaces are the vesicles of pollen cells and the acrosome of male gametes; in addition, zinc is also found in the cytosol. Again the extensive involvement of this metal in hydrolytic (digestive) enzymes clearly requires that these are not active in the cells to avoid their self-digestion; hence this zinc and its proteins are in vesicles and are rejected at some stage to extracellular space. Bound iron is usually concentrated in special membranes, such as those of mitochondrial and thylakoid organelles, and, when inside the cytoplasm of cells, much is in the form of ferritin, which can be considered to be a special compartment of its own, where a protein wraps around $Fe(OH)_3$ solids like a membrane. Bound copper and perhaps molybdenum are largely extracellular (see Chapters 15 and 17), and some of their metallo-enzymes are found between double membranes, as in the periplasmic space of bacteria, e.g. in *E. coli* (see Fig. 3.1(b)), or organelles.

Amongst non-metals, the essential elements chlorine and silicon are largely rejected from cells, while elements such as phosphorus, sulphur (as sulphide but not sulphate), and selenium are accumulated. There is also a group of elements which have little or no biological function and which biology seems to exclude. They are the group III and IV elements, such as Al, Sc, La, Ge, Ti, (Sn), and Pb, all of which could have been accumulated if a useful function had been found for them. Surprisingly, some cells take up large amounts of barium or strontium and deposit them in quite distinct vesicles as sulphates.

The different elements are moved in very different ways to different places by the use of metabolic energy and all these movements are kinetically controlled, so that their spatial distribution in a cell has to be described by a map of local, energized, steady-state concentrations, selective for each element, which relates to the 'wanted' function. This distribution is part of evolutionary 'design' and is contrary to expectation from thermodynamic pressures. The fact that the distributions are energized means that they are time-dependent, being susceptible to fluctuations of environmental conditions, such as light intensity cycles, oxygen tension variations, availability of food, reproductive or other binding cycles, and triggering of activity. The fluctuating movement of elements and its adjustment is much more obvious for some elements than for others. Thus, H^+, Na^+, K^+, Ca^{2+}, Mg^{2+}, Cl^-, or HPO_4^{2-} are mobile, or

at most semistatic in the cases of Ca^{2+}, Mg^{2+}, HPO_4^{2-}, or even H^+, while other elements, namely some transition metals such as iron, are largely, static as ML complexes and their function does indeed rest, very largely, in a fixed compartment or even in a localized surface zone of a membrane or of a cell interior.

From what has been said it becomes more and more evident that, if the full use of the elements in biology is to be appreciated, we must grapple with the description of the nature of biological space and with the disposition of elements in that space, for it is this organization which causes the flow of material as well as the uptake, storage, and transduction of energy. Since these are time-dependent phenomena, space and time relationships are essential knowledge in biology. Our immediate need is to describe biological space more thoroughly. We shall do so by reference to single cells, but in complex organisms, such as man, the element distribution is organ- as well as cell-specific. The intractable problem which this imposes will be approached only here and there in the present chapter.

3.4 The role and nature of membranes

Fig. 3.5 A schematic representation of the relationship between the cytoplasmic membrane, an internal membrane (vesicle), and the protein network. One series of proteins, e.g. spectrins, lies under the outer membrane, bound to it, while another series, here called tubulins but including a variety of filamentous structures, crisscrosses the cell and holds vesicles in place. The vesicle membrane may form part of the outer membrane reversibly.

Biological systems have some obvious ways of dividing elements into spatial patterns using membranes:

(1) capture in aqueous phases by membrane enclosures;

(2) capture by proteins at localities either on or in membranes or filaments;

(3) precipitation of salts in vesicles.

We now look more closely at the nature of the membranes related to all three. Biological membranes not only divide space but they determine mobility (in the membrane) and incorporate a number of energy-using devices that can move elements from one compartment to another either as simple ions, small chelates, protein complexes, or even crystallites. The energy-using devices are proteins which act as pumps fixed in the membrane or in a system of membranes, whereby the membrane can even bud into the cell to form a new vesicle or coalesce with an existing membrane to create a new compartment (see P_2 in Fig. 3.4).

However, a cytoplasmic membrane is not just a lipid barrier, but is part of a total cellular structure through a network of proteins in the cell. These proteins lie both under the surface of the lipid membrane as a network and perpendicular to it in the form of filaments of which tubulin is one example. In Fig. 3.5 we give a scheme which may be useful in thinking about this description. Notice that the proteins are tension fibres and have directional quality; the shape of the membrane is to some degree decided by the tension in the protein filaments. However, the system is also interactive, through these filaments, with internal vesicles (see Fig. 3.5 and Chapter 7).

Obviously due to the asymmetry of tension from the internal filaments, the outer membrane has a varying radius of curvature (Fig. 3.6), and from this fact

Sphere
(some bacteria?)

Cylinder
nerves

Box-like
plants, muscles, fibroblasts

Fig. 3.6 Some possible cell and organelle shapes. X is the nucleus and the cell (top) may have polarity. Differential protein distributions are shown by hatching. Filaments spread from the nucleus in eukaryotes to give shape.

alone there will be a lateral distribution of its chemical contents. This follows from the fact that all components in the membrane migrate to lower surface tension overall. Different parts of the membrane accumulate macromolecules plus their combined elements differentially; hence, they are ordered phases but with no repeats in the manner of a crystal: outside surfaces differ from inside surfaces and left is distinguished from right. Cell or organelle membranes are therefore asymmetric three-dimensional phases which undergo shape changes from one moment to another according to the local energization of tension.

Now, some cells confine the cytoplasmic membrane within an outer wall-like structure, for example many bacterial and plant cells. The shapes of their outer membrane may well be controlled by this outer wall, although complex invaginations are frequently observed. Other cells have a very flexible outer coat of polysaccharide and their shapes are open to external as well as internal forces. Their membranes are affected by the forces on the outer polymers, i.e. *connective tissue*, and the cell shape is now dependent on a steady-state relationship between the local pressures produced by both the internal and the external matrices. Such cells often show long extensions as villi or dendrites, as in nerve cells. The cell and its outer membrane are then interactive with the internal milieu and the external environment. This is frequently true for animal cells. It is against this background that we study the nature of membrane phases themselves and element distributions in them; detailed examples are given in Section 3.5.

3.5 Different types of membrane

The membrane described so far is the cytoplasmic membrane of higher cells or that of the organelles, mitochondria, and chloroplasts. These membranes are connected to internal filaments (Fig. 3.5). Inside higher cells there are quite different vesicle membranes (Figs 3.1(a) and 3.4) which have no internal protein fibres. It is often even difficult to be sure whether the internal space behind these membranes connects with the extracellular space or not. In some cases this does not happen, e.g. the Golgi apparatus, but the various so-called reticula may or may not be so connected. Again there are vesicles which are formed from the outer membrane and which introduce external molecules or particles into the cell by phagocytosis and other internal vesicles which reject chemicals from cells. Both of their membranes can be formed from local zones of the outer membrane (Fig. 3.5) and again the common feature of the *internal* spaces inside these membranes is that they do not have a fibrillar proteinaceous structure. They are limited by the forces imposed on them by the *internal* filaments of the cytoplasm, e.g. tubulin (Fig. 3.5), by the pumping of molecules into them, and only rarely by internal constructions not attached to the membrane. These may be based on polysaccharides or glycoproteins. The latter biopolymers are often related to export processes, to biomineralization, or to the sending of messages.

To deal with such complexity one must again try to divide the subject matter, i.e. the description of limitations on distribution of elements by different membranes, and we have chosen the following items:

(1) internal and external biopolymer assemblies related to membranes;

(2) lateral membrane organization and in-membrane flow;

(3) transmembrane organization and flow;

(4) the position of vesicles in cells.

3.5.1 Internal and external particles and their relationship to membranes

Large particles, many of which are unrestricted by membranes, assemble both inside and outside cells. Here we include not just the filamentous structures described in Section 3.5, but also 'free-floating' units, such as enzymes associated with particular sequences of reactions, e.g. fatty acid synthesis (Fig. 3.7). Inside cells, the most familiar large particles are those formed by RNA and DNA, i.e. ribosomes and nuclei (Fig. 3.7). These phosphate-rich polymers are held together by special basic proteins, e.g. histones, by magnesium ions, and by some monovalent ions. The DNA is connected to some of the filaments and is confined in eukaryotes by a special 'nuclear' membrane. One kind of RNA (ribosomal) is often connected to certain reticula membranes. In the course of a cell cycle these particles may form and divide; hence we know that once again we are examining a dynamic set of particles with an interaction with the membrane via filaments. The positioning of particles is very different in prokaryotes and eukaryotes.

As already stated, apart from the large particles there are, both inside and outside cells, large filamentous structures. Inside the cell the structures are made only from proteins and typical of these are the tubulins, vincins, and actin/myosin filaments (see Table 3.1). They have associated with them localized enzymes which assist in energy-driven changes of the filaments (Fig. 3.8). Outside the cells, associated glycoproteins of great complexity and length are to be found; they form both a surround for single-cell organisms and also the soft tissue matrix which holds together, often with the support of biominerals, the multicellular organisms. Typical proteins are collagens, but there are many others, including the proteoglycans. However, many of the glycoproteins are sulphated, as in myelinated tissues of nerves. Sulphated polymers are not found inside cells. Very many of the intra- and extracellular filaments are firmly attached to membranes (Fig. 3.7).

The chemical elements associated with these structures are highly specific; inside cells magnesium (and probably zinc) is deeply involved with the fibrous

Fig. 3.7 A representation of some large particles in a cell. Some are free floating in the cytoplasm and some are in or on membranes. They are distributed in different membrane-limited compartments. The nucleus has a membrane in eukaryotes only and in these cells the energy capture devices are in organelles.

■ ATP/ADP exchanger

Fig. 3.8 The localized transfer of an energy-carrying chemical, creatine phosphate (CrP), from a mitochondrial cristae (Mito cristae) membrane (Fig. 3.2) organization to a fibril organization for contraction. Two different creatine kinase (CrK) enzymes act in different locations.

Table 3.1 Some filamentous structures of cells*

Filament	Function
Actomyosin	Contractile protein which can be used to shape a cell or to exert force between cell units
Tubulin	Reversibly formed filament for structure and vesicle transfer and cell shape
Intermediate filaments	So-called because of their intermediate thickness: function uncertain
Collagen	Filamentous structure outside cells—connective tissue

* See Fig. 3.5.

structures, their organization, energization, and synthesis; outside the cells, the major element involved is probably calcium. Most importantly, the actual synthesis and degradation of the extracellular matrices of many if not all cells is associated with the enzymatic activity of *extracellular* copper oxidases and zinc proteases. The constant balance of filament production and cross-linking, i.e. energized synthesis, and degradation both inside and outside cells is all part of the dynamic particulate organization of multicellular organisms much as it is a part of the division cycles of cells (see Chapter 21). Growth and repair are constant activities.

Filaments are organized in space, so that they give a cell vectorial quality within the aqueous matrix. The cell's metabolism governs their formation, but also exposes them to mechanical stress and strain, which is the basis of contractile action as in muscle. This action, which is directional, is again related to patterns of element flow.

If a cell is to maintain shape for a period and then adjust to another shape, it follows that the filamentous network inside as well as outside the cell must have only temporary stability. The way in which this could be managed in cells is shown in Chapter 7, Fig. 7.14.

3.5.2 Lateral membrane organization and in-membrane flow

A biological membrane has to be described as a three-dimensional object with proteins embedded in it or protruding from it (see Fig. 3.9). Consideration of the inside/outside organization and ordering across the membrane is very important for the discussion of electron diffusion and ion or molecule movement *across* membranes. But the lateral organization of proteins must also be taken into account and this leads to the consideration of the *curvature* of the membrane (Figs. 3.6 and 3.9). (This in turn is associated with the capture

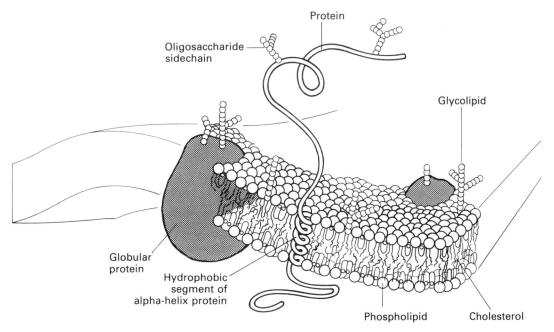

Fig. 3.9 The organization of proteins in membranes.

of certain proteins and to the flow of vesicles as already described; see Fig. 3.4). It is perhaps unnecessary to stress that all these aspects relate directly to the localization of the metals in the cell space since many of the proteins in the membranes are indeed metalloproteins. We will examine more closely this question by considering briefly three specific examples of membranes: that of the thylakoid of the chloroplast; that of the mitochondria; and that of certain epithelial cells.

The thylakoid membranes have a rather large set of incorporated proteins, largely metalloproteins, which occupy more than 50 per cent of the membrane volume; they include the whole apparatus of two photosystems, PS I and PS II, and the energy-capturing device, the ATP-synthetase. The relation of the different units is normally represented as a diagram—the Z diagram of photosynthesis which is supposed to indicate a path of electron flow (see Chapter 5, Fig. 5.3). Today much more is known of the topology of the membrane than this diagram describes (see Figs 3.3 and 3.10). For example, it has been found that, after exposure to light, the photosystem PS I and the ATP-synthetase are in non-oppressed, curved, and exposed regions of the membrane, while PS II is in the oppressed region. This means that there are different microcircuits of electron and proton flow and of ATP generation laterally disposed along the membrane as well as across it (Fig. 3.10). The exact location of each metalloprotein involved (and consequently each metal) is not known, but it is suspected that many of metalloproteins face the aqueous phase of the chloroplast and not that of the thylakoid. The centre of the membrane itself is full of magnesium chlorophyll and the lateral organization places manganese (apparently bound to a 30 000 Da protein) in the oppressed regions (PS II), but iron is more evenly distributed. The case of plastocyanin, the only copper protein involved, is curious since it is found inside the thylakoid, perhaps assisting the electronic communication between PS I and PS II. This may be due to the fact that besides having the right redox potential this protein can withstand the low pH values (of the order of 4) generated in this compartment, whereas other metalloproteins may dissociate there.

The case of the mitochondria is similar, but here the lateral organization is not yet known with the same clarity although the proteins themselves have been more thoroughly studied. Figures 3.2 and 3.11 summarize the present knowledge of the essential structural features of the mitochondrial membrane system and again it is found that a large part of it is made up of proteins which

Fig. 3.10 A description of the distribution of particles, PS I and PS II (photosystems), and proteins in and on thylakoid membranes of chloroplasts. Q represents quinone and QH_2 hydroquinone.

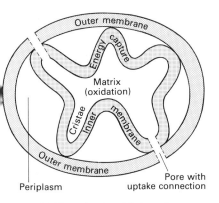

Fig. 3.11 The mitochondria have two membranes. Many of their proteins are made outside the cell cytoplasm and then are placed in different compartments of the organelle. The membranes of the organelle have shape and different proteins are also positioned in lateral organization. The distribution of charge is also differential both across and along membranes.

are overall in excess of lipids. Besides the proteins of the interiors of the membranes in Figs 3.10 and 3.11, many other intraorganelle proteins exist, not bound inside the membrane but bound to it and to other proteins, extending more than 100 Å both ways from the 40 Å-thick lipid bilayer. One feature of this organization is that the inward movement of proteins, and perhaps of small molecules ML, is across the zone where the outer and inner membranes come together (Fig. 3.11). There is a special transfer unit here, for iron complexes and vitamin B_{12} (Co), among other elements. We return again to mitochondrial organization when the tunnelling of energy is described in Chapter 5.

The time-dependent organization of these organelles is related to the light intensity (thylakoids) and oxygen concentration (mitochondria); both factors cause the organelles to be energized and it is this energization that generates the lateral positioning of the different proteins, as well as that of the ATP-synthetase. Note that here it is not radial filaments which cause organization but interactions between proteins in membranes. Effectively these changes are massive invaginations of the inner membrane and similar invaginations are often seen in prokaryote cytoplasmic membranes. This lateral organization occurs together with vertical (transmembrane) movements of ions locally, so that ion gradients occur *across* and *along* membranes. As referred to before, the cations include Na^+, H^+, Mg^{2+} and the anions include OH^-, Cl^-, and HPO_4^{2-}; their lateral gradients cause ionic currents along membrane surfaces. Similar local eddy currents are well-known on nerve surfaces, but are probably common to all cells and organelle surfaces once energized. We discuss these features in more detail in Chapter 5.

Another and very different example of the differential placing of elements in proteins in a cell with a particular shape is illustrated in Fig. 3.12. There we show an epithelial cell. The bottom of the cell, the basolateral membrane, has one set of pumps for moving ions into and across the cell while the finger-like protrusions at the top have quite a different complement of pumps moving material in or out of this region. Flow from top to bottom of the cell is inevitable and no doubt there are special internal carrier systems to assist the transport. It is particularly important to observe the distinction between invaginations (Fig. 3.11) and extensive villi (Fig. 3.12). The shape of the epithelial cell is decided by long filaments inside the villi.

We next need to examine the lateral membrane organization in relation to the extremely anisotropic distribution of the large particles and the matrix of filaments inside and outside the cell. The interaction between the two is inevitable since energy production (ATP) is laterally different and it is ATP which energizes fibres. The tension in the filamentous structure is then connected to cell and vesicle morphology or shape and the membrane is a pattern of differentiated activity. External to the cell there will therefore also be energized organization. Quite generally, the outside of a cell becomes a pattern of energy flow of ions and tensions in filaments and this pattern could well be part of the morphogenic field, i.e. the force which decides the interrelationship of cells to one another. Of course, transmembrane currents are also related to those fields.

We have stressed in this section the organization of proteins, and elements in proteins, in membranes, but we shall see in Chapter 7 that the transverse and lateral distributions of lipids and polysaccharides are equally important.

Microvilli localized enzymes
Alkali phosphatase
Sucrase
Peptidase

Basal/lateral membrane

Na/K ATPase
Ca pump

Epithelial cell

Fig. 3.12 The distributions of some proteins in highly-shaped epithelial cells.

3.5.3 Transmembrane organization and flow

Transmembrane organization is immediately apparent in cells from the fact that Na^+ and Ca^{2+} are largely pumped out of cells or into vesicles, whereas K^+ and (Mg^{2+}) are retained inside, so that, in a steady state, the Ca^{2+} gradient across the cytoplasmic and many vesicle membranes is about 10 000. Examples of vesicles that can hold high concentrations of Ca^{2+} are the sarcoplasmic reticulum of muscles (in fact reticula, generally), the calcizomes, and the various synaptic vesicles of nerves; it is frequently the case that these types of vesicles also contain high concentrations of calcium-binding proteins, e.g. calsequestrin and chromagranin A. When a local electrical pulse (Na^+/K^+) reaches a 'gate' for calcium, which corresponds to the opening of a hole in a membrane, this ion flows suddenly into the cell and triggers a predetermined reaction. Subsequently, it must be removed by pumping through the same or, more frequently, another hole; hence a current of Ca^{2+} always flows along a path in a cell, not just in and out of it, and the size of this current is time-dependent. This activity is localized by the position of the channels and pumps in the membranes. While all the simple cations have gradients across cells so do many anions and, while chloride and sulphate are concentrated outside cells, phosphate is usually concentrated inside.

These many ion gradients, localized or generally across cell membranes, imply that there are transmembrane asymmetries of large molecules, some of which will be partly dependent upon the gradients themselves. In Fig. 3.13 we show a receptor on the outside connected to an internal calcium-binding protein (CAM), a directional pump, attachment points for filaments, and a locally placed vesicle. All these features are related to transmembrane differences in protein constructs, in lipid distribution, and in polysaccharide positioning. In particular, virtually all polysaccharides are on the outside of membranes. In Chapter 7 we shall discuss these distributions in more detail.

In parallel with the protein systems in membranes for the movement of calcium are particles for the transport of many other species and elements. The

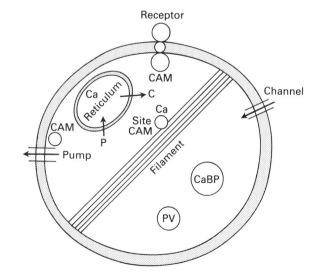

Fig. 3.13 The ways in which calcium distribution is affected and the ways in which its distribution affects cell organization. Calcium is pulsed into the cytoplasm by chemical or electrical stimulation through channels (C). It is distributed or retained by proteins, parvalbumin (PV) and calbindin (CaBP). It affects the cell through action on enzymes and filaments via calmodulins (CAM) and is removed by pumps (P). The disposition of many if not all these units is organized.

case of hydrogen (and other non-metals) is different from those of the alkali or the alkaline-earth ions; in its flow across membranes hydrogen can move as an ion or in the covalently bound form and, as such, it is pumped in many cells. Examples of hydrogen-containing covalent compounds pumped into cells are various simple sugars, carboxylic acids, or the coenzyme NADH. Inside some compartments it can be oxidized or ionized to the proton and so it may exist at a given free ionic level (as H_3O^+), different in each compartment, as is easily recognized by pH measurements. It also exists in many bound states which *do not* equilibrate with the ionic state (e.g. NADH) and this can be recognized by measuring the effective redox potential by a redox electrode or a reducible dye. Notice that the fact that hydrogen is common in many organic molecules where it is covalently bound must not be allowed to hide from us that these are HL compounds comparable with ML compounds.

To finalize these comments it is worthwhile to point out that, while many elements are moved about in proteins and others are transferred by the pumping of simple cations or anions, there is also the possibility of concentrating given elements in particular parts of the space by the movement of small ion chelates. Major examples are those of Fe(citrate), Fe(haem), Fe(hydroxamic acids), Co(vitamin B_{12}). These metal ion 'carriers' are similar to coenzymes in many respects and there must be receptors that can recognize such molecules specifically and then transport them across membranes in a time-dependent process. Two or more forms of one element then act as different kinetic entities even in the same pool, e.g. haem iron acts separately from free Fe^{2+}, see Table 6.4. Thus, the movement of some elements can take place in different forms; iron for instance, enters cells as an iron complex, is processed by insertion catalysts in mitochondria, and much of it returns to the cytoplasm as iron–porphyrin. Similarly, phosphorus enters as free phosphate, is processed by a capture catalyst, ATP-synthetase, and remains as ATP. Both of these activities still leave the mitochondria with a quite high concentration of the free elemental forms—iron as a cation and phosphate as an anion. Other elements, such as calcium and manganese, are pumped into the mitochondria out of the cytoplasm so they too are considerably higher in this compartment, rather than in the cytoplasm, particularly of the eukaryotes. Although it is not possible, yet, to determine the concentrations of *free* Cu, Fe, and Mn, the redox elements, in the cytoplasm of eukaryotes, it is likely that they will be present in very low concentration to protect the exposed DNA from oxidative mutation. This problem does not arise with zinc and this is one of the reasons why this element is used almost exclusively in cytoplasmic acid–base catalysts.

A special case which is worth mentioning here is that of vanadium which is accumulated in the vacuoles of specialized cells in the blood of certain tunicates; recent results show that vanadium enters the cell as vanadate through the phosphate channel and is reduced to VO^{2+} and V^{3+} apparently by a set of pigments related to catechol called tunichromes. In these cationic states vanadium cannot leave the cells and accumulates there, apparently against the concentration gradient.

3.5.4 Flow of vesicles in cells

Whereas the compartmental handling of the elements Na, K, Mg, Ca, Cl, S, P, C, N, H, and, probably, Mo and Si, in advanced cells, and of all these elements

plus Fe and Mn in prokaryotes and organelles, such as mitochondria, is often through the use of directed pumps (either for the element in some simple ionic state or as a small molecule, such as Fe-citrate or ATP), another device for the handling of the transition elements in higher eukaryotic cells is through the use of specialized complexing proteins transported in vesicles. Three possibilities can be considered for the flow of vesicles: one from the outer membrane of a cell to the large endosomal vesicular compartment (Fig. 3.4); a second from the Golgi membrane to the outer membrane (Fig. 3.14); and a third of flow along the tubulin structures running inside a cell (Fig. 3.15). In the first case, the outer membrane, which has receptors incorporated in it—the so-called 'coated pits'—can pick up certain metalloproteins and then fold around the receptor/metalloprotein pair. On closing off the membrane it phagocytoses the coated pit (see Fig. 3.4). A new compartment is thereby formed, which flows to a large endosomal vesicular compartment where the membrane becomes divested from the metalloprotein and reforms the vesicle that flows back to the outer membrane. Since a portion of external fluid is also carried in the process, metal ions abundant in that fluid can also be transferred to the endosome, e.g. calcium, but this is not their normal method of transport.

The second circulation scheme involves the synthesis of proteins by ribosomes on membranes such that the proteins are excreted into an internal vesicle, e.g. into the Golgi apparatus (Fig. 3.14). The Golgi buds off vesicles continuously. These vesicles flow to the outer membrane, combine with it, and reject the proteins which are to be used in the external fluids of the organism, e.g. the lymph- or bloodstream. They can also be discharged into *internal* compartments, as in the case of lysosomes, peroxisomes, etc. One should recall

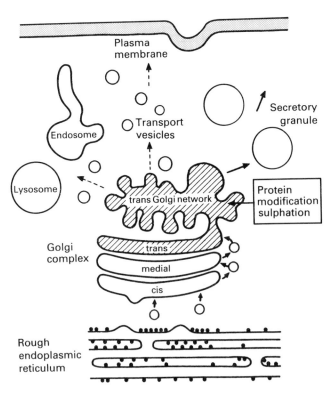

Fig. 3.14 The Golgi and the rough endoplasmic reticulum with extensive variation in protein and particle distributions and a variety of curvatures in the membranes. Some vesicles budding from the Golgi are shown, contrast Fig. 3.4. The positioning of the vesicles is controlled by filaments which are not shown.

(a) Neurofilaments are produced in the nucleus and incorporated into the axon to form part of the vesicle transport system. Actin and the neurotransmitters are carried on this system in vesicles to the synapse and the growth cone where the actin is incorporated and the neurotransmitters are released.

(b) In the Alzheimer neuron, the neurofilaments have formed insoluble tangles around the nucleus. Axonal transport is stopped leading to neurotransmitted deficiencies at the synapse. Actin is no longer transported and collects in vesicles around the nucleus to form Hirano bodies.

Fig. 3.15 A comparison of the transport processes taking place in (a) a normal and (b) a diseased neuron illustrating vesicle incorporation and motion.

that much of what these internal compartments enclose is from 'outer' spaces, i.e. regions that are in many respects similar to the surroundings of the cells. The exported proteins may be designed to pick up a certain metal and then become the agonists of the receptors, to be again captured by a cell. Examples of this behaviour are those of transferrin (uptake of iron and manganese) and of the albumins (uptake of zinc and perhaps copper), both with known receptors, and that of the cobalamin carrier protein (uptake of cobalt), for which a receptor has been postulated.

Axonal transport along tubulin fibres is another device for the movement of vesicles. By careful time-lapse photography it has recently been shown that there is a continuous two-way traffic of different types of vesicle to the end and back of long cellular fingers (Fig. 3.15). It is considered that acetylcholine and adrenalin vesicles in nerve termini are moved in this way. (Interestingly, in the brain there are many nerve endings with vesicles which contain zinc at quite high levels. These zinc vesicle are very noticeable at the ends of the mossy fibres of the hippocampus.) This flow of vesicles along tubulins may well be the mode of transfer of silica, or other minerals, to the growth points of hard structures (see Chapter 20).

Failure of the transport of vesicles along tubulin filaments leads to an aberrant organization and the cell fails. This may be the case in Alzheimer's disease (Fig. 3.15) which has been associated with the incorrect distribution of elements such as Al, Si, and Ca.

Since the flow of protein-containing vesicles is a general process for the handling of the transition elements in their transport between compartments of higher eukaryotic cells, including the uptake and rejection of some of them, it is essential to recognize that the production of *receptor* proteins and of *export* proteins controls the spatial distribution of these elements. Since different cells produce different proteins, the pattern of cellular uptake of elements must be organ-specific; for example, liver absorbs much iron, but white muscle does not. It is equally important to realize that there is a time dependence of protein production; hence the uptake of transition metals can also be conditioned by need on certain occasions. Receptors for iron, for example, increase when cell division is stimulated and cancer cells are rich in

such receptors for much longer periods of time than normal cells, a fact that suggests interesting lines of intervention to control their proliferation.

Only in passing do we have space to refer to the obvious flow of many cells from place to place in, for example, the bloodstream.

3.6 Special solution conditions in vesicles

As to the nature of the internal vesicle compartments three generalizations seem to apply: they are more acidic (pH < 6); they are more oxidative; and they have a higher calcium content. The acidity arises from the fact that proton ATP-ases (pumps) are located in the membranes in such a way that they generate acid in the vesicle, and it may well be that these ATP-ases are associated with the 'coated pits'. (It is known (see Section 3.5.2) that proton ATP-ases accumulate in curved membrane regions of thylakoids and it is possible that they are responsible for the generation of the curved surfaces in these organelles and in the coated pits, favouring the formation of vesicles and their flow to and from the cytoplasm.)

The high redox potential arises from the absence of reducing materials and their enzymes; in lysosomes and peroxisomes, for instance, iron is held as Fe(III), not Fe(II), and manganese can be raised to the Mn(III) state. The Golgi apparatus retains sulphur as sulphate in contrast with the sulphide of the cytoplasm.

The high calcium content probably arises from the enclosing of the external fluids, but also from the activity of calcium ATP-ases which pump this element out of cells and into vesicles, since the outer membrane of the cells inverts on formation of a vesicle which then behaves as an inverted cell. This is the reason why the internal fluids of vesicles resemble the external fluids of the cell concerned.

The importance of the acidity and of the redox potential of the vesicle phases must not be missed; we have repeatedly mentioned that these factors can control the availability of certain ligands and their form; hence they introduce a limitation on the type of enzymes that can function in them. Zinc and indeed most *divalent* transition metal–protein complexes lose stability around pH 5 to 6, so that they are of no value in vesicles such as lysosomes, for example. The *trivalent* metal protein complexes can be made stable at a much lower pH and may be used in such conditions. Acid phosphatases in lysosomes, for instance, are Mn(III) and Fe(III) proteins, probably formed in these vesicles by apoproteins exported by the Golgi system.

It is also the acidity that releases the metals from their carrying proteins and releases the carrier from the receptor. The carrier returns to the outside of the cells, but the metal stays inside. An example is that of iron which is transported externally as transferrin, captured, and transported inside cells by a specific carrier and then used or stored, when in excess, on a special storage protein, ferritin, to which we have already referred. It is the escape of ferritin which leads to iron deposition external to cells (see Fig. 12.4).

The redox potential is also of the utmost consequence, as stressed before, and we will only note here that iron proteins are not very useful oxidases in external solutions or inside vesicles since the dissociation of iron caused by low

pH is made irreversible by the oxidation of Fe(II) to free Fe(III). It is, therefore, the copper proteins which are of greatest value as redox enzymes *outside* cells and inside vesicles, since neither Cu(I) or Cu(II) will exchange from metalloproteins due to the high intrinsic stability of their complexes.

The further aspect of vesicles, to which we will devote an entire chapter, Chapter 20, is that of the formation of inorganic deposits in cell compartments. Indeed, gross deposition of certain elements in certain specialized vesicles can lead to the formation of inorganic mineral phases, especially of Fe, Si, Ca, and P. These precipitates can be amorphous or crystalline, but may be given precise form ('pellets') by the proteins in the vesicles and, when exported, may be packed in regular solid structures, a process again favoured by other proteins that assist deposition of those solids. Bone, teeth, and shells are formed this way, as well as other solid structures, such as the gravity and magnetic sensors used by some species of animals. Curiously, this process is prone to loss of control in higher organisms, particularly in old age, so that the laying down of both iron and calcium minerals in an incorrect manner, especially in humans, is a major cause of ill health. Atherosclerosis, cataracts, haemosiderosis, etc. are all diseases related to incorrect movements of minerals, e.g. deposition in veins and arteries. Even Alzheimer's disease is associated with mineral deposits. Biological systems operate very close to saturation limits of many salts!

3.7 Co-operativity of separations and localizations

We have seen that cell shape is connected to tubulin filament growth and we know that the shape will lead to segregation of units in membranes. If these two processes, tubulin growth and segregation, share a common energization device, then they become co-operative (or antico-operative). The positioning of vesicles is constrained by the filamentous, tubulin, structure of a cell; they move along the filaments, sometimes budding from outer membranes, sometimes coalescing with them. Clearly, the fate of vesicles is associated with the state of fibrils and outer membranes and some co-ordination of activity is obviously required. It is highly probable that this co-ordination is dependent on the states of calcium and of phosphate (see Chapter 21).

3.8 Summary of metal ion positioning

We see now that the *position* of a metal ion, including its concentration gradient, in bound and unbound forms is controlled and must be absolutely critical to an understanding of its function in biology, but we return to this theme in Chapter 6. The picture is likely to be confusing until the distribution is established more firmly. To give those readers who have little acquaintance with the level of organizational complexity of biology a feeling for the situation

we provide in Fig. 3.16 an indication of some of the zones into which different elements are placed. Table 3.2 is an illustration of how a variety of elements can occur in external or vesicle compartments of multicellular organisms. We turn, finally, to a brief description of multicellular organization.

Fig. 3.16 A picture of some of the distributions of elements in eukaryotic cells. Iron is often in membranes while Cl^-, Ca^{2+}, and Na^+ are concentrated away from the cytoplasm; K^+ and Mg^{2+} concentrate inside cells, Mn^{2+} is to be found in vesicles, but copper is more usual outside cells. Zinc is everywhere and cobalt is rare everywhere. Aluminium may be rejected. P = a protein; Ch = chelating agent. Compare Fig. 2.16.

Table 3.2 Proteins (apoproteins) for external retention of metals

Caeruloplasmin	Cu	Oxidase in blood
Haemocyanin	Cu	Oxygen carrier in blood
Alkaline phosphatase	Zn	External surface of membranes
Xanthine oxidase	Mo	Oxidase in milk
Acid phosphatase	Fe, Mn	In lysosomes
Catalase	Fe	In peroxisomes
Hydroxylase (adrenal)	Cu	In chromaffin granule vesicles
Plastocyanin	Cu	In thylakoids
Urease	Ni	In plant vacuoles

3.9 Symbiosis and multicellular systems

The stress on the divisions within single cells must not be allowed to hide the differences in cells which are part of the same organism or are locked in symbiotic union. The most obvious peculiarity is the positioning of the nitrogen fixation system and, therefore, of much molybdenum, in root nodule bacteria while there is little or none of this activity or of molybdenum in roots. Symbiosis is very common (it may even be one of the main sources of evolution) and there are a multitude of ways of dividing activities and elements which parallel the division between mitochondria, chloroplasts, and cell

cytoplasm. In fact, mitochondria and chloroplasts may have resulted from the capture of one type of organism by another.

The creation of cells with different purposes in a single organism implies that element distribution is unique to a cell type and this is known to be the case. The most obvious is the distinction between red blood cells (iron mostly) and white cells (more zinc and little iron). We shall not discuss this problem here and will return to it in Chapter 21.

3.10 Spatial distribution of non-metals

The spatial distribution of non-metals in some simple forms such as H_2O, CO_2, N_2, and O_2 is not limited by membranes, by binding, or by charge. In essence, these inert molecules form the liquid and gaseous atmospheres in which biology has developed. Here we are hardly concerned with these forms of the non-metals since they are not fixed in space; neither is it interesting to look at H, C, N, O in biopolymers since they are combined everywhere. Spatial distribution of H, C, N, and O becomes interesting when and where their small inert molecules are converted to reactive forms, i.e. where they are activated. It is now well known that N_2 goes immediately to NH_3, H_2O goes to bound hydride (NADPH), bound O atoms, and O_2 which is lost to the atmosphere, CO_2 goes to HCHO on its way down to $-CH_3$, etc. Once in reactive form, it becomes essential to move the non-metals about in such a way that they only react when and where they are required. In part this is managed by handling these elements in kinetically trapped covalent molecules, e.g. C/H in the carboxylic acids of the citric acid cycle (mitochondria) or the sugar phosphates of glycolysis (cytoplasm), N in amino acids or nucleotide bases (cytoplasm), etc. which are limited in their movements by membranes. In addition their distribution is managed, while still holding the elements in very water soluble forms so that they diffuse in any one aqueous compartment, e.g. C/H in sugars, by energizing their transport between compartments. This transport is observed to be by pumps or exchangers which move them through membranes, and it is very specific. For these reasons we become more concerned with the localization and transformation of specific compounds where the elements are incorporated into kinetically inert or somewhat labile forms as, for example, in C–H, rather than with the localization of elements H, C, N, or O as non-metals. This means that, for example, carboxylates of the Krebs cycle, amino acids, and phosphates of glycolysis, although they all contain C and H, often belong to quite separate C and H pools either in space and/or in kinetic connection. This is well known in metabolic and synthetic pathways of the biological systems of course. Yet, despite this separateness of the metabolic pathways, they must all relate to one another for several reasons: first, all pathways have come from CO_2, H_2O, N_2—a few small primary molecules; second, they will all be degraded to these same molecules in an orderly manner so that energy and material is not wasted; third, the metabolic pools cannot be independent since amounts of proteins, fats, polynucleic acids, etc. are required in rather strictly controlled ratios in order to build an organism. The demand is then for an intercommunicating network

based on additional kinetically controlled units which interact with the metabolic pathways as distributors and controls, i.e. small basic carriers of the pools of $-H$, $-CH_3$, $-NH_2$, $-COCH_3$, $-OPO_3^{2-}$, etc. Naturally, these carriers can only distribute the basic units properly if they also act as (allosteric) controls over the networks of metabolic pathways. The essential carriers of the non-metals are the mobile coenzymes (see Section 2.17), and it is distribution of these carriers, frequently related to vitamins, to which we now turn. While doing this we must note which coenzymes are in a particular compartment and which go between compartments. Some coenzymes are not mobile (Table 3.3).

Table 3.3 Transport and reaction systems for non-metals

Transported Units	Mobile coenzyme (immobile)
H_2/reducing equivalents	NADH, NADPH, (flavin), QH_2 glutathione
CH_3-	Coenzyme M, S-adenosyl methionine (B_{12} coenzyme)
CO_2	(Biotin)
CH_3CO-	Coenzyme-A (lipoic acid) (F-430, Ni)
CH_3CHO	(Thiamin)
SO_4^{2-}	PAPS
PO_4^{3-}	NTP, ATP, etc. P_2, acyl phosphate
Sugars	UDP
NH_3	Glutamine (pyridoxal phosphate)

3.11 The spatial transfer of H, C, N, S: mobile and fixed coenzymes

3.11.1 H (and O) transport

There is no intracellular carrier for elemental oxygen, O, and it may well be that, since water is everywhere and dioxygen is a highly energized, mobile, species, extracellular distribution to cells only is necessary in multicellular organisms, being managed by a variety of dioxygen carriers, and the movement in cells is not controlled. The distribution of oxygen atoms after the element has been incorporated in specific localities by oxidases is then in combined form with carbon, for example in acetate, and oxygen atoms themselves are not transferred from one carbon to another. Even $-OH$ as such does not have a carrier though it can be rearranged on carbon skeletons by isomerases. The distribution of H itself is managed by a series of carriers at different redox potentials which do not equilibrate directly with one another. At the low potential extreme (-0.5 V) there are the carriers NADH and

NADPH which pick up H from organic molecules, such as malate, in aqueous phases. They transfer H, site selectively, to a variety of acceptors for synthetic purposes (NADPH) or to the membrane of the degradative respiratory cytochrome chain for energy production (NADH). Once the reducing equivalents enter the cytochrome chain, they remain membrane-bound or -associated until they react with O_2 to give H_2O, or perhaps in some systems with sulphur to give H_2S. While in the membrane two varieties of coenzyme handle H: fixed site reactivity is generated by flavins (see Fig. 3.1); mobile transport of H is managed in the membrane phase by quinones, QH_2. Before passing the reducing H to dioxygen or sulphur, energy is retained in proton gradients (see Chapter 5). Remember that here we wish to observe the location and timed movement of H as an element, i.e. H^+, H^-, and $H^·$ (including covalently held H).

Now, the H– transfer systems at -0.5 V (NADH, NADPH) in water are quite separate, in a kinetic sense, not only from the QH_2 pools in membranes (0.0 V) but also from the external and internal redox couples which govern the equilibration of hydrogen reactions at somewhat higher potentials. Outside cells ascorbate, for example, is involved in maintaining a redox system with a potential above 0.0 V, while in the cytoplasm of cells the glutathione couple transports H at close to the same potential, 0.0 V. Because the couples NADPH/NADP, NADH/NAD, QH_2/Q, and 2RSH/R–S–S–R do not intercommunicate directly there is not a redox state of the cytoplasm in a cell, but at least four pairs of redox potential buffers acting on distinctly opposite series of different metabolic pools. The separation is maintained despite the fact that three of these pools lie in the same aqueous phase. The fact that there is no direct intercommunication between pools allows the energy of different C–H bonds to be used without equilibration and loss of energy as heat, and it allows specific networks of synthesis and degradation. A general transport of H between all systems would have been disastrous. The movement of H is then not just a matter of catalysis of substrates but of selective transport in selected pathways. The positioning of systems for hydrogen reactions in vesicles, on membranes, and on filaments is clearly critical. (*All this activity is critically dependent upon the correct location of inorganic catalytic ions too.*)

Now we have described the simplest form of transfer of H and before going forward to look at carbon, phosphorus, and nitrogen transfer we refer to H-transfer in groups very briefly. Just as acetyl, CH_3CO, is an effective way of transferring O it is also an effective way of transferring H (and C). Looking at the coenzymes in Table 3.3 we note that there are other coenzymes which transfer H with carbon, as CH_3- for example, and this building of organic molecules from fragments in different pathways is central to metabolism.

3.11.2 Phosphate transport

Unlike O, H, and C, phosphate itself cannot gain direct access to cells and in this respect it is to be likened to inorganic cations. However, once it has been transferred across membranes by pumps it becomes locked into many kinetically stable, although not very stable, inorganic and organic compounds but it does not undergo redox state change. Note also that phosphate is not only transferred in small fragments but in larger molecular units, e.g. nucleotides.

The situation may next be compared with that of the hydrogen pools in the cytoplasm. The first step of synthesis involving phosphate is of ATP and then of GTP, UTP, CTP and other 'high' energy forms of phosphate from ATP. The exchange of energized phosphate is controlled such that each pool communicates to some extent with a separate reaction path, but also more directly to one another than the different forms of bound H. The networks of phosphorylation can then be separately blocked, although the pool of phosphate 'energy' is interlocked. This will become of consequence in later chapters.

The transfer of phosphate in different metabolic pools is also managed by a whole range of specific enzymes which recognize specific molecular species as is true for C, H, O, and N. Phosphorylated compounds apart from nucleotides are seldom moved across membranes.

3.11.3 Carbon transport

Carbon fragments, C_1 and C_2, are transported by very different sets of coenzymes which diffuse in the cytoplasm or operate at fixed sites. Thus, coenzyme A, methylmethionine CoA, vitamin B_{12}, and lipoic acid, all transport carbon. Notice that the carrier for acetyl-groups, CoA, is not a carrier for methyl groups, which methionyl CoA is. Methylation and acetylation must not be confused; they are different ways of moving C/H/O in combination. The interested chemist should take a biochemist's wall chart of metabolism and follow the way in which, say, C_1 fragments are introduced from CO_2, follow the path to CH_3- groups, and then follow their subsequent distribution. The carbon has then to go into C_2, C_3, etc. units. C_1 incorporation into CH_3CO- fragments is an essential step. (Note in this connection the special pathways using fixed Co and Ni coenzymes in archaebacteria; Chapter 16.)

The further incorporation of larger carbon units is managed differently for the individual chemicals. Fatty acid fragments are built up from acetyl CoA by successive C_2 (acetyl) additions which are retained by CoA. The control of the movement of units is very sophisticated. In particular one part of CoA, pantothenic acid, is suitable for swinging between a CH_3CO- uptake point and a delivery point during the addition of acetyl to a growing $(CH_2)_n$ chain. The end product is the synthesis of lipids. Other forms of movement of carbon include the varieties of carriers for fragments such as amino acids, saccharides, isopentyl groups, and so on and some of these systems are in special compartments, e.g. the Golgi apparatus for polysaccharide synthesis.

3.11.4 Nitrogen transport

Nitrogen is transported mainly as glutamine, which is the precursor of the amine group in most amino acids, and as pyridoxal. The amino acids are transported by tRNAs to the ribosomes for protein synthesis. There is a quite separate transfer of $-NH_2$ in the formation of lysine; in this case the coenzyme is pyridoxal. All this transport is confined to the cytoplasm.

Nitrogen is not incorporated in biological polymers outside the cytoplasm, unlike carbon or sulphur, both of which are also moved around in the Golgi apparatus.

3.11.5 Sulphur transport

Sulphur is transferred as sulphate using the coenzyme PAPS. The transfer is confined to the Golgi apparatus where sulphation occurs. Now, while sulphate is a dominant form of anionic sulphur outside cells, inside cells it is present largely as sulphide. Thus, the catalysts for sulphate reduction must be retained outside the cell cytoplasm. The transfer of sulphide is not understood but it may involve small Fe/S clusters that are mobile.

3.12 Summary of non-metal transport

As we have seen, the essential difference between metals and non-metals is increased kinetic inertness so that for non-metals chemical pathways are readily separated by placing catalysts in addition to using specific pumps. Many coenzymes are specially selected organic molecules which carry the units of H^-, HPO_4^{2-}, CH_3^-, CH_3CO-, etc. in an activated condition so that later catalytic transfer is easy. Examples are ATP(HPO_4^{2-}), PAPS (SO_4^{2-}), and $-S-(COCH_3)$ in acetyl coenzyme A. However, there are also coenzymes that are located at fixed points in proteins, in contrast with those which are free and can traverse water or membranes on carriers; this is the case for ATP which is required to enter or leave a large number of vesicle and organelle systems.

We conclude that biological non-metals, like the biological metals, may well be compartmentalized in very highly organized distribution networks. The fact that locally molecules may diffuse randomly in one phase does not mean that they are not following spatial paths out of the phase and into a designated second one. One general restriction is the membrane and another is the existence of points of reaction or transfer. The facts that, in membranes, pumps can be positioned specifically and, in aqueous phases, catalysts can be located on networks of filaments, means that distribution can be controlled. While it is easy to follow elements such as Ca^{2+}, which does not alter its chemical form as an ion, it is more difficult to follow carbon or nitrogen or even iron because of their many possible oxidation states in kinetically controlled conditions. Even so, there are many kinds of devices controlling transfer to local organelles, reticula, and vesicles. We have seen that the vesicles themselves are transported in given ways too. The systematic movement of transmitter molecules to the ends of nerve fibres is not different from the movement and exocytosis of silica rods to form the *lorica* outside the cell of some protozoa. We shall return again and again to the dynamic organization of elements in biology.

Further reading

This chapter only contains material which is in many biochemistry textbooks. The books which follow are suggested for general reading. While reading these books we ask the reader to note which elements are involved in any process described and where the process occurs in cells.

Armstrong, F. B. (1989). *Biochemistry* (3rd edn). Oxford University Press, Oxford.

Bergethin, P. R. and Simons, E. R. (1990). *Biophysical chemistry—molecules to membranes*. Springer-Verlag, Berlin.

Darnell, J., Lodish, H., and Baltimore, D. (1990). *Molecular cell biology* (2nd edn). Scientific American Books, New York.

Stryer, L. (1988). *Biochemistry* (2nd edn). W. H. Freman & Co., New York.

4

Kinetic considerations of chemical reactions, catalysis, and control

4.1	Introduction	95
4.2	Chemical transformations	96
4.3	The nature of acid–base reactions: hydrolysis and condensation	97
4.4	The hydrolysis of proteins, RNA, DNA, and other polymers	99
4.5	The nature of ion flow	100
4.6	On/off reactions: control systems	103
4.7	Electron transfer reactions: electronics	105
4.8	Redox potentials of complexes	108
4.9	Atom transfer reactions	109
4.10	Group transfer	111
4.11	The nature of transition-metal centres in catalysis	112
4.12	Free radical reactions	114
4.13	Sizes of atoms, stereochemistry, and reaction paths	115
4.14	The creation of local small spaces: mechanical devices for transfer	116
4.15	Summary	118

4.1 Introduction

In Chapters 1 to 3 we have described the chemical elements of life, the principles behind their chemical speciation, and the observed compartmentalization of biological systems. Now, in their special places, the elements are used in a functional sense, e.g. to provide structure and to assist chemical transformation. Before we can look at their functional value in *biology* we need to be acquainted with their *chemical kinetic* potential and we shall do this in two steps. In this chapter we consider the basic rules of chemical transformations, i.e. the kinetic controls and barriers for ion and compound reactions which can be utilized by biology and see how they are dealt with

locally, but we do not wish to involve biological considerations in any depth. Then, together with the basic rules for the thermodynamic combination of the elements and for their distribution in compartments, outlined in the previous chapters, we can turn, in Chapter 5, to the possible ways in which energy can be put into biological systems in order to see how, in principle, the full functional potential of the chemical elements in biology, including catalysis, control, signalling, and structure, can be realized in the appropriate place and at particular times. This is not the same problem as that of the ways of developing the functional potential of elements in chemistry, since in chemistry we have a much wider range of variation of solvent, of pressure, of temperature, of phase removal, and so on. In biology there is instead an energized organization in the control over the distribution of elements. Fundamentally, of course, there is no difference from the deployment of elements in compounds by man and his spatial and energetic manipulation of that chemistry by, for example, phase transfer and temperature controls, e.g. in a chemical factory. Biology is just so much more particularized and complex on a microscale since all activities are in the same cell or organism, but, in essence, properly developed biology (and chemistry) seek to ensure that every atom has the correct partners and controls, in the appropriate zone, for function at the right time.

4.2 Chemical transformations

In this book there are only a few basic kinetic aspects of chemistry which we have a need to analyse in relation to biological activity, namely, those referring to:

(1) acid–base reactions: hydrolysis;

(2) the nature of ion flow: signalling and control, electrolytics;

(3) electron-transfer reactions, electronics;

(4) atom-transfer reactions (redox reactions);

(5) group-transfer reactions;

(6) free radical reactions.

These aspects will be considered in the next sections, but some preliminary considerations on catalysis in general are necessary.

First of all, catalysis by itself can only assist a downhill reaction, so that we shall need to consider how to incorporate or retain energy in a catalysed system. In part this means that the catalyst will not just activate a molecule, but will protect it from adventitious attack. The need for such organized and energized catalysts is seen if we consider the synthesis of any condensation polymer (high-energy reaction in water) as opposed to its degradation in digestion.

There must be catalysts for the forward and back reactions which are quite distinct, differently controlled, and occur in different places. The polymerization reactions require energy (e.g. from pyrophosphate) and a series of catalysts, Cat (2), which are placed inside cells. (Note that DNA and RNA are synthesized in the nucleus, proteins in ribosomes, and polysaccharides in the Golgi.) The opposed reactions need to be very largely prevented in cells, but

accelerated in digestion, for example, which occurs outside cells. In this case, catalysts (1) and (2) are enzymes differently placed and stored in space but with one fundamental feature in common: they are acid–base catalysts and not redox catalysts. In acid–base catalysis there is a rearrangement of atoms without changes of oxidation state. Once again there is space/time dependence in these catalytic acts. A very similar problem concerns the direction of redox reactions also involved in synthesis and degradation.

There is one group of biological polymeric molecules which we have not mentioned above–the lipids. The fully reduced carbon chains of biology are generated by condensation followed by reduction

$$CH_3COR + CH_3COOH \longrightarrow CH_3COCH_2COR + H_2O,$$
$$CH_3COCH_2COR + 2H_2 \longrightarrow CH_3CH_2CH_2COR + H_2O,$$

where R is a retaining centre on an enzyme. Naturally, two different catalysts are required, acid–base in the first reaction and redox in the second. Notice that these and all other biological polymers are formed by removing water.

This very simple introduction to some living processes has shown that in biology two basic reactions must proceed at *controlled* rates and in chosen directions: (1) acid–base; and (2) redox reactions. We need to note the character of these reactions, to know how to control them, i.e. to catalyse and to stop them, and then to see that they go in a specific direction, which requires energy.

4.3 The nature of acid–base reactions: hydrolysis and condensation

Consider the reversible reaction of ester formation

$$RCOOH + R'OH \rightleftharpoons RCOOR' + H_2O. \tag{4.1}$$

The carbon atoms remain associated with the same numbers of oxygen and hydrogen atoms—there is no change of oxidation state—but the atoms are re-arranged. The need for catalysis arises from the stability of the different carbon–hydrogen, oxygen–hydrogen, and carbon–oxygen bonds. To get reaction, the R'OH group must attack the $>$CO group of the acid, and naïvely we can see that this will be assisted by higher energy states, such as

$$\tag{4.2}$$

when we can write that the C^+ is attacked by the negative polarity of the oxygen atom of R'OH to give

$$\tag{4.3}$$

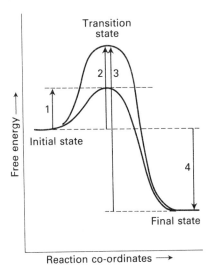

Fig. 4.1 The effect of a catalyst is to lower the energy of intermediates or transition states relative to the uncatalysed pathway. 1: Free energy of activation of catalysed reaction; 2: free energy of activation of forward reaction (uncatalysed); 3: free energy of activation of reverse reaction (uncatalysed); 4: overall free-energy change of reaction.

Fig. 4.2 (Lower) the order of hydrolysis of a simple ester by a series of Lewis acids and (upper) the order of hydrolysis by the same Lewis acids substituted in a zinc enzyme, carbonic anhydrase. Obviously, Lewis acid strength is not dominant in enzymes.

Overall loss of water (condensation) gives the product in both cases. The pattern is the same for all condensations mentioned above. Immediately we observe that catalysts must act by binding a molecule so as to favour vulnerable states (Fig. 4.1), the intermediates, and transition states. One way to achieve this is to induce strain in the substrate simply by deforming bonds, while using binding energy to induce that strain. This induced strain is, in principle, independent of the chemical nature of the binding in that it is just a mechanical effect. To strain a molecule, a large number of matched contacts is useful and surfaces of proteins (large molecules) are ideal for straining *large* substrates. If there is an electrostatic component of the binding, it can be used to help to induce charge separation (see eqn (4.2)) as well as to induce mechanical strain. Such a catalyst is an acid–base catalyst since acid and base centres are distinguished by their affinity for negatively charged atoms (acids are positive centres) or positively charged atoms (bases are negative centres). Clearly, the proton or a metal ion with high negative charge affinity, e.g. Zn^{2+}, will be a good acid, and an anion, such as hydroxide or thiolate, will be a good base. In order to control condensation and hydrolysis, biology has developed a series of acid–base catalysts within proteins, i.e. enzymes. We shall see that the useful character of inorganic elements in these catalysts arises especially where the substrate is a *small* molecule or when the attack needs to be indiscriminate and fast. The metal ion is a site of great attacking power which is useful in both synthesis and degradation (Table 6.3).

Now, acid attack depends on the electronic deforming (polarizing) power of an ion as well as on its electrostatic binding (it is permissible to refer to this as partial covalent bond formation). In Chapter 2 we observed that the Irving–Williams series, i.e.

$$(Mg^{2+}, Ca^{2+}) < Mn^{2+} < Fe^{2+} < Co^{2+} < Ni^{2+} < Cu^{2+} > Zn^{2+},$$

was a series of decreasing radius, increasing electrostatic effect, and increasing electron affinity, polarizing effect, or covalence.

This order is the observed order of acid catalysis by these ions in model acid–base reactions (see Fig. 4.2); however, in biology, the order, when metals are available or bound in the same enzyme site, is never this one (Fig. 4.2) and is different in different compartments, due to control over concentrations of these cations, and in different enzymes, due to selective incorporation and control of conformation, as we shall see in the following sections.

Amongst bases, the common attack through –OH, –SH, or –NH may give rise to relatively longer-lasting 'covalent' intermediates; the problem then is not just attack, but the leaving ability to give products. In both respects proteins are 'designed' to assist the organic catalytic acts. The bases chosen by biology are not always conventional, but are often the conjugate base of a weak proton acid which is especially reactive within the protein (enzyme) matrix; a good example is the –OH group of an alcohol, e.g. serine.

This analysis of acid–base catalysed reactions has avoided the detailed description of the precise way in which a base like RS^- or an acid like Zn^{2+} can polarize an ester bond. The obvious mode is the formation of an intermediate so that in the reaction the base RS^- actually becomes acylated and the acid Zn^{2+} becomes complexed to the attacking anion (OH^- or RO^- above). The required property of the catalyst is that it must be a good attacking group, able to form bonds quickly, and must also be a good leaving

group, i.e. it must form unstable bonds. Note that there is an obvious barrier to these reactions (Fig. 4.1), which needs to be overcome; within proteins there are special ways of increasing the attacking power of atoms or groups of atoms, for example, zinc, and alcohol –OH groups (of serine), which we shall describe in Chapter 7 under the name of the 'entatic state'.

While this description is clearly satisfactory for a downhill acid–base hydrolysis, it does not permit the reaction to go uphill in energy. All esters and amides are unstable in water and a way of protecting the active site for their syntheses and many of their other reactions has to be found, which avoids attack by the solvent. For uphill syntheses energy input is necessary, e.g. in the synthesis of pyrophosphate from two phosphates. We return to group transfer below and to energy input in Chapter 5.

Combinations of acids and bases in correct spatial organization opposite substrates obviously improves catalytic power since the activation event involves polarization of at least one bond, i.e. there are two centres to attack (eqn (4.2)). Again the use of two acid centres, e.g. metal ions, can be valuable since one can bind the substrate in a useful orientation while the other generates attack either directly or through a bound base, e.g. hydroxide. This point is made since it has been observed that several hydrolytic and synthetizing enzymes have two metal ions at the active site, e.g. zinc and magnesium in alkaline phosphatase and two zinc ions in phospholipase C. Model catalysts have been used to show the great enhancements of reactivity of two metal ions at active sites over one. We shall not describe the mechanisms of acid–base catalysis by enzymes in more detail until we have examined limitations of biology in Chapter 6.

4.4 The hydrolysis of proteins, RNA, DNA, and other polymers

While a major concern of acid–base catalysis is with small molecules, (glycolysis, and amide and ester hydrolyses), a section is added here to make sure that we remember a peculiarity of biological systems. The 'stable' polymers of biology, which we have described in Chapters 2 and 3 as a framework for the reactions and movement of the elements, are constantly turned over. The turnover, comprising synthesis and degradation, is an effective way of generating a steady state concentration of a polymer. Acid–base catalysts control both these rates, as well as the supply of monomers and energy for synthesis. It will then be important to notice the parts of space in which the polymers and their acid–base catalysts are placed. Obviously some zones are mainly digestive, e.g. the lysosomes, some are synthesizing, e.g. the ribosomes and the nuclei, and some are apparently neutral, e.g. the cytoplasm.

There are two factors which modify any general picture of hydrolysis rates of polymers from the above account of model reactions. One is the strength of the fold, which varies from almost random coil to tightly folded and cross-linked polymers, and the second is the location of the hydrolytic catalysts. The random coil polymers are relatively easily digested, even in the cytoplasm, but

the cross-linked proteins are not. However, the stabilizing cross-linking of proteins is reversible due to release of metals, redox reactions that break –S–S– bridges and so on. There is then a steady state of protein concentration dependent on X, where X can be a metal ion or a redox process giving an –S–S– bridge. The scheme shows that there may be little free apoprotein in a solution of a protein, a cross-linking agent, and a protease. This situation is made the more secure by the feedback control to DNA or RNA over protein synthesis by X, e.g. a metal ion, and its uptake. The limitation on free protein is a very important control over selective binding (Chapter 2). Similar considerations apply to RNA: messenger RNA is continuously made and digested; ribosomal RNA and DNA are in protected states.

Now, the degradation process and the synthetic process need catalysts which must be kept apart, i.e. zoned, for obvious reasons (see Chapter 3). Proteins which are not strongly stabilized do not last long in the bloodstream of animals, for example, since there are proteases present everywhere outside cells. A major activator of catalysts of hydrolysis is external calcium. When excessive calcium enters a cell it even switches on degradation in the cytoplasm, which corresponds to cell suicide. Magnesium is much more concerned with the hydrolysis of phosphate-containing polymers, and again this is rapid outside cells, e.g. in digestion of DNA and RNA.

Some zones are designed to be oxidative. Here –S–S– bridge formation, as well as calcium cross-links, protect certain proteins from hydrolysis, e.g. outside cells.

The function of the proton in acid-catalysed enzymic reactions is more to the fore in zones which are acidic, pH < 7. The obvious ones are parts of the extracellular digestive tract of multicellular organisms and lysosomal vesicles. Such zones often have *active* metal-containing enzymes for hydrolysis too. Clearly, we cannot speak of general principles of acid–base or redox catalysis in biology without reference to a compartment and its content at a given time. Just as we shall find that certain elements, Zn, Ca, and Mg, are overwhelmingly used in acid–base catalysis, so we shall find that certain compartments are for acid–base reactions and not redox reactions.

4.5 The nature of ion flow

4.5.1 Signals and controls

Acidic and basic ions are obviously important in the catalysis of chemical change, but they are equally important in signalling and control. The transfer of information from site to site in biology necessitates a signalling system. This need be no more than the passage of current, the flow of charge, and, in particular, the flow of ions. However, the kinetics of electrostatic flow, referring now to movement along or across cell membranes and cellular systems, must be selective, as in any transmission system, so that there must be allowed and forbidden flow paths for different biological ions. Speed of flow is also essential. Moreover, there has to be a receptor for the message, which eventually will be transcribed into a chemical change. All this means that, in

addition to the rate of movement from site to site, the rates of binding and release of the messenger ions have to be analysed. We then must consider a series of steps where an ion A flows to, binds, and then leaves a receptor B.

(1) A flows to B (rate of diffusion);
(2) A binds to B→AB (rate of association);
(3) AB undergoes conformation change (rearrangement rate);
(4) A leaves B→A + B (dissociation rate).

All these rate processes are not just conventional catalyses of acid–base reactions; they involve transport in space and time, i.e. they can be vectorial. However, quite frequently, we can relate rates of diffusion and/or rates of association to steps in acid–base catalysis.

For example, the movement of the ion A may well require the release of water from a solvation sphere of a cation M^+ or an anion X^-. The leaving of water is an acid–base dissociation

$$H_2O \to M^+ \longrightarrow H_2O \text{ (base)} + M^+ \text{ (acid)},$$
$$OH_2 \to X^- \longrightarrow H_2O \text{ (acid)} + X^- \text{ (base)}.$$

These reactions and their reverses may well be required for action at a receptor or as an ion passes through a membrane and could well be vitally important for ion selection. We shall then need to know the speed of water release since it will be important in message transmission, in control, in catalysis and so on.

Finally, binding involves mechanical rearrangement of the receptor ligand B (e.g. a protein) when bound to A as AB. We must observe the kinetics of rearrangement of molecules here. The distribution rates of strain energy are involved (step (3)), since the metal binding pulls on a protein. The control of surfaces and the control of mobility gives proteins extreme value as channels, gates, triggers, and so on.

There is little doubt that H^+ and OH^- are the most useful acid and base, respectively, not only in catalysis but in ion migration since they have the fastest diffusion rates. They are very valuable in biology, but we must remember that the biological pH is 7.0 which is extremely restrictive for the concentration of these species. Function in biology does not follow easily from chemical principles without very careful examination of the restrictions which do not allow wide variation of parameters. Since the physiological pH is close to neutrality (~ 7), H^+ and OH^- are 10^6 to 10^5 times less available than Na^+ and K^+ and, therefore, despite their higher mobility, they are at least 10^3 times less useful as charge carriers. For this reason, K^+, Na^+, Cl^-, and even Ca^{2+} are preferred in biology's electrolytic circuits.

We turn now to some details of acid–base reactions in ion flow.

4.5.2 Ion migration rates: electrolytics

The migration of an ion in a membrane and, more especially, its movement to and from water will be partly dependent on the rate at which water leaves and enters the co-ordination sphere of the ion. For anions this rate is almost invariably very fast, of the order of 10^9–10^{10} per second. For cations the rate

decreases from this limit for H^+, Na^+, and K^+, down via the large divalent and trivalent ions, Ba^{2+}, Sr^{2+}, Ca^{2+}, and La^{3+} (10^7–10^9 per second), to the smaller divalent ions Mg^{2+} (10^5 per second), to the smaller highly charged ions, Al^{3+} (1 per second) (see Fig. 4.3). The rates of this exchange are one major factor, apart from concentration, in the selection of Cl^-, Na^+, K^+, and Ca^{2+} as biological ion current carriers. The second factor is binding, which is minimal for Cl^-, Na^+, and K^+, so that on/off rates to organic matrices remain very fast (see Section 4.6).

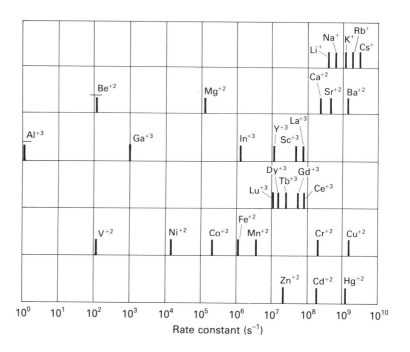

Fig. 4.3 Characteristic rate constants (s^{-1}) for substitution of inner-sphere H_2O of various aquo ions. *Note.* The substitution rates of water in complexes $ML(H_2O)_n$ will depend on the symmetry of the complex. (Reproduced with permission from Frey, C. M. and Stuehr, J. (1974). Kinetics of metal ion interactions with nucleotides and base free phosphates. In *Metal ions in biological systems*, Vol. 1 (ed. H. Sigel), p. 69. Marcel Dekker, New York.)

Two factors clearly need to be carefully distinguished now in the discussion of the selective passage of ions across membranes. There may be thermo-dynamic barriers to entry into the channels, as discussed in Chapter 2, but there can also be differences in rates of diffusion in a channel. When we measure currents, i.e. permeabilities, through membranes in the presence of mixtures of ions, we study a combination of these factors which are locked in the permeability ratios (Table 4.1), which differ for channels designed to allow passage of Na^+ predominantly or K^+ predominantly. Many of the channels appear to be of considerably larger diameter than the bare ions and it could well be that some of the channel walls are lined with water molecules (Fig. 2.11); these water molecules reorganize differently as different ions pass through them (see Chapter 8 for more details).

In order to establish a net flow it is necessary to generate an initial out-of-balance configuration of ion concentration or of field, for example, across membranes. To achieve the concentration or field gradient, some charged species, ions, must be pumped across the membrane. The rate of pumping depends on the rate of energy input and the capacity of the pump. The design

Table 4.1 Permeability ratios for two types of channel

	Li^+	Na^+	K^+
Sodium channel	0.95	1.00	0.05
Potassium channel	0.02	0.02	1.00

Fig. 4.4 A diagram of the interior of part of the redox protein, cytochrome *c*, to illustrate the coupling of proton networks in proteins with electron transfer to their metal ions. On redox reaction, Fe(II) to Fe(III), there is a change in the hydrogen-bond network associated with Gly-41, Asn-52, and Thr-78. Such changes could alter rates of proton flow through the network. * represents a water molecule.

of pumps for H^+, Na^+, K^+, Ca^{2+}, and Mg^{2+} amongst many other moieties is a critical acid–base kinetic problem in biology (Chapter 8). The selectivity of the rate of pumping is quite different from the thermodynamic selectivity of ions entering channels, although it too depends in part on interactions between ions and water molecules. The actual transfer step in a pump requires also a conformation change of a protein, which is of necessity slow, about 10^2 to $10^3 \ s^{-1}$, and this motion of the protein may or may not be allowed, depending on the size of the ion in the pump channel.

A very special case of ion migration is that of the proton since it can diffuse in an H-bond network by rotational/translational diffusion of water molecules. This mode of very rapid diffusion is characterized by a hop distance of some 0.5 to 1.0 Å (Table 4.2); the equivalent motion of OH^- is easily written. Proton channels readily exclude all other ions since no other ion can follow such an H-bond path, Fig. 2.11. An important consideration will be the coupling of proton channel movement to electron movement in proton pumps which, if they are to conserve energy, must be 'gated' (see Section 4.7 and Chapter 5). An impression of a proton network within a protein is given in Fig. 4.4 which shows the interior of cytochrome *c*, an electron transfer protein.

There is one other type of current carrier: small organic cations, such as acetylcholine (AcH) and glutamate, which can be very selectively recognized by receptors to build a specific current chain, e.g.

$$\text{Message } Na^+/K^+ \longrightarrow \text{Release of X (AcH)} \longrightarrow \text{Message } Na^+/K^+.$$
$$\text{(Cell A)} \qquad\qquad\qquad\qquad\qquad\qquad\qquad\qquad \text{(Cell B)}$$

Table 4.2 Proton and electron transport

	Protons	Electrons
Hop distance (Å)	0.5 to 1.0 Å	<15 Å
Site	Bases	Metal ions, aromatics
Activation energy	pK$_a$ change	Redox energy, ΔE
	Relaxation, R	Relaxation, R
	Rotational diffusion	

Here X is an organic carrier of current which selects for a special B-type cell from amongst many others. It is the specificity of these current carriers which makes slow complicated (parallel) biological circuits so different from the fast electronic circuits of man, e.g. in computers. However, the use of the slower ion movements allows diversity in the current carriers themselves and each current carrier can function specifically at any point in the circuit since it can be differentially recognized. This allows very sophisticated information distribution.

4.6 On/off reactions: control systems

Control demands that, as well as diffusing rapidly, the messenger system should bind, change the conformation at the site of binding, and subsequently should leave as rapidly as possible so as to allow the system to relax after information has been received. Thus we require k_{off} to be as large as is

compatible with the ability to switch the device. Now, the energy capacity to switch a protein conformation will depend on binding strength, where the binding constant is

$$K = \frac{k_{on}}{k_{off}}.$$

Biological switches need to be fast and their speed appears to be controlled by the time scale limit at which proteins can rearrange since proteins must have reasonably stable structures. This limit seems to be close to 10^{-3} s. Thus k_{off} should ideally be 10^3 s^{-1}, when K can be no larger than about 10^6 to 10^7 since k_{on} is diffusion-limited at 10^9–10^{10} s^{-1} for all ions or molecules. Ions able to bind with a k_{on} as high as possible and K around 10^6 make therefore the fastest and consequently most powerful controls. The Ca^{2+} ion meets these criteria wonderfully well, while K^+ and Na^+ fail on grounds of poor binding, except in ring chelates, where on and off rates are slow, Mg^{2+} fails on the grounds of its slow on and off reactions, and Ni^{2+} has even slower on rates. Therefore calcium is the universal biological trigger. We shall see how these on/off reactions communicate through protein matrices in Chapters 7 and 10.

For slower controls, resembling hormones rather than transmitters, there are several possible candidates: Mg^{2+} (on rate slow (10^5 s^{-1}) weak binding but relatively high concentration (mM range)), Zn^{2+} (on rate faster but stronger binding, i.e. much slower off rate, and present in lower concentrations as a free ion), Fe^{2+} and Mn^{2+} (on rates fast, stronger binding, but present in lower concentrations and redox active). Clearly, such elements are not useful in triggering (10^{-3} s time scale) but they can be used in regulation (10^3 s time scale). The very strong binding of ions such as Cu^{2+} and Fe^{3+} makes them almost permanent parts of a system once bound, so they are of lower value in either process. Table 4.3 summarizes some of the important on/off rates.

Amongst anions, phosphate and phosphate derivatives match calcium in control function since they diffuse rapidly and can bind quite well. (In a simple-minded sense, HPO_4^{2-} is negative calcium.) The organic ions are like acetylcholine in current selectivity (see Section 4.5.2) but, like organic

Table 4.3 Idealized binding and rate constants

Ion	k_{on} (s^{-1})	k_{off} (s^{-1})	K	Function
Na$^+$, K$^+$, Cl$^-$	$>10^9$	$>10^6$	$<10^3$	Electrolytic message
Ca^{2+}	10^9	10^3	10^6	Mechanical trigger
Mg^{2+}	10^5	10^2	10^3	Phosphate transfer
Zn^{2+}, Fe^{2+}	10^8	10^{-2}	10^{10}	'Hormone' communication
Cu^{2+}	$>10^8$	$<10^{-7}$	$>10^{15}$	No exchange
C, H, N, O	Covalent, enzymic control			Structure building
HPO$_4^{2-}$ (RPO$_4$)	10^9	10^6(?)	10^3	Trigger
Protein PO$_4$ (phosphorylation)	Slow	Slow	Weak	Kinetic control faster than C, N, O bond breaking
H$^+$	$>10^{10}$	Slow to 10^{10}	Huge range	Catalysis, energy store
OH$^-$	$>10^{10}$	Slow to 10^{10}	Huge range	Catalysis

molecules generally, they can be metabolized and it is hard to maintain their concentrations steady, especially outside cells, without wasting metabolic energy. In this case, it is the hydrolysis of a weak ester or anhydride which is used for control. In Chapter 21 we look at the general way in which calcium plus phosphate compounds inside cells act co-operatively in control devices.

While discussing molecular ion movement we must not forget the diffusion and binding of larger organic ions, even up to the size of proteins. It appears to be a general observation that barriers to on/off reactions, apart from binding constants, are related to the polarities of the interacting surfaces. Polar surfaces come together and part rapidly; non-polar surfaces stick together in water.

4.7 Electron transfer reactions: electronics

Electron transfer depends on the approach of an electron donor to an acceptor. There are two cases:

(1) long-range or outer sphere electron transfer, when the co-ordination sphere of both donor and acceptor is altered to only a minor degree during reaction;

(2) inner sphere electron transfer, where the donor and acceptor form a direct chemical bond by changing their co-ordination spheres.

In biology, the first case is used more generally and especially for making electronic conductors, i.e. biological wires. Direct electron transfer in an inner sphere complex is often reserved for difficult chemical transformations, i.e. for catalytic steps in enzymes.

Long-range electron migration in biology is via hop conduction, i.e. the electron goes from one metal ion to another over a limited distance. A satisfactory theory of the rates of electron transfer has been developed, particularly by Marcus. The Marcus equation is

$$\ln k = \ln \nu_{ET} \frac{(\lambda - \Delta E_m)^2}{4 k_B \lambda T}$$

where ΔE_m is the redox potential difference between source and sink; λ is the relaxation energy due to the small motions required before electron transfer; k_B is the Boltzmann constant; and ν_{ET} is the limiting rate at infinite temperature T.

We must then consider ΔE_m, the redox potential difference between the source and the sink of the electrons, which is the driving force, and the relaxation energy, λ. Obviously, the only chemical elements suitable for electron transfer devices must have at least two oxidation states (Fig. 1.9). Hidden in the term ν_{ET} is the very important dependence on distance of the overall rate.

The effect of distance on unimolecular electron transfer rates has been clearly demonstrated in model complexes of many metals held apart on a rigid organic framework (see the 'Further reading' section). The rate constants fall

by a factor of approximately 10^2 for each 3 Å increase in distance; hence it is expected (and found) that hops of 15 Å are at the upper limit of usefulness in biology. These inferences from models have been thoroughly demonstrated in the electron transfer of the reaction centre of photosynthetic bacteria. Figure 4.5 illustrates this in photosystem I in which electrons excited at the chlorophyll dimer $(BChl)_2$ hop successively to other chlorins (BCl) or phaeophytins (BPh) and then to quinones, Q, while new electrons are fed in from cytochromes, Cyt. The medium, here a variety of protein groups, appears to have no influence and all redox reaction potentials are below $+0.6$ V. Photosystem II is very similar except in the section at very high oxidizing potentials between $(Chl)_2$ and the Mn (for O_2 production) when a tyrosine radical intervenes. Tyrosine, tryptophan, or histidine radicals can only be really useful in redox reactions in the range above $+0.75$ V. The study of almost every electron hop in this wiring diagram follows expectation based on model reactions.

We shall consider later in more detail the value of inserting unsaturated groups in the chain between, but not bound to, the metal ions. Bound unsaturated ligands, e.g. porphyrins, spread the orbitals of the metal, but keep it protected from substitution reactions. The remaining restrictions on the rate of electron transfer are hidden in relaxation energies λ, given a fixed redox potential difference which decides the direction of flow.

All electron transfer processes involve a change in charge locally. There is then a change in the electrostatic interaction with the environment and in certain circumstances, e.g. a bare ion going from $M^{2+} \rightarrow M^{3+}$ in water, there will be a considerable change in bond energy and in the entropy of the water. This would require a high reorganization energy hindering transfer, since the states, source and sink, must be thermally adjusted before the electron can move. The term λ would then be caused to be disadvantageously large; apparently this is observed in some model reactions. Biological electron transfer need not suffer from this problem because the electron transfer centres

Fig. 4.5 The electron transfer scheme of the bacterial reaction centre equivalent to photosystem I. Cyt, cytochrome; BChl, chlorophyll; Q, menaquinone; BPh, phaeophytin.

can be held relatively rigidly in the correct structures to make optimal electron transfer in proteins and out of water. We shall look at their structures later, since it is the nature of the protein matrix which is central to biological electron transfer.

In a protein, the change from one oxidation state to another, involving relaxation of the surrounds of the electron transfer site, could therefore be simply vibrational adjustments. Alternatively, it could involve quite major conformational movements, but in this case energy is used and properties other than simple electron conduction can be achieved and the coupled conformational change has to occur separately in time from electron transfer. The former situation at a series of hop centres would generate a 'wire' where electron transfer is not coupled to any other process; the latter case would correspond to a coupled process in which electron transfer was not permitted unless a certain conformational change had occurred or was followed immediately by a major relaxation of nuclei. The major relaxation could then cause other events, such as movement of ions (see Fig. 4.4). This reaction is called 'gating' and the transfer of energy is called 'energy transduction' (Chapter 5).

The above applies to all those reactions in which the co-ordination sphere of the metal ion is not changed. Obviously, then, it applies particularly to the metal ions which have all their co-ordination sites filled by ligands which are sidechains of proteins. In the case of open-sided metal ions there is the possibility that the non-protein ligand, water or any other ligand, could be lost, and this change in structure prior to and/or following electron transfer means that the relaxation term may again be large so that coupling to these steps can occur. We find that wiring 'circuits' for rapid electron transfer in biology are built from the closed co-ordination-sphere metal ions and the rate is limited at the end of a chain by open-sided catalysts (Fig. 4.6). They are the ones which bind dioxygen, hydrogen peroxide, or even dihydrogen so as to couple electron transfer to reactions which effectively are the energy generators of an electronic (chemical) battery (see Chapter 5). This applies to a relatively highly organized system—almost a solid state device—but there is also the possibility of molecular diffusion-based devices, e.g. liquid three-, two-, or one-dimensional molecular flows, for carrying electrons between sites. All these possibilities are put to use in biology.

As to the material used, it is curious to note that man makes his wires from metals such as copper; biology makes hop conductors from metal ions embedded in insulating material, the proteins. The choice of metal is not constrained except by the availability of two oxidation states at the appropriate potential. Nearly all transition metals in a number of oxidation states could be so manipulated, but it is observed that iron is virtually the only metal used. It could well be that this was an accidental result of the availability of iron in the early evolution of life. Besides iron, only copper is used, and much less commonly, in aerobic systems, which were a later development.

Before concluding we must note that a control protein or a coupling device must have a physical character different from a catalyst in that the former must change shape easily. In Chapter 7 we return to these problems.

Fig. 4.6 The mitochondrial respiratory chain. (a) The sequence of redox centres given in the diagram is based on the E values obtained with de-energized heart mitochondria. Cytochrome c links the cytochrome c reductase and cytochrome c oxidase complexes. (b) The arrangement of cytochrome c with respect to the positions of cytochrome reductase and cytochrome c oxidase is not well defined. The reductase and oxidase complexes span the inner mitochondrial membrane with cytochrome c loosely attached to the outside of the membrane. The open-sided reaction centres are at the beginning (RH) and end (O_2). a and b refer to different cytochromes.

4.8 Redox potentials of complexes

It is useful to discuss the thermodynamic redox potential change associated with electron transfer here since it is the driving force referred to in the preceding section. The redox potentials are given by the gradients of lines in Figs 1.7 and 1.8 and are those for free ions in aqueous solutions at pH = 7 and in the absence of other ligands. If we give this redox potential the symbol E°, then in the presence of ligands the value changes. Consider two free aquo ions M^{II} and M^{III}; if they both form complexes ML with the same ligand, the redox potential changes from that of the hydrated ions by the strength of binding in ML

$$E = E^\circ + \frac{RT}{\mathscr{F}} \ln K^{II}/K^{III}$$

where K^{II} and K^{III} are the stability constants of $M^{II}L$ and $M^{III}L$, respectively. The use of a great variety of ligands to manipulate redox potentials is of the greatest importance since, for example, the assembly of metal ion complexes into a series within a membrane (Fig. 4.6) is the way in which biology makes a wire. The wire has the best current flow with least loss of energy if the redox

potentials of the couples in the wire are ordered and the rate of electron transfer is also controlled by this difference (see Section 4.7). By adequate selection of ligands L, usually proteins, the redox potential of iron in different couples may extend from -0.5 to $+0.4$ V.

Using model systems it has been easy to show how increase in σ-donor power of ligands tends to decrease the redox potential of a transition-metal couple. On the other hand, increase in π-acceptor power of ligands tends to stabilize lower oxidation states and increase the redox potential. The use of saturated and unsaturated ligands based on nitrogen donors readily illustrates these points, but it is not so easy to explain the properties of, e.g. P- and As-based donors in inorganic chemistry, since in these cases the π/σ effects are very mixed. Finally, we must consider a series of ligands which induce spin-state changes in, say, the Fe^{2+} and Fe^{3+} ions. Models show that, since the energy for spin-pairing is less for Fe^{2+} than for Fe^{3+}, increase in π-bonding together with σ-bonding will drive Fe^{2+} low-spin first and, therefore, raise the redox potential as the total ligand field strength increases, but at still stronger ligand fields the Fe^{3+} state becomes low-spin and the redox potential falls (see Chapter 2). The example of the iron porphyrin proteins illustrates the point (Fig. 13.3).

The redox potentials of organic molecules in the electron transfer chains are matched with those of the metal ions (see Fig. 4.6). At low potential, pyridine nucleotide and flavin coenzymes have potentials around -0.5 to -0.3 V, i.e. near those of Fe_n/S_n couples; at intermediate potentials, around 0.0 V, there are quinone coenzyme couples. Of course, dioxygen has a higher potential, $+0.8$ V at pH $=7$, and is associated with haem and copper couples. A feature of these three non-metal systems is that redox change is necessarily associated with acid–base protonation reactions:

$$Q + 2e + 2H^+ \longrightarrow QH_2,$$
$$\tfrac{1}{2}O_2 + 2e + 2H^+ \longrightarrow H_2O,$$
$$Flavin + e + H^+ \longrightarrow Flavin\ H^{\cdot},$$
$$NAD^+ + 2e + H^+ \longrightarrow NADH.$$

This covalent hydrogen binding is fundamental to the energy capture devices.

In enzymic reactions, the oxidation of organic molecules can be achieved so as to give organic radicals of high redox potential using couples such as FeO(IV)/Fe(III) (about $+1.0$ V). Groups which can then be oxidized are phenols and indoles and these oxidations are greatly used to detoxify, to protect, and to strengthen connective tissue. It is also possible to use the oxidation of these groups to radicals reversibly when they are part of a side-chain of an amino acid in a protein (Table 4.4). They may be quite generally valuable in electron transfer through proteins at high redox potential.

Table 4.4 Organic radicals in enzymes

Enzyme	Radical
Ribonucleotide reductase	Tyrosine, $-CH_2^{\cdot}$ of B_{12} enzyme
Photo-reaction centre	Tyrosine, phaeophytin, chlorophyll
Cytochrome c peroxidase	Tryptophan
Galactose oxidase	Tyrosine
Prostaglandin H synthetase	Tyrosine

4.9 Atom transfer reactions

While acid–base and electron transfer catalyses are major activities within cells, a third equally important process is oxidation and reduction by atom transfer; an obvious need in biology is to transform CO_2 to $HCHO + H_2O$, N_2 to NH_3, and so on. The catalytic act may well be the uptake on one site of an

atom or ion, H^-, O, etc., which is obtained from, say, H_2 (NADPH) or O_2 (FeO in an oxidase), and its transfer to a carbon centre, e.g.

$$CH_4 + O \longrightarrow CH_3OH.$$

When an oxygen atom or H^- are added to a compound we term this a 'two-electron oxidation or reduction' (by atom transfer). Some other examples are

$$ClO^- + SO_3^{2-} \longrightarrow Cl^- + SO_4^{2-}$$

and

These reactions require the activation of oxygen or hydrogen and the usual mechanisms in chemistry require the creation of an intermediate such as FeO or CoO or Fe–H, Ir–H, etc. on the surfaces of metals or in organometallic compounds. These metal intermediates are often seen in biological systems too, but we must observe the very extensive use of non-metal coenzymes described in Chapter 3. Very frequently, the use of a metal is associated with very small redox substrates as in acid–base catalysts (see Table 6.3).

In order to make atom transfer (as opposed to electron transfer) easier, two conditions must be met: (1) the central atom must be able to make and break its immediate atomic links, i.e. it is an inner sphere reaction; and (2) the change of the central atom formal oxidation state has to be two. A typical reaction would be

$$ClO_4^- + R_2Se \longrightarrow R_2SeO + ClO_3^-.$$

Referring to oxygen atom transfer we find that this is difficult in the first row of the periodic table, but easier down each group given that redox potentials permit the reaction (see Fig. 1.8). Thus SeO_2 gives O more readily than does SO_2. Elements confined to one oxidation state do not take part in these reactions, so that they are possible properties of C, N, (O), S, Se, and a variety of transition metals, but not of Si, P, and Zn, for example. Notice that there is a potential danger with the transition-metal ions since they are likely to confuse one- and two-electron reaction paths; the Fe and Se peroxidases (Chapters 12 and 18) are good examples to contrast for their different value.

A typical two-electron transfer in hydrogen chemistry is H^- transfer, and once again a shift of two in effective oxidation state is required. For these reactions biological systems resort to complex coenzymes of the pyridine nucleotide and flavin classes or to the use of transition metals. Here biology has not generated many of the equivalent electron-rich series of metal compounds for catalysing these reactions which are well-known in organometallic chemistry; the best it does is to use nickel or iron sulphides, as we shall see.

Many very fine catalysts made by man use quite heavy metals, e.g. Rh, Ir, Pt, which biology has not managed to obtain. Instead, biological systems have manipulated the first-row transition metal ions and especially their spin states to induce reactivity (Fig. 4.7 and see Section 4.11). This heightened reactivity in the catalyst atoms is required for the activation of O_2, NO_3^-, NO_2^-, SO_4^{2-}, and H_2 among others.

Fig. 4.7 The ranges of available oxidation and spin states of different elements which, together with their energies (redox potentials), generate different efficacies in catalytic acts. Biology has selected elements on this basis but by incorporating the elements in special ligands and proteins has enhanced their functions both in chemistry and spatial effect.

While the atom addition reactions of such reagents as dioxygen and dihydrogen are clearly described, by definition, as oxidation and reduction, the further classification of reactions is not so simple. 'Oxidation' is still adequate for the description of reactions of good electrophilic reagents, such as the halogens and sulphur, but is not so easily applied to species based on carbon or nitrogen. For example, if two ethylene molecules combine to form cyclobutane we do not call this oxidation and we do not say that the oxidation state of carbon is changed, but we do describe the reaction of ethylene with dioxygen in these terms since a cyclic ethylene peroxide is formed. The problem is well-known (and irresolvable) and arises from the definition of oxidation states. We shall not attempt any such classification of reactions here, but observe that all reactions are catalysed by re-adjustment of electron density, although it is not always obvious which electron adjustments have been made, due to a variety of factors. Similarly, the transfer of groups, like the transfer of O-atoms, is often loosely described by referring to types of 'oxidative' reactions of catalysts.

4.10 Group transfer

Atoms other than H or O which can be transferred usefully are the halogens, as X^+, but a much more profitable series of reactions for biological systems is the transfer of groups such as CH_3^+, (CH_3^-), CH_3CO^+, and, in fact, any such similar ion (or unit). The preparation of group transfer agents is also a major preoccupation of organic and organometallic chemists; most of this chemistry is today outside the reach of biology since it requires non-aqueous solvents and inert atmospheres, but it may well have been that in very primitive organisms this path was used (see Section 4.11). Instead, biology has now a series of coenzymes and metal centres as well as side-chains of amino acids for such reactions; as expected it is often heavy elements, especially sulphur, which are used because they make good attacking and leaving groups, e.g. in coenzyme A (Fig. 2.15), but lighter atoms such as N can also be activated, especially in unsaturated molecules, to perform these functions, e.g. in biotin or imidazole. For those who wish to think more deeply about such processes and relate acid–base and redox functions of atomic centres it is of interest to compare the release and uptake of H^+ with that of CH_3^+ or CH_3CO^+ in detail. Here we give a brief summary only, referring readers to books on bio-organic reactions.

It is conventional to separate reactions such as

$$H_2PO_4^- \longrightarrow H^+ + HPO_4^{2-} \tag{4.4}$$

from

$$CH_3COPO_4^{2-} \longrightarrow CH_3CO^+ + PO_4^{3-} \tag{4.5}$$

and from

$$CH_3COSR \longrightarrow CH_3CO^+ + RS^-, \tag{4.6}$$

although they are all electron-pair reactions. In fact, it must be generally true that a leaving base in an acid–base reaction such as (4.4), which has a pK_a in

the range 4–12, will function as a positive ion group transfer reagent. Obvious examples are imidazole, phosphate, thiolate, and carboxylate. The big difference in considerations of catalysed group transfer is that while proton and simple ion reactions of type (4.4) come very rapidly to thermodynamic equilibrium, reactions (4.5) and (4.6) do not, so that kinetic pathways, dependent on catalysts, decide which reaction will occur. The mobile coenzymes of biology for group transfer are usually based on reaction (4.5) or (4.6) (cases of CoA, lipoic acid, NTP) or upon a well designed aromatic moiety which does not equilibrate rapidly, e.g. NADH in hydride transfer.

In summary, non-metal atoms suitable for group transfer are particularly S and Se, followed by N, while, amongst metals, Mo, Mn, V, and Fe are more suitable for transferring O and Co and Ni are more useful for the transfer of H^- or CH_3^-.

4.11 The nature of transition-metal centres in catalysis

Metal — ligand

$d_{x^2-y^2}$

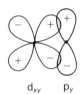

d_{xy} p_y

Fig. 4.8 Simple picture of the overlap between metal d orbitals and ligand p orbitals to give σ- and π- interactions. d_{z^2} orbitals also give σ-bonds; d_{yz} and d_{xz} give π-bonds.

In describing metal ion functions we have not looked at all their kinetic capabilities; we have only examined the Lewis acid properties of the metal ions, which are a function of the energy of unoccupied exposed orbitals of any element or ion. This term is additional to electrostatic polarization due simply to the charge on an ion. High-affinity hidden orbitals are not effective, but the degree of their exposure depends critically on the metal in question, on the other ligands of the metal, on the spin states generated in a complex (Fig. 4.7), and on the spatial characteristics of the complex. When considering ions such as Mg^{2+}, Ca^{2+}, and Zn^{2+} some of these considerations are of little importance since these ions have a hidden spherical core, but in the transition metals the more exposed 3d core orbitals are polarized, as in Fig. 4.8, for octahedral symmetry. The two less stable orbitals are now in the direction of the donor lone pairs of a ligand, while the three more stable orbitals can form π-bonds to it. Naturally, it is the lower three which are occupied preferentially by the ions' own electrons. Usually, in biological systems, we only need to discuss ions with five or more 3d electrons, i.e. metal ions from Mn(II), d^5, onwards, which tend to be π-donors from the highest occupied molecular orbitals (HOMO), e.g. d_{xy}, but σ-acceptors to the lowest unoccupied molecular orbitals (LUMO), e.g. $d_{x^2-y^2}$. Examples are low-spin $3d^6$ (Fe(II), Co(III)). As the number of 3d electrons increases further, the σ-orbitals become occupied and the metal ion can become a σ-donor of two electrons (Ni(II) d^8, Co(I) d^8). Thus, a metal ion is not just a σ-acid, but can be a π-donor, or a σ-base. Clearly, while acidic functions are closely paralleled by the activity of the proton, which shares with transition metal ions the property of acidity not just due to electrostatics, other properties are quite different. Hence, the metal ion, while binding via σ-acceptor properties, can totally adjust the reactivity of the bound group by acting as a π-donor. These 'basic' properties of transition metals allow them to bind and to activate unsaturated small substrates such as CO, O_2, NO, etc. The 'basic' properties can be called into use after the acidic properties of the metal ion in that, after forming the acid–base complex, the metal centre (with a lower formal charge) may alter its role and act as a π-base to unsaturated ligands. In this way, a reaction can

proceed from a planar d^8 system to a pentagonal prism and then to an octahedral system, e.g.

with subsequent atom rearrangements. Such reactions are well known on metal centres in organometallic chemistry.

In the description of these reactions and their catalysis a rather loose language has again developed. For example, if we consider the reaction of Ni(II) in a square planar geometry as a d^8 ion and add to it, e.g. Cl_2, such that the nickel becomes octahedral, we may describe this as 'oxidative addition' since Ni can now be seen as a Ni(IV) d^6 ion with two chloride ligands. By the same reasoning, the reverse reaction is 'reductive elimination'. It is clear that many syntheses and degradations can proceed by these pathways. Notice that in reactions such as these oxidative additions nickel may well have been the element of choice, especially early in evolution. Reference to metallo-organic chemistry texts may lead one to think that some form of organometallic catalysis pre-dated life.

The above description has concentrated upon octahedral complexes but a similar description applies to other structures. Additionally, the 3d shell acts as a partial protection for free radical intermediates; indeed, while all half-filled orbitals, e.g. in high-spin Mn(II), $3d^5$, are free radicals, they only give rise to free-radical reactions readily if the d orbitals are exposed. Electrons are clearly more reactive in σ than in π orbitals so that $3d^7$ (low-spin) cobalt in vitamin B_{12} and $3d^9$ copper cations in enzymes readily initiate radical pathways.

In this same context we note that too rigid a limitation on conformational change at a centre could prove to be disadvantageous to catalysis. Most reactions require that the catalyst flows through a series of states as the substrate is transformed into the products. On the other hand fast simple electron transfer reactions require this structural switching to be minimal. It follows that cations which are extremely stabilized by polarization in a particular geometry are not likely to be of great functional value. Here we can compare ions such as Fe^{3+} (d^5) and Zn^{2+} (d^{10}) with Cr^{3+} (d^3) and Ni^{2+} (d^8) noting that the first two, which do not have strong preferences for co-ordination symmetry, differ strikingly from the second pair, which are strongly constrained in octahedral sites. There is here a correlation with the absence of Ni^{2+} and Cr^{3+} from aerobic biology and the suspicion arises that they are deliberately avoided, even rejected. Given that this is only true of aerobic conditions, there is the further probability that, in the presence of sulphide, nickel chemistry is much more flexible, but this sulphide chemistry is not open to Cr(III). Chromium is not found in archaebacteria.

4.12 Free radical reactions

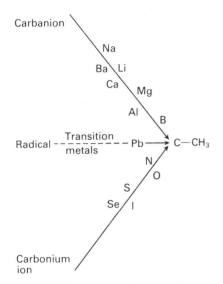

Fig. 4.9 The relative control over the type of leaving group derived from CH_3 or H, i.e. X^+, X^-, and X^\cdot. Carbanions obviously come from A-group metal methyls and carbonium ions from S, Se, I. Radicals can be generated by elements more able to undergo one-electron reactions and they may be the lowest pathway of dissociation for atoms of closely similar electronegativity, e.g. H_3C-H.

The major part of the above mechanistic account of the central reactions of chemistry, except the discussion of electron transfer, concentrates on two-electron reactions, e.g. transfer of ions such as H^-, CH_3^+, CH_3^-, OH^-, and H^+ or of units such as CO and O_2 inserted into molecules. There remains the possibility that, for example, CH_3^\cdot will be transferred. Figure 4.9 illustrates the likelihood of radical transfer from potential catalyst elements as against CH_3^+ or CH_3^- transfer. We see that as polarity is removed from the substrate so CH_3^\cdot transfer (and similarly H^\cdot, etc. transfer) becomes more probable. This is not to say that radical pathways become generally easier; rather ionic pathways become very difficult. Radical pathways will be assisted by an intermediate catalyst if the central catalytic atom can undergo one-electron reactions easily. In this case, there are three possibilities:

(1) the use of transition-metal ions;

(2) the use of aromatic molecules such as quinones and flavins and, with more difficulty, tyrosine and tryptophan;

(3) the use of heavy elements, where ionization energy differences are small, e.g. lead.

Rather than elaborate here, we simply state that certain reactions such as those of H_2, $R-(CH_2)_n-CH_3$, terpenes, and even sugars, i.e. molecules obviously not open to ionic reactions, do take place more readily through radical pathways. The usual metal ions involved in the catalysed steps are Mn, Fe, Co, Ni, and Cu; Mo and Se are more usually employed in two-electron reactions. The situation will be discussed in several later chapters (see Table 4.5).

We observe again the need to consider other factors than just catalysis. A biological system, especially DNA, is very vulnerable to free radicals; hence cytoplasmic catalyses must not release products which are free radicals. Reactions are then controlled in space (see Section 4.14) and partly by localization of the catalyst. Having dealt with the main kinetic mechanism we need now to make some additional points which apply to most, if not all, the different reaction paths mentioned.

Table 4.5 Free radical systems in biology

Value	System	Elements
Free radicals for polymerization	Construction of cuticles, lignins, etc. Protective pigmentation	Cu
Detoxification (removal)	(a) In cytoplasm: superoxide dismutase (b) Outside cells: ascorbate oxidase (c) In membranes: radical reactions	Cu, Mn, Fe Cu (Vitamin E)
Protection	Reactions of leucocytes in vesicles	(Flavins)
Special reactions in cells (local traps)	Sugar reductions (for DNA) Sugar rearrangements P-450 oxidations	Co, Fe, Mn Co, Fe, Mn Fe

4.13 Sizes of atoms, stereochemistry, and reaction paths

The intermediates of reactions may be formed by dissociative or associative paths. It is generally thought that, for a given geometry of the initial compound, the dissociative path is favoured by the smaller centres, while larger centres favour associative intermediates. Given that bond strengths are usually greater for smaller atoms the implications are that:

(1) Reactions of saturated small atoms, e.g. carbon, are slow.

(2) Reactions of saturated larger atoms can go by associative mechanisms, e.g. Si, P, S.

Examination of standard texts provides many examples.

The dissociative and associative paths involve different stereochemical

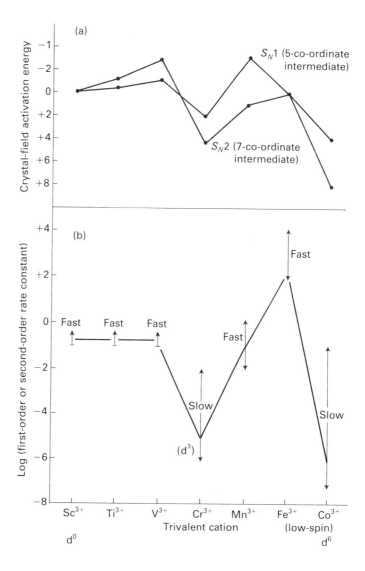

Fig. 4.10 The crystal field stabilization energy on going from a 6-co-ordinate complex to either a 5-co-ordinate (dissociative reaction) or a 7-co-ordinate (associative reaction) intermediate. S_N1 and S_N2 denote nucleophilic substitutions via unimolecular or bimolecular reaction paths, respectively. (b) An impression of observed rates of substitution showing how the energy in (a) affects reactivity depending on the number of d-electrons in the ion.

demands and here the polarization of the inner core of electrons (3d) of the transition-metal ions plays a role. Figure 4.10 shows the crystal-field energy changes for trivalent ions from Sc^{3+} to Co^{3+} for the two steps (the crystal-field energy change parallels the change of polarization of the 3d core). The bottom part of the figure shows the observed ranges of rate constants for these ions in reactions by either pathway. It is obvious that some metal ions, e.g. Co^{3+}, cannot be used as catalyst centres since they do not bind or release groups easily, while others are excellent, e.g. Fe^{3+}.

4.14 The creation of local small spaces: mechanical devices for transfer reactions

Discussions of mechanism in inorganic and organic chemistry usually leave aside mechanical devices; the reactants are seen as just diffusing together. Biological catalysts can behave just in this way, acting as a relatively rigid recognition surface for partners. This is the standard approach to digestive enzymes, proteases, for example. A very different reaction route is also found in which the process of binding alters the catalyst in a mechanical sense; a simple example is a hinge-bending enzyme (Chapter 7). The reaction sequence, where E represents the enzyme, A and B the reactants, and C and D the products, is

(1) $\quad E + A \longrightarrow E'A;$

(2) $\quad E + B \longrightarrow E''B;$

(3) $\quad E'A + B \longrightarrow E'''AB;$

(4) $\quad E''B + A \longrightarrow E'''AB;$

(5) $\quad E'''AB \longrightarrow E''''CD;$

(6) $\quad E''''CD \longrightarrow C + D + E.$

The different enzyme conformational states are represented by dashes.

The last step (6) could be re-written as a series of steps to follow steps (1) to (4) replacing C and D for A and B, respectively. The series of reactions shows that the enzyme is a mobile polymer. If steps (1) and (3) only and not (2) and (4) are permitted, there is a required order of addition of reactants to the catalyst. If the way back from E'''' can be arranged to be via conformations very different from E', E'', and E''', then the enzyme cycles rather than retraces its steps, and energy is required in such a one-way flow (see Chapter 5). Pumps are of this kind; there is a big difference between a two state (open \rightleftarrows shut) reversible equilibrating system and a multistate cycle, which can be energy driven to avoid reversibility that allows synthesis of unstable compounds, as discussed in Section 4.2.

Now, let us examine what can be achieved by avoiding a random set of conformations. If we take the case of the desired reaction

$$ATP + X \rightleftharpoons XP + ADP$$

where ATP is just an organic triphosphate, A–P–P–P, this is in competition with hydrolysis, which is energetically much more favourable

$$ATP + H_2O \longrightarrow P_i + ADP$$

where P_i is free inorganic phosphate. Hence, the binding of ATP must not be such that it is activated without the presence of X. It may also happen that the reaction

$$XP + H_2O \longrightarrow X + P_i$$

is favourable, so neither the reactant ATP nor the product XP must be bound initially in an activated state. The required transfer of P from ATP to X or of P from XP to ADP must follow only if X and ATP or XP and ADP are together and activated after a conformation change forms a local pocket which eliminates water. Since ATP is an energy store, the reaction retains energy in the products instead of losing it to hydrolysis. The above case is one of random substrate addition; an example is that of phosphoglycerate kinase. Consider next the case of attack by a reagent which, no matter what the conformation of the protein, is very aggressive, e.g. oxygen bound to iron, not in a carrier protein, such as haemoglobin, but in a detoxifying aggressive oxidase, such as cytochrome P-450. Oxidative attack on the protein itself is now the risk, but this possibility can be removed if step (1), the binding of the substrate A to be attacked, is required before B. In this case, when O_2 is bound, previous binding of A allows O_2 to see A only in its site of binding and protects the enzyme from O_2, i.e. steps (1) and (3) are allowed, but steps (2) and (4) are forbidden. This is a required order mechanism which opens a space for a reagent.

These mechanical movements in proteins require 'hinges'. Later we shall attempt to show that the enzymes which have hinges can use β-sheets for their recognition sites, but that the hinges themselves are based on α-helices (Chapter 7).

We turn next to the final mechanical trick using helical motions—the irreversible ratchet. If during a series of reactions there is an unstable intermediate then the destruction of the intermediate loses energy from the reaction system. It is possible then to have two kinds of pump; one which transfers energy from one system to another without deliberate loss (a balanced pump) and another which during its cycle loses energy. The second will not run reversibly. Examples are given in Chapter 8. The movement of ions in channels may well depend on such mechanical properties of proteins and so may the synthesis of ATP from proton gradients. Finally, this is the mode of transfer of the energy of ATP to the filamentous systems which maintain tension.

So far in this chapter we have considered kinetics as a problem in a free aqueous solution, so that reaction rates are expressed in terms of rate constants and concentrations per unit volume. In fact, as seen before, a biological cell is criss-crossed by fibrous proteins and each cell has many membranes; hence molecules attached to the fibres or membranes can diffuse along a one-dimensional path, while those in a membrane can diffuse on a two-dimensional matrix. Because of the attachment of the fibres to the membrane there are many ways in which movement can be controlled. At one time such ideas were considered fanciful, but it is clear today that, as well as channelled diffusion, there are such phenomena as the diffusion of proteins along DNA and in membranes and the diffusion of vesicles along fibres in nerve cells. It is also likely that ion diffusion on charged surfaces is extremely important.

4.15 Summary

This chapter has described in brief outline the kinetic constraints and controls in chemical systems with only brief reference to biology to keep the topic in the context of the book.

The kinetically valuable features of the use of elements other than C, H, N, O, S, and P which we have discussed are:

(1) strong acid–base catalysis, e.g. by Zn^{2+};

(2) rapid mobility with weak binding in electrostatic currents (Na^+, K^+, Cl^-);

(3) rapid exchange with moderate binding (Ca^{2+}, Mg^{2+});

(4) electron transfer using Mn, Fe, Co, Ni, Cu, Mo;

(5) ease of atom transfer from Se, Mo and to some degree from Mn, Fe, Co, Ni, Cu;

(6) potential for free radical reactions (many transition metals).

We must next look at an area where biological chemistry is particularly unlike normal chemistry. This lies in the use of energy for synthesis, for organisation and to maintain flow, which is the subject of Chapter 5. We shall then be able to return to the way in which essential elements are exploited functionally in biology and, in Chapter 6, we shall show how the chemical functions of the elements, now not just the catalytic activity, but also the ability to transfer electrolytic charge and electrons, and to participate in structures, are utilized in biology. This returns us to the way in which elements are partitioned in space. It might be difficult for a chemist to say why he should choose iron rather than copper as an oxidation catalyst in a given process since both could, in principle, do the same job. Biological systems use both iron and copper in oxidative catalysis, but their roles are very well separated in space so that their total function is very different. This is equally true of acid–base catalysts where Zn^{2+} is used as a general acid with the proton, but we find that for particular reactions Mg^{2+} is preferred, e.g. phosphate transfer, while other metal ions are most important in special zones, Fe^{3+} and Mn^{3+} in lysosomes, and Ca^{2+} outside cells. Again Na^+ and K^+ are biologically totally distinct and are not similar ions.

Further reading

There are many standard textbooks which describe the kinetics of organic compounds and catalysis by enzymes.

Atkins, P. W. (1990). *Physical chemistry* (4th edn). Oxford University Press, Oxford.

Basolo, F. and Pearson, R. G. (1958). *Mechanisms of inorganic reactions*. Wiley, New York.

Deisenhofer, J. and Michel, H. (1989). The photosynthetic reaction centre. *Chemica Scripta*, **29**, 205–20.

Johnson, M. K., King, R. B., Kurtz, D. M., Kutal, C., Norton, M. L., and Scott, R. A. (ed.) (1990). *Electron transfer in biology and the solid state*, Advances in Chemistry Series, No. 226. American Chemical Society, Washington, DC.

Marcus, R. A. (1965). Theory of electron transfer. *Journal of Chemical Physics*, **43**, 2654.

Norman, R. O. C. and Hill, H. A. O. (ed.) (1986). *Inorganic and organic radicals*, Proceedings of a discussion. The Royal Society, London.

Packer, L. (ed.) (1986). *Protons and water*, Methods in Enzymology, Vol. 22. Academic Press, New York.

Taube, H. (1970). *Electron transfer reactions in solution*. Academic Press, London.

Note: *Structure and Bonding* (1991) **75** contains a series of articles concerning electron transfer in proteins, pp. 1–224.

5

Energy in biological systems and hydrogen biochemistry

5.1	Introduction	120
5.2	Biological systems and light	122
5.3	Oxidative energy: mitochondria	124
5.4	Proton migration coupled to redox reaction	126
5.5	The coupling of gradients to ATP formation	126
5.6	Anaerobes in the dark	128
5.7	Compartments, energy, and metabolism	131
5.8	Organization and tension	132
5.9	Motion of organisms	133
5.10	Local storage of elements	134
5.11	The evolution of the compartments for energy generation	135
5.12	Early sources of energy	135
5.13	Sulphur compounds	136
5.14	Feedback	136
5.15	Flow of material	136
5.16	The role of the proton in homeostasis	137
5.17	The proton and redox potentials	138

5.1 Introduction

Energy can only be contained by biological systems in the same way as man contains it in chemicals and in physical storage devices. Some of the obvious static forms follow.

(1) **Electrical storage.** Storage due to charge separation, as in a condenser (electron store), or at a boundary between aqueous phases where diffusion, which would neutralize charge, is prevented, i.e. individual ion stores.

(2) **Concentration storage.** Here a compound be it a salt, NaCl, or an organic

molecule is held in solution and prevented from diffusion by a membrane. Such junctions occur in electrolytic concentration cells.

(3) **Light energy storage (capture).** Light is absorbed locally in an excited pigmented layer before going to heat or another form of energy, e.g. electrical, chemical, conformational, or mechanical energy.

(4) **Chemical bond energy storage.** Some combinations of chemicals are less favourable than others as is illustrated by the use of fuels by man. The combination of C with H(CH_4) separately from O (as O_2) is less favourable than bound C with O and bound H with O. The basis of the petrol engine is

$$CH_4 + 2O_2 \longrightarrow CO_2 + 2H_2O + \text{energy}.$$

The redox states of carbon and other elements in Figs 1.7 and 1.8 are plotted on an energy scale; this scale of chemical energy ignores combinations of elements other than with H and with O and refers to fixed unit concentration conditions for each, i.e. unit gas pressure or unit concentration in water. We obtain the energy equivalent of a chemical change from differences along the ordinates of these figures; the slope is the chemical potential of unit change of redox state, i.e. the redox potential. The difference in redox potential between two couples is an energy source. Life obtains energy from chemicals in many different ways.

(5) **Energy storage in a field.** Life on Earth is in a gravitational field and movement away from the Earth, i.e. in this field, costs energy. Therefore objects held high above the Earth store energy even if this energy is usually trivial, and ignored in chemistry. It is not trivial in physics and biology. Of much greater importance are the electric fields in biology.

(6) **Energy due to tension storage (mechanical energy).** Storage of energy in a stretched rubber band is an obvious example. Biological systems are full of deformable string-like objects, e.g. muscle fibres and ligaments, and very similar fibres (filaments) are present in most cells. We can extend this description to two-dimensional surfaces, e.g. of balloons. All cell membranes or containing walls of biology are in balloon-like tension, e.g. the skin. Deformation energy is often contained in the confined conformational or configurational energy of polymers.

It would certainly be useful if we could give quantitative expression to some of these energy stores. Using ΔG to sum the free energy change, that is the available energy for work on going from one state to another, we define ΔG° of a chemical as its energy in a defined state relative to its component elements in defined elemental states in which they sense no field, are not in tension except being under unit pressure, are at unit concentration (if in solution), and are at 25°C. Change of concentration to a value [M], where M is any element, different from unity causes a change in ΔG given by the expression

$$\Delta G = \Delta G^\circ + RT \ln[M].$$

Changes of other factors extend the equation to fields and tension

$$\Delta G = \Delta G^\circ + RT \ln[M] + \text{energy due to physical effects}.$$

(A high negative ΔG represents a highly stable state by convention.) Change of

chemical combination is then considered in the term ΔG°, which contains the sum of the internal energies of all the chemical bonds in the molecule relative to those in the separated elements.

Biological systems accumulate energy in order to create or develop ordered chemical constructs, so they have to adjust ΔG°, [M], and the physical condition of a large number of chemical elements within compounds. The starting condition of the elements before they enter biological systems differs from element to element (see Chapter 1). Most required non-metals outside biology are present as stable and available simple gases, N_2, O_2, and CO_2, or as molecules or ions in solution at reasonable concentration, H_2O, HCO_3^-, NO_3^-, SO_4^{2-}, and Cl^-. These are basic food stuffs for many unicellular systems and plants. This high availability applies to some metals too, e.g. Na^+, K^+, Mg^{2+}, and Ca^{2+}, but for the rest of non-metals and many metals their concentration, [M], is very low. The biological problem is to get these elements out of these relatively stable conditions (large negative ΔG values) into the constructs we observe in biology. This is achieved in various ways as described in the following sections.

5.2 Biological systems and light

Biology depends very largely on sunlight energy, $h\nu$ (Fig. 5.1). Light is absorbed by membrane pigments, e.g. green (Mg) chlorophyll, and, as in a man-made photocell, this energy is converted to a local gradient of electronic charge, ψ, that is an electronic potential energy. Chemically, this is seen as a change of oxidation state between, say, two compounds of hydrogen, or of two metal ions near D and D_1 or D_2 (Fig. 5.1) on opposite sides of a membrane, e.g. of a thylakoid. One centre D becomes more positive, the other more negative. The electron/positive charge separation can then be converted to freely diffusing entities by moving the charges to unbound molecules or ions in the two solutions on opposite sides of the membrane (Fig. 5.1). The usual case is the transfer of charge from or to hydrogen, giving H^+ from bound H^- (NADH) on one side and giving OH^- and bound H^+ (QH_2) on the other. We can write a simple symbolic equation where 'free' may represent two local or

Fig. 5.1 The reactions of photosystem I. Light drives electrons and hence the separation of charge across a membrane to give an electrical potential, ψ. The separated charges can rest on metal ions or organic molecules. At the edges of the membrane the charges can migrate to the aqueous phase to give a proton gradient and the chemical species generated can be used to drive organic reactions. Note the metals used. Q, quinone.

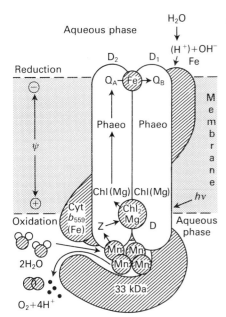

Fig. 5.2 The reactions of photosystem II. Light drives electrons and hence the separation of charge across a membrane to give an electrical potential, ψ. Because the system works at higher redox potentials than photosystem I the charge separation is used to generate dioxygen as a product as well as a proton gradient and reducing equivalents. Note the metals used. Chl, chlorophyll; Phaeo, phaeophytin; Q, quinone; Z, a tyrosine radical.

Fig. 5.3 The combination of photosystem I and photosystem II shown on a redox energy scale and indicating their interaction (but see Chapter 3). Z, a tyrosine radical; P and P*, ground and excited states of reaction centre chlorophyll; Cyt, cytochrome; PC, plastocyanin; Fd, ferredoxin; Q, quinone. Note the different uses of four metal ions. Fe–S$_A$, Fe–S$_B$, and Fe–S$_R$ are different iron–sulphur proteins. LHC is a light harvesting system, FNR is ferredoxin reductase.

delocalized pools in the membrane or in the aqueous phases and where the subscripts in and out represent relative separation in space

$$\text{Light } (h\nu) \longrightarrow \left| \begin{array}{c} \text{bound } \psi \\ M^+_{in} + M^-_{out} \end{array} \right| \xrightarrow{\text{Q + NADH}} \begin{array}{c} \text{free} \\ H^+_{in} + OH^-_{out} \end{array}.$$

Notice that the switch is from physical energy as light, to a physical field (electronic potential) localized on atoms, to a chemical transformation (ΔG° change), and finally to a concentration change of H^- (NADH) and H^+ ($RT \ln[M]$ change); Fig. 5.1). We may now use QH_2 (H^-) to reduce CO_2 and N_2, for example, to compounds for synthesis, e.g. the sugars of DNA, the ammonia for amino acids, and so on. That is a further, now favourable, ΔG° change. Intermediate small molecules are required by all metabolic systems and these intermediates already hold elements at a considerable positive chemical free energy; making polymers increases still further the positive ΔG° and the adverse $RT \ln[M]$, and an extra source of energy is required (see Section 5.5). Alternatively, CO_2 could be reduced to CH_4 or heavier hydrocarbons, e.g. oils and stored.

In addition to using light, $h\nu$, for direct reductive metabolism, chloroplasts use M^+ produced in a second photosystem as an oxidizing agent, e.g. for the production of dioxygen of the air which is a 'waste' energized product or rather a remote biological energy store (Section 5.3) and at this stage of the discussion (Fig. 5.2) energy goes into the oxidation states ΔG° of the element oxygen relative to its normal condition in water. Manganese is the metal used to make dioxygen from water in biology; hence oxidation states of manganese are the intermediates (Figs 5.2 and 5.3), i.e. the reaction $4M + h\nu \rightarrow (4M^+)$ is used to drive the change $2Mn^{2+} \rightarrow 2Mn^{4+}$ followed by

$$2Mn^{4+} + 2H_2O \longrightarrow 2Mn^{2+} + 4H^+ + O_2.$$

The storage of O_2 in the air and CH_4 (oil) in the ground (an alternative reduced form of CO_2 is $(CHOH)_n$, cellulose, e.g. in plants) by the combination of the two photosystems (see Fig. 5.3) is just separated ΔG° storage of oxidizing and reducing compounds in chemicals, as far as living systems are concerned, since they can combust (eat) oils or cellulose. The reaction of O_2 with these reduced chemicals gives a second supply of oxidative energy.

$$[CHOH] + O_2 \longrightarrow CO_2 + H_2O + energy,$$
$$CH_4 + 2O_2 \longrightarrow CO_2 + 2H_2O + energy.$$

As well as energy stored in ATP, see Section 5.5, the reactions produce intermediates for chemical synthesis either by plants or animals. The combination of the two photosystems, Figs 5.1 and 5.2, is shown schematically in Fig. 5.3 but their real spatial distribution is complicated (see Chapter 3).

5.3 Oxidative energy: mitochondria

The reducing compounds, e.g. bound H_2 in organic compounds, and the oxidizing compounds made from light can be applied to opposite sides of a separate (mitochondrial) membrane, which is different from the chloroplast membrane used in light energy capture, and their reactions then can lead back to electrical ψ (electron gradients) (Figs 5.4 and 1.2). This is done by back-reacting the CHOH compounds to give H^- and then M^- and back-reacting the O_2 to give M^+. Even light could be generated in this way but it is not; usually the energy goes elsewhere even to heat. The two modes of energy generation, light and oxidation of hydrocarbons, are closely similar in that both produce gradients of potential and then of H^+ and both use rather similar components in membranes (Fig. 5.5).

At this stage we have shown how biology transforms sunlight, $h\nu$, (or chemical reaction energy) to energized chemical combinations or to electronic or ion gradients. Now let us go forward to the use of the electronic energy ψ provided by light or by the reactions of O_2 and H_2 in a slightly different way. For example, we can write

$$4M^+ + 4 \text{ bound } XH \longrightarrow 4M + 4 \text{ bound } X + 4H^+ \text{ (bound or free)},$$
$$2H_2O + 4M^- + O_2 \longrightarrow 4M + 4OH^- \text{ (bound or free)}$$

and let the two reactions occur on opposite sides of a membrane through the placement of their respective catalysts. We have then created a pH gradient across the membrane (Figs 5.1 to 5.6). In principle, we can imagine that a clever device could be made such that H^+ is allowed to move back across the membrane by a new path down its gradients while moving forward Na^+, K^+, Ca^{2+}, or molecules across the membrane (see Fig. 5.6). This is called an exchanger and it makes storage gradients of any ion or chemical. Biology is observed to make electrolytic gradients in this way and it can also be used to accumulate organic molecules (food) in cells. We have followed the processes as shown and we can see that light can be used to alter electrical gradients and ΔG° (chemical storage), and now it can be used in uptake of elements or compounds (concentration storage).

Aerobic

Fig. 5.4 A schematic representation illustrating a biological fuel cell. The reaction shown is the burning of dihydrogen with dioxygen while capturing the energy of the reaction in a proton gradient across a lipid membrane. The initial reaction of dihydrogen is to generate protons isolated in the phase (in) and a flow of electrons in a biological wire, Fe proteins b. Due to the fall in redox potential at the various couples the electrons flow to the protein c which carries electrons in the outer aqueous phase to a second particle (FeCu), proteins a. Here the electrons combine with dioxygen and water to produce hydroxide ions in the (out) phase. The named components are those found in biology but the reaction of H_2 and O_2 as shown is not observed. In fact a more complex reaction scheme is found in biology between oxidizable organic molecules such as acetate and dioxygen but the diagram contains the essence of the biological problem. Note the value of the membrane as a diffusion barrier. a, b, and c are cytochromes.

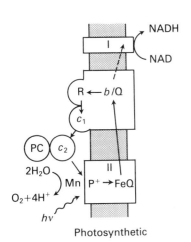

Photosynthetic

Fig. 5.5 The parallels between the electron transfer, middle sections, of photosynthesis, below, and of oxidative phosphorylation, above, are illustrated. a, b, and c represent cytochromes a, b, and c respectively, R is a high potential iron protein, P is a photocentre, and Q are quinones. Details are not necessary here.

Fig. 5.6 An interpretation of energy transduction of biology in terms of electrical circuit symbols. Initially a photocell or fuel cell unit is used in the construction of a source of potential energy. The current of electrons flows to a storage unit, ψ. The storage of electronic charge can be converted to electrolytic gradients in aqueous phases and these new storages (condensors) can interconvert (exchangers). Alternatively, energy from any of the condensers can be discharged to produce chemicals, e.g. ATP, and the chemical energy of ATP forms a second fuel cell using hydrolytic, not redox energy. The points X can be connected in the membrane as shown or outside the membrane in the aqueous phase.

5.4 Proton migration coupled to redox reaction

The flow of electrons in biology must be coupled to processes which conserve the energy for reactions such as synthesis. We know that the overall steps are

Light or O_2/XH reactions→redox (electronic) energy→proton gradients.

In the above devices the proton gradients were formed by the different redox reactions at the opposite sides of a membrane but it has also been found that the electron flow can be coupled directly to electrolytic flow of protons, literally transferring protons (pumping) across the membrane. Such movement requires a highly selective (irreversible or gated) motion of protons. In essence we know the different characteristics of proton and electron flow (Table 4.2): the electron hop must be accomplished by a conformational change which moves the proton. Protons move 0.5–1.0 Å at a time by rotational/translational diffusion from molecular centre to molecular centre involving H-bond breaking and making. It must be the case then that change of oxidation state (or metal ions) must induce a conformational change which makes and breaks a hydrogen-bond network in a cyclic manner since the flow must be gated (Fig. 4.4). We shall maintain that such transmission of force, the opening and closing of gates, movement at hinges, is controlled by protein conformational changes which are due to interlinked helical movements in proteins (see Fig. 5.11 in Section 5.8).

5.5 The coupling of gradients to ATP formation

Apart from the creation of gradients we have also described a mechanism for accumulating energy in compounds, i.e. in $\Delta G°$ of oxidized and reduced elements which are ordered in space across membranes. The basis is oxidation–reduction reactions and it is achieved by the organized assembly of essential elements (Fe and Mn, particularly) in membranes, as in Figs 5.1, 5.2, and 5.4; this mechanism is common to most if not all living systems. We still need another procedure since biological polymers are not made by redox reactions but by condensation reactions which remove water (see Chapter 12)

$$n[\text{HOROH}] \longrightarrow [\text{OR}]_n + n\text{H}_2\text{O}.$$

We know that this is achieved in biology by using a very small condensation polymer, pyrophosphate $P_2O_7^{4-}$ (in ATP) (see Fig. 5.7), which can be represented as $P \sim P$ where \sim stresses the unstable bond.

Pyrophosphate can drive condensation of all biopolymerizations since it is a drying agent, e.g.

$$P_2O_7^{4-} + 2(\text{HOROH}) \longrightarrow \text{HOROROH} + 2\text{HPO}_4^{2-}$$

written here for sugar reactions, but a very similar reaction can be written for amino-acid or nucleotide polymerizations. Thus the free energy change $\Delta G°$

Fig. 5.7 The pathway of energized protons generated by the photocell or the fuel cell of Figs 5.1–5.4 are led to the ATP synthetase unit F_1 where ATP is made from ADP and P_i. The nature of the F_0 unit is to provide a proton channel and a transducer of the energy from proton gradients to stored mechanical energy which in turn drives the states of F_1 so that it absorbs ADP + P_i but releases ATP in a cyclic series of events. Follow the arrows from left to right but note that the system is reversible. X^- is a carboxylate anion of a protein side chain. The left hand channel is a generalized picture of the production of proton gradients.

for pyrophosphate going to two phosphates is great enough to drive the polymerization, but where does pyrophosphate (ATP) come from? It has been clearly shown by many research workers that the reaction

$$2HPO_4^{2-} \longrightarrow P_2O_7^{4-} + H_2O$$

is driven by the electrolytic (pH) gradient of H^+/OH^- made across biological membranes described in Section 5.3. We shall illustrate the principles behind this process later but the basic transformation takes place in a membrane. After the charge separation has generated ψ, a membrane potential, or a pH gradient confined by a membrane, charge in the form of H^+ is allowed to flow back down a special channel for H^+ ions (Fig. 5.7). As it flows through the channel, the H^+ ions use their energy, i.e. ψ or ΔpH, to drive conformational changes in the proteins of the channel (i.e. gradient energy goes to mechanical energy) and, in turn, this mechanical energy is transduced into chemical bond energy forcing the condensation of two inorganic phosphates to give pyrophosphate

$$P_i + P_i \longrightarrow P_2 \quad \text{(see Fig. 5.7)}$$

Actually in biology this condensation is of a special diphosphate, adenosine diphosphate (ADP) with P_i to form adenosine triphosphate (ATP)

$$ADP + P_i \longrightarrow ATP.$$

Furthermore Mg^{2+} is required for the reaction.

The H^+ flows in a circuit driven by light or redox reaction (Fig. 5.6) and connects systems dissipating this energy in the production of gradients of M (or foodstuffs) and in the production of the condensation reagent ATP. The equivalence of the energy storage modes is shown in Fig. 5.6. A detailed picture of the mechanical device of Fig. 5.7 is given in Section 5.8.

Two points remain. The pyrophosphate $\Delta G°$ energy can also be used in the reverse sense to drive gradients, since this is just the reverse of its synthesis. Indeed pyrophosphate (ATP) is often used to accumulate and distribute elements and compounds all over the cell (see Fig. 5.6 and Chapter 6), but the chemical energy can also be transferred to other nucleotides, GTP and UTP (generally written NTP), which are used selectively in different metabolic pathways for the same general purposes as ATP. This selective use of phosphate energy parallels the production of NADPH for synthesis but of NADH for degradation in (high-energy) hydride transfer, Chapter 3.

The second point to observe is that in the above account the inorganic ions and elements dominate the several processes subsequent to the absorption of light. This chemistry is of redox changes of metal ion (electronic potentials and currents), changes in ion concentration gradients (electrolytic potentials and currents), and the polymerization of phosphate. It is for this reason that Fig. 1.3 was provided so early in this book. Naturally, without proteins the organization could not be built but without the inorganic elements the energy capture could not operate. Dominant also in all this biological chemistry is the function of hydrogen as the go-between from light and electrical energy to pyrophosphate; hydrogen is indeed the only element that can carry out this role (see Chapter 1, Section 1.2).

5.6 Anaerobes in the dark

Some biological systems do not use light and do not use dioxygen in energy generation. If we are to understand these anaerobic energy sources in the dark then we must start from a compound with one given chemical energy, $\Delta G°$, and transform it without changing the average redox states of carbon but producing energy. The usual source is the reaction of sugars, e.g.

$$[\text{HCHO}]_n \longrightarrow n/3\ \text{CH}_3\text{CHOHCOOH (lactic acid)}$$

which does not involve a general oxidation or reduction of carbon, i.e. although there is an internal oxidation/reduction of this element its average oxidation number remains the same (here it is zero).

The process can be looked upon, schematically, as

$$\begin{array}{ccc} -\text{CH}-\text{CH}-\text{CH}- & \longrightarrow & -\text{CH}_2-\text{CH}-\text{CO}_2\text{H} \\ \quad|\qquad|\qquad| & & \qquad\qquad| \\ \ \text{OH}\quad\text{OH}\quad\text{OH} & & \qquad\ \text{OH} \end{array}$$

Here, every third carbon is oxidized and one carbon is reduced. If we take the oxidation states of all carbons in glucose to be zero, the two carbons of lactic acid involved in the reaction are in oxidation states -3 and $+3$, respectively. This is then an internal disproportionation, oxidation–reduction reaction, driven by a favourable $\Delta G°$ change.

Indeed, if we look at the differences in $\Delta G°$ between the reactants and the products in this type of reaction (Fig. 1.7), one finds that lactic acid must lie lower than $[\text{HCHO}]_n$ and, therefore, this reaction of sugars can be used to make pyrophosphate (ATP) in a coupled series of reactions involving phosphorylated intermediates, as shown in Fig. 5.8. This series of organic chemical transformations is called 'glycolysis'.

Given that the phosphorylation of intermediates and the formation of ATP require energy, i.e.

$$\text{P}-\text{OH} + \text{HO}-\text{P} \xrightarrow{+\Delta G°} \text{P}-\text{O}-\text{P} + \text{H}_2\text{O}$$

$$(\text{ADP} + \text{P}_i \xrightarrow{+\Delta G°} \text{ATP} + \text{H}_2\text{O}),$$

and, since the free energy change on going from $[\text{HCHO}]_n$ to $\text{CH}_3\text{CHOHCOOH}$ is small, the final yield of ATP is rather limited—just two ATP molecules per molecule of glucose going to pyruvate (see Fig. 5.8).

Notice how poor an energy source this is compared to the aerobic process (in the respiratory chain)

$$[\text{HCHO}]_n + n\text{O}_2 \longrightarrow n\text{CO}_2 + n\text{H}_2\text{O}$$

which requires the different set of transformations of Fig. 5.9, called the 'citric acid cycle', leading to a proton gradient that is able to generate a much larger amount of ATP, now a total of 38 ATP molecules for each molecule of glucose transformed. In Table 5.1 we compare the yield of the two processes, which is considerably different, but notice also the entirely different chemistry and inorganic elements required (Figs 5.8 and 5.9).

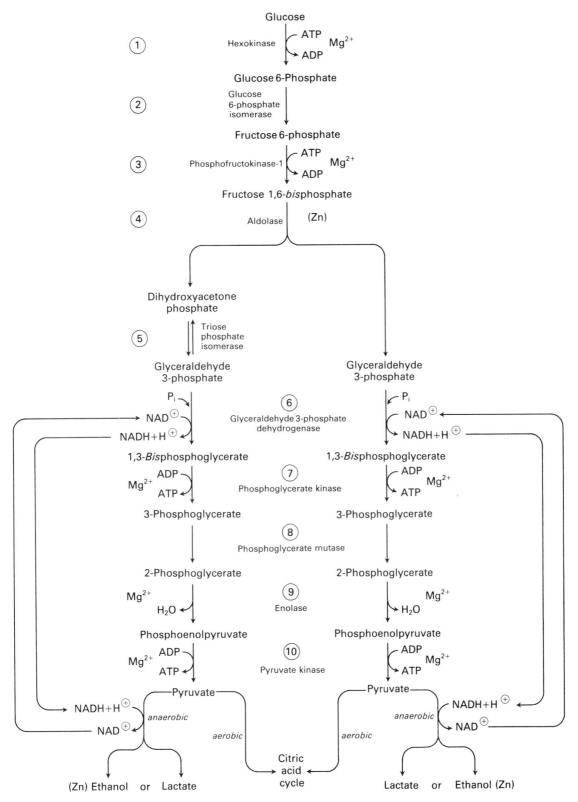

Fig. 5.8 The anaerobic breakdown of carbohydrates in the cytoplasm of cells. Notice the uses of the inorganic elements Mg and Zn while control is exercised by Ca and phosphates as shown in Chapter 10. Details of the pathway are readily available in bochemistry textbooks.

Table 5.1 The production of metabolic energy—a comparison of the anaerobic and the aerobic processes

Processes	Anaerobic glycolysis	Glycolysis+ respiration
Free energy change in the reaction (kcal)	47	686
Number of ATP molecules produced	2	38
Free energy change stored in ATP (kcal)	15.6	296
Yield of conversion (%)	33	43.2

There are reasons to believe that glycolysis was a very early metabolic pathway in the history of life. Whether it preceded energy from transformation of minerals is a moot point and there are, of course, other ways of obtaining energy from changes in chemical states of several elements. Bacteria are

Fig. 5.9 The aerobic breakdown of carbohydrates in the mitochondria of eukaryotes. Notice the uses of the inorganic elements, Fe, Cu, Mg, and Mn, while control is exercised by Ca, see Chapter 10. Details of the pathway are readily available in biochemistry textbooks.

known which use nitrogen or sulphur cycles, redox reactions of iron and manganese, and a multiplicity of carbon sources (see Section 5.12).

Now we can examine other metabolic cycles or reaction pathways in the same way looking for their connection to energy generation or use, noting where they occur in the cell, and observing the functional roles of inorganic elements in catalysis and control apart from their direct participation in reactants. The reader is asked to do this for himself so that he gains the feeling for the unavoidable interaction between pathways, controls, and energy. It is important that the essential involvement of many elements is observed and for this reason we give a reminder scheme of those elements which are most prominent in different functions, Fig. 5.10. In this book we wish to keep the inorganic elements in the forefront of the mind.

Fig. 5.10 The interrelationship of the functions of the elements is given here to stress the intimate connection between them and energy. This diagram is only an aid to thinking about the ways in which energy and metabolism have to be linked frequently using inorganic chemical functions. C, N, O are excluded.

5.7 Compartments, energy, and metabolism

One of the features of complex cells, eukaryotes, is the separation of metabolic pathways in compartments. The most obvious are the three we have been examining, photosynthesis in chloroplasts, oxidative phosphorylation in mitochondria, and glycolysis in the cytoplasm. They all synthesize ATP for the further metabolic steps which generally consume energy either just in ion or molecule transport through membranes or in molecular syntheses. So far in this account ATP was considered to act only on the cytoplasmic side of the membrane. Virtually all compartments have pumps associated with them, but compartments such as the Golgi apparatus (Fig. 3.14) engage in glycosylation (polysaccharide synthesis) and sulphation, for example, which are processes that require energy; hence the nucleotide triphosphates themselves must be pumped into the Golgi. There are other cell compartments which are degradative, e.g. lysosomes and peroxisomes, whose names explain their

functions. They have no need of internal energy from pyrophosphate. Yet others, such as the outside of cells and vesicles are just deposits, even of solids such as $CaCO_3$, SiO_2, $FeO(OH)$, and so on. These different compartments may be energized only by their own reactions, not by ATP, but they are set at a pH or a redox potential poise which is different from that of the cytoplasm. Lysosomes are usually at pH < 6, as are plant vacuoles, and ATP is used externally to pump protons and other ions into them. When we look at the outside of cells there is no pyrophosphate; hence energy must be obtained in a quite different way from transferred chemicals, i.e. from chemical $\Delta G°$ directly (see Chapter 21). All the compartments contain special proteins so energy is always used in the transport of proteins through membranes since all proteins originate from ribosomes.

5.8 Organization and tension

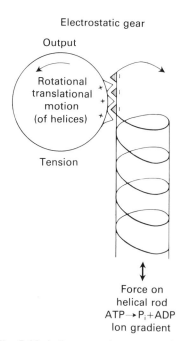

Electrostatic gear

Output

Rotational translational motion (of helices)

Tension

Force on helical rod
ATP→P_i+ADP
Ion gradient

Fig. 5.11 A diagrammatic representation of the general way in which drive is transferred from one helical system to another. The pumps in membranes as well as mechanical drive in muscles depend on the adjustment of protein/protein surfaces due to the flow of chemical bond or ion gradient energy.

Now we have seen how biology can obtain energy from different sources and in many different forms. A final form is tension in filaments that controls the shape of organisms and allows movement. A shaped object in biology is usually an energized object. Once again it is the hydrolysis of pyrophosphate (ATP) which is used to drive conformational and configuration changes in polymers to cause constantly held tension and to allow movement. Note that Mg^{2+} ions are always used in these ATP reactions.

Tension in biological polymers can be achieved in different ways. The first is by the use of mechanical displacement of units accompanied by conformational changes in polymers; this is called the sliding filament model of muscle action shown in Fig. 10.12 and generalized in Fig. 5.11. Another elementary device of this kind is seen in rubber where stretching restricts configurational entropy. A second mode for the creation of tension is to drive (reversible) polymerization in selected directions; the diagram of polymerization of tubulin monomers shows the basic idea (Fig. 7.14). The polymer is built from shaped units and therefore it can only change length in a given direction and as it does so it stretches or relaxes the membrane to which it is connected: the tension develops in the containing membrane and any fibrous molecules attached to this membrane. An elastic rubber is very different in that it can be deformed in all directions. Polymerization with a direction may well be very significant in defining biological shape and it is involved in many processes such as cell division and nerve cell extension. This, in turn, may be the basis of memory—a record of external observations reconstructed in a spatial internal pattern, the brain.

A very important role in the transfer of energy from one form to another is played by intermediate mechanical storage devices. An example is the transfer of the energy of a proton gradient to a chemical bond in ATP. The general picture of the events is as follows (see Figs 5.12 and 5.7, which show an ATP-synthetase enzyme as consisting of two parts). A bundle of protein helices, F_0, in a membrane (see Fig 2.11) is arranged in such a way that protons can bind with considerable affinity but they cannot pass through the bundle in this state so as to cross the membrane. When the binding sites are loaded by, say, three

Fig. 5.12 A set of intermediates in the transfer of energy from an ion gradient to the synthesis of ATP or the reverse. Curiously, in some sense, the steps of transfer of energy in the drive of muscle or bacterial flagellae may be similar to that in pumps. (See Fig. 5.11.)

protons, the binding is such as to force considerable re-adjustment of the helix packings, $F_0''H_n^+$, where the electrostatic energy stored initially in $F_0'H_n^+$ in the membrane has been changed now in large part into mechanical energy. The helices rotate with respect to one another and slip to some extent. This structural change moves the protons so that they can now be released across the membrane (their binding energy is lowered) while energy is conserved in the conformational tension in the helices. The relaxation, movement of F_0'' to its original state F_0 via an intermediate state, is coupled to a groove opening reaction in a second protein F_1 which releases bound ATP from its separated protein domains (Figs 5.7 and 5.12). The bound ATP had been formed initially from ADP + P by the higher binding of ATP and without net energy input, $F_0F_1 \rightarrow F_0'F_1'$ ATP (Fig. 5.12). Thus the mechanical release of ATP by reduction of binding and not permitting reversal of reaction to ADP + P is driven by an electrolytical/mechanical coupling in the helices of the membrane. This use of helix/helix movements in membranes is probably common to channels, gates, pumps, and exchangers in many membranes. An excellent example is the seven-helical bundle of bacteriorhodopsin.

The discussion of tension and organization is inevitably connected to static shape, which we leave until Chapters 20 and 21, and to dynamic motion.

5.9 Motion of organisms

The motion of organisms requires what is effectively an engine. The development from whip-lash flagellae of unicellular organisms to muscle movement cannot be covered here nor can we describe the apparatus in any detail. We need to note certain features of it, however. Energy is required and dissipated as heat and in large part this energy comes from ATP or from ion gradients. The machinery to which it is applied must be a mechanical device which moves in a controlled manner. The devices are not all the same but they depend upon cyclical operations as in simple man-made pumps and engines. Finally, they must have controls such that they can drive vectorially and with adjustable speeds. The involvement of various inorganic ions, such as Mg^{2+}, H^+, Ca^{2+}, and phosphates, like the machinery itself, is still the subject of much research and it is therefore only possible here to lead the reader who is interested to references.

5.10 Local storage of elements

Once energy has been devoted to constructing a biological cell in the form of compartments, the next step in our discussion is to see how it is used to trap elements and molecules, including proteins, in these compartments. Only if the elements are properly placed in a physical (Chapter 3) as well as a chemical sense (Chapter 2) will the biological system work. The discussion is, of course, strangely circular in that without organization no new organization can be generated, but we do not ask here how the first ordering (physical or chemical) was achieved and became a biological system needing all the components of Fig. 1.3.

The different energy storage systems of biology are steady state stores maintained by the constant activity of cells. For example, the concentration of ATP is kept very closely constant at a fixed value, slightly different in different cells. All the storage devices in this kinetic scheme can only remain in a fixed relationship, and thereby cellular activity and shape remain constant, if they all communicate amongst themselves. The cell must then have an information network (always energized) which maintains this homeostasis (Fig. 5.10). We shall see that calcium, the proton, and phosphate play major roles, but that they are all maintained in steady states far from equilibrium with their surroundings. While calcium is held in concentration gradients, i.e. a $RT \ln[M]$ energy store, and phosphate is largely held in kinetically stable but thermodynamically unstable chemical bond systems, i.e. a ΔG° store, the proton (or hydrogen) is held in a variety of $RT \ln[M]$, ψ, and ΔG° stores. The wonder of hydrogen chemistry is this easy switching from $RT \ln[M]$ to ΔG° energy or to ψ field energy and the reverse.

There is, therefore, the following connection of energy types in a homeostatic state

Light→[electronic⇋electrolytic⇋mechanical⇋chemical $(P_2O_7^{4-})$]→heat.

This is the essence of biological energy capture and use which remains largely related to inorganic element properties since

Light capture	→(Magnesium) chlorophyll proteins,
Electron transfer	→Iron, copper proteins
Electrolytic current	→Proton flow (Na^+, K^+, Cl^-),
Condensation	→Phosphate to pyrophosphate
	(magnesium required).

While stressing these aspects of energy capture and use, we must not overlook the fact that the electron transfer is also used directly in biological chemistry so that the separation of oxidizing from reducing equivalents, starting from water (or H_2S), causes the reduction or oxidation of metal ions or that of compounds of C, N, etc. It is only after these stages that organic syntheses can start.

5.11 The evolution of the compartments for energy generation

In this chapter we have concentrated on the three most important present-day sources of energy for life—photosynthesis (photon capture and photophosphorylation), oxidative phosphorylation, and glycolysis. These processes occur in many different species. As life developed from prokaryotes to eukaryotes the membrane-dependent process moved from the cytoplasmic membrane into the organelles of higher plants and animals, i.e. into chloroplasts (photosynthesis) and mitochondria (oxidative phosphorylation), while glycolysis remained in the cytoplasm. Note again that only glycolysis does not involve oxidation reactions. There is a belief (and some evidence) that both the organelles are remnants of captured prokaryote cells placed inside anaerobic eukaryotes. The isolation of the energy production machinery into the chloroplasts and mitochondria has been accompanied by segregated functional metal ion roles in both synthesis and degradation: the capture of CO_2 occurs in chloroplasts but degradation occurs in the citric acid cycle in mitochondria and there are obvious advantages in the close association of these redox pathways and of the energy transduction machinery. Here we note again that such compartmentalization requires a separation of key chemical elements in catalysts and controls which must therefore be utilized in somewhat different manner in prokaryotes and eukaryotes.

There is also very good evidence that the present-day use of light to generate dioxygen did not start until some one-third of the time in the development of living systems had passed (Fig. 1.11). The first use of light to capture energy and drive chemistry is thought to have been in a system similar to photosystem I (Fig. 5.1) but it is not necessary to assume that the same kind of pigments were present. The interesting feature of the earliest photosystem was that it probably produced sulphur from H_2S rather than dioxygen from H_2O. It is postulated that sulphide and sulphur chemistry had a dominant role until the evolution of photosystem II. Before this use of light there must have been even earlier energy sources.

5.12 Early sources of energy

Energy capture could not have started as it is seen today since any of the three processes require organization. We can imagine that the action of light could develop compounds, i.e. change $\Delta G°$ on the surface of the earth, and many accounts of such processes exist. We then need to imagine how an organization evolved to manage the use of this extra $\Delta G°$. Again many accounts exist which vary from the production of vesicles to the use of clay or other minerals. Such matters deeply concern the biological chemistry of the elements, but their experimental foundation remains insecure. Rather than discuss the origin of life we concentrate on some of its known forms and mention in a final paragraph of this chapter some presumed early forms of life.

In the deep ocean vents and in anaerobic muds there are bacteria which have a rather different metabolism from that described above; they are the

archaebacteria and related forms, which depend on sulphur cycles rather than oxygen. They can utilize CH_4 and CO_2 as a carbon source and perhaps they are able to use chemical transformations of compounds which formed in unstable states, i.e. with high positive $\Delta G°$, as direct sources of energy from the environment. While they introduce new metabolic pathways and different chemical elements are involved, e.g. nickel, the principles of energy assimilation and use are the same.

5.13 Sulphur compounds

Although not so obviously involved in the bioenergetics of cells today there is one other common molecular construct of high positive $\Delta G°$, i.e. the thioesters. It is noticeable that many syntheses, all based on the transfer of acyl fragments, go via thioacyl units (see Chapters 3 and 6). There has been the suggestion that the earliest bioenergetic controls worked through this 'high-energy' bond, but since this is somewhat speculative we leave further analysis of it to Chapter 19.

5.14 Feedback

Once again we must stress the obvious fact that since much of biology requires energy this must be distributed but not used randomly or wastefully; the energy status must be a form of homeostasis of gradients and chemical bond energy. There then must be feedback control to all the units involved in the energy generating system. Since, ultimately, the source of internal information is coded in DNA, it is to DNA that this feedback must go and from which it must come, so that DNA is not just a sequence for casual read-out but is a timed controlled code responding to the states and energies of the cell. Feedback requires that there are messengers, no matter if it is feedback all the way to DNA, to RNA, or to the chemical pathways of metabolism. We shall be very concerned to find elements or very simple compounds which act as messengers in cells and others, if any, which act outside cells or between the outside and the inside of cells in order to use energy efficiently.

5.15 Flow of material

In biological systems different forms of energy are captured and then successively transduced by flow of material from the local site of creation to the site of utilization. When this flow is of ions or electrons it is seen as currents. Currents require energy just to maintain them. In part this is kinetic energy of flow and it constantly requires boosting since energy is continuously lost as heat. By making pathways, hop electron conductors and ion channels, the currents are not random but create patterns under diffusion control. The flow

**Proton current
in hyphae**

of uncharged particles is equally important since their creation and destruction can lead, through localized catalysis (see Fig. 6.4), to patterns of chemicals and patterns of mechanical stress inside cells. Morphology and growth may well be linked to these patterns of element and compound flow. We have to attempt to picture this complex set of interacting flows of material, for it is inevitable that, in a system far from equilibrium, flow (current) takes on an importance equal to the static local potential which drives it (see Chapters 6 and 21).

5.16 The role of the proton in homeostasis

The proton connects the bound covalent chemistry of H_2O, CH_4, $[CHOH]_n$, i.e. of water, fatty organic compounds, and water-soluble organic compounds, to the synthesis of ATP; the unique character of the oxidation states of free and combined hydrogen, oxygen, and carbon generates this possibility. Now we need to notice more specifically the role of the proton as it travels from its source either in membranes or on the surface of membranes at high local concentration or high chemical energy, i.e. the bound proton, to its sink in a reaction with OH^- to form water (Figs 1.2 and 5.9). The interesting feature is that the very nature of biological molecules, often phosphates and nitrogen bases with pK_a values (log [proton binding constants]) close to seven, means that the proton is not only in communication with the bioenergetic systems mentioned above but that at the pH of biological systems, $pH = 7$, it affects virtually all biological polymers as its concentration is changed. The proton is an ideal element for such an intercommunicating role between biological localities since it travels rapidly to many parts of a cell through the Grotthus mechanism. This mechanism, which involves rotational/translational diffusion, is open to the proton alone since hydrogen is everywhere in water and biological polymers. One universal steady state homeostasis of biology is via the proton.

This is not contradictory with the existence of proton concentration gradients in cells; in fact each different vesicular system may well be separately buffered at a different pH. Moreover, there are extremely acidic zones in some organisms or organs, e.g. in man's stomach. The local proton gradients have received much attention in the literature since they must exist whenever sources and sinks are in different places connected by flow (see Chapter 21). It may be too that there are pathways of flow along membrane surfaces.

More importantly, in order that the connection from bound covalent H to protons can then go forward to ATP formation, protons must flow in defined channels which are not open to other ions. The selectivity is not fully understood but we may assume that it is the small size of the proton and its ability to dehydrate from $H_9O_4^+$ to bound H^+, a complete dehydration not open to any other ion on combination with one Lewis base, which allow it to move via a series of hops using Lewis bases, perhaps even organized water molecules (Fig. 4.4). Now, as well as flowing down a pH gradient, protons can be driven, i.e. pumped, against a gradient to give an acid concentration locally.

Once a proton gradient is formed, it can be coupled to a movement of another ion or molecule by exchange.

The major purpose of this short section on the proton, which repeats some early comments and which should be taken together with Section 1.2 and the remarks on bioenergetics in this chapter, is to stress its universal role. It is a central element in homeostasis (Chapter 21) but it must not be thought of as being involved in simple buffering. The pH is maintained by metabolism (energy input) and both directly and by feedback it interacts with oxidation reduction reactions and not just with acid–base metabolism.

5.17 The proton and redox potentials

The concepts of the redox potential and the pH of a chemical or biological compartment are deeply interlocked since the proton can also pick up electrons to give reduced species. Redox potentials in water are defined relative to a standard pH and for most purposes we use the redox potentials against the hydrogen electrode at $pH = 0$. The actual potential of the H_2/H^+ couple in water at $pH = 7$ is then -0.42 V and that of dioxygen/H_2O is $+0.8$ V. Virtually all biological cells have a cytoplasm which operates at $pH = 7$; in almost any environment, it is as if Nature had selected a buffer with as low an OH^- and H^+ concentration as possible. High pH values risk hydrolysis and hydroxide precipitations, but lower pH values reduce absorption of CO_2, so that it could well be that $pH = 7$ is a carefully selected value for synthesis and stability.

The effect of $pH = 7$ on redox potentials is easily calculable from $pH = 0$ values in tables and we observe that inside biological cells the effective redox potential range is low due to the concentration of reducing materials. In fact, all cells appear to have chemicals in the cytoplasm that are not in equilibrium with one another but in at least two constrained kinetic pools—one of which gives a redox potential range at about -0.5 V (NADH/NAD), i.e. close to the H_2/H^+ potential at $pH = 7$, and a second at a potential of around 0.0 V (glutathione), which is close to the highest potential of compounds of carbon bound to oxygen and hydrogen. These conditions have probably existed from very early life forms and they have a profound effect on elements which undergo redox change. Figure 1.8 shows that, apart from carbon, only the chemistry of one other non-metal element falls in this range—that of sulphur (see Section 5.13).

All the above analysis points firmly to the fact that even if the H^+ concentration may equilibrate rapidly within some compartments, it does not equilibrate with the chemical potential of bound H in the vast majority of compounds, e.g. H_2, CH_4, $[CHOH]_n$, NADH, glutathione, and so on. This lack of equilibration which is also true of OH^- (H_2O) and O_2 means that local acidity and redox potential are independent and can be varied according to the catalysts and transport systems in specific zones. Lack of equilibration does not mean absence of communication. All systems must interact. In Chapter 21 we shall ask whether elements other than hydrogen assist in the communication networks in cells between redox and hydrolytic metabolism.

Further reading

The general ideas on the nature of energy capture in biological systems are to be found in standard books such as the following.

Cramer, W. A. and Knaff, D. (1990). *Energy transduction in biological membranes.* Springer-Verlag, Berlin.

Deisenhofer, J., Epp, O., Miki, K., Huber, R., and Michel, H. (1985). Structure of the photosynthetic reaction centre. *Nature,* **318**, 618.

Harold, F. M. (1986). *The vital force—a study of bioenergetics.* W. H. Freeman & Co., New York.

Henderson, R., Baldwin, J. M., Ceska, T. A., Zemlin, F., Beckman, E., and Downing, K. H. (1990). Proton pumps in bacteriorhodopsin. *Journal of Molecular Biology,* **213**, 899–929.

Khorana, H. G. (1988). Bacteriorhodopsin and proton translocation. *Journal of Biological Chemistry,* **263**, 7439.

Nicholls, D. G. (1982). *Bioenergetics.* Academic Press, London.

Stryer, L. (1988). *Biochemistry* (2nd edn). W. H. Freeman & Co., New York.

Williams, R. J. P. (1985). In *The enzymes of biological membranes* (ed. A. N. Martonosi), Vol. 4, pp. 71–110. Plenum Press, New York.

Williams, R. J. P. (1978). The multifarious couplings of energy transduction. *Biochimica Biophysica Acta,* **505**, 1–44.

The development of the theories of the involvement of proton gradients follows from two articles.

Mitchell, P. (1961). Coupling of phosphorylation to electron and hydrogen transfer by a chemi-osmotic mechanism. *Nature,* **191**, 144–8.

Williams, R. J. P. (1961). The functions of chains of catalysts. *Journal of Theoretical Biology,* **1**, 1–13.

Surveys of many topics mentioned in this chapter appear in the special volumes of *Biochimica Biophysica Acta,* labelled 'Bioenergetics'.

References to origins of life will be found in

Cairns-Smith, A. G. (1985). *Seven clues to the origin of life.* Cambridge University Press, Cambridge.

Mason, S. F. (1991). *Chemical evolution.* Oxford University Press, Oxford.

6

The functional value of the chemical elements in biological systems

6.1	Introduction	140
6.2	Major chemical properties of elements in aqueous solutions	141
6.3	Biochemical functions of the chemical elements	143
6.4	The living process	146
6.5	The chemical flow in biology	149
6.6	Organization and flow	156
6.7	The integration of activity	160
6.8	Conclusion	160

6.1 Introduction

Chapters 1–3 have shown the ways in which biological systems accumulate elements both through chemical speciation and through separation in compartments. Chapter 4 illustrated the potential kinetic restrictions and advantages of the use of various elements in biological reactions while Chapter 5 described the different ways of capturing and using energy. We have mentioned that not only can the elements be energized through their concentrations, e.g. by uptake into compartments, but that many of the molecular forms in which they appear in biology are thermodynamically unstable, e.g. phosphate in ATP. While in no way suggesting that biological systems can invent a chemistry totally foreign to *in vitro* laboratory chemistry, these considerations do suggest that biological chemistry might be unorthodox. This goes beyond the obvious feature that biological organic chemistry, unlike most of man's organic chemistry, is carried out in the presence of water, to the idea that the 'elemental' chemistry in biology is sometimes not easily appreciated from *in vitro* behaviour. In this chapter, largely an overview of points made in the previous five chapters, we set out briefly to look at the conventional functional value of the elements in chemistry and biochemistry, and then, after looking at the additional major requirements for a system to be

living, i.e. biological, we ask to what degree the selection of an element for a particular biological task is a necessary consequence of its chemical quality and to what extent there are biological peculiarities. It will be assumed that a peculiarity is the result of evolutionary drive toward the best chemical systems for a living organism. Since we do not understand living systems in depth the discussion will be diffuse, but we hope it will give some essential insight.

6.2 Major chemical properties of elements in aqueous solutions

A very brief general summary of the chemical properties of metals and non-metals, a reminder of what was discussed in Chapters 2 and 4, is conveniently given here.

Taking metals first, it may be said that the great advantages of their chemistry in aqueous solution are:

(1) the transport of positive charge by stable ions;

(2) ready, often strong, and selective interactions with organic ligands;

(3) easy bond breaking of ionic bonds in low positive oxidation states;

(4) considerable polarizing power of the ions of high electron affinity, i.e. Lewis acidity in later transition and B-group metals;

(5) readily adjustable bond directions and lengths of some but not all elements;

(6) easy change of stable oxidation states of some transition elements, i.e. easy one-electron chemistry;

(7) in high oxidation states, some properties closely related to those of non-metals, e.g. slower ligand exchange but still readily activated exchange for atom transfer;

(8) good π-electron donor properties of heavier elements in low oxidation states.

The advantages of non-metal chemistry in water are in quite different directions:

(1) the transport of negative charge by stable anions, e.g. Cl^- and HPO_4^{2-};

(2) strong kinetic control of homolytic and heterolytic breaking and making of covalent bonds which are thermodynamically unstable;

(3) weak but selective interaction between their anions and cations, including organic cationic sidechains of proteins;

(4) high electronegativity, especially of the light non-metals, providing good Lewis bases;

(5) more fixed bond lengths and bond angles, giving stereochemistries of high precision, especially for light elements;

(6) difficult single unit changes of oxidation state and thus a restricted (but important) radical chemistry since one-electron non-metal reactions are rarely easy or easily controlled.

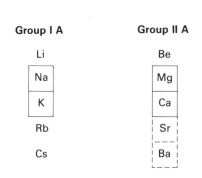

Group I A

Li

Na

K

Rb

Cs

Group II A

Be

Mg

Ca

Sr

Ba

Non-redox non-metals

B

Si P

Group II B

Zn

Cd

Hg

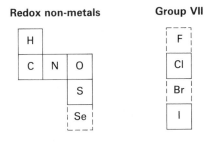

Redox non-metals

H

C N O

S

Se

Group VII

F

Cl

Br

I

Redox metals

Cr Mn Fe Co Ni Cu

Mo

Table 6.1 Examples of chemical reactivities

Extremely difficult reactions	Slow reactions	Fast reactions
$[CH_2]_n$ reactions in all classes	Ester hydrolysis, amide hydrolysis	Dissociation of ion pairs, M^+X^- and $M^{2+}X^-$
Rearrangement of sugars Inversion of configuration at carbon atoms	Many cis–trans conformational changes	Proton reactions of strong and moderately weak acids
Atom (group) transfer from weakly polar compounds	Atom (group) transfer from more polar compounds	Electron transfer M_1 to M_2 across short distances

Undoubtedly the reader will wish to add one or two points to these two lists to make them more comprehensive.

It is now obvious that, in the presence of water, non-metals are the basis of most thermodynamically unstable but kinetically stable compounds, i.e. all major polymers and substrates, while metals are the best electron-transfer agents, the best local catalyst atoms, and, amongst the conducting devices via ionic diffusion, they provide a multitude of agents. These statements are independent from biology but we see that biology needs both non-metals and metals because it has to build structures, requires catalysts, and needs communication systems. While it is extremely dangerous to overgeneralize, we can provide a guide table of chemical reactivity from these observations (see Table 6.1) and we can ask what kind of elements do we need to take part in or to activate these reactions. In addition, Fig. 6.1 gives an easy-to-read overview of some of the different aqueous solution chemistries of the *early elements* in the periodic table, and it is clear that we can get all sorts of covalent and ionic molecules but inadequate catalysts. If good catalyst centres are required we need to turn to some of the heavier elements and observe also that in every group of the periodic table heavier elements react faster than lighter elements in the same type of compound.

All the above are points made from knowledge strictly from *chemistry*; the immediate next step in this discussion is to use the *biochemical* data obtained by analysis usually of *extracted* compounds to get a first impression of how the elements are used functionally in biology, but note that biochemical analysis is to some degree destructive of biological systems. This discussion cannot take into account to any degree that a biological system is an organized, energized, system *in permanent flow*.

Fig. 6.1 A simple division of part of the periodic table stressing the separation in biology between elements: (1) those which are found as cations from those found as anions; (2) those which have a variety of redox states from those that do not; and (3) those which exchange ligands or groups rapidly from the first co-ordination sphere from those that do not. B and Si are intermediate in many ways.

6.3 Biochemical functions of the chemical elements

It is useful at this stage to summarize the biochemical roles that the different elements can perform (Table 6.2). A systematic examination of these roles by biochemical methods has been made possible by remarkable changes in our knowledge over the last 30 years. More powerful methods of chemical analysis have been developed and practice and theory have progressed enough to allow detailed investigation of the structure, dynamics, and reactions of many isolated large molecules. Naturally, most of the biochemical molecules are organic compounds of the bulk elements—carbon, hydrogen, oxygen, and nitrogen—sometimes with sulphur and phosphate groups and in a few cases with some covalently combined minor non-metal elements, e.g. selenium. These elements are required for the major monomers and polymers of biology, that is, the central constructs of life. However, in catalytically active enzymes we often find other (inorganic) elements; in fact, some one-third of all enzymes need metal ions. There are also biochemical materials that are fully or predominantly inorganic, e.g. bones, teeth, shells, etc. Furthermore, there are quite high, and precisely maintained, concentrations of a variety of simple inorganic ions throughout biological systems. Some of these are essential as

Table 6.2 Main biochemical functions of the chemical elements

General function	Elements	Chemical form	Examples of specific functions
Structural functions (biological compounds and support structures)	H, O, C, N, P, S, Si, B, F, Ca, (Mg), (Zn)	Combined in organic compounds or sparingly soluble inorganic compounds	Components of biological molecules; formation of tissues, membranes, skeletons, shells, teeth, internal structures, etc.
Electrochemical functions	H, Na, K, Cl, (Mg), (Ca), HPO_4^{2-}	As free ions	Transmission of messages (nerves); production of metabolic energy
Mechanical effects	Ca, (Mg), HPO_4^{2-}	As free ions exchanging with bound ions	Triggering of muscle contraction; lysis of vesicles
Acid–base catalysis	Zn, (Ni), (Fe), (Mn)	Combined in enzymes	Food digestion (Zn); hydrolysis of urea (Ni); phosphate removal in acid media (Fe, Mn)
Redox catalysis	Fe, Cu, Mn, Mo, Se, (Co), (Ni), (V)	Combined in enzymes	Reactions with oxygen (Fe, Cu); oxygen evolution (Mn); nitrogen fixation (Mo); inhibition of lipid peroxidation (Se); reduction of nucleotides (Co); reactions with H_2 (Ni); bromoperoxidase activity (V)
Various specific functions	Mg	In chlorophyll	Light harvesting in photosynthesis
	Fe, Cu	In proteins	Transport of oxygen
	I	Special covalent compounds	Hormonal action
	Fe, Ca, Si	Mineral compounds	Sensors (magnetic, gravitational)
	O_2, N_2, CO_2	Gases	Flotation (fish)

information carriers and in the co-ordination of activities. One thesis of this book is that biochemistry cannot be considered as just a part of organic chemistry, i.e. the chemistry of C, H, N, O, etc., since combinations of these elements alone do not represent the analytical make-up observed in a living system; on the contrary, life seems to have evolved from a much wider use of elements in the periodic table. A few examples of the uses of inorganic elements, most of which have been referrred to in the previous chapters, clearly show some, not all, of these functions, as revealed by biochemical analysis (Table 6.2).

(1) The hard structures of animals are largely mineral, most frequently calcium salts (see Table 20.1 for examples of biological minerals). Of course, this is not the case for the plastic cuticles of insects or of the woody stems of many plants.

(2) Electrolytic connection in and between cells, e.g. nerve messages, rests largely on the concentration gradients of sodium, potassium, and calcium amongst metal ions and chloride amongst non-metal ions.

(3) Phosphate transfer, essential in bioenergetics, is almost invariably linked with magnesium reactions in enzymes.

(4) Mechanical changes in cells are frequently initiated by calcium concentration changes.

(5) External triggering in digestion, blood-clotting, fertilization, muscle contraction, cell–surface interactions, etc. also depend largely on calcium concentration changes.

(6) The capture of light in photosynthesis depends on a magnesium-containing pigment, chlorophyll.

(7) Electron transfer in energy transduction is very largely dependent on iron proteins.

(8) Oxidative metabolism is almost always dependent on iron and copper catalysts.

(9) The capture of combined nitrogen from nitrate or dinitrogen relies amost totally on a molybdenum-containing catalyst centre, though there is some use of vanadium.

(10) The release of oxygen from water, that maintains our present atmosphere, is absolutely dependent on manganese (part of an O_2-evolving enzyme).

(11) Transport of oxygen requires iron or copper proteins.

(12) The synthesis of some DNA and RNA, the activation of water as a leaving or attacking unit in condensation, and hydrolysis reactions, etc. depend on the intervention of more than 150 zinc (and magnesium) enzymes involved in biological reactions. (DNA synthesis also requires either cobalt, manganese, or iron in ribonucleotide reductase.)

(13) Bound pyrophosphate (ATP) is the major carrier of chemical free energy for driving reactions.

(14) The role of the free proton is dominant as an intermediate in energy metabolism.

Again the reader will want to add other points, many of which we hope to cover from Chapter 8 onwards.

Glancing at the above list of activities of the various elements one is struck first by the extent of the total involvement of inorganic elements in structural and mechanical functions; electrochemical control; acid–base and redox catalysis; generation, transmission, and storage of energy; transmission of messages and triggering effects; and so on, very much in line with the chemical possibilities for individual elements (compare Fig. 6.1 and Table 6.2). However, there is remarkably little in common between the functions of different elements (see Table 6.3). Such a complete separation of element function is not expected from chemistry judged directly from the periodic table, since neighbouring chemical elements in the table are often similar in properties, although some of these specialized biochemical roles are more or less obvious. For example, copper, zinc, and iron are good Lewis acids, and each is used to some degree in biology; however, at first sight puzzlingly, zinc has been selected overwhelmingly above all others for this function. Again potassium is largely inside cells, and sodium outside cells, so there must be good *biological* (not chemical or biochemical) reasons for such an individualized choice which biochemical analysis by itself does not reveal. Some of these features will have arisen, of course, since biological evolution, but not chemistry, has been very restricted by the distribution of the elements themselves in the universe and on earth (see Chapter 1).

This analysis of choice of an element for a given function is complicated since we have seen in the previous chapters that an element may be present in living systems in different locations, different oxidation states, and different forms, with distinct properties and functions (Table 6.4). For instance, iron (II) is different from iron (III) and haem iron behaves as a separate entity from

Table 6.3 Some specific metal ion catalyses

Small-molecule reactant	Metal ion	Examples
Glycols, ribose	Co in B_{12}	Rearrangements, reduction
CO_2, H_2O	Zn	Carbonic anhydrase
Phosphate esters	Zn	Alkaline phosphatase
	Fe, Mn	Acid phosphatase
N_2	Mo(Fe)(V)	Nitrogenase
NO_3^-	Mo	Nitrate reductase
SO_4^{2-}	Mo	Sulphate reductase
CH_4, H_2	Ni(Fe)	Methanogenesis
$O_2 \rightarrow H_2O$	Fe	Cytochrome oxidase
	Cu	Laccase, oxidases
O-insertion from O_2 (high redox potential)	Fe	Cytochrome P-450
SO_3^{2-}, NO_2^-	Fe	Reductases
$H_2O \rightarrow O_2$	Mn	Oxygen-generating system of photosynthesis
H_2O_2/Cl^-, Br^-, I^-	Fe(Se)(V)(Mn)	Catalase, peroxidases
H_2O/urea	Ni	Urease
Small peptides and esters	Zn	Hormone control (peptidases)
Ethanol	Zn	Dehydrogenase

Table 6.4 Examples of separated functional pools of one element

Element	Pool 1	Pool 2
Mg	Free ions	Chlorophyll
Fe	Free ions	Haem
S	–S–S– and RS$^-$	Sulphate
Ni	Free ions	F-430
Zn	Free ions	Enzyme sites
H	NADH (degradation)	NADPH (synthesis)
P	ATP (glycolysis)	UTP (glycosylation)

both in many different ways. This means that any attempt to relate biological functions in living systems to biochemical analysis of elements must take into account specific non-exchanging complex species in physiochemical environments. However, no matter how complicated we make this type of analytical description, there are obviously some overriding principles of living systems which are still missing. One such problem is the fact that we usually find metalloproteins of living systems with a stoichiometric ratio, but apparently no excess apoprotein and only a limited fixed excess of free metal. Another problem is the observation that the ratios of all the elements in a cell are quite fixed. A third problem is that a living system is always in flow within a highly specific structure.

6.4 The living process

We shall take it that all life known to us is based on similar physical and chemical features though of very different complexity. The major physical features largely described in previous chapters are:

(1) Life is a controlled (chemical) *growth* process at pH = 7.0, temperature 300 ± 25 K.

(2) Growth requires *energy*.

(3) Life exists in a series of aqueous and membranous isolated *compartments*, which it makes (see Table 6.5).

(4) Living systems are *organized*, develop, and then fail. The system *self-assembles* and *self-degrades*.

(5) Such systems must *reproduce* to remain extant.

(6) Inevitably they grow in association with a wide variety of *elements in the environment*.

(7) When they grow they grow in *well-defined* ways; there is *morphological* growth and development.

We can also make some further restrictions and largely chemical statements which must be kept in mind.

(1) Living processes occur in fixed internal, aqueous, *salt* solutions which are retained by cells and their membranes.

(2) Life must capture energy in small molecule chemicals, i.e. within *unstable compounds*, or in *gradients* in order to generate its larger compounds.

(3) Life depends on the continuous *synthesis and degradation* of unstable water-soluble as well as organic, lipid-soluble, *polymers*. (While we now know that reproduction today is dependent on DNA, we do not have to consider this finding as fundamental; in principle, there could be other polymers which would carry genetic information. Similar statements can be made about the functions of RNA, proteins, polysaccharides, and fats. The necessary feature is just a controlled, reproducible set of polymers which are of considerable but limited kinetic stability in water and which require energy to produce and reproduce them faithfully.)

(4) The initial function of energy capture must be to take up or make the appropriate *unstable monomers* for the synthesis of polymers, which also occurs in water. There then has to be a balanced set of monomers and polymers; hence, ions, molecules, and polymers are all in a *homeostatic* relationship.

(5) The overall composition of a living system contains elements in closely fixed ratios.

(6) The synthesized molecules must be placed in an *organized array* because *directed flow* of material is essential.

(7) Control of growth requires a complex set of *messages* and *feedback controls*, which again demand flow which is constantly maintained.

(8) Series of organized *catalysts* are necessary to interconvert chemicals at room temperature and this requires capture of various elements.

(9) Life has evolved with time and the chemistry of the elements in life has been and still is an evolving development. This is necessary to adapt to environmental changes.

Once again the reader may wish to add further points, but the essence is clear.

While the biochemical analysis in the previous section revealed some of the refinement of the chemistry, and this will need to be examined carefully when we describe the biological chemistry of specific elements in Chapters 8–19, there are in addition definite new features in living systems. Outstanding are the organization at any one time (Table 6.5) and the time-dependence of the whole flow system, synthesis/degradation, and specific growth (Tables 6.6 and 6.7). There is clearly some integrating activity which has a large energy input. The function of an element or a compound in biology has to be seen in the context of where and when it combines and how it moves, is used, and is removed. In an effort to see how these features must force a very separate and almost individual chemistry upon each of the elements (already seen above to some extent by biochemical analysis) we shall examine some of the major requirements to build an organized system which can develop and reproduce. Such a system must have a regulated stability and a channelled dynamics at any time and in this organization there must be integration of function (see Fig. 6.2). In this figure all the activity pockets must feed back to one another and directed organization is of the very essence of biology. We illustrate this point first with the nature of biological polymers and here it will be necessary to repeat in part some points made in previous chapters.

Table 6.5 Compartments and reactions in eukaryotes

Compartment	Reactions	Elements
Cytoplasm	Synthesis of DNA, RNA proteins	Mg, Zn, K
	Glycolysis	Ca, Mg
Mitochondria	Oxidative phosphorylation Citric acid cycle	Fe, (Cu), (Ca), (Mg)
Chloroplasts	Photophosphorylation	Mg (chlorin), Fe(Cu)
	Carbon assimilation	Mg
Sarcoplasmic reticulum	None(?)	(Ca)
Golgi apparatus	Glycosylation Sulphation	Mn
Endoplasmic reticulum	$2RS^- \rightarrow R\text{--}S\text{--}S\text{--}R$	(?) (Ca)
Vacuoles	Urea hydrolysis	Ni, (Mn)
Peroxisomes	Oxidative degradation	Fe, Mn
Circulating fluids	Hydrolysis oxidations	Zn, Ca, (Na) Cu

Table 6.6 Some basic element flow units of biology

Metabolite (element)	Reactant and control carriers (coenzymes)	Protein-catalysed reactions
Electron (e^-)	Metal ions, glutathione	Many, e.g. energy conversion, oxysugar reductase
Hydride (H)	Pyridine nucleotide (NAD)	Many, e.g. energy conversion
Methyl (C)	Methionine	Methylation
Acetate (C) (and other carbon fragments)	Coenzyme A	Fatty-acid synthesis, etc.
Phosphate (P)	Adenosine (guanosine) triphosphate (ATP)	Many, e.g. glycolysis and oxidative phosphorylation, DNA, and RNA synthesis
Nitrogen (N)	Pyridoxal phosphate	NH_3 exchange
Amino acid (N)	Adenosine (guanosine) monophosphate (AMP)	Ribosomal synthesis of proteins
Sugars (C, H, O)	Uridine diphosphate, dolicol phosphate	Many, e.g. glycoprotein formation
Nucleotides (C, N, O)	Nucleotide phosphates	RNA/DNA synthesis in nucleus
Hydride/electron (H^-/e)	Coenzyme Q (quinone)	Energy conversion (membrane)
Sulphate (S)	Adenosine phosphate-sulphate (APS)	Sulphonation
Energy (ΔG)	Triphosphates	Energy utilization
Light quanta ($h\nu$)	Chlorophyll	Energy capture (membrane)

Although there are carriers for H, C, P, N, S, and e^-, there are no simple carriers for O. Rather, this element is carried in carbon compounds or H_2O or O_2. Note different carriers for different oxidation states.

Table 6.7 Examples of diffusion-controlled zones

Process under control	Control of diffusion
Oxidative phosphorylation	Membrane curvature (mitochondrial)
Photo-capture	Membrane curvature PS I/PS II (separation of O_2 release)
Creatine kinase exchange	Muscle and mitochondrial relationship
Na^+/K^+ channelling	Nodes of Ranvier

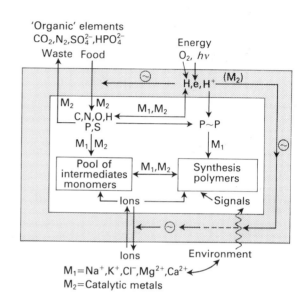

Fig. 6.2 A very simplified picture of the multitude of cellular activities in a prokaryote which have to be in mutual communication. The role of inorganic elements is stressed. This figure will be used again in Chapter 21.

6.5 The chemical flow in biology

6.5.1 The synthesis and degradation of the polymers of life

We ask first 'Is there a common feature amongst the biological polymers?' Chemically, the answer is that all the biological polymers are condensation polymers; indeed biological systems rarely use free radical polymerization since this is very difficult to control, while it is relatively easy to carry out condensation reactions in a very controlled step-by-step manner

$$A + B \longrightarrow H_2O + AB \xrightarrow{+C} ABC + H_2O.$$

Man achieves this in solid state synthesis of peptides and polynucleotides following Nature. The reactions are reversible also in a step-by-step manner and this is part of control (Chapter 4).

None of the condensation reactions is spontaneous, of course, so it is clear that they need some driving power. It is equally clear that biological systems have a need for a wide variety of catalysts for these syntheses. In order to appreciate these requirements we must see the close interrelationship of the whole activity. The overriding need is to build a set of polymers of four major types from simple monomers:

(1) polynucleotides from phosphate and sugars (and bases);

(2) polypeptides from amino acids;

(3) polysaccharides from sugars;

(4) hydrocarbons from acetic acid, followed by reduction.

There are three immediately perplexing features: (1) the polymers are produced in balanced ratios, yet the exact mode of production is not the same for each, e.g. for proteins and DNA; (2) the products appear in position in space; (3) all the reactions use the same initial sources of energy. All the reactions are acid–base reactions, so that the catalytic elements involved are usually Mg, Ca, and Zn, and various triphosphates are used for energy. While some separation of the different syntheses is managed by the differential uses of ATP, UTP, CTP, and GTP in spatial organizations, there remains the need for integrated feedback control. Furthermore, each set of the four polymers is processed in a specific machinery located in spatial compartments: polynucleotides in the nucleus, polyamino acids in ribosomes, polysaccharides in the Golgi, and fats in fatty acid synthesis particles. Communication between all these compartments is obviously required.

These features and the observed stoichiometric ratios of bound coenzyme in enzymes and of metals in metalloproteins, as well as the general absence of excess of apoprotein (or metal ion or free coenzyme, e.g. flavin) indicate that the synthesis must have a degradative counterpart, for example,

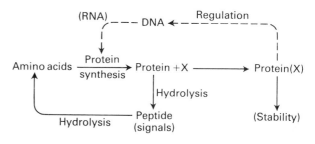

where X is either a metal ion, such as Zn, or a coenzyme such as flavin, and the broken arrows to and from DNA represent instructions or feedback of information.

Synthesis and degradation lie at the heart of flow within the turnover of all macromolecules and feedback is clearly essential for balanced growth,

Table 6.8 Feedback in polymer synthesis

Polymer	Some feedback systems
DNA (nucleus)	Transcarbamylase (Zn), ribonucleotide reductase (Fe) (Mn) (Co)
RNA (nucleus)	Controlled by similar chemistry to DNA and by DNA
Proteins (ribosomes)	Control at the level of ribosomes and DNA by regulating genes, hormones, Fe^{2+}, Zn^{2+}, etc.
Polysaccharides (Golgi)	Pumped supply of monomers from cytoplasm, Mn^{2+}
Glycolysis/glucogenesis	c-AMP, Ca^{2+}, ATP controlled proteins
Fatty acids	Redox controls sharing ATP, cAMP, Ca^{2+} with glycolysis, glucogenesis, and the citric acid cycle series of enzymes

All syntheses are controlled by material and energy supply and such ions as HPO_4^{2-} and Mg^{2+}.

Table 6.8. Moreover, the feedback is not just in one pool, e.g. proteins, but must be between pools, RNA and proteins, to maintain stability.

We now summarize our knowledge of the organization of the flow of polymers:

(1) They are synthesized in specific localities.

(2) Starting from a locality they are transported in a particular manner across membranes, or in vesicles, or they are prevented from such movements.

(3) They are deposited in a region assisted by local recognition.

(4) They are turned over by destruction locally.

(5) There is linked to them a communication network which relays back to the synthetic machinery information about excesses or deficiencies (see Table 6.8).

(6) The information in the communication network is not made from polymers but much of the structural network itself, as well as the biological support structures, are so made.

(7) The polymers form the genetic code in DNA, the transcription of the code, RNA, the protein framework and the filaments of the cell, and also the localized catalysts. The catalysts then locate monomer activity.

(8) The lipid polymers form the membranes which make all functional compartments (Table 6.5).

The polymers are all produced from monomers and we therefore turn to the ways in which monomers are synthesized and degraded.

6.5.2 Synthesis and degradation of monomers

The monomers are all formed initially from very small molecules made of at most two elements H_2O, CO_2, N_2, PO_4^{3-}, SO_4^{2-}, O_2 (see Fig. 6.2). Several of these very small molecules are first converted to more reduced states, e.g. $HCHO$, NH_3, H_2S, etc., some in combined form. (In passing, note again that these reactions are reductions which can only be carried out at reasonable rates by *metal ion* catalysts; see Chapter 4.) Condensation reactions next yield more complex molecular species. The production of these monomers is written here in formal, not real, reactions

$$6\,HCH(OH)_2 \longrightarrow C_6H_{12}O_6 + 6\,H_2O$$
$$HCH(OH)_2 + NH_3 \longrightarrow HCH(NH_2)\,OH + H_2O,$$
$$HCH(OH)_2 + HPO_4^{2-} \longrightarrow HCHOH(OPO_3)^{2-} + H_2O.$$

In this way amides, esters, and ethers are made. It is also possible to go on to make more complex small compounds and in series of reactions (shown in all biochemical textbooks and on wall charts in biochemical laboratories) to go further to amino acids, nucleotides, saccharides, and ketones, i.e. to the monomers for polymer syntheses (see Fig. 6.3). Some steps require further reactions and again and again catalysts are required. The catalysts differ, but the initial elementary components are the same for all the monomers and the process, oxidation/reduction followed by condensation, is fixed.

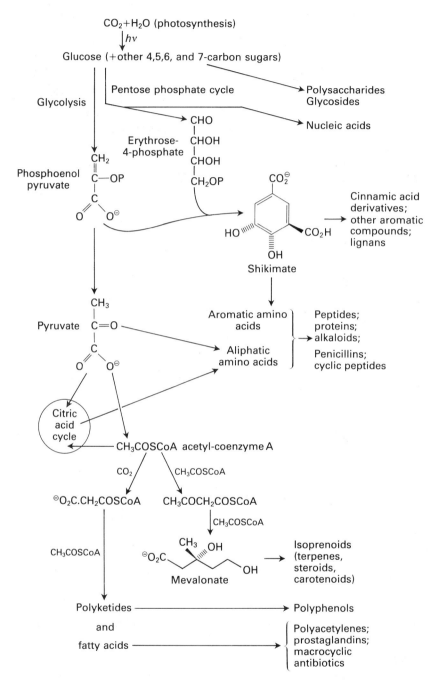

Fig. 6.3 A sketch of some of the required balanced connections in the substrate chemistry of cells. It is a revealing exercise to label the arrows with required metal ions for catalysis. (See the use of this figure in Chapter 21.)

In order to carry out these reactions in an orderly controlled way the elements H, C, N, P, S must be packaged and sent to controlled destinations. One way cells achieve differential packaging, for example, of reducing equivalents, is through the action of the freely mobile coenzyme carriers NADH, NADPH, and glutathione. NADH is involved in the degradation of small substrates in both glycolysis and the citrate cycle, while NADPH carries hydrogen for incorporation in the synthetic paths of glucogenesis and photosynthesis. The path of incorporation is further split since glutathione as well as NADPH delivers hydrogen in specified paths of reductive synthesis. As mentioned before there is a parallel here with the differential transport and use of phosphate in different NTP molecules, of iron as free ion or as haem iron, and of magnesium in chlorophyll or free magnesium ions (Table 6.4).

Now monomers, indeed substrates, occupy a curious position in that they are the source of synthesis but also very frequently the source of energy. The digestion of polymers (outside cell) was kept separate from their synthesis (inside cells) for very good reasons. The consumption of the monomers for other purposes than syntheses has also to occur inside cells and this consumption can give both energy and new intermediates. Clearly, the activities must also be separated; thus, we find oxidative degradation in mitochondria and glycolysis in the cytoplasm well separated by membranes just as was the handling of reducing equivalents well separated from synthesis by different coenzymes. The advantages of compartments also for transmitters and hormones and for stores of digestive enzymes becomes obvious. We repeat that all compartments must remain in communication with one another (Fig. 6.4).

The implication is clearly that in order to achieve balanced synthesis the relative amounts of monomers flowing in different synthetic and degradative paths must be balanced and, as for the polymers, feedback control is essential from the different monomer pathways of Fig. 6.3. However, we also require a balanced supply of chemical elements of a greater variety than those used to make monomers for synthesis.

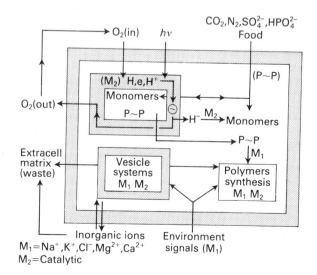

Fig. 6.4 A very simplified picture of a eukaryotic cell illustrating some of the complexity and sophistication needed to maintain a network of balanced interactions in a multicompartment cell. This figure will be used again in Chapter 21.

6.5.3 The uptake and loss of elements

As commented in previous chapters, apart from incorporation into organic molecules, a large number of inorganic elements are used in communication, control, regulation, and catalysis and many of these elements have to be obtained from outside the cell. There is then a balance utilizing many devices including synthesis of reagents and pumping into compartments but all the time the elements are held in steady states of elevated or depressed concentration levels which are critical for all functions of cells (see Chapter 21). Here we include carefully controlled salt solutions containing Na^+, K^+, Mg^{2+}, Ca^{2+}, Cl^-, and so on both inside and outside the cell, e.g. in circulating fluids. All these ions are pumped and clearly the pumps must be under feedback control and linked to one another by exchangers.

The need to make special chemicals in special places means also that their catalysts have to be positioned, but we have seen in Chapter 4 that the catalyst for small molecules has to contain activating metal ions. Clearly, to make the catalyst enzymes the metal ion transport has to be as specialized as the polymer and monomer transport, and, as for the monomers, the metals must react in the required place. Not only is this the case but it is also necessary to ensure that redox active metals, e.g. Cu and Fe, do not get into certain other spaces for they, as ions, catalyse simultaneously both oxidation and hydrolysis indiscriminately. They must not interfere with the chemistry of zinc. There have to be 'metal-free' zones as well as prescribed metal-activated zones, so that detoxification can also be managed. The reason for the intensive element selection in Table 6.3 is becoming clear; each catalytic element is identified with idiosyncratic sets of reactions in a compartment so as to keep control partly through the secondary use of the element in communication systems but partly to avoid confusion in reaction pathways.

6.5.4 The flow of energy in compounds

While building up a picture of the distribution of polymers, monomers, and of the chemical elements, we have insisted that all the polymers are unstable and have a limited kinetic stability and this is true of all the monomers too. In fact the whole biological set of organic chemicals is unstable to hydrolysis, oxidation, and diffusion to equilibrate concentration gradients, so that unless energy is constantly put into each part of the system it will degrade cell by cell, polymer by polymer, monomer by monomer, gradient by gradient. The whole biological set of chemicals is an evolved energized group with a timed life from synthetic to degradative steps connected to energy. Of necessity, the use of H, C, O, N, P, and S as elements and, indeed, of all inorganic elements is built into this covalent and kinetically regulated, energized, system. Another way of visualizing the situation is to note that the major cycles and pathways of primary metabolism are both synthetic and degradative, e.g. glycolysis, glucogenesis, and the citric acid cycle (Fig. 6.3). Degradation in one section of a pathway, giving energy, leads to components for synthesis in another to achieve balance in the production of all the many monomers and four types of polymers. It is required then that always the system retains stored *energy* which is generated largely as NTP in particular parts of space. Thus, energy produced locally must flow and this flow has to be guided (Table 6.9). It must be continuously captured and lost: light always passes eventually to heat.

Table 6.9 The uses of pyrophosphate in four coenzymes

Coenzymes	Uses
ATP	Pumps of ions and most molecules Actomyosin tension generation Many kinases. Assimilation of CO_2, N_2. Fatty acid and glycolytic paths of metabolism
GTP	Protein synthesis, tubulin growth, many kinases (G-proteins)
UTP	Synthesis of polysaccharides
CTP	Choline synthesis

The complexity of the uses of energy, the fact that the whole system is constantly multiplying and growing or repairing, that it can maintain shape or alter it, and that the several operations are timed require a continuous monitoring, feedback, communication and energy network. We have then, in every respect a flowing energized chemistry within a dynamic organization. It is this operation with which our description of the biological chemistry of the elements must come to terms. When the energy flow stops, the order and the system disintegrate and death occurs.

Energy transfer, as we showed in Chapter 5, is often in the form of NTP, phosphate transfer. Now, phosphate transfer is not involved just in syntheses of polymers, in energizing stress fibres, in pumping ions and molecules, but, through the process of phosphorylation of proteins, it brings about major controls and regulations. Indeed, some one-third of in-cell proteins are phosphorylated and dephosphorylated regularly. The kinases which bring about phosphorylation interact mainly with ATP and GTP. However, the GTP and ATP connections are distinct (Table 6.9), just as NADPH and NADH transfer H in separate paths. Thus phosphate *movement* is a fundamental part of energy transfer, communication, and synthesis. All must be in homeostatic balance.

6.5.5 Self-organization and flow patterns

The self-synthesis of an organism depends further on the mutual recognition of the polymers so as to build nuclei, ribosomes, and so on. While so doing they also pick up metal ions and organic cofactors so as to assemble the catalytic machinery of the cell. These polymers are forced into folded states by their sequences (Chapter 7) and we need to know how these sequences are related to activity and flow. We also need to observe that their surfaces come together to give the organization but that much organization has to be relatively flexible since the systems move and grow. (We shall see the importance of electrostatic interactions in Chapter 7.)

The flow we have described so far has concerned the movement of polymers, monomers, elements, and energy through aqueous phases and across membranes into compartments. As we pointed out in Chapter 2, the compartments contain organized systems due to the ways in which filamentous structures self-assemble both along membranes and perpendicu-

lar to them, so that a cell has a shape. Now when such anisotropic organization evolved, flow of material and energy became more complex and this is especially true once organisms gained control of the external solutions of cells and built cell–cell organization. The flow ceased to be random and developed into defined patterns. An obvious example is the flow of blood in the human body but, at a more microscopic level, flow will be caused by the anisotropic distribution of active systems along all membranes. Clearly, this flow becomes part of the information carrying machinery and part of the self-assembly so that the flow and the organization are an interactive whole. The growth tip of a fungus is a fine illustrative example (Fig. 6.5).

Fig. 6.5 Representation of the morphogenic changes taking place in the regenerating tip of *Acetabularia*. The sequence (a) through (e) shows the morphogenic changes from the initial state to the beginning of whorl formation. We wish to uncover how the development of the tip is related to the distribution of proteins in the membrane and of currents of ions which are known to flow at the tip, see Chapter 21.

Developing region

Region of no growth

Composition of the membrane is changing while current flow around the surface changes

(a) (b) (c) (d) (e)

We go forward therefore in the next paragraphs to look for the relationship between the fundamental *flow* of energy and the flow of material *within the self-organized system*. Energy capture, organized information and activity, like the chicken and the egg, had to develop simultaneously. We shall see that this means that 'organic' and 'inorganic' chemistry had to come together to make a living system since we have already demonstrated that organic materials on their own do not and could not have generated many of the flow systems we observe in catalysis and communication. We turn to stress some of these aspects at the risk of being repetitive. In the following section we start discussion in the opposite sense from above, i.e. from the smallest units, and build forward, in marked contrast to the start from polymers which we have just undertaken. The stress will now be on flow within observed organizations of very elementary charges. The reason for including this section is to link the material flows discussed so far to the electronic and electrolytic flow which is so fundamental to biology today.

6.6 Organization and flow

6.6.1 Patterns of electron flow: electrical potentials

The movement of electrons in energy capture and the manipulation of electronic energy need metal ions in either chemistry or biology but, while in chemistry there is a rather wide choice, the primary metal ion of biology for this purpose is iron. This choice is probably an historical accident since, although iron was readily available at the time of the origin of life, it is not so

today. Man makes electronic devices, e.g. wires, from many different metals and semi-metals, and makes hop semiconductors out of very many simple compounds, but biology cannot. As mentioned before, virtually all wires in biology are made from hop-conductor *iron proteins*. When we look at these proteins more closely we find that not all the iron is held in the same way: some is in different Fe_nS_n clusters and some in different haem complexes. The question, therefore, is not just why iron was selected to the exclusion of other metal ions but why a particular type of iron complex in a particular protein of well conserved sequence was chosen, rather than some other combination. We must always remember that a function has to be performed in a given place (and perhaps at a given time); hence, the protein has developed so as to position the element and to keep it there without exchange, as well as to modulate its use through its redox potential. In other words, energy has been used to prepare proteins which force the metal ions (or other groups) into special energized states and sites relative to the places for the absorption of light or redox reactions, so that flow of electrons is driven in a defined spatial way (see Figs 4.5 and 4.6). This direction of flow ensures that many membranes carry asymmetric charge distribution and thence an electrical potential.

These photochemical and redox electron transfer pathways must also be controlled, since otherwise all biological systems would burn; in other words, the immediate products of electron transfer must feed back so as to prevent excessive drive to reduction plus production of free O_2 by light in chloroplasts, or excessive oxidation using O_2 in mitochondria. Neither system is reversible so that excessive concentration of reducing equivalents does not lead to reduction of O_2 by combustion and to emission of light by chloroplasts, just as excessive oxidative pressure does not lead to the generation of O_2 by mitochondria. Instead, there are ways of building up back-pressure against electron flow through the establishment of membrane potentials and there are also feedback inhibitors at various levels, including the ATP-synthetases. All of these features require organized assembly (see Chapter 3 for the description of chloroplasts and mitochondria). Electron flow is connected, of course, to proton levels in the fluids surrounding the bioenergetic membranes but the apparatus involving electron flow is also controlled by levels of Cl^-, Ca^{2+}, and by phosphorylation, for example, and it must contain switches. The organization is indeed very sophisticated.

The next step in handling energy is the connection from electronic to electrolytic power sources. This is a necessity since free radical polymerization and free radical reactions in general are to be avoided.

6.6.2 Patterns of proton gradients and flow electrolytic potentials

The manipulation of electrolytic energy in space needs ions which flow readily and are stable. The obvious choices are the highly available Na^+, K^+, and Cl^- ions. The not so obvious choice at $pH = 7$ is H^+, yet, as explained in Chapters 1 and 5, H^+ is the one readily mobile cation which can be reduced to covalently bonded H and bound H^-, i.e. *connected* to electron flow. Hydrogen can carry both positive charge as protons and negative charge (electrons) in compounds and it therefore *connects electronic and electrolytic circuits. No other ion can act*

in this simple fashion. Moreover, as H^+, it binds strongly and so can bring about conformational switches, i.e. it acts in controls and transduction. (The parallel here is with ions such as Mg^{2+} and Ca^{2+} which have no redox properties.) *Hydrogen ions* are essential and there is no possible substitute for this element in this function but its flow is tightly organized and controlled and there is much more to this use of H^+ than the creation of an osmotic gradient. Overridingly this means that the *location* of proton pumps and channels is controlled, as it is for other ions, and again its routes of flow must be gated. Again, the flow of protons is *switched* on and off by the feedback from proton concentration *and* by that of other products of energy transduction which must also be correctly placed.

6.6.3 Patterns of ion gradients and flow: communications

The electrolytic power of proton gradients is used to pump in some chemicals, to reject others, and to make ATP, the first kinetically stable but energized covalent chemical association. These are necessary steps for ensuring dynamic osmotic stability, developing signalling, and promoting polymer synthesis. Protons can also be changed back to reducing power. We take these one at a time (Fig. 6.2).

The pumping of ions or molecules, X, by H^+ across membranes, e.g. by exchanges of the kinds H^+/X^+, H^+/X^-, or H^+/X, is thermodynamically possible across all membranes. This is a major biological activity and cellular systems develop internal osmotic and electrical stress (electrolytic ion potentials) partly in this way. These tensions must be controlled by homeostasis. This can best be managed by manipulation of the pumping of the stable readily available ions, such as Na^+, K^+, and Cl^-; consequently, these very elements, sodium, potassium, and chlorine, become essential components of biological stability. The fact that they have developed later into current carriers in nerves and so on is, at first, a side issue but, obviously, as these ions do not bind well and cannot be chemically transformed, they are the best electrolytes available for carrying currents. Selection between the ions here is based on straightforward chemical principles, but *the positioning and selectivity* of the channels and pumps is a critical feature due to particular proteins in a controlled organization (see Chapter 8). Cations, not anions, dominate the process because there are more unreactive cations available. Subsequent to the case of the simplest ions came the development of organic molecule exchange and signalling.

While we observed the need for both degradative and synthetic paths for the covalent molecules, monomers, and polymers, we now must understand that there is an inevitable constant dissipation and a constant rebuilding of ion gradients. This constant electrolytic activity is an absolute requirement, e.g. in brain, where we readily observe a variety of current waves at any time. It is not so noticeable in muscle and other cells, but it is there in all cells. Equivalent to the energy going into synthesis of covalent molecules is the energy going into pumping of ion gradients; this means that there is a constant electron flow to generate the proton gradients and a constant proton flow to generate ATP and then the ion gradients. All these continuous flows are a necessary part of life; much as there is 'organic' life chemistry—synthesis and degradation—so there

Circulating flows of ions in Chara cells

(Chloroplast)
hv energy $+ H_2O \longrightarrow$ Dioxygen

ATP \longleftarrow Proton gradients Reducing equivalents

(Mitochondria)

is 'inorganic' life chemistry—flow of electrons and ions. Clearly the two, the making and breaking of somewhat kinetically stable covalent bonds and the development and dissipative loss of gradients, have to be connected since once again they share the same energy source (Fig. 6.4). The first connection lies with the chemistry of hydrogen; the second lies with the initial polymerization and depolymerization of pyrophosphate from these gradients:

$$\text{Synthesis} \rightleftharpoons \text{Pyrophosphate(ATP)} \rightleftharpoons \text{Ion gradients.}$$

Between the different activities and, therefore, also between the different catalysts which govern the activities there has to be continuous, spatially organized, feedback communication. A major control role lies with the simplest cations, Ca^{2+}, Mg^{2+}, Zn^{2+}, and so on and the slightly more complicated anions largely based on phosphate. The fact that additional sophistication can arise from the use of more complex cations such as acetylcholine and adrenalin and then of a range of transmitters and hormones must not be allowed to hide the basic requirement for a communication network which is organized and continuously active, that is, *in flow*. We shall have to find how this vast variety of evolved communication networks based on ions, coenzymes, transmitters, hormones, tension, and so on is integrated.

We look at this complexity in more detail in Chapter 21 after we gain a sense of the integrated feedback circuitry which distinguishes biological from biochemical but we stress again that each biological part is locally *organized* by an interaction of the ion gradients, the organization within the membranes, and their connection to filaments and their tension.

6.6.4 Tension patterns and flow

At this stage we must remember the features observed in Chapter 3 concerning the location of filaments, vesicles, and membrane enclosures, in cells and the membrane curvature, i.e. *their imposed shapes*. It is these anisotropic constructions which decide the localities of the pumps and channels for the movements of ions and molecules. As a consequence they dictate patterns of flow of all molecular species, but patterns of flow imply patterns of concentration and gradients. Now, ion concentration gradients lead back to control mechanisms of many kinds; note that the patterns can be of H^+, Na^+, K^+, Ca^{2+}, and of phosphates, such as ATP. There is then a connection between reactions such as polymerization actin and tubulin, for example, and ion concentrations and, in turn, these growths of filaments redetermine cell shape until some balance is achieved. On the other hand, all these polymerizations are also governed by the binding and reactions (phosphorylation) of triphosphates, e.g. tubulin by GTP and actomyosin by ATP, and once again there is a feedback circuit connecting many simple species, Ca^{2+}, Mg^{2+}, Na^+, K^+, H^+, and many forms of phosphate.

The cell membrane constantly adjusts its shape so that the homeostasis is dynamic. Now this leads to a constant communication between the inside and the outside of the cell which is reflected in the shape or morphology.

6.7 The integration of activity

We have looked at living systems in two ways; in Section 6.5 we have looked at the incorporation of elements into effectively static, steady state structures, starting from polymers and monomers (C, H, N, O, S), going to specific elements (Mn, Fe, Cu, Zn, etc.), and proceeding to the creation of organized structures. All the processes described used energy. In Section 6.6 we started from the organization and looked at the flow of electrons (Fe, Cu, H), ions (Na, K, Mg, Ca), and energy (P, H) within it. We finish with the convincing impression that the essence of life is continuous change, growth, and with it death, not steady state, since it is inevitable that the living system responds to the timed modifications and development of its surrounds. Growth is then an integrated response to external as well as internal forces.

Integration requires much more than synthesis and degradation even within internal order; it requires communication, maintenance of organization during growth, and the ability to respond to the outside world. While simple elements in particular forms, free or as compounds, are necessary for the building of structure, catalysis, and energy transmission, communication networks demand quite other properties (Table 6.10), and it is essential that we see how the communication network in and around cells and organs is managed.

Table 6.10 The properties required in communication systems

Communication system	Required property
A. Message transfer (fast)	Directional guided motion: Na^+, K^+, Cl^+ in tubular structures, channels, gates, and nerve cells. Reversible
B. Message impact (no memory) (optimal speed)	Motion as above followed by binding at specific sites, release and fast recovery: Ca^{2+}, cAMP, IP_3, H^+. Reversible
C. Message impact (memory) (speed not essential. Fidelity)	'Permanent' trace giving new systems following (A) and (B). Growth of organic structure stimulated by messengers with long-term binding: hormones, phosphorylation, and heavy metal binding. Irreversible

This table should be taken in conjunction with Table 4.3.

6.8 Conclusion

6.8.1 The biological selection of elements

We have found a satisfactory way of appreciating the interrelations of the chemistry and biochemistry of elements and molecules in living systems from very detailed analytical observations, Fig. 6.6, but this leaves us still distant from an understanding of the specific functional roles of either the elements or

Source of energy (light or redox reaction) →	e (Fe,Cu,Mo) → NADH	H⁺/H → Ion gradients	ATP	Synthesis (polymerization) \| Enzyme triggering

Fig. 6.6 Interrelation of some functions of the elements in biology. Controls are largely excluded.

Redox chemistry of substrates	Energy transfer, signalling and chemomechanical transduction	Acid–base chemistry of substrates
Fixed sites	Mobile and semi-mobile elements	Fixed sites (largely)
Fe,Cu,Mn,Mo,Ni (Co),(Se),(S)	Na⁺,K⁺,Mg²⁺,Ca²⁺,H⁺,HPO₄²⁻, Cl⁻,Fe(II),Mn(II),(Zn²⁺)	Zn²⁺,Ni²⁺, Fe(III),Mn(III) Mg²⁺

the molecules within *biological* systems. A simple example illustrates the point. There is no obvious *chemical* reason why an oxidative catalyst for a given reaction should be made from copper rather than from iron. While in biology it is found, now and then, that either copper or iron can be used for a particular function, e.g. dioxygen transport or electron transport, it is also clear that, generally, copper and iron are used in quite different ways. If we are to understand the biological functions of the elements, we must look more closely at the selected roles the major 25 elements play against the background of the chemical and biochemical principles discussed in Chapters 1–5 and against the background of evolution, i.e. the way living systems have become organized. We must also look therefore at their selective association with biological polymers. In turn this demands that we give in Chapter 7 a summary account of the nature of these polymers. We shall be particularly concerned to uncover the dynamics of the polymers within structures since only through such knowledge can flow be understood. If we are correct, flow is of the essence of biological chemistry.

6.8.2 The evolution of the chemistry of life

Nobody doubts that it is the polymers of life which have evolved and we shall see how this evolution meets demands for structure, catalyses, and communication, but it is not often noticed that the utilized chemistry of the elements has also evolved coincidentally and that this evolution gives biological chemistry a historical context quite absent from normal chemistry. One major change was the advent of a predominantly oxidative atmosphere of dioxygen some 10^9 years ago (Fig. 1.11). Associated with this change and interactive with it, have been very great changes in availability and hence in the utilization of, especially, Fe, Cu, Ni, Zn, and Ca, while other metals almost disappeared from biology, e.g. V. When we say that the polymers of life have evolved, this does not imply that it is the basic use of C, H, O, and N in organic molecules which has evolved, but rather the way sequences have changed in DNA, RNA, proteins, and so on, and this is, at least in part, to take advantage of the change in the availability of the other elements. We can only see this combined evolution by inspecting the way in which different elements are used in different organisms and in different biological niches, while observing their

partner proteins which have in part evolved to match the evolving potential
the elemental chemistry. This evolution also generates novel chemistry of tk
elements. One major purpose of Chapters 8–20 is to see the value and use
each element in biology before we look again at the integrated nature of tk
activity of the elements in Chapter 21.

Further reading

This chapter is of a very general nature and is little more than an attempt t
make a brief summary of the content of the preceding chapters and of tk
central themes dealt with in textbooks of general cell biochemistry an
physiology. The only novel features are the stresses on the roles of tk
inorganic elements and on the dynamic balances of all the units, inorganic an
organic, in the energized living organism. Rather than give references here w
hope the reader will use those at the ends of the previous and followin
chapters.

7

The role of biological macromolecules and polymers

Part A Proteins and nucleic acids 163

Part B Polysaccharides and lipids 193

A Proteins and nucleic acids

7.1 Introduction 163
7.2 Protein composition and basic structure 164
7.3 The protein fold and the internal motions in solution 171
7.4 The amino-acid composition of specific proteins 173
7.5 Structural proteins and mechanical devices 178
7.6 Matching of proteins and organic and inorganic ions 178
7.7 Enzymes 180
7.8 States of metal ions in proteins 182
7.9 Summary of proteins 185
7.10 Nucleic-acid composition and outline structure 185
7.11 Metal-ion binding to polynucleotides 188
7.12 Nucleic acids and proteins 190
7.13 Polymer synthesis and degradation 191

7.1 Introduction

In the previous chapters we described the elements, the physical–chemical basis of their uptake, their movements and distribution inside and outside cells, and the chemical properties on which their functions are based. In the discussion it became clear that most biological elements, particularly the transition metals and many non-metals, have their differential effects through binding to proteins, either permanently, in catalysis, electron transfer, etc., or transitorily, in activation, control, or triggering. Even sodium and potassium pass through protein channels in membranes. For this reason, proteins may be considered, from the point of view of inorganic biological chemistry, to be the most important biological macromolecules. Although we have already stressed the co-operative action of metals and proteins and referred to the need

for mutual adjustments on binding, we did not consider in any detail the dynamic protein structures which have developed to fulfil the various functions they perform and we limited our discussion to the several (static) conditions required for effective and selective metal binding by proteins. The importance of adjustment or conformational mobility is becoming more and more evident and justifies the inclusion in this text of a large part of a chapter dedicated to the analysis of the structure and dynamics of proteins, an analysis in sufficient depth to enable the understanding of their functional specialization even when metals are not involved, at least directly. While stressing the role of these polymers, other biological macromolecules, nucleic acids, polysaccharides, and membrane lipids, must not be overlooked. It is true that they hold a less prominent position in the context of the biochemistry of the elements due to the fact that their functions are of lower importance or are less known. Nevertheless, these biopolymers are overwhelmingly important biological molecules and an effort has been made later in this chapter to consider their role in an analogous perspective. We have previously mentioned the role of *polysaccharides*, especially combined with proteins, in assisting the recognition of the cell environment and we have also discussed in some detail how the bi-layer *lipid* biological membranes divide space, restrict the flow of materials into and out of cells and cell vesicles, and assist the organization of chains of metalloproteins and enzymes embedded in them (Chapter 3). The other class of polymers, *nucleic acids*, are, of course, determinant of the synthesis of proteins partly through their interaction with metals, particularly Mg^{2+}. There are also control proteins which allow the influence of metal ions to reach the DNA and RNA through their structures (regulatory systems; see Section 7.11). These aspects are well enough known to show that (inorganic) chemists would be well-advised to pay attention to all these biopolymers and to take into account also their particular structural and dynamical features.

There is a further reason for an examination of all these biological polymers. It is more and more necessary to see cellular or organism activity in some integrated way rather than as isolated units. This has become apparent in our discussion of the functions of the elements. The integration of the elemental activities depends on organization which is based on the biological polymers considered in an interactive manner with the energized distribution of the elements. In this chapter we shall not be concerned just with isolated proteins, polynucleotides, lipids, and polysaccharides, but we shall try to describe their total functional potential and the integration with the chemistry of the elements. We shall start each section on proteins, polynucleotides, polysaccharides, and lipids, respectively, with a simple description of isolated polymers and then we shall look at their roles in organized systems.

7.2 Protein composition and basic structure

As is well known, proteins are linear chains of α-amino acids connected by peptide bonds that usually exceed 50 units in length. In fact, *enzymes* are usually above 100 amino acids in length. Shorter chains are referred to as polypeptides and several of these have important control (hormonal)

functions. There are some 20 natural amino acids (see Table 7.1, which also shows their potential for binding metal ions), and some modified versions of them also appear through, for example, methylation, carboxylation, or hydroxylation (Table 7.2). The amino acids can be divided roughly into four classes as shown in Table 7.3, and their relative presence in a protein largely conditions the structure and dynamics of that protein.

The amino acids are incorporated into linear sequences in various proteins and the vast array of proteins arises from the huge number of possible combinations. The exact sequence of the amino acids corresponds to the 'primary structure' of the corresponding protein and this sequence largely determines all the other forms of internal association which lead to the so-called 'secondary', 'tertiary', and 'quaternary' structures of proteins. The 'secondary' structure (helices (A), parallel (B), and anti-parallel (C) β-sheets,

(A)

(B)

(C)

and β-bends) arise from hydrogen bonding between the peptide links, i.e. the –CO–NH– units. The particular structure which arises is related to the nature of the amino-acid sidechains. The 'tertiary' structures arise from further folding due to interactions between sidechains, e.g. hydrophobic or electrostatic interactions or formation of covalent disulphide –S–S–bridges, and are, of course, specifically dependent upon the nature of the sidechains (in later sections we will see that other types of interactions, e.g. bonding to cofactors, metal ions, etc., can also lead to specific tertiary structures). The 'quaternary' structure refers to the form of association between several complete protein subunits, usually by specific surface interactions. Figures 7.1–7.3 exemplify most of the types of structures, but see Fig. 7.13 for a β-sheet barrel.

These structures correspond to a static picture of the proteins, i.e. to a certain particular conformation for each one, but we are further interested in conformational changes and structural dynamics since it is through these processes and properties that each protein acts during its function, whether this is catalysis, control, triggering, or some other. A simple list of some important classes of mobility is given in Table 7.4.

To understand the corresponding relationship to functional mechanisms it is necessary first to clarify the meaning of 'conformational change' in the context of protein dynamics and for this we must realize that proteins are not rigid bodies. Indeed, even the most structured proteins show some mobility and there are others that are indeed very mobile. As a matter of fact, the range of 'mobility' (a concept in which we include all the rate processes within

Table 7.1 The natural protein amino acids [*]

Name	3-letter symbol	1-letter symbol	Sidechain	Character	Metal-binding
Aspartic acid	Asp	D	$HOOC-CH_2-CH\begin{smallmatrix}COO^-\\NH_3^+\end{smallmatrix}$	Acid	M^{2+}, M^{3+}
Glutamic acid	Glu	E	$HOOC-CH_2-CH_2-CH\begin{smallmatrix}COO^-\\NH_3^+\end{smallmatrix}$	Acid	M^{2+}
Tyrosine	Tyr	Y	$H-O-\bigcirc-CH_2-CH\begin{smallmatrix}COO^-\\NH_3^+\end{smallmatrix}$	Neutral	Mn^{3+}, Fe^{3+}
Alanine	Ala	A	$CH_3-CH\begin{smallmatrix}COO^-\\NH_3^+\end{smallmatrix}$	Neutral	—
Asparagine	Asn	N	$H_2N-CO-CH_2-CH\begin{smallmatrix}COO^-\\NH_3^+\end{smallmatrix}$	Neutral	—
Cysteine	Cys	C	$HS-CH_2-CH\begin{smallmatrix}COO^-\\NH_3^+\end{smallmatrix}$	Neutral	Zn, Cu Fe, Ni Mo
Glutamine	Gln	Q	$H_2N-CO-CH_2-CH_2-CH\begin{smallmatrix}COO^-\\NH_3^+\end{smallmatrix}$	Neutral	—
Serine	Ser	S	$HO-CH_2-CH\begin{smallmatrix}COO^-\\NH_3^+\end{smallmatrix}$	Neutral	M^+?
Threonine	Thr	T	$CH_3-CH(OH)-CH\begin{smallmatrix}COO^-\\NH_3^+\end{smallmatrix}$	Neutral	M^+?
Histidine	His	H	$\begin{smallmatrix}HC=C-CH_2-CH\\^+HN\quad NH\\C\\H\end{smallmatrix}\begin{smallmatrix}COO^-\\NH_3^+\end{smallmatrix}$	Neutral Basic	Zn, Cu Mn, Fe, Ni †
Arginine	Arg	R	$H_2N-C(=NH_2^+)-NH-CH_2-CH_2-CH_2-CH\begin{smallmatrix}COO^-\\NH_3^+\end{smallmatrix}$	Basic	†
Lysine	Lys	K	$H_3\overset{+}{N}-CH_2-CH_2-CH_2-CH_2-CH\begin{smallmatrix}COO^-\\NH_3^+\end{smallmatrix}$	Basic	†
Glycine	Gly	G	$H-CH\begin{smallmatrix}COO^-\\NH_3^+\end{smallmatrix}$	Non-polar hydrophobic	—
Isoleucine	Ile	I	$CH_3-CH_2-CH(CH_3)-CH\begin{smallmatrix}COO^-\\NH_3^+\end{smallmatrix}$	Non-polar hydrophobic	—

Table 7.1 *continued*

Name	3-letter symbol	1-letter symbol	Sidechain	Character	Metal-binding
Leucine	Leu	L	CH_3—CH—CH_2—CH, CH_3, COO^-, NH_3^+	Non-polar hydrophobic	—
Methionine	Met	M	CH_3—S—CH_2—CH_2—CH, COO^-, NH_3^+	Non-polar hydrophobic	Cu, Fe
Phenylalanine	Phe	F	⬡—CH_2—CH, COO^-, NH_3^+	Non-polar hydrophobic	—
Proline	Pro	P	CH_2, H_2C, H_2C, NH_2^+, C, COO^-, H	Non-polar hydrophobic	—
Tryptophan	Trp	W	indole—CH_2—CH, COO^-, NH_3^+	Non-polar hydrophobic	—
Valine	Val	V	CH_3—CH—CH, CH_3, COO^-, NH_3^+	Non-polar hydrophobic	—

NB. Carbonyl groups of main- or sidechain may bind M.
* Ionic forms, predominant under physiological conditions at pH 6–7. † Anion-binding.

Table 7.2 Some modified amino acids

Amino acid	Metal-binding	Metal involvement in synthesis
γ-Carboxy-glutamate (Gla)	Ca^{2+}	None
Hydroxy-aspartate	Ca^{2+}	Fe?
Hydroxyproline	(Bone)	Fe
Hydroxylysine	(Bone)	Fe, Cu
(Methyl)$_n$lysine	None	None

NB. Several amino acids can be phosphorylated and some can be glycoxylated.

Table 7.3 Classes of amino acids

Charged hydrophilic	Intermediate (hydrophilic)	Hydrophobic	'Structural'
Lysine	Serine	Leucine, isoleucine	Glycine
Arginine	Histidine	Valine, alanine	Proline
Glutamate	Glutamine	Phenylalanine	
Aspartate	Asparagine	Threonine, tyrosine	
(Histidine$^+$)		Tryptophan, (proline),	
		Cysteine, methionine	

proteins) is such that at one extreme a change of conformation may mean something like a phase change, the so-called R (Relaxed) ⇆ T (Tense) switch, or a change in molecular structure like a cis ⇌ trans isomerization (relatively slow processes), while at the other extreme it can only be described as a fast equilibration over many conformers of similar energy or, in terms of statistical thermodynamics, a variation in the population of states within a large ensemble, whose appraisal requires the calculation of the corresponding partition functions.

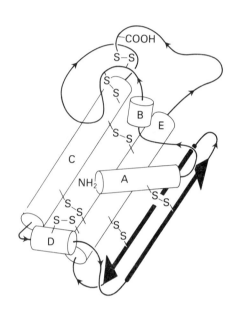

Fig. 7.1 An outline structure of a protein, here phospholipase A_2, showing α-helical runs of amino acids as cylinders, A to E, and anti-parallel β-sheet runs as heavy black arrows. Disulphide cross-links are shown, and runs of no secondary structure appear as thin lines. The structure is relatively immobile, and binds calcium in a constrained loop (see Chapter 10). (Reproduced with permission from Professor J. Drenth.)

Table 7.4 Mobility in proteins

Unit of mobility	Example
Sidechain	
Aromatic ring	Flipping of ring
	Flapping of ring
Aliphatic chain	Rotation about several bonds possible
Aliphatic ring	Cis/trans isomerism (proline)
Segment	Motion within β-bends
	Motion of termini
	Concerted motion of helices
	Partial unfolding
Whole protein	Unfolding/folding
	Rotational tumbling
	Diffusion free or on a membrane surface or on DNA, etc.

Figure 7.4 exemplifies the two extreme situations and an intermediate one. In Fig. 7.4(a) there is only one ground state (free energy minimum) for the ordered (native) and one for the disordered (denatured) form; in Fig. 7.4(c) there is no sharp distinction between ordered and disordered forms; in Fig. 7.4(b) there are several conformations associated both with the partly ordered and disordered forms which are clearly separated by energy barriers but differ little in energy. Binding of molecules or metal ions can switch the fold which is most stable.

It should be noted that even when the change of conformation is (apparently) just from one conformation to another, as in the case of Fig. 7.4(a), corresponding to a very structured and rigid protein, it is highly improbable that the change is a single-step process. The probable nature of such a readjustment is that an initial change at one point in the structure spreads through small steps into the protein to other distant regions. This has been confirmed by many studies of mobility either of the protein backbone or its sidechains, using various techniques, with particular emphasis on nuclear magnetic resonance (NMR). One of the most interesting results of these studies was the establishment of definite correlations between the conforma-

Fig. 7.2 The structure of a simple helical protein, troponin C and see calmodulin (Chapter 10). Note the absence of any extensive β-structure and that the helices, A to G, are relatively mobile. The calcium sites I to IV have mobility, but in crystals only the bottom two are occupied. (After M. W. G. James and C. E. Bugg.)

Fig. 7.3 (a) The subunit of haemoglobin. (b) Quaternary structure of haemoglobin from four multihelical peptide chains ($\alpha_2\beta_2$). (After M. F. Perutz.) The structure can undergo considerable conformational changes. Note that it contains no β-sheet.

(a)

(b)

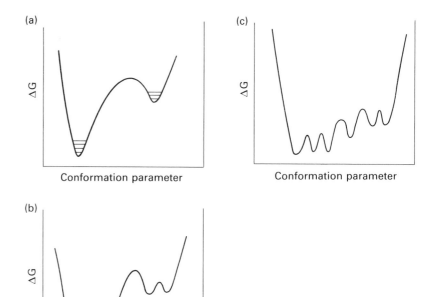

Fig. 7.4 The relative values of the free energy of different folds, ΔG, and its dependence on conformation for different theoretical classes of protein. In (a) there is only one deep free energy minimum for the ordered (native) and one for the disordered (denatured) form. In (b) local motion within these two states gives rise to several conformations associated with the ordered and disordered forms. In (c) there is no sharp distinction between ordered and disordered forms. The abscissae are labelled 'conformation parameter' to indicate that a change from one state to another is made by altering the fold. The horizontal lines in (a) illustrate states which are associated with motions of sidechains. Examples are
(a) phospholipase A_2,
(b) calmodulin,
(c) metallothionein.

Table 7.5 Protein conformational motions related to known or possible functions

Functional property	Protein motion
Ion movement in channel	Very fast local bond movement around an ion 'site'
Electron transfer	Fast relaxation of conformation around site: small changes
Enzyme catalysis	Movement of sidechains with substrate transformation: fast binding adjustments (induced fit)
Allosteric co-operativity	Segmental shift, especially of helices: relatively slow
Triggering by ions	Segmental shifts, especially of helices: intermediate rate
Binding of surfaces	Adjustment of surface groups: fast
Two-dimensional diffusion on membranes or DNA	Adjustment of surface groups, e.g. lysine: fast, helix adjustments: intermediate or slow
Folding	The whole structure is dependent on metal ion binding in some cases: relatively slow
Covalent attachment	Phosphorylation is particularly important: intermediate rate of local change

Fast here implies $>10^5$ s^{-1}, intermediate 10 to 10^5 s^{-1}, and slow <10 s^{-1}.

tion motions of several proteins and their known or possible functions (see Table 7.5 for various examples). The number and nature of the different cases studied is sufficient to confirm that not only structure (a static concept) but also conformational changes and their rates, i.e. structural dynamics, are essential data to understand fully protein function. The discussion can be extended from the highly structured proteins all the way to random coil proteins to which we return.

7.3 The protein fold and the internal motions in solution

Our knowledge of protein folds comes mainly from X-ray diffraction studies of protein crystals, but it has been shown in recent years for several globular proteins that the main chain fold seen in the crystals is very similar to that seen in solution using NMR methods. Very recently, the essential structures of proteins up to 12 000 mol. wt. have been completely determined in solution. Some uncertainty exists with regard to the dispositions of sidechains, even in those proteins which form crystals (and many will not), and fluctuations are found to occur about one determined average structure, forming a family with related conformations of sidechains. In crystals these can be measured by crystallographic B-factors, which in effect measure vibrational/rotational motion, but it is clear that there is much more mobility in solution.

The mobility of the protein chain in solution increases continuously with the temperature for all proteins, but the rate of this increase varies from case to case. It can also vary within a protein from segment to segment and is influenced by the pH, ionic strength, and, of course, by binding to other molecules or ions. Naturally, the active sites of enzymes must have relatively low mobility, since they must recognize a specific substrate, and the surfaces in contact with water must be much more mobile than the interior due to the interactions with mobile solvent and both inorganic and organic ions. It is usually the case that enzyme sites are pockets, concave surfaces, where, naturally enough, conformational changes are much more limited than on the open convex surfaces. Limited mobility is an aid to specific recognition, visualized in the lock and key concept, but mobility is required to aid speed of recognition, hand in glove fitting.

The influence of the temperature is easy to understand, but it may take different forms as illustrated in Fig. 7.5. In this figure two classes of proteins are considered: for one of them, A, there is a small increase in mobility (corresponding to a decrease in order) until a large sudden co-operative change is observed and the protein is denatured. This behaviour is typical of globular proteins, e.g. lysozyme (see also Fig. 7.4(a)). For the second protein, B, there is a gradual increase in mobility but no sharp change is observed at any temperature. Calmodulin, free from calcium, is an example of this type of behaviour. This class of protein is obviously represented at the opposite

Fig. 7.5 The variation of order with temperature in two classes of protein, where order is defined as unity for a protein which occupies only one, the most stable, conformation and is zero for a random coil. In class A there is increasing but very local disorder with temperature until the onset of melting. In class B, disorder increases continuously and is more widespread at all low temperatures. A third class of protein (not shown) will have very little order at all temperatures of biological interest.

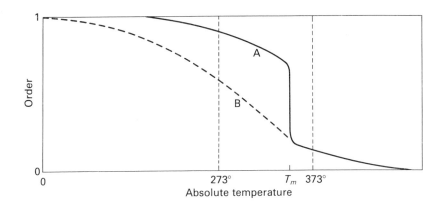

Table 7.6 Some proteins of little tertiary structure studied by NMR

Protein	Characteristics in solution
Chromogranin A	Loose cross-linking net
Calcium-binding proteins	Loosely structured proteins especially at $T > 50°C$ with marked changes on binding calcium
Glycophorin	Membrane protein with little or no tertiary structure in solution
Phosvitin and tooth phosphoproteins	Structural protein with no tertiary fold
Histones	Bind to DNA; in solution some segments are not folded
Lipoxygenase-I	Region remote from active site behaves as random coil
Myelin sheath protein	Poorly structured in aqueous solution
Glucagon	Very nearly a random coil in solution
Metallothionein	No fold as apoprotein: folded with metals
Osteocalcin	Little or no fold in the absence of calcium
Cytochrome c	Little or no fold in the absence of haem

extreme by Fig. 7.4(c). Finally, there is a group of proteins which does not fold at room temperature (Table 7.6).

As to the sidechain groups of proteins we expect that there will be rapid rotation of all methyl groups, flipping rotation of aromatic and of some aliphatic groups, flapping motions of groups such as proline and coenzymes, and swinging motions of some larger groups such as tryptophan (see Fig. 7.6).

Fig. 7.6 Some internal motions of protein sidechains.

Fig. 7.7 The fold of cytochrome *c* showing the haem (black rectangle) linked by two thioether bridges (S′) into the chain. The shaded residues have very restricted motion while the hatched residues are more mobile (see also Fig. 4.4).

Naturally, the motions of all these groups will be more or less limited by the folding of the protein and their location in the chain; accordingly, the motion activation energies vary from 5 to over 100 kJ/mol^{-1} (Fig. 7.7).

We are led to an important conclusion. *The state of proteins and a biological assembly of proteins and other polymers is dependent on the coexisting physical and chemical conditions: there is no fixed structure.* The degree of rigidity (fold) is only partly determined by amino-acid composition.

7.4 The amino-acid composition of specific proteins

The amino acids can be divided in four classes as shown in Table 7.3 and it is to be expected that the relative percentages of each group in each protein, particularly of the charged (hydrophilic) and of the hydrophobic amino acids, will have a profound influence on its properties. Indeed, a high percentage of charged amino acids is likely to lower the stability of folded structures, increase hydration, and increase the mobility of the proteins (which can be further increased if extra hydrophilic chains are incorporated into the proteins through glycosylation). Thus, we may suppose generally that there will be a relationship between the protein-chain mobility and the percentage of hydrophilic groups, but we must also ask about those proteins which contain large numbers of oppositely charged amino acids and are virtually neutral—will they fold tightly and be rigid or will they be more mobile? A high relative percentage of hydrophobic amino acids will, of course, have an influence opposite to that of charged amino acids and one expects to find this in the most structured proteins. However, should such proteins become embedded in membranes, the variety of states open to them may well change dramatically. Again binding to metal ions may constrain mobility and protect proteins from

hydrolysis. It is clear that much of evolution to a function is dependent on the development of protein composition and sequences.

The ratio of hydrophobic to charged amino acids may therefore be examined as a first approximation to judge the character of the proteins in water. Table 7.7 gives the amino acid composition of a series of proteins from the well structured globular enzymes at the top of the table to the non-structured, almost random coil, control or structural proteins at the bottom, while Fig. 7.8 shows the distribution of those proteins with different ratios of hydrophobic to charged amino acids. The separation is clear at the extremes: to the right one finds the less mobile; to the left the most mobile proteins. Enzymes are, accordingly, placed to the right, whereas most control proteins are to the left side. Inhibitor proteins fall together with enzymes; they are rigid blockers. Some proteins, however, appear in unexpected positions, e.g. the cytochromes, which have a content of charged amino acids apparently higher than expected for the function they perform and are located at the left side of Fig. 7.8.

The case of cytochrome c (Fig. 7.7), is common to other proteins that strongly bind some cofactor such as haem, flavin, or even some sterols. Without the cofactor they may be only loosely folded, but when cofactors bind they become tightly folded, i.e. the cofactor acts as a strong folding agent and may be considered as equivalent to several hydrophobic groups contributing to the shaping of hydrophobic platforms within the protein structure. Folding will be even tighter when stabilized by chemical cross-links, as happens with some haem proteins such as cytochrome c (Fig. 7.7). It may be that fitting large units inside proteins requires a mobile sequence.

Now it must be noted that proteins are not uniform in charge or hydrophobic group distribution and may have some segments that are

Table 7.7 Amino-acid composition (per cent) of some proteins and corresponding hydrophobic/charged amino-acid ratios

Protein	Charged (C)	Hydrophobic (H)	Ratio H/C	Pro+gly
Elastase (pig)	10	53	5.3	13
Trypsinogen (cattle)	13	47	4.3	15
Papain (papaya)	16	52	3.2	18
Ribonuclease (pig)	20	48	2.4	9
Alcohol dehydrogenase (horse)	22	54	2.5	17
Lysozyme (man)	23	51	2.2	10
Immunoglobulin, kappa chain	19	51	2.6	11
Collagen (cattle) a-l-chain	13	46	3.5	54
Lac repressor (E. coli)	18	54	3.0	10
Azurin (P. fluorescens)	21	50	2.4	12
Flavodoxin (P. elsdenii)	22	52	2.5	15
Myoglobin (horse)	27	46	1.7	12
Cytochrome c (man)*	30	46	1.5	16
Cytochrome c_3 (D. gigas)*	27	43	1.6	14
Myelin membrane protein*	25	38	1.5	22
Histone H.2 (trout)*	27	48	1.7	14
Parvalbumin (hake)*	31	50	1.6	11
Ribosomal protein (E. coli)*	33	44	1.3	10
Troponin-C (rabbit)*	40	42	1.1	9
Chromogranin A*	~44	41	~0.9	10

* Proteins which are partially or not folded unless bound by cofactor.

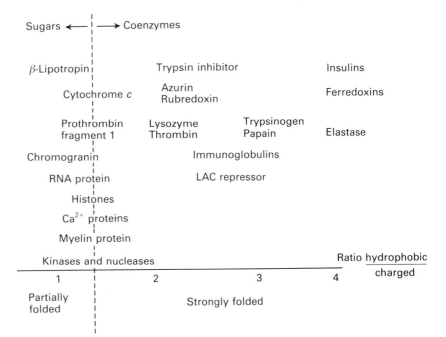

Fig. 7.8 The distribution of proteins with different ratios of hydrophobic to charged amino acids. Note the division between most enzymes, to the right, and most structural or control proteins, to the left. Proteins to the right of the dotted line have much lower mobility than those to the left.

predominantly hydrophobic and others that are hydrophilic, which merely states that sequence as well as composition is important. Structurally, such proteins are expected to have two kinds of behaviour: the hydrophobic parts of the sequences can fold, and the folded part may well be an enzyme, leaving a long and more random external segment. Kinases are bi-lobed and have highly charged, more mobile, central linker regions, for example, and these features may explain the apparent exceptional behaviour to the general distribution of enzymes in Fig. 7.8. Since their primary effect is to transfer phosphate from ATP, kinases must bind not only this species but also substrate and cations such as Mg^{2+} (see Fig. 9.3). Conformational control could well be as represented in Fig. 7.9(a) in which a V-shaped bi-lobed structure can have a hinge-bending action on substrate binding (see Fig. 7.9(b) for details). This movement brings together the two substrates on sites X and Y and the catalytic group Z, thus promoting the reaction between X and Y; the original conformation is re-established after reaction and the products are expelled. (Hinge bending of this kind is a functional necessity if reagents have to be protected during reaction.) Another example of this hinge-bending movement in catalysis is that of cytochrome P-450, when the substrates, some organic molecule RH, e.g. a sterol, and O_2 (and the catalytic group is the haem) are brought together through closing of the V-shaped enzyme molecule. The products, H_2O and ROH, are expelled after the reaction, and the V-shaped structure opens again in a redox cycle for binding a new set of centres of the substrates (see Fig. 7.9(c)).

Of course, these considerations apply in aqueous solution and other factors must be considered in lipid bilayers. Sequences of some 20 hydrophobic amino acids are usually associated with helices inserted in membranes and may alternate with hydrophilic sequences making for channels of helices in the

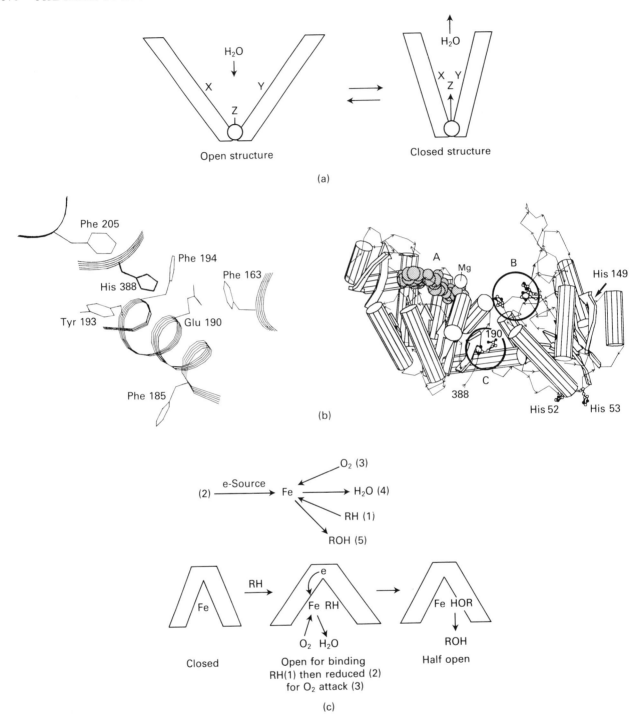

Fig. 7.9 (a) A schematic representation of kinases showing the V-shaped structures before substrate binding and the hinge action on substrate binding. The latter brings together the two substrates on site X and Y and the catalytic group Z. Water is expelled in the closed form. Much conformational mobility is required in such an enzyme near the hinge. (b) The open structure of phosphoglycerate kinase (right) showing ATP.Mg (shaded) binding on one β-sheet domain at A; the binding site for 3-phosphoglycerate at circle B with three histidines on the other β-domain; and the looser hinge region at circle C. C is shown in more detail to the left of (b). (After C. C. F. Blake and H. C. Watson.) Helices carrying Tyr 190 and His 52 are part of the more mobile hinge region, C. (c) The proposed opening and shutting of P-450. The enzyme is not cross-linked and is helical. Reaction occurs in the half-shut state and via a further open state when product is lost. The reaction step order, numbered, is shown above.

bilayers (Chapter 8). Segmental mobility in membranes will require hydrophobic contacts between helices and the functions of charged amino acids will now be associated with active sites for ion-binding, e.g. H^+, Na^+, Ca^{2+}.

We must now enquire about the role of the 'structural' amino acids, glycine and proline, since those of intermediate character need no particular comment. Most proteins in Table 7.7 contain 9–17 per cent glycine plus proline and these amino acids frequently occur at turns. A much higher glycine plus proline content is found in collagen, and proteins with such high ratios of the two amino acids are often termed 'collagen-like'; they have little tertiary structure and form special quaternary structures, e.g. long multiple interwoven helices (Fig. 7.10). In collagen there is incorporated the hydroxyproline and hydroxylysine of Table 7.2, and the latter is made into cross-links by the action of metallo-enzymes. Such filaments are not enzymes, not control devices, but structural elements inside and outside cells. Smaller groups of prolines now next to residues larger than glycine, even such as alanine or even an acidic amino acid, do not allow triple helix formation. In this case the effect of a series of prolines with other amino acids can be to create a somewhat mobile multijointed rod. In the rod a very limited number of conformers are attainable so that it can act as a swinging arm for the transfer of chemicals. Indeed, this is the way in which fatty acid synthetases are presumed to work as one $-CH_2CO-$ unit after another is added to a growing chain. There are much more sophisticated ways of analysing protein sequences than those given here. The difficulty in using them is that they are designed to show what type of fold a protein should have as opposed to whether it folds or not. It is very important, of course, to know if a protein is a simple set of helices or is composed of a β-sheet or a β-sheet linked by helices. Some amino acids, e.g. aspartic acid, favour sheets while others (usually larger), e.g. glutamic acid and lysine, favour helices. There is a general suggestion that the value of α-helical

Fig. 7.10 Model of the collagen triple helix with the structure of the intermolecular H-bonds. Three left-handed single strands wind themselves around one another in a right-handed spiral. For steric reasons every third residue must be glycine.

proteins lies in their segmental rod-like motion, i.e. in the relay of motion, while the value of β-sheets (with helices) lies in their ability to define a receptor or donor site very precisely. The latter are generally the basis of enzymes. Notice that increase in selectivity must go with decreased mobility and slower response.

7.5 Structural proteins and mechanical devices

Because of their lack of a precisely defined function those proteins that are not enzymes are least discussed. In fact, through their mechanical, not chemical, properties they control cell shape, contractile systems, cell division, and so on. Frequently, these proteins are made from single- or multiple-strand helices. They extend as intricate networks both in the body of the cytoplasm, e.g. actomyosin and other filaments, and under the cell membrane (Fig. 3.5). They may pass through the membrane in places. We leave aside the contact with the outside matrices until we have described polysaccharides. Inside cells they contact and hold in place organelles, vesicles, and the nucleus. As a consequence of the composition and nature of these proteins, different cell types have different shapes and different roles. Some cells have multiple long protruding villi which move easily, e.g. epithelial cells, some are elongated and relatively rigid, e.g. muscle cells, and some have discoid shape, e.g. red blood cells. These proteins isolated on their own do not show their full functional value since: (1) they connect to one another; (2) in the cell they are in energized 'stretched' states; and (3) they are controlled by the exact ionic medium. It is through these proteins that there is mechanical communication between in-cell, in-membrane, and external events (see Figs 7.25 and 7.30). They usually have both anion, e.g. ATP, and cation, Mg^{2+} and Ca^{2+}, binding sites.

7.6 Matching of proteins and organic and inorganic ions

So far the examination of proteins in terms of their amino-acid composition indicates that there are several major factors affecting the ability to maintain a fold: one is the ratio of hydrophobic to charged amino acids; another is the percentage of glycine plus proline; a third is the local sequence; and a fourth is the ability to form helices or sheets. The problem of the theory of the fold is clearly very complicated and has defied accurate analysis to date. Naturally, the possiblity of cross-linking, e.g. through –S–S– bridges, is also a major factor to consider that depends upon the amino-acid composition (Table 7.8). Not directly dependent upon the amino-acid composition, although related to it, one must also consider the possible effects of the interaction of opposite charges in a buried hydrophobic region and, particularly important, binding to ions, prosthetic groups, and other cofactors, which also give cross-linking and may lead to the folding of even heavily charged proteins (Table 7.9). We shall see that calcium and, to a lesser degree, zinc are extremely important in controlling and stabilizing folds.

Table 7.8 Some –S–S– bridged proteins and their mobilities

Protein	Number of bridges	Mobility
Lysozyme	4	Relatively stable fold, $\alpha\beta$ type; high internal mobility; hinge motion
Phospholipase A_2	7	Stable fold, $\alpha\beta$ type; high internal mobility
Prothrombin kringle	2	More mobile fold; higher internal mobility
Toxins EGF-peptide proteinase inhibitors	4	Restricted internal mobility, β-type

NB. In all cases loops are the most mobile regions.

Table 7.9 Metal cross-linking agents

Metal	Protein	Cross-link
Zn	Alcohol dehydrogenase	$(RS^-)_4$
Zn	Carbonic anhydrase	$(imid)_3$
Fe	Rubredoxin	$(RS^-)_4$
FeS_2Fe	Ferredoxin	$(RS^-)_4$
Fe_4S_4	Ferredoxin	$(RS^-)_4$
Fe	Cytochrome c	(imid) (thioether)
Cu	Azurin	(imid) (RS^-) (thioether)
Ca	Parvalbumin	$(-COO^-)_4$

At the same time, as the charges in the sequence of the proteins interact with one another internally and with salts, on the surface of the protein they interact with other proteins, RNA, DNA, membranes, and polysaccharides. Where these polymer–polymer interactions are electrostatic they are very sensitive to salts, either through general salt effects or through specific ion-binding. This means that many functions of proteins can be regulated by altering general salt levels or by changing the concentration of particular ions. Biological cells, e.g. nerves or muscle, are then capable of regulation by dissipating controlled ion gradients, directly or via membrane potentials, especially by control of the flow of K^+, Na^+, H^+, Ca^{2+}, and Mg^{2+} amongst cations, and of Cl^-, HCO_3^-, $HOPO_3^{2-}$, and OH^- amongst anions. It is becoming increasingly clear that these ions act by changing the relative conformation of the mobile proteins. Interestingly, the cations are found to be in competition with polyamines or polyamine-like surfaces built from small basic proteins, e.g. histones. These interactions are very important in nuclei (DNA) and in ribosomes (RNA). We note, in particular, the competition between many different charged species which must exist at the surface of membranes, at several points along highly charged polymers such as DNA and RNA, and on the highly charged structural proteins. It is inherent in this interaction that rolling contacts between the polymers rather than exact fitting of rigid shapes are made to allow rapidity of structural change.

Although the roles of calcium are frequently stressed, the importance of a variety of other ions, organic and inorganic, must not be overlooked. Many effectors, transmitters or hormones, are small, highly mobile, *organic* molecules with charged end-groups, e.g. acetylcholine, γ-aminobutyric acid, adrenalin, cAMP, etc., and the binding of these small molecules is often extremely strong. The same can be said for various drugs and toxins, e.g. chlorpromazine and the neurotoxin from Cape cobra, that also have charged centres. It would seem that, especially through the use of organic cations and anions with particular shapes, biological systems have at their disposal subtle ways of adjusting charge–charge interactions and, through these, various metabolic processes, but very frequently these binding reactions affect or are affected by inorganic ions. *Since all the inorganic ion concentrations are energized (Chapter 5) so are the states of all proteins in cells.*

7.7 Enzymes

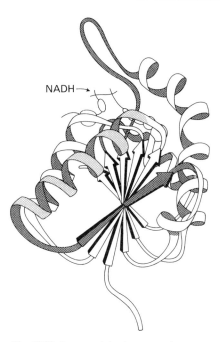

Enzymes generally have the following features: (1) they select for binding and action from all the molecules in their own compartment down to the level of one type of molecule in a thousand; (2) they act as catalysts often giving catalytic factors of as much as 10^{10}; (3) their surfaces are prochiral. While these are general rules, several enzymes, especially digestive enzymes, are much less selective. Given the above discussion of mobility and selectivity opposite a variety of structures we cannot be surprised that most enzyme sites are in concave zones on the surface (grooves) of folds based on β-sheets since this is an ideal way in which to develop selective and strong binding, i.e. ensure recognition (Fig. 7.11).

In this chapter we do not attempt to explain the action of any particular enzyme but wish to draw attention to the general problem of catalysis by folded proteins, especially those containing metal ions. We refer to metal ions not just because they fit the general content of the book but because by their nature they allow a more detailed examination of the activity of a reaction centre, using a huge variety of spectroscopic tools, than is possible for enzymes lacking metal ions. We can see the nature of the complexity through the following simple description.

The substrate or substrates are bound and in this process a large number of interactions come into play. The semi-mobile enzyme surface is adjusted to make particular long- and short-range bindings but there will also be repulsive energy changes and changes in hydration which will result in considerable entropy alterations involving the enzyme itself and the substrates. Going forward from the binding step, which may have distorted the substrate(s), toward the transition state, i.e. towards product(s) (Haldane–Pauling postulates), we need to consider the attacking groups which will form the intermediates (Fig. 7.12 and see Fig. 4.1). In hydrolytic reactions (acid/base)

Fig. 7.11 Lactate dehydrogenase has an NADH-binding domain which is shown as a β-structure with connecting helices, after M. G. Rossman.

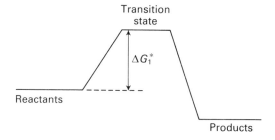

Fig. 7.12 Above, the normal activation free energy, ΔG_1^*, in a diagram for a one-step reaction (see Chapter 4). Underneath is the catalysed path for a normal catalyst with reduced activation free energy, ΔG_2^* ($<\Delta G_1^*$), and the path for an entatic catalyst, ΔG_3^*, which shows that the catalyst site is 'energized' at all stages of the reaction relative to a normal catalyst by the strain energy (X) imposed by the protein fold on the catalytic groups.

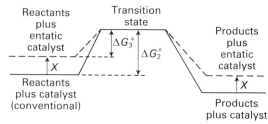

these are usually polar centres such as basic RS^- or RO^- (either protonated or not) or acid proton-carrying centres. A proton transfer from enzyme to substrate or the reverse is a very usual step in such a reaction path and it is often possible to analyse but difficult to describe the intermediates. It has been observed that the active site-attacking groups are usually more aggressive than the common run of RS^- or RO^- sidechains or protons in proteins which are not at active sites. To explain this a series of reasons has been invoked involving solvation, energy distribution of bonds, and so on (see Section 7.8.2). Even using crystallographic structures it has proved to be very difficult to give precise details of these activation steps. Much as in solution reaction kinetics, the solvent, here the surface of the enzyme in the concave enzyme pockets of active sites, can behave peculiarly in mobility as well as directly as a catalyst. Again, there are many points of contact between protein and substrate and no doubt strain is induced in large substrates. Throughout this description we have had great difficulty in following the processes which we describe because our spectroscopic tools are limited. (Notice that no reference has yet been made to leaving groups simply because they must be the same as those which we call attacking groups since we can write the reactions of a catalyst in terms of either the forward or the back reactions; it is not useful to make a catalyst from a group which becomes unable to leave an intermediate.)

Perhaps the most obvious step for the purposes of this book is to look next at those enzyme reactions which require metal ions as opposed to those that do not. Two simplifying features emerge. The substrates are often very small molecules and the states of metals can be followed by spectral means. Table 6.3 supplies a list of very small substrates and their enzymes. One feature of biological (and man-made) catalysts is that all small molecules, such as N_2, CH_4, NO, CO, H_2O_2, and, generally speaking, O_2, are activated by metal ions. Such small molecules can hardly be activated except by electronic perturbation through metal ions since they cannot bind to any, never mind many, organic sidechains on the surfaces of proteins. Metal ions are the most aggressive reagents, i.e. attacking groups, in the chemical (biological) armoury. Many of them are excellent Lewis acids since they can carry two or more positive charges on one small atom and remain of high electron affinity; this is a direct consequence especially of the underlying 'd'-shells of electrons. They can also create radicals easily due to their one-electron chemistry—again a property of the 'd'-shells. One essential function of metal ion chemistry is then that metals give very localized attacking power to a protein (see Chapter 4). In passing, notice that there is a risk too if this attacking power is not extremely well controlled since the metal/substrate complex might attack the protein; here lies the importance of localization of metal ions. Larger substrates do not, in general, need to have enzymes with very powerful attacking groups since bond strain can be induced by multiple points of attachment using the binding energy.

A factor which is particularly useful in the study of metallo-enzymes is that the metal can be replaced often without loss of activity, see Fig. 4.2. This site-specific substitution can introduce a spectroscopic probe, e.g. Co(II) for Zn(II), or extend the range of techniques used, e.g. lanthanide ions for calcium or Mn(II) ions for magnesium. Such isomorphous replacements, compare D for H, are extremely powerful.

7.8 States of metal ions in proteins

7.8.1 General aspects

Early in this book we stressed that metal ions in a protein, especially an enzyme, had properties modulated by the protein or coenzyme organic moiety to which it is bound. Where the modulation of electronic or geometric state was a distortion of these states from those normally observed in inorganic chemistry, the modulation being obviously of functional value, Vallee and Williams (see 'Further reading') described the metal ion as being in an 'entatic' (under tension) state. One example is sufficient here, since the point will arise in several later chapters.

To clarify this concept of 'entatic' state one may imagine that the metallo-enzyme in its ground state has a kind of permanently 'energized' catalytic atom, so that, as well as lowering the energy of the transition state for the catalysed reactants as with ordinary catalysts, the 'starting' energy of the metal protein ground-state complex is at a level closer to that of the catalyst's own transition state through adjustment of the bound metal. Figure 7.12 compares the two situations for normal and entatic catalysts. An example is given in the next section.

7.8.2 The copper blue centres

The nature of the deeply blue coloured centres of such proteins as azurin and plastocyanin (Fig. 7.13) has been the subject of intensive examination by very many methods that we do not intend to summarize here, but it very quickly became apparent (1960–65) that the co-ordination geometries of the copper(II) blue sites were unusual; their outstanding properties were the high redox potentials, the curious absorption spectra, and the intriguing electron paramagnetic resonance (EPR) signals. The prediction was made that these sites had a strained geometry for Cu(II) so as to reduce the rearrangement energy on going from Cu(II) to Cu(I); it then became apparent that the protein was designed for binding Cu(I) rather than Cu(II). These views were beautifully proven by the X-ray crystal structure determinations of Freeman and Jensen, who established that the protein fold was a β-barrel which formed the binding site and which was unaffected by removal or addition of the metal; in other words, the metal is captured in a hole in the protein (Fig. 7.13) and the hole generates a strained entatic state for Cu(II). All subsequent work in solution really establishes this position in detail. The design ensures that the co-ordination sphere and the metal form an excellent electron transfer site of controlled potential and minimal relaxation energy. The complex formed by the isolated strained site itself would surely have had a very high electron transfer rate, say 10^9 s^{-1}, to a second small donor/acceptor, but for the fact that it is somewhat hidden in the protein by an enclosing strand of peptide (see Chapter 4). In fact, this very fast rate of electron transfer is sacrificed to gain: (1) a large electron transfer distance, about 10–15 Å, the beginnings of the making of a spatial construction like a wire, but based on a sequence of hops; (2) an external surface near the electron transfer site to provide a recognition zone for a very limited number of protein partners, to or from which electrons

Fig. 7.13 The β-barrel fold of the copper protein, plastocyanin, which is a rigid structure designed to fit Cu(I) but not Cu(II) which becomes entatic in the rigid site (see Fig. 7.12). Note the difference from the protein mechanical rack of haemoglobin which is made from helices which can move and share strain with the metal ion. (After H. C. Freeman and L. Jensen.) The filled squares and circles show binding zones for donor or acceptor proteins.

are required to go (Table 13.1); (3) a protected site away from such molecules as O_2 and CN^-. Thus the electron transfer rate becomes optimized together with other functions at a rate constant (protein–protein) of around $10^4 \ s^{-1}$. The proteins control one other factor: the location of these copper blue proteins in an appropriate cell compartment and assembly. The most important point to note here is that the protein is a β-barrel and it is this that confers the rigidity and supports the entatic state of the metal site. Some further examples are given in Table 7.10. While we concentrate on metal ions in Table 7.10, sidechains of proteins or organic cofactors can be forced into similarly activated environments by the effect of the fold and its energy demands, e.g. tyrosine free radicals. We stress that it is the virtue of the metal ions that they allow detailed knowledge of the energetics of enzyme sites.

From what has been said earlier it must be clear that immobile β-structures will support the entatic state most readily. When we come to very helical proteins (not usually enzymes) we find a somewhat different situation in that the metal-binding and the apoprotein fold are not compatible so that strain, relieved somewhat by helix/helix motions, is induced in both on binding, e.g. in calmodulin. This is a rack action. A combination of an entatic state and a rack is possible as in haemoglobin where the porphyrin takes the place of the rigid β-sheet. The porphyrin holds high-spin iron, Fe(II), in an entatic (out-of-plane) state but strain at the metal is relieved on binding O_2; the iron goes low-spin and moves into the plane of the haem, while by rack action the strain

Table 7.10 Examples of entatic states in metallo-enzymes and metalloproteins

Metallo-enzyme	Metal	Donors	Structure of active site
Carboxypeptidase	Zn(II)	2 imidazoles 1 carboxylate 1 water	Pentaco-ordinated, with bidentate carboxylate
Alcohol dehydrogenase	Zn(II)	1 imidazole 2 thiolates 1 water	Distorted tetrahedral
'Blue copper' site, e.g. in plastocyanin	Cu(II) Cu(I)	2 histidines 1 cysteine 1 methionine	Distorted tetrahedral, with a very long methionine–copper(II) bond
Cytochrome c	Fe(II) Fe(III)	Porphyrin 1 imidazole 1 methionine	Distorted octahedron, strained axial bonds in Fe(III); see also cytochrome b
Some ferredoxins	Fe(II, III)	4 cysteines 4 inorganic sulphur atoms (S^{2-})	Distorted cubane type cluster (Fe_4S_4); a great variety of structures
Vitamin B_{12}	Co(III)	Corrin 1 benzimidazole	Open-sided distorted octahedron (the cobalt–carbon bond is broken when cobalt is Co(II))

is induced in the protein helices. There must be no confusion between entatic strain and rack action.

7.8.3 Protein/protein assemblies and metal ions

When proteins come together they may form a well defined structure, as in the reaction centre for photosynthesis in Fig. 4.5, and in haemoglobin, or a loose association. The value of the well defined organization is that it gives very specific direction to the reactions of reactants and products, for example, to electron flow. In particular, buried metal ions for electron transfer can be in organized arrays. The value of looser association is that assembly and disassembly are rapid. The need for fast exchange between proteins is obvious in such functions as:

(1) scavenging, e.g. cytochrome c scavenges the periplasmic space for electrons from several electron donor enzymes (Fig. 5.5);

(2) flexible development and contraction of villi, e.g. tubulin forms rapidly reversible filamentous structures (Fig. 7.14), which stretch out or contract the membranes (Fig. 3.5);

(3) control such as that exerted by calmodulin.

The different types of organization and its rigidity depend upon the type of amino acid and metal ion involved at the junction. Oppositely charged amino acids on two protein surfaces are likely to generate highly mobile electrostatic joints while hydrophobic matching is more likely to give rise to more rigid structures. Amongst metal ions we must expect that inside cells the role of magnesium will be important at protein contacts, whereas outside cells calcium has the dominant functional significance and provides greater mobility.

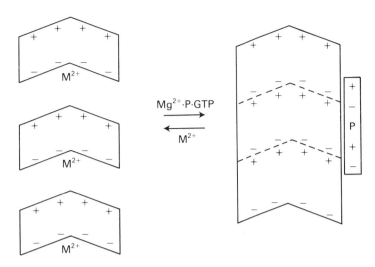

Fig. 7.14 The assembly of tubulin monomers, probably quite rigid units, into filaments due to the presence of specific concentrations of metal ions, e.g. Mg^{2+}, and GTP, then followed by hydrolysis of GTP (energy dissipation). Dissociation may be aided by Ca^{2+}. The whole assembly may require additional proteins, P.

7.9 Summary of proteins

In the first part of this chapter we have tried to describe proteins and the way in which proteins and inorganic elements combine functionally; we emphasized that it is the protein which provides *dynamic structure* within the organization of biology, such that various devices, e.g. channels (Na, K), filaments (Mg, Ca), control systems (Ca, Fe, Zn), enzymes (many elements), and so on, can be made and can evolve. However, it should always be kept in mind that: (1) the whole organization of a protein is energized by the physical conditions, temperature and pressure, by the input of energy due to chemical gradients, i.e. electrostatic fields, and by chemical binding of cations, especially Mg^{2+} and Ca^{2+}, as well as of phosphate; and (2) the protein's position in space is also energized. Chapters 13–20 will amplify this combination and organization of elements and polymers.

7.10 Nucleic-acid composition and outline structure

The linear polymeric structures of deoxyribonucleic acid (DNA) and ribonucleic acid (RNA) are closely related. They are both condensation polymers of ribose phosphates which carry four purine and pyrimidine bases. The difference in the backbones lies only in the reduction of one, 2', $CHOH$ group of the ribose of RNA to CH_2 in DNA, but this does appear to make double-helical strands more possible in DNA. This extremely important reduction is controlled by one of the three metal ions, Fe(III), Mn(III), or Co(III), in a quite extraordinary set of enzymes which generate a free radical pathway in the cytoplasm of cells (see Chapter 12). The backbones of the polymers carry the bases shown in Fig. 7.15 and these sequences in DNA are the basis of the genetic code. Of course, there is a close relationship between the bases of any RNA derived from a DNA by transcription and the bases of

Fig. 7.15 A short stretch of single-strand DNA giving the formulae of four bases.

the DNA itself. DNA has both the function of replication and of RNA production and the enzymes which carry out the polymerization reactions are often, but not always, zinc enzymes. Moreover, the production of the bases is under the control of a zinc-requiring enzyme, aspartate transcarbamylase. The RNA is translated to proteins via the ribosome machinery and magnesium enzymes play a considerable role in this process. Finally, the control of transcription and translation rests with regulatory transcription and translational factor proteins, many of which are zinc metalloproteins (*not* enzymes) (Table 7.11). In this part of the chapter we shall not be concerned with any of these metal-containing proteins but with the polymeric nucleic acids themselves and their metal ion and possibly small anion-binding sites. We shall look at the nature of the nucleotide monomers more closely while

Table 7.11 Metal enzymes related to DNA and RNA production

Enzyme	Metal ion	Function
Ribonucleotide reductase	Fe, Mn, Co(B$_{12}$)	Ribose→deoxyribose
Transcriptional factors	Zn	Control DNA→RNA
RNA polymerases (also reverse transcriptase)	Zn	Condensation polymerization
Various NTP-utilizing systems	Mg	Bioenergetic driving forces

describing saccharides since the basic monomer of RNA and DNA is a phosphated saccharide.

Some of the known structures of DNA and RNA are illustrated in Fig. 7.16. The structures have a very different character from that of proteins in that they do not fold easily since they are polyelectrolytes. However, DNA must be able to assemble and disassemble quaternary double-helical structures rapidly in order to replicate. Single-stranded DNA (RNA) is a very mobile polymer both at the phosphate and at the ribose. The grooves of DNA provide some selective specificity since the groove width and chemical structure are dependent on the local base sequences. The grooves are recognized by helical sections of proteins and by a great variety of chemicals (drugs) some of which are metal chelates. Specific base reactions are known, for example, with anticancer drugs such as alkylating agents, and reagents such as copper or iron complexes can be chosen so that they bind in the groove and generate destructive reactions of great value in the foot-printing of DNA.

In the double-helical structures there is little mobility of the sidechains since they are all cross-linked by H-bonds but, in principle, the five-membered ribose rings can flip between conformers very easily and the obvious way to pull apart the two chains is to use a zipper-like action employing the mobility of both the rings and phosphate ester joints.

(*Note.* In what follows we shall often refer to *poly*nucleotides, synonymously with nucleic acids, to avoid confusion with *mono*nucleotides.)

(a)

(b)

Fig. 7.16 (a) The fold of DNA. (After A. Rich.) (b) The outline of alternating double helices and loops in a stretch of ribosomal RNA (see Fig. 7.17 for detail).

7.11 Metal-ion binding to polynucleotides

The outstanding features of a single-stranded system (note RNA in particular), assuming a random fold, as a potential centre for ion binding can be simply listed.

1. *The diesterphosphate with the sugar moiety.* This is a very weak centre and it is doubtful if there will be binding of effective affinity by one such site to any cation. The charges are then simply screened by long-range effects familiar in all polyelectrolyte solutions, i.e. by Na^+, K^+, Mg^{2+} in biology. However, organic cations, such as spermine, behave differently (Chapter 9).

2. *The nitrogen centres of the bases.* Again, in a random polymer, these are not likely to be binding sites for cations except for reactions with very heavy metals such as Hg^{2+}. Aromatic bases of low pK_a are poor donors and most transition metals can only form stable complexes with them through chelation.

Thus, free single-stranded polynucleotides are of little interest to us in the context of the present book. However, such single-stranded polymers can bind to proteins and become locally structured, so metal ion chemistry again becomes involved and undoubtedly this is the case in the complex units of ribosomes (see Section 7.12).

The centres on double-stranded helical DNA and single-stranded RNA folded into double helical segments are similar, but now the organized surfaces become of interest since the double strand (Fig. 7.16) has potential sites for metal-ion interaction. Thus the major and minor grooves, which have base-specific widths, expose both the ribose phosphate moieties and parts of the bases. These organized centres of charge and N-donors (bases) can bind metal ions, but the phosphate anions can also bind polyamines and small sequences of single-strand polynucleotides (i.e. anions), so that the exact form of a double helix structure and its curvature may well be controlled by molecules or ions in the grooves. We shall turn to the competition and co-operative effects between them later.

There are potential co-operative structures which are not double-stranded but can be created irregularly along the chains if the chain alternates between double and single (or triple) strands. Such a structure is seen in transfer RNA (tRNA) (Fig. 7.17), in larger RNA (Fig. 7.16), and is the basis of the cruciform of DNA (Fig. 7.18). The stability of these folded forms is again cation-dependent. The co-operativity here involves directly Mg^{2+} or polyamine binding. These structures may well be critical in the expression of function.

7.11.1 Cavities and metal-ion binding in DNA

The reader is reminded of the discussion of cavity size and cation binding which was given in Chapter 2. There we saw that there were many factors controlling binding to a cavity which we could simplify under the headings:

(1) the cavity size, treating it as an empty circle or hollow sphere of given radius;

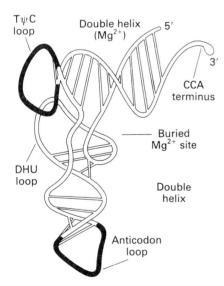

Fig. 7.17 Schematic diagram of the three-dimensional structure of a tRNA. (After A. Klug and A. Rich.)

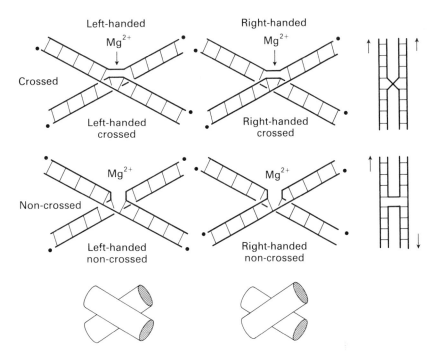

Fig. 7.18 Possible topology of strands and helices in the stacked X structure of the four-way DNA junction. The strands can be drawn as non-crossed or crossed, generating an antiparallel or parallel alignment of the sequences respectively (indicated by the arrows). Because of the non-planarity of the X, the helices can cross in either a right-handed or a left-handed form (the convention for the handedness indicated is based on the definition of dihedral angles). The helices at the bottom are stippled. The combination of these variables generates four possible structures. The 5′ termini of strands are indicated by filled circles. (After a diagram by Dr D. Lilley.)

(2) the lining atoms of the cavity which act as donors;

(3) the support structure behind the cavity.

Now it is known that DNA(RNA) must be relatively mobile in all its interesting biological forms since, for the most part, these polymers must pass through a reading (translational, transcriptional) machinery; the construction as phosphate ribose esters aids this mobility. The sizes of the cavities these polymers can form, as seen in grooves (Fig. 7.16) and crucifixes, are unlikely to have strict limitations; hence many cations and polyamines can bind but not very specifically or strongly. In cells, however, only Mg^{2+} and polyamines are present in sufficient quantity (Chapter 9). We need to consider equilibria of the kind

$$\left.\begin{array}{l}\text{Open structure with}\\\text{large cavity for}\\\text{very large cation}\end{array}\right\} \quad \underset{\text{polyamine}}{\overset{Mg^{2+}}{\rightleftharpoons}} \quad \left\{\begin{array}{l}\text{Closed structure with}\\\text{small cavity for}\\\text{small cation}\end{array}\right.$$

To a first approximation Mg^{2+} appears to vary but little in concentration in cells; if this variation can be neglected, then the degree of folding DNA/RNA is controlled by the production of polyamines such as spermines and histones. While Mg^{2+} tends to force nucleic acids towards crucifix forms, both spermine and histones appear to stabilize condensed states of regular DNA and, since Na^+ and K^+ do not bind, they stabilize very open ionic surfaces. In all of these interactions the relative selectivity of interaction with small molecules and metal ions is very slight. There is then little question of enzyme activity except for RNA (with RNA as substrate) where the fold has peculiarities (Fig. 7.16). Specificity develops only over very considerable surface area, e.g. in binding of promoter proteins. We return to these problems in Chapter 9 which treats magnesium biochemistry.

The bases in the folded forms offer the possibility of reaction (binding) to many heavy metal ions. This field of research was strikingly awakened by Rosenberg's discovery of the cisplatinum drugs. Here binding is to N(7) of guanidine bases. Undoubtedly many drugs in this and related families remain to be found in man's attempt to fight cancer and viral diseases. At the same time we must notice that other heavy metal ions will also bind to the bases, but this represents damage to DNA and could either be useful or cause mutational activity which must include cancer itself. Here the relationship between the value of drugs and the dangers of chemicals is close.

7.11.2 DNA sequences and mutation

The DNA sequences contain the coded information for the cell. It is the development of these sequences that lies behind evolution as we understand it. The rearrangement of sequences and the changes of the bases by mutation are well recognized mechanisms for the trial and error search by biological systems in their efforts to evolve greater 'fitness'. The details of the involvement of simple elements in these changes of sequence are not clear but it is known that elements such as Mn, Fe, and Cu through their oxidizing actions can cause DNA breakdown and then mutations. It appears that all biological systems reduce exposure of their DNA to damaging reagents such as $O_2^{\cdot-}$, H_2O_2, and OH^{\cdot}, produced by these oxidation catalysts, to a large degree. In this respect eukaryotes have the most powerful controls in that they use a superoxide dismutase in the cytoplasm which equilibrates only with an extremely low level of metal ions (Cu, Zn) while the prokaryotes use other enzymes (Fe, Mn) which involve a higher risk. Moreover, eukaryotes keep the reactions of H_2O_2 in special vesicles. In multicellular organisms the protection against mutation and therefore against cancer has a high priority. Development of complex organisms is often extremely slow allowing control.

7.12 Nucleic acids and proteins

Major particles in cells are nuclei (DNA) and ribosomes (RNA). They contain folded forms of both proteins and polynucleotides but it is the proteins that control the shapes of nucleosomes, DNA units of about 200 base pairs, and ribosomes, RNA combinations of high molecular weight. Each of these units has associated with it an unknown complement of cations, especially Mg^{2+}. There are present also two special classes of proteins which can activate the transcription or translation machinery; these proteins bind in the grooves of DNA but hold RNA in a complex form (Fig. 7.19). For these activities there needs to be a series of factors and enzymes and we shall note the role of iron and of zinc in the DNA transcription systems (Table 11.4) and haem in the RNA translations. It is obviously necessary to have a communication network amongst all the systems. This is especially obvious in cell division when new proteins and new DNA have to be built. The division process itself requires filamentous structures of yet other proteins to be made mechanically active at a precise time. The filaments bind both to the DNA units, chromosomes, and

Fig. 7.19 The way in which a helical set of two protein segments could fit into the grooves of DNA. Note the value of mobile helices. Several metal-binding regulatory proteins may be of this kind (see also the discussion of calmodulin, Section 10.7). The structure is from the works of B. W. Matthews and T. A. Steitz.

to the membranes. There is a deep involvement of the simultaneous directed use of energy and the completion of a synthetic cycle. How is this integration managed? We do not know yet, but Mg^{2+} and NTP must be involved.

Again we repeat that it is necessary to realize that the organization of DNA and RNA is not rigid but energized and dynamic. There are general structural principles, but these are temperature-, pressure-, electric-field-, and ion-concentration-dependent and are linked into the organization of proteins and, as we shall see, of membranes and even of polysaccharides. DNA in particular has to be a dynamic structure, more like a river than a lake, since it has to replicate.

7.13 Polymer synthesis and degradation

An important part of our understanding of biological systems concerns the rates of synthesis or repair of biopolymers and the rates of their degradation. The polymers are produced in response to messages; in fact, even the monomers are not wastefully generated. A full appreciation of the interactive system which is required to generate such control is possible only when we see that: (1) because DNA is a double-stranded helix, adenosine and thymine as well as cytidine and guanine must be produced in appropriate amounts; (2) similar statements can be made about proteins and RNA: all polymers demand a somewhat equal production of a great variety of monomers; (3) no polymer can be made without generation of appropriate amounts of ATP, GTP, etc. for synthetic energy. Again associated elements must be collected in correct proportions, so there has to be over-riding chemical integration within the system as well as an appropriate means of generating particular polymers and the physicochemical conditions of their existence. In a steady growing system this implies a homeostatic set of rates of production and degradation which can only be achieved if feedback systems regulate the synthesis by start

and stop signals and/or if the excesses to requirement are destroyed by a feedback self-hydrolysis. All this activity demands controlled flow of information. Some very simple examples show the delicate nature of this balance.

1. Photosynthesis requires that the absorption of light should provide a *balanced* supply of energy *and* of reduced carbon.

2. Oxidative phosphorylation is coupled through the citric acid cycle not just to ATP (energy) but to the synthesis of amino acids for protein synthesis.

3. Messenger RNA (mRNA) for a given protein must have a well maintained level. Ribosomal RNA (rRNA) and transfer RNA (tRNA) are much less significant in this respect, but tRNAs have to be produced in amounts commensurate with the use of the amino acids.

4. The glycolytic pathways are coupled to the production of monomers not only for proteins but also for DNA, RNA, fats, and polysaccharides.

It is the mixture of functions in which any major small molecule or element of biology is involved which demands an intense communication and exchange network *inside cells*. (Extracellular activity is equally co-ordinated and examined later.) We shall ask, especially in Chapter 21, can we see as yet what are the underlying co-ordinating units? Are they the mineral elements to a large extent since it is these elements which control major electrostatic and chemical gradients as well as catalysis?

When we turn in the next section to the final two types of biological polymers, fatty acids $(CH_2)_n$ and polysaccharides, then the synthesis becomes less exacting and their direct role in control appears to be less, but, especially in the case of their phospholipids, the problem is not now referable to the properties of single molecules but only to *lipid phases*. The lipids form the solvent of a phase, much as water molecules do, and the polysaccharides are even more difficult to describe in structural molecular terms. Since they are polymers we describe them first in terms of molecules but we shall see that phase concepts are of great value in the analysis of their functions.

B Polysaccharides and lipids

7.14 Introduction to polysaccharides 193

7.15 Backbones of biological polymers 195

7.16 Sidechains: physical properties 197

7.17 The information content of biopolymers 198

7.18 Space-filling 199

7.19 Chain branching in polysaccharides 201

7.20 Polysaccharide sidechains: chemical nature 201

7.21 Functions of biological polymers and oligomers 202

7.22 Larger random polysaccharides 204

7.23 The cation interactions 210

7.24 Conclusions about polysaccharides 210

7.25 Properties of lipids 210

7.26 The connection between lipids, hormones, and coenzymes 216

7.27 Summary of biopolymers 217

7.14 Introduction to polysaccharides

Polysaccharides in living systems (Table 7.12) have five main roles: (1) in the formation of plant *skeletons* (cellulose) and cell walls; (2) in connective tissue in animals; (3) in generating recognizable features on external proteins or on the outer surfaces of cells; (4) in mucus, adhesive and cohesive fluids, etc., many of which either aid or inhibit flow; and (5) in storage. In the present chapter we will be concerned mostly with the first four roles.

In most organisms the outside (exo- and ectocellular) regions of the cell have the carbohydrates associated with proteins (i.e. glycoproteins and proteoglycans) and lipids (i.e. glycolipids) or in hydrated polysaccharide gels. In unicellular organisms the combination of these macromolecules, polypeptides and polysaccharides, gives rise to cell walls, protective polymers, receptors, and enzymes, both in membranes and in the periplasmic and extracellular spaces. In multicellular organisms, while the same diversity exists, the mixed (hetero) and homo polymers also form the structural network which holds the organization together (Table 7.12). (For a review on saccharide structures and functions see 'Further reading'.) The primary structure of these polysaccharides is often very complex since their constituent monosacccharides can be linked in many different ways. Consequently, the potential information encoded into a polysaccharide via its primary sequence and three-dimensional structure is considerable. We shall find that the

Table 7.12 Important polysaccharides of some living organisms

Organisms	Polysaccharides	Composition	Biological function	Functional groups other than —OH
Algae	Fucoidan	D-glucose, D-mannose, D,L-galactose, L-rhamnose, L-fucose, L-xylose, L-arabinose, D-glucuronic acids, D-galacturonic acid, L-3,6-anhydro-galactose	Extracellular mucilage	(a) —COO$^-$ (c) —O—SO$_3^-$
	Cellulose	β-(1→4)-D-glucan	Cell wall	—
	Xylan		Cell wall	—
	Mannan		Cell wall	—
	Alginic acid	D-mannuronic acid	Intercellular space in marine algae	—COO$^-$
Higher plants	Cellulose	β(1→4)D-glucan	Cell wall	—
	Hemicellulos	β(1→4)D-xylan, xyloglucan, arabinogalactan, arabinoxylan, etc.	Cell wall	—
	Pectin	α(1→4)D-galacturonan, arabinogalactane, rhamnogalacturonan, etc.	Cell wall of young tissues	—COO$^-$
	Gums	Uronic acid, hexoses, pentoses	Secretion products	—COO$^-$
	Mucilages	Galacturonic acid	Cell wall, nutrient reserve	—COO$^-$
Animals	Chitin	β(1→4)N-acetylglucosamine	Invertebrate skeletons	(a) —NH—COCH$_3$
	Glycogen	α(1→4)D-glucose with ramifications in α(1→6)	Nutrient reserve	—
	Hyaluronic acid	D-glucuronic acid, N-acetyl-D-glucosamine	Cell cohesion	(a) —NH—COCH$_3$ (b) —COO$^-$
	Chondroitin	D-glucuronic acid, N-acetyl-D-galactosamine	Cartilage tissue	(a) —NH—COCH$_3$ (b) —COO$^-$
	Chondroitin sulphate	Idem with galactosamine sulphated at positions 4 or 6	Cartilage tissue	(a) —NH—COCH$_3$ (b) —COO$^-$ (c) —O—SO$_3^-$
	Glycoproteins	Mannose, galactose, glucuronic acid, glucosamine, galactosamine, sialic acid	Cartilage tissue (many others)	(b) —COO$^-$ (d) —NH$_2$

NB. Some of the polysaccharides may be phosphorylated. They, like carboxylated derivatives, bind metal ions.

interactions of the different polysaccharides with inorganic ions are quite different from those with proteins and that these interactions have a selective nature of their own.

The aim in this section is not to comment on the nature of the saccharide-associated proteins or lipids but rather to pose questions about the nature of the saccharides as polymers in an attempt to relate their functional significance to their composition via their structures and the dynamics of these structures. It is useful to start, however, by comparing polysaccharides with proteins and nucleic acids, the biological polymers already described. This will supply a summary of similarities and differences between the three major classes of biological polymers while introducing the polysaccharides. Only when their different features are clear will the association with different chemical elements become understandable.

7.15 Backbones of biological polymers

7.15.1 Proteins

The backbone of proteins is formed from the link –[–RCH–CO–NH]–. The peptide group –CO–NH– is relatively rigid and fixed in a plane so that protein dynamics and folding concern the flexibility about the unit –NH–RCH–CO.

7.15.2 Nucleic acids

The backbones of polynucleotides consist either of 3′, 5′-ribose phosphate or its 2′-deoxy analogue, i.e. they are polysaccharides with special sidechains and condensed via phosphates. Here, the unit is much larger and has much higher flexibility in that there is effectively free rotation at C-5′ and at each of the ester oxygens of the phosphorus, as well as conformational mobility within the five-membered ribose ring since various endo–exo forms easily interconvert.

7.15.3 Polysaccharides

Polysaccharide backbones are generally based on the condensation of five- and rigid six-membered cyclic oxide rings (furanose and pyranose, respectively); note again how large these units are (Fig. 7.20). Here different linkages are possible, unlike for proteins and DNA, where all the linkages in a molecule are of one type. Note that the anomeric carbon (C1 for most monosaccharides, C2 for sialic acids) is involved in the linkages (e.g. 1→2, 1→3, 1→4, or 1→6 links). This fixes the position at this site of the oxygen atom with respect to the ring. If this were not the case, the ring would open at the anomeric position to yield a set of isomers with the OH group directed either axially or equatorially with free interchange, which would make the unit useless as a specific recognizable group. As the anomeric oxygen may lie axial or equatorial to the plane of the ring, there arises the possibility of two isomers for each monosaccharide, an α-form (O-axial to the ring) and a β-form (O-equatorial to the ring).

In the case of glycosidic (i.e. ether) links from any of the oxygens 1, 2, 3, 4, and 6 to any other such oxygen of another monomer there is flexibility around the C–O–C unit but this is limited by the steric properties of the carbon atom links to two other carbon atoms. There is greater flexibility via the 6 linkages since here the ether bond is in the structure (hexose ring)–CH_2–O–(hexose ring) and this not only introduces rotamers of its own but also increases the freedom around the ether oxygen (compare the phosphates of DNA).

From the above, one might expect that chain flexibility would follow the order

$$\text{Single-strand DNA and RNA} > 1{\rightarrow}6 \text{ saccharide} > \text{proteins} \\ = 1{\rightarrow}(2, 3, 4) \text{ saccharide}$$

with the DNA and RNA molecules having the greatest flexibility for a single chain due to the phosphate linkages. However, we must also observe that in proteins every *third* chain atom is a point of considerable mobility, while in the polysaccharides mobility at two or three consecutive atoms is followed by a

Maltose (β-form)
(4-O-α-D-glucopyranosyl-β-D-gluco-pyranose)

Glucose (α) Glucose (β)

Lactose (α-form)
(4-O-β-D-galactopyranosyl-α-D-glucopyranose)

Galactose (β) Glucose (α)

Sucrose
(α-D-glucopyranosyl-β-D-fructo-furanoside)

Glucose (α) Fructose (β)

Fig. 7.20 Some disaccharides showing the variety of features discussed in the text.

relatively large and rigid saccharide unit. Finally, there is the effect upon conformation of the α or β configuration at C-1 (Fig. 7.20). Generally, β1→4 structures are associated with more rigid elements than α1→4 structures (compare cellulose and chitin with glycogen and starch; Fig. 7.21). Interestingly, one finds that the core regions of complex class oligosaccharides are composed of β1→4 linked regions (see Section 7.19).

The immediate point concerning polysaccharide chains is the diversity of chain links, the large volume occupied by each link compared to its sidechains, and the peculiarities of the high mobilities in some of them. Clearly, the differences in character must lead to very different polymer properties. One obvious point, however, is that the *backbone* of all the polymers except proteins is usually a part of the interactive surface with small molecules or ions.

(a)

Cellulose

β-D-Glucose

(b)

Chitin

Fig. 7.21 (a) Cellulose; (b) chitin; (c) starch; note branching.

(c)

Glycogen or starch

7.16 Sidechains: physical properties

7.16.1 Proteins

The sidechains of proteins are characterized by a variety of physical and chemical properties from some 20 amino acids with one common flexibility characteristic apparent (Table 7.1). Almost all are based on

$$\alpha \quad \diagdown \atop \diagup \text{CH–CH}_2\text{—X units}$$

and the construction of most amino acids is to provide mobility as well as extension. The X group can provide very water-soluble units, when, if they are all of this kind, the protein tends to be extended in water, or very water-insoluble groups when it tends to fold. The huge variety of X in an n-mer peptide (i.e. a protein) generates the rich variety of the fold in lipid or in water. This variety also permits every kind of strength and mobility in the fold since X can be more or less hydrophobic on average. Overwhelmingly, proteins are folded structures of relatively low internal mobility due to their sidechains, which occur close together. A vast range of specific interactions with ions and molecules is generated by the sidechains and the fold (see Section 7.6).

7.16.2 Nucleic acids

In contrast to proteins, the sidechains of the nucleic acids are few, five major kinds, and very similar. They have no internal flexibility, being aromatic ring systems which can only rotate with much restriction about sugar-base bonds and are, to a fair degree, equally hydrophobic. All five are found attached to the hydrophilic pentose phosphate polymer; consequently, there is one type of basic fold, the double helix, with bends and crucifix structures now and then. This double helix, although very stable, can be modified by weak, non-selective, ion binding as we have seen above. Note that diversity both in proteins and nucleic acids is achieved by unique *linear* coupling in sequences. However, the repetitive local structures of polynucleotides clearly generates less selective surfaces.

7.16.3 Polysaccharides

Polysaccharides are quite variable in sidechain structure but the sidechains are relatively small (see Table 7.12 and Fig. 7.22). Generally, the sidechains help to make the polymer more water-soluble. Moreover, they are not so mobile as in proteins in the sense that they are not connected via CH_2-units to the main chain, the single exception being the β–CH_2OH unit attached to carbon 5 in pyranoses. Of the two faces to each saccharide ring, one is generally more hydrophilic than the other due to the greater number of hydroxyl sidechains on one side of the ring than on the other; this is especially marked in monosaccharides such as mannose. The hydrophobicity of a polysaccharide is then based on *one face* of the hexose ring, i.e. is a property of the backbone and

Fig. 7.22 (a) A typical saccharide bound to lipids. (b) A typical branched saccharide bound to lipids, R_1 and R_2.

not of the sidechains, in marked contrast with those of proteins and polynucleotides. Thus, on the whole, any fold of a polysaccharide is based on weak main-chain main-chain interactions, and tertiary structures are not strong. It is the weakness of the tertiary, and quaternary (and even secondary) structures of many saccharides (but not all) that make them a very special set of biopolymers, often lacking structural definition and therefore having very different roles from proteins and polynucleotides. Above all, it is the interaction with solvent water which dominates their fluctuating conformations and their extreme mobility. Note that their diversity lies now in sequence and *branching* (which was not present in the other polymers). They are such then, that they can be recognized by large surfaces, but do not supply *recognition* centres for small molecules or ions.

At this point it is instructive to consider the variety of polymers and oligomers that can be built from the simple building blocks.

7.17 The information content of biopolymers

The relative complexities of the various biopolymers can be gauged by considering the number of different structures of a given n-mer that could occur. Following such considerations, including the different ways of linking saccharides, it is clear that the inherent complexity of polysaccharides is much greater than that of the other types of biopolymer and that even short sequences have high degeneracy of sequence. If one considers branched structures, the number of possible sequences is enormous and, therefore, the number of recognizable sequence patterns is huge so that we could consider this to be an information capability. Note that one refers to theoretical

information capability and there is no suggestion that genetic information is encoded directly in the primary structure of polysaccharides, as it is with nucleic acids. Rather, the large number of possible sugars (see Table 7.12) indicates the wealth of opportunity for constructing polysaccharides with different shapes.

7.18 Space-filling

7.18.1 Proteins

The ability of a linear polymer to fill space depends on the flexibility of the joints and the sizes of the units. It is notable that proteins are polymers of short units. For this reason, a complete turn can be made in a small space so that the α-helices, β-sheets, and β-turns, assisted by main-chain H-bonds, can form and fill space easily in very effective ways, leaving surfaces to generate properties and little or no room for water in the interior. Their folds are admirably suited to provide:

(1) almost rigid platforms for recognition, e.g. enzyme active sites and receptors mostly based on β-sheets;

(2) stable rod-like elements of considerable length based on α-helices, e.g. the mechanical transmission rods of biology;

(3) local energized regions generated by the strength of the whole fold;

(4) effectively, a three-dimensional form of information from a relatively rigid folded linear sequence;

(5) internally co-operative structures (compare sodium chloride crystals) such that the majority change structure (phase) by melting;

(6) concave as well as convex surfaces.

7.18.2 Nucleic acids

Nucleic acids are obviously based on quite another principle in that the unit is an anion. A polyribosephosphate will then resist tertiary and quarternary folding but, as with proteins, the double helix interior of the bases of DNA and RNA in dimeric units contains little or no water. Detailed DNA and RNA folding is controlled by base sidechains or added cations of various types, e.g. Mg^{2+}, polyamines, and histones in the helix grooves. Calcium is not present in cells. The consequences of the requirements for sidechain to sidechain H-bonding and for the hydration of the soluble ribose phosphate are such that, by themselves, these polymers tend to form long mobile strings. They are a *linear* form of information since, in large part, they have few of the properties given for proteins above. The linear information includes small local variations in groove size and considerable differences in H-bonding capacity. Both DNA and RNA can recognize (large) molecules locally using considerable surface area but small ions only rather generally. RNA, being single-stranded in part, also interacts with proteins to give ribosomes while DNA interacts with histones to give nucleosomes.

Because of the way in which it fills space, RNA is capable of forming weak catalytic sites of low selectivity only, called ribozymes, since there is little diversity or attacking power on its surfaces.

7.18.3 Polysaccharides

Polysaccharides have an interesting position relative to these other two classes of polymers insofar as space-filling is concerned since, even as unbranched polymers, they have rather a large (fat) rigid-single unit, the hexose ring, which in polymeric form tends to give an extended chain due to its considerable hydrophilicity. A likely secondary structure open to the backbone is the extended H-bonded fibril which, when built from *homo*-polysaccharides, X_n, or $[XY]_n$, can yield strong structures, e.g. celluloses from $(1 \rightarrow 4)$ β-glucose (Table 7.12). There appears to be no way in which a *tertiary* fold would then be generated, since the exteriors of these structures are very hydrophilic (see, for example, cellulose). Branched polymers have a shape even more open to contact with water, so that they tend to form swollen gels, starch, i.e. $(1 \rightarrow 4)$ α-glucose with $1 \rightarrow 6$ branches (Fig. 7.21). However, three factors, introduced earlier, suggest that we should not be looking for simple general structural principles concerning the occupation of space by polysaccharides. They are:

(1) chain branching varieties;

(2) alternating chain linking (i.e. they are frequently heteropolymers within the backbone);

(3) the nature of their sidechains.

It appears that these are polymers in which there is a very low level of fold co-operativity but are such that a great variety of types and shapes of polymers can be constructed and each polymer may well be highly mobile. While the above limitations on these constructions makes them of reduced value in, for example, the generation of recognition surfaces by folding, as found in proteins, they become a form of recognizable information through variations in sequence, connectivity, and branching. Together, *these variations provide quite new ways of generating information in three dimensions without a rigid fold, i.e. by hand in glove, not lock and key, fitting.* Regarding their potential functions, perhaps for the very reason that they are limited in their folding, they do not have local regions which generate enzyme *activity*, neither can they generate entatic states of metal ions. It is clear that the great advantage of the saccharide polymers is that they can build large *passive open zones* of little fixed shape of either great flexibility or at least of limited rigidity outside cells and often connected to cell membranes. These possibilities are greatly increased by chain branching which is absent in the other polymers. (The image is of tree-like forms moving in the wind.) We must look for functional significance in properties associated with for example (viscous) flow as well as recognizability. The surrounds of a cell in an organization demand almost free flow of food, etc. to it through what must be an open mesh. In particular, since the polysaccharides are outside cells, we shall have to observe effects of ionic strength (Na^+, K^+, Cl^-) and specific interactions with calcium, on these features.

7.19 Chain branching in polysaccharides

Some simple points should be clarified. First, chain branching can provide either a flexible or a relatively inflexible connection (e.g. 1→6 and 1→3 linkages, respectively). A common question is the significance of rigidity and mobility at branching points. In this regard, one can consider the forms of trees and umbrellas. Rigid branching after starting from a linear trunk provides the most economical way of generating a large surface while covering the ground from which the trunk emerges. More mobile terminal features allow rapid adjustment on contact. Note that a branched saccharide can then:

(1) protect a large part of the surface of a protein or a membrane due to the size of the monomer;

(2) provide a large exposed surface, especially on a membrane, carrying recognizable features but using few residues (Fig. 7.23);

Fig. 7.23 Sequence of N-linked oligosaccharide from soybean agglutinin; note the positions of β-(1–4) links giving some rigidity and of α-(1–6) links giving great mobility. The core is common to many glycoproteins.

(3) bear functional units, without defined folding based on a sequence.

Furthermore, branching in a concentrated group of large polymers is effectively a cross-link which reduces freedom of movement and creates entanglements, such as occur in gums, saliva, slime, and so on. Additional entanglements will occur through Ca^{2+} binding.

7.20 Polysaccharide sidechains: chemical nature

The most common polysaccharide sidechains are the fixed –OH groups, the 6 CH_2OH, and the $NHCOCH_3$ group (Table 7.12 and Fig. 7.21). These units are obviously highly effective H-bonding centres and allow solvation as well as intra- and extra- chain–chain interactions. Other common units are anions $-CO_2^-$ and $-OSO_3^-$ which repel one another; these units will be discussed later since they become of major importance in weak cation binding. The few hydrophobic units include O-methyl groups. Now, the distribution of the sidechains is unique to a given saccharide and one can ask what properties are conferred by, for example, N-acetylation. It must also be remembered that, unlike proteins, the sidechains can be attached at several points around the hexose ring, so that once again three-dimensional information can be developed without folding.

A special feature of the $-NHCOCH_3$ group of some saccharides is that it permits in homopolymers the formation of secondary structure using sidechain H-bonding. Some of the most rigid saccharides are based on the β-sheet formed from the antiparallel strands of these [NH–CO] units which

can be compared to the β-sheet of proteins. The β-sheet and the α-helix are perhaps equally rigid, but the (curved) β-sheet effectively forms a strong three-dimensional rigid surface while the helix is effectively a one-dimensional rod with flexibility between rods as with proteins. Note that these polymers are homopolymers or of very little monomer diversity.

Now that the three major classes of polymers have been described we can look at their functional value in biology.

7.21 Functions of biological polymers and oligomers

7.21.1 General Notes

The major functions of polysaccharides as opposed to other polymers can be classified roughly as in Tables 7.12 and 7.13.

We need to stress the complementary nature of these polymers through their very different characters and relationships to one another which we have just explored. The proteins are ideal for folded nearly solid state structures, for relatively rigid mechanical devices (sliding filaments, jointed rods), and for chemical binding of a diversity of kinds in relatively fixed conformations. Random protein structures are rare. DNA and RNA are relatively ineffective in all these roles and carry translatable information. In all of their polymers there is much mobility of strands and double helices. Polysaccharides, which are polymers of lower fidelity, are used for making space more usable through the generation of loose 'random' networks or for making very selective recognizable sequences. When they are used for the building of frameworks, they are of very simple usually homopolymer kinds, e.g. celluloses (see Table 7.13).

Table 7.13 Important polysaccharides of some living organisms

Function	Saccharides	Other biopolymers
Stores of genetic information	No	DNA, RNA
Catalytic surfaces	None known	Protein–enzymes/RNA
Structural supports	Cellulose	Collagen, silk
Recognition devices	Oligosaccharides?	Antibodies
Recognizable variations	Oligosaccharides	Protein surfaces
Physical (mechanical) devices	Proteoglycans	Muscle proteins
Protective devices	Lignins	Coat proteins
Energy storage	Glycogen	Seed proteins

7.21.2 Roles of polysaccharides: relatively rigid structures

The roles of homo- and heteropolysaccharides as rigid structural elements in cells are well known and have been extensively reviewed elsewhere. These structures include cellulose, hemicelluloses (in plants), teichoic acids, cell-wall peptidoglycans, extracellular and capsular polysaccharides in bacteria, chitin (invertebrates), and fungal and algal cell-wall saccharides. Among these polysaccharides there are those that have an ability to have a rigid three-

dimensional structure such as cellulose. It is sometimes the case that these structures leave small voids, to be filled by cations such as calcium (Chapter 10), which stabilize the structure, or large voids to be filled by hydrated silica as in many plants (see Chapter 20) which make composites.

Turning to other structural roles, carbohydrates are known to influence the chemical and physical properties of glycoproteins; these properties include three-dimensional structure and susceptibility to degradative enzymes and heat. For example, the activity of 'antifreeze' glycoproteins, e.g. those which inhibit the formation of ice crystals in the blood of arctic fish, is destroyed upon modification of its carbohydrates. For many other glycoproteins, however, removal or alteration of the carbohydrates does not result in qualitative change of their functional capability but rather in a change of physical stability.

Carbohydrates may also induce conformational changes in glycoproteins; this has been shown in fibronectin (a molecule responsible for anchoring and guiding the migration of cells), the three-dimensional structure of which is altered by both heparin and heparin sulphate (the carbohydrate to which it normally binds). This saccharide-induced change in three-dimensional structure is thought to play a role in maintaining the ability of the molecule to generate a cell-binding domain.

7.21.3 Recognizability of short mobile polysaccharide structures

The cell surfaces of multicellular organisms are equipped with protein receptors and ligands for the recognition of free molecules, specific substrates, and the surfaces of other cells. Such machinery is also found within cells and is involved, for example, in the intracellular routing of proteins. Numerous studies have suggested that polysaccharides (specifically, branched oligosaccharides) participate in such recognition events. These phenomena can be classified as cell–molecule, cell–substratum, cell–cell interactions, and intracellular routing of glycoproteins. Other phenomena of importance to saccharide function include species- and organ-specific glycosylation, microheterogeneity, and site-specific glycosylation. Once again, polysaccharide/polysaccharide interactions are frequently mediated by cations and it is *calcium* which demonstrates this association; these effects become particularly important in cell–cell interactions. These interactions are not to be seen through such images as die/mould or lock/key fitting but rather as mobile fingers of a hand (branched chain) fitting into a mobile glove and remaining mobile. There is no loss in selectivity in this fitting providing the number of points of contact is large.

Involvement of cell-surface oligosaccharides has been shown in species-specific aggregation in several biological systems, including formation of sponges and slime moulds, fertilization in marine algae and invertebrates, the mating reactions of compatible yeasts, genesis and morphology in sea urchin and mouse embryos, interaction of malaria causing parasites within host red blood cells, and the recognition of pathogens and symbionts. Surface carbohydrates are also involved in the binding of cells of infective agents, for example, bacteria and viruses. Targeting of a parasite to host cells has been

shown in *Giardia lamblia* in which a specific surface lectin, activated by secretions from the host's duodenum (the site of parasitic invasion), allows binding to the appropriate host cells.

We now turn quickly to other classes of polysaccharides in order to illustrate the huge variety of 'non-structural' functions they fulfil, especially outside cells. However, the associated roles of other polymers, especially proteins, must not be forgotten; in fact, it is most frequently the combination of proteins with polysaccharides, the glycoproteins, to which we shall be referring, and all the time we must remember the ionic atmosphere.

7.22 Larger random polysaccharides

7.22.1 Non-interactive polysaccharides

In descriptions of biopolymers we are now used to the notion of structure as in the chemical crystalline state, e.g. the structure of proteins. This is valid for certain types of polysaccharide, but another useful starting point for a quite different type of polysaccharide is the perfect-gas or dilute-solution model of random motions. The appropriate structure/dynamic approach is given in texts on synthetic polymers such as polystyrene. The following analysis applies to any random polymer whether it is a protein, a nucleic acid, or a polysaccharide and to branched or straight chains which are contained in a vessel which need not be rigid, e.g. a balloon or a skin. One can start from concepts such as the osmotic pressure, π, where one writes, without reference to molecular structure,

$$\pi = RTc \tag{7.1}$$

where c is the concentration of solute in moles per litre, R is the gas constant, and T is temperature. Next, we take into account either repulsive or attractive interactions, with solvent and between molecules, in the second and further virial expansion terms, A

$$\pi = RT(c + Ac^2 \ldots). \tag{7.2}$$

The very fact that the polymers are made of large monomeric units causes there to be an excluded zone (often called the excluded volume) and the factor A also includes this effect. Looking at this more carefully, we see that, for a polymer molecule, the 'concentration' depends on the ratio of molecular volumes of individual polymers to the volume of a molecule of solvent. This ratio is not fixed for polymer molecules, since there are internal variations in volume with concentration; hence the amount of free solvent varies depending on the degree of swelling of the polymers. These two, swelling and solvent activity, are interactive terms so that there are additional terms in the virial expansion due to the internal constructions in single polymer molecules. The restriction on space and then the solvation of the polymer is limited by its free volume at a fixed concentration.

A parallel starting point for the dynamics of molecules in solution is the viscosity of a polymer solution. Ideally,

$$\eta = \eta_s(1 + c\eta) \tag{7.3}$$

where η_s is the solvent viscosity. Once again, it is necessary to take into account the internal and external interactions between polymer molecules and solvent and, assuming them to be in equilibrium, we expand the viscosity equation

$$\eta = \eta_s(1 + c\eta + Bc^2\eta^2 + \ldots) \qquad (7.4)$$

where the higher order terms, B, etc., will be related to internal and external interaction dynamics with the solvent. Unfortunately, there is insufficient space in this article to deal with the properties of the mutual flow of polymers and solvent which is so very important in biological movements, e.g. on the outside of aquatic animals or in the alimentary canals of all animals. (All these terms depend also on the inorganic contents of the solutions as we shall see.)

Now, when the solutions become concentrated the polymers lose freedom of movement except locally due to entanglements from which they cannot escape. (The properties of entangled non-interacting polysaccharides, for example, in mucous secretions, is described by theories such as the tube theory of Doi and Edwards, see Further reading.) These entangled polymers which are held to or within a floating frame (e.g. a cell membrane or a skin) will be restricted in volume by any surface they meet. In other words, work is done when the polysaccharide is constrained to a volume smaller than the random walk volume; in effect, a constrained random polysaccharide measures available volume through its internal pressure (eqn (7.1)).

7.22.2 Random extensions of glycoproteins: cell surface packing

Another way in which expanded regions of polysaccharides attached to proteins in membranes may be responsible for, or related to, volume restrictions is through spatial packing which relates to shapes of cells. For a spherical cell the surface tension of the membrane is uniform, so that packing of molecules in the membrane is only radially different, i.e. from inside to outside. In the case of a non-spherical membrane with regions of different curvature local surface tension would be different if the composition of the membrane were uniform. As a consequence, components of the membrane sort themselves so as to reduce the differences. This differential packing of molecules in curved and straight membrane zones is shown in Fig. 7.24.

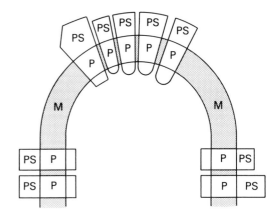

Fig. 7.24 Membrane surface tension. The polysaccharides, PS, bound to proteins, P, are shown defining the curvature of a cell surface by virtue of the way in which their size and shape dominate packing in the membrane, M.

Conversely, the presence of differently shaped molecules can *cause* the curvature. Packing and shape now involve an interaction between external packing forces and the tensions of filamentous structures internal to the cell in contact with the surface (Fig. 7.25). The function of different polysaccharides which are on the outer surface of membrane proteins could then be to assist localization of protein activities in zones along the inside as well as the outside of the membrane; all these associations will be affected by cation binding if the surfaces are negatively charged.

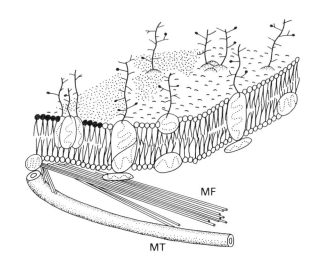

Fig. 7.25 Membrane protein networks: MT, microtubules; MF microfilaments. The precise organization is thought to be critically dependent on the level of Ca^{2+} and Mg^{2+} ions in the cells and in the surrounding medium.

7.22.3 Polysaccharides showing chain–chain repulsion or attraction

In this section we shall consider sol and gel structures of polysaccharides. A limitation away from a random coil can be based on repulsive or attractive energies as well as on simple space-filling considerations. Examples include:

(1) anionic linear polysaccharides (repulsions) (NB. Note that DNA and RNA are really in this class);

(2) linear polysaccharides interacting through H-bonds (attractions).

We shall only look briefly at hydrophobic forces as these do not appear to be of such dominant importance in polysaccharides as compared with proteins and nucleic acids. Each new force is an extra term in the virial expansion of eqns (7.1) and (7.2) but it also introduces some structural chemical specificity, much of which now depends on selected sidechains.

7.22.4 Anionic linear polymers

A repeated anionic unit in a linear polymer will tend to stretch out the molecule (cf. DNA). Such a structure will resist collapse of its volume, as for a random polymer, but that volume is now proportional also to the increase in chain/chain electrostatic repulsion. This repulsion pushes the polysaccharide away from any other negatively charged surface and the polysaccharide will

sense long-range electrostatic fields as well as short-range volume constraints. However, if the polymer is in the shape of a rod, it can sense direction too.

It is not too difficult to see that, if we build units of a large number of negatively charged but otherwise random polysaccharide chains on to a (protein) backbone and then concentrate the units, a very open highly hydrated mesh will be formed (Fig. 7.26). This mesh, usually of chondroitin sulphates, provides a relatively fixed volume through which molecules can diffuse rapidly to and from cells—it is somewhat like a thixotropic ion-exchange gel in that it can change shape considerably, but through attachment and containment, not inordinately. These sol/gel structures are of decisive

Fig. 7.26 (a) Structural formulas of the repeating disaccharide units of some major anionic sugars which are part of connective tissue. (b) Structure of the large aggregating chondroitin sulphate proteoglycan from cartilage. The domain arrangement of the molecules is shown (i), with the three globular domains (G1, G2, and G3), the two chondroitin sulphate domains (CS1 and CS2), and the keratan-sulphate-rich region (KS). The aggregate that forms on interaction with hyaluronate (HA) and which is stabilized by the link protein is shown in (ii).

Fig. 7.27 A typical structural diagram of gel/sol changes with ionic concentrations.

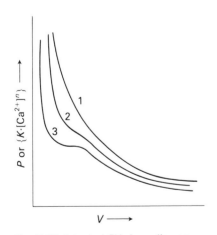

Fig. 7.28 A typical *PV* phase diagram substituting calcium ion concentrations for *P* and the molar volume of a polysaccharide for *V*. The ideal case is the top curve (1). The bottom curve (3) illustrates phase separation. *K* is the binding constant.

importance for multicellular systems since it is within them that cells are active even to the point of division.

Now just because these polysaccharides are *anionic* gels, or sols, they are extremely dependent upon the inorganic salt medium in which they reside. Two types of interaction affect their volume:

(1) the general ion effects (Debye–Hückel–Onsager);

(2) the specific binding (cation) effects.

It is here that the inorganic chemistry of biology again becomes important.

General theories of ion interaction with polyelectrolytes have been developed (see Kotyk *et al.* in Further reading). These interactions can cause collapse of a widely dispersed solution of an anionic polymer, a sol, down to a very small volume, a gel, over a very small concentration range of cation (Fig. 7.27). Furthermore, like all other attractive forces, they give rise to equations of state with terms parallel to those for condensable non-ideal gases (eqn (7.2)). This is illustrated in Fig. 7.28 and we note that such condensations can be selective to the calcium ion in particular (Fig. 7.29).

So far, these polymers have been considered as free units; now one must ask what value they could have when bound to a surface, i.e. as in Fig. 7.25. Polymers (such as these) will not be very useful as sensors of external conditions on the surface of single cells if the cell is a sphere and membrane-bound molecules can move over the surface such that the cell remains isotropic. In such a case, escape of the polymer from repulsion by a surface (of the same charge) could be managed by diffusion of the anionic polymer along the membrane. A rearrangement of glycoproteins would occur but the cell itself might not be repulsed. A more rigid anisotropic cell is more interesting. For example, the red blood cell has a discoidal shape that could not arise if the membrane was under isotropic constraints. In fact, the shape relates to

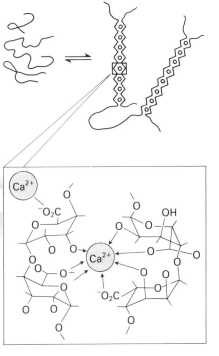

Fig. 7.29 Schematic representation of association of polyglucoronate sequences by chelation of Ca^{2+}, the 'egg-box model'. Oxygen atoms co-ordinated to calcium are shown by arrows.

differential radii of curvature at different points which is due to internal fibrillar structures (Fig. 7.25). This, in turn, means that there will be differential features along the membrane and that these features are directly related to curvature (Fig. 7.24). In essence, a shaped membrane can relieve new tension applied to it by re-arrangement of molecules. When such a cell strikes a constraint, the interaction cannot easily be relieved by protein diffusion round the cell surface. Instead, a pressure is communicated *into* the cell, i.e. *signalling* will occur (see Fig. 7.30). A message enters the cell when a mechanical pressure or an electrical repulsion is exerted first on the random (charged) polysaccharides and then, say, via a membrane protein helix (adjustable in an up and down sense), is transferred to the intracellular protein trigger connected to the fibrils. In effect, the polysaccharide acts as an entropy sensor. The cell will now change shape. These processes may perhaps be assisted by metabolic events communicated to stress fibre systems including the opening and closing of channels in the membranes (see Fig. 7.30). The cell then moves its membrane away from the constraint. By this mechanism, a red cell under pressure from the flow of blood, for example, could squeeze into a capillary smaller in diameter than the initial size of the cell itself. As stated earlier, this type of sensor can be directional, especially if the sensor is rod-like and carries charge. The message could also consist of the inward flow of cations, especially Na^+ and Ca^{2+}, if the rearranged membrane proteins form a channel (Fig. 7.30).

The interaction with a surface may also be attractive due to the attachment between negative surfaces and negative polysaccharides through the mediation of calcium ions. Extension (but not retraction) of membranes into pseudopods is aided by such interaction. Cell–cell interactions made in this way generate communication networks between cells; these are positive communications similar, in essence, to those generated by repulsion, as described earlier in this section. In part, this binding is related to the sol–gel changes produced by ions (see Fig. 7.27). Many cells move along surfaces by using flowing contact through charged polysaccharide/charged surface contacts involving calcium.

In practice, one observes many carboxylated and sulphated polysaccharides with a wide variety of composition giving the same function (e.g. chondroitin sulphates). Some of these functions are those typical of non-biological polymers, e.g. adhesives, flow-assisting fluids, etc., but here one should stress the value of the polymers as volume and field sensors and note that these are properties usually not associated with proteins.

Fig. 7.30 Push–pull mechanisms. The structure on the left represents a glycoprotein connected to a single helix device (push–pull signal) while on the right the connection is to a multihelix channel signal. In the first case, force on the polysaccharide restricts its volume; hence it exerts a pressure on the helix, e.g. glycophorin. An alternative mechanism is for the pressure or binding agent, e.g. an antibody, to bring together two molecules in the membrane so as to create an ion channel with series of helices in the membrane (Chapter 3).

7.23 The cation interactions

At this stage of the discussion it is clear that polysaccharides were 'designed' to interact with cations to very different degrees so that the full potential value of their structural variety was developed. Their selectivity can be based especially upon the incorporation of anionic sidechains. There are three ways of proceeding: (1) very low density of such sidechains, as in celluloses, with virtually no role for cations but a potential interaction with hydrated silica particles, multiple charged positive surfaces, or possibly crystalline calcium salts which tend to carry low negative charge; (2) increasing anionic density due to sulphation or formation of diester phosphates rather than carboxylation where such polymer chains repel and have poor binding interactions with cations, hence the polymers are of great value in that they can provide a very open gel through which molecules can diffuse to and from cells; (3) increasing anionic density but now including carboxylate and finally mono-ester phosphate centres. These latter centres interact more strongly with cations (see Chapter 10) and the polysaccharide chains condense around the cations; important examples include the roles of Na^+ and K^+ (general cation effect) and of Mg^{2+} and Ca^{2+} (specific cation effect) in the stabilization of walls of cells, of connective tissue, of cell/cell interactions, and so on. We shall look again at this problem of immense consequence in bio-inorganic interactions in later chapters, Chapters 8–10 especially.

7.24 Conclusions about polysaccharides

It is difficult to draw a simple conclusion. We have effectively stated that polysaccharides must often be described by equations of state similar to those used for liquids due to their high mobility. We cannot discuss their properties in terms of molecular structures except through the full development of statistical mechanics. Yet, as we have said earlier, equations of state rather than structures apply equally to some proteins and nucleic acids and, as we shall see next, to membranes. These apparently separate groups of polymers are actually all knitted together (see Figs 7.19–7.24, for example). Moreover, they are interactive with the aqueous phase which is energized through its ion content. This means that, like the other polymers of biology, the whole of their activity is feedback-linked to the energization of inorganic contents.

7.25 Properties of lipids

7.25.1 Lipids in membranes as macromolecular assemblies

Lipid molecules in themselves are polymers based on such chains as $-(CH_2)_n-$ and, obviously, such chains do not become involved with the binding of inorganic elements; they are very loosely held to one another and the membranes of biology are liquid rather than solid barriers between aqueous

solutions. The headgroups of the lipids are polar and are exposed to ions. However, for the most part we shall not be concerned with single molecules of fatty acids or phospholipids but with the separated membrane phase which they form. This phase has a repeating pattern of headgroups in water, such as saccharide $-CO_2^-$ or $-(RO)_2PO_2^-$, and such headgroup repeats are reminiscent of the grooves of DNA and have the same interest for us. We shall therefore treat the membrane surface as if it were a two-dimensional, very large, mobile, fluctuating macromolecule carrying a considerable electrostatic potential and proceed to the discussion of membrane organization and dynamics and the effect of cations and anions upon them. First, we must see the complications in such an approach by examining the molecules involved in some more detail.

7.25.2 Composition and general functions of lipids

As is now clear, the basic framework of the biological membranes (see Chapter 3) consists of double layers of lipid molecules held together mostly by hydrophobic van der Waal's forces (see Fig. 7.31). Embedded in these double layers one finds several 'integral' proteins (i.e. non-dissociable without disruption of the membrane) with carbohydrate chains attached to them and protruding from their external, extracellular surface, as also happens with the glycolipids discussed later in this section. The relative percentages of these three components of membranes (lipids, proteins, and carbohydrates) vary considerably from one membrane type to another; usually one finds 25–65 per cent lipids, 25–62 per cent proteins, and about 10 per cent carbohydrates. Figure 7.25, which is similar to Fig. 3.9, is the accepted model of the biological membrane, based on that proposed in 1972 by Singer and Nicholson (see 'Further reading'), in which the various components are represented. The arrangement is not static but in some regions the proteins are highly organized in complexes. Besides the 'integral' proteins, membranes also contain 'peripheral' proteins and glycoproteins associated with their internal and external surfaces respectively (see Figs 7.24 and 7.25).

All membrane lipids are amphipatic, i.e. they have a hydrophilic head group and a hydrophobic (double) tail in their molecules. Most of them are phosphoglycerides, i.e. derivatives of glycerol with two of its three hydroxyl groups esterified with fatty acids and the third esterified with a phosphorylated molecule which may be choline, ethanolamine, serine, or inositol. The fatty acid residues are long hydrocarbon chains, generally with 12 to 18 carbon atoms, which can be saturated or unsaturated, usually with one or two double bonds (see Fig. 7.32 and Fig. 7.33). These membrane lipids are, therefore, diglycerides, and differ from storage fats (non-membrane lipids) that are triglycerides, i.e. have the three hydroxyl groups of glycerol esterified with fatty acids. The other lipids that occur in biological membranes are cholesterol (a disc-shaped molecule with four fused rings –A, B, C, D–, a short hydrocarbon tail, and an hydroxyl group attached to ring A (see Fig. 7.33) and some derivatives of sphingosine, an amino alcohol with a long unsaturated C_{15} hydrocarbon chain in which the amino group is linked to a fatty acid residue. When the hydrophilic head group of these lipids is phosphorylcholine the resulting derivative is sphingomyelin; when it is a carbohydrate one obtains a glycolipid, either a cerebroside (with a single saccharide) or a

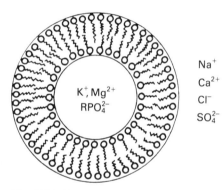

Fig. 7.31 (Below) A diagram of the bilipid layer membrane of a vesicle or a cell with (above) a typical lipid, phosphatidylcholine. The ions surrounding and inside a cell are shown but the distribution is reversed in vesicles.

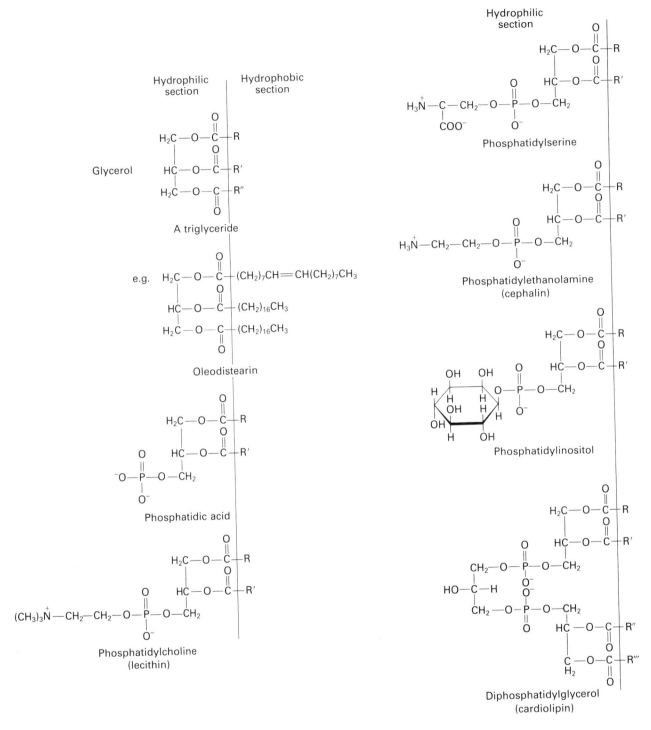

Fig. 7.32 Derivatives of glycerol-forming lipids for membranes.

Table 7.14 Some important molecular characteristics of biolipids

Characteristic	Property affected
Unsaturation	Stiffness, packing, melting point
Branched chain	Stiffness, packing, etc.
Chain length	Packing, solubility, condensation
Ether link	Lower hydration
Polysaccharide chain	Recognition, space-filling
Anion groups	Cation (proton) binding
Cationic groups	Anion binding
Protein attachment	Membrane function, light capture
Cholesterol inclusion	Stiffness, packing

ganglioside (with an oligosaccharide) (see Fig. 7.22). It is clear that the different lipids confer different properties on a membrane (Table 7.14).

All these lipids occur in variable amounts in different membranes, (Fig. 7.33 and Table 7.15) but the glycolipids are always in the outer parts of the bilayers—compare the 'integral' proteins. Furthermore, the lipids of the two layers *are always different*; for instance, in the red blood cell membrane, the outer layer contains more phosphatidylcholine and sphingomyelin, but the inner layer contains more phosphatidylethanolamine and phosphatidylserine. This asymmetry may be related to the difference in available surface of outer and inner monolayers to accommodate the charged headgroups of the lipids—note the difference in volume of the headgroups $-NH_4^+$ and $-N(CH_3)_4^+$ which occur in the four lipids just mentioned; hence the smaller groups occur in the lipids of the inner layer (Fig. 7.31). The difference is then related to the curvature of the membrane. The distribution may also be a requirement to avoid unequal surface tensions and lateral pressures in the membranes that might affect the translational movement of the proteins which span the membrane. Indeed, unless constrained from moving, the membrane proteins of eukaryotic cells can diffuse in the lipid membrane and this movement can be fast, e.g. proteins may go from one extreme of a cell to the other in only a few minutes. The same is, of course, true for the lipid molecules in each layer. Finally, the unequal distribution may be related to the chemical, especially the ion, content of the inner and outer aqueous phases.

Fig. 7.33 The content of different lipids in the human red cell membrane.

Table 7.15 The different amounts of different lipids and proteins in different membranes

Fraction	Molar ratios		Ganglioside (μmol/mg · **N**)
	Cholesterol: phospholipids	Cerebrosides: phospholipids	
Cerebral membranes			
Myelin (large fragments)	1.03	0.5	0.07
Myelin (small fragments)	0.73	0.25	0.18
External synaptosome membrane	0.72	0.05	0.27
Synaptic vesicles	0.41	Very low	0.07
Microsomes	0.48	0.09	0.14
Mitochondria	0.15	0.03	0.03
Other membranes			
Erythrocytes (plasma membrane)	1.17	—	—
Intestine (brush borders)	1.11	—	—
Liver (plasma membrane)	0.56	—	—

Mobility of the proteins in the membrane is an important factor and we shall describe its importance in Chapter 8. Before that, it is useful to call attention to the fact that membrane proteins can be classified into two types according to their shapes within the hydrophobic interior of the membranes; one has rod-like tightly coiled α-helices of hydrophobic amino acids with their sidechains projecting outward from the helices; the other has, essentially, a globular structure within the membrane hydrophobic region. Usually, the rod-like proteins span the cell membrane and their external segments in eukaryotic cells always have several hydrophilic oligosaccharide chains bonded to various backbone amino acids (nine types of saccharides in up to 15 sugar chains); the internal fragment, in contact with the cytoplasm, has a short tail of hydrophilic amino acids but no saccharide molecules (Figs 7.24 and 7.30).

7.25.3 Membrane functions and their lateral anisotropic organization

An artificial bi-layer membrane of one single phospholipid is expected to be spherical and homogeneous (Fig. 7.31); natural membranes, however, are far more complex and some parts of them may be different from others, not only in their composition, as was seen in the previous section, but also in their properties and reactions (Chapter 3). One can refer to this feature as microheterogeneity, microzoning, or anisotropy and this applies both along and across membranes. Some examples of localized reactions have already been commented upon, e.g. those associated with receptors and coated pits (see Chapter 3), but it should also be realized that this anisotropy is time-dependent since the membranes are rather fluid and their components are generally mobile. Fluidity is, of course, an important property since it conditions, to a considerable extent, the movement of materials within

membranes and hence that of the proteins embedded in them. Membranes are then structured liquid phases obeying complicated equations of the kind (7.2) and (7.4). Fluidity must be maintained homeostatically and this is achieved not only by general temperature control but by changes in the lipid composition of membranes. Since unsaturated fatty acids and fatty acids of shorter chains have lower melting points, as the temperature drops more of these acids are incorporated in the membranes and the opposite is found if the temperature increases. Cholesterol has a complementary regulatory effect, causing higher melting phospholipids to become more mobile and lower melting phospholipids to become less mobile; *in vivo*, at physiological temperature, it increases viscosity and restricts the mobility of the phospholipids. Naturally, since the inner surface of the membranes has a higher percentage of unsaturated fatty acids, it probably is more fluid unless its content of cholesterol is higher to counterbalance the effect of the fatty acids. All this means that fluidity is an all-important property that must be under metabolic control. (Cholesterol is such an unusual component of membranes that we must ask about its appearance in evolution, section 7.26.)

The mobility of the membrane proteins (or lipids, in some cases) may also be affected by the addition of 'ligands' to the membranes, e.g. Ca^{2+}, protonated polyamines, or other peripheral proteins. In some cases this may lead to the formation of 'clusters', 'patches', or 'caps' (see Fig. 7.34).

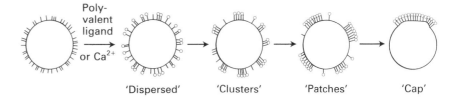

Fig. 7.34 Formation of clusters, patches, or caps driven by cations, e.g. Ca^{2+}, or cationic organic compounds.

7.25.4 Ion association with lipid surfaces

Since membrane surfaces are negative they will accumulate cations through both a general salt effect and specific binding. These are very hard to distinguish since the types of group exposed on the membrane surface, O-donors, have little selective binding for different cations. We expect binding to be important, however, since the charge density on the surface of a membrane is high. We also expect polyamines or surfaces which closely resemble polyamines, e.g. those of polylysine and cytochrome c, to bind well to these membranes. As in the case of DNA a competition exists between organic and inorganic cations for the surfaces

$$M^{2+}\text{surface} + \text{polyamine} \rightleftharpoons \text{polyamine surface} + M^{2+}.$$

The modulation of the interaction, as in the case of DNA, can be managed: (1) by altering M^{2+} (pumps); (2) by synthesis of the polyamine; and (3) by modifying the surface, for example, by phosphorylation.

7.25.5 The influences of selected inorganic elements

We have seen that ions are pumped across the membranes so that the lipid

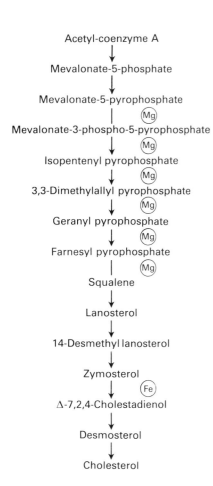

Cholesterol biosynthetic pathway

bilayer faces a gradient from the cytoplasm of high K^+ (Mg^{2+}) and phosphate anions, which is opposed by high Na^+, Ca^{2+}, (H^+), and Cl^- ions in both the outer bathing solution and in most vesicles (Fig. 7.31). These energized gradients can interact with the lipids, so that the distribution of lipids across a membrane is energized by the chemical potentials (not just electric potentials) at the surfaces. The strength of the interaction is very like that discussed in Section 7.22. Undoubtedly, the most effective ion involved in this interaction is the calcium ion. Most of the charged saccharides attached to the lipids also carry negative charge and as such they assist calcium interaction; we show this schematically in Fig. 7.34. Insofar as there is lateral discrimination of ionic concentrations due to steady-state flow of ions along membranes there will be lateral organization of lipid molecules also. Another form of energization is phosphorylation of the proteins on the inside of the bilayer; for example such phosphorylation is known to control the organization of protein complexes in the thylakoid membrane.

We assume that there is no direct binding of trace elements or of monovalent ions to the membranes.

7.25.6 Potentials on membrane surfaces

The fact that positive ions are unequally distributed in the aqueous phases on either side of membranes gives an inner to outer *solution* potential across membranes (see Chapter 8). Quite different, though interactive, is the potential which arises from the (anion) surface charge disposed differentially across the 50 Å of the membrane. The interaction of this *surface* potential with the *free* (not bound) ions in solution is usually handled by Gouy–Chapman theory (see Kotyk *et al.*, in Further reading), so that there is a summed potential. If the membrane is not isotropic, local areas have local surface potentials and we must then always be aware of the possible development of local eddy currents through holes in membranes while a steady state is maintained by ion pumping.

7.26 The connection between lipids, hormones, and coenzymes

7.26.1 Cholesterol

The lipids are not just used to form membranes. Cholesterol is an example. This molecule (see Fig. 7.33) is present in animal but not plant or prokaryote membranes. It requires dioxygen for its synthesis from the lipid squalene utilizing an iron enzyme (see the diagram of the 'cholesterol biosynthetic pathway' in the margin). As stated, its function, like that of the variation of the degree of unsaturation in the fatty acid chains of lipids, is to modulate the fluidity of animal membranes. However, it is important in a second respect in that it is the starting material for the steroid hormones. The further hydroxylations required also demand iron enzymes.

7.26.2 Essential fatty acids

Animals cannot synthesize all fatty acids. Of particular interest are the linoleic acids, which they require in their diet. Their release as free fatty acid glycerides from inositol phosphate derivatives is an extremely important mode of signalling associated with Ca^{2+} and kinase C. They are also precursors of arachidonic acid and then of the prostaglandins. The syntheses of the latter require iron enzymes and dioxygen.

There is a grossly intriguing feature of the secondary metabolism described in this section, namely, that it is so dependent on dioxygen and iron. Behind the hormonal activities of sterols and prostaglandins lies a deep level of homeostatic and morphological control, that of iron.

A further important point not to be missed in the context of this book is the way many essential molecules have interlinked syntheses, for example, related to isopentenyl pyrophosphate, the reactions of which require magnesium, see opposite, which leads to ubiquinone, cholesterol, and dolichol.

7.27 Summary of biopolymers

In section 7.21.1 we summarized the complementary nature of the three groups of large biological polymers, proteins, nucleic acids, and polysaccharides based upon a description of the way in which their monomers were built into sequences and the structures which then developed, see Table 7.12. We then turned to situations in which the properties of the polymers, overwhelmingly the polysaccharides, were described better by bulk, not molecular, properties and equations of state. This approach was also found to be suitable in large measure for the description of fatty acids in lipid membranes. The fatty acid polymers separate zones of action and create an organic phase as well as separated aqueous compartments for reactions, but they cannot be usefully described through structures of isolated molecules.

The biological polymers are invariably energized in more than one way. They are energized in the obvious sense that their chemical bonds are unstable to hydrolysis. Additionally, the polymer is energized by the way it is placed and contained in space—polysaccharides outside cells, for example. In the local space in which it is found the polymer is obviously in a gravitational field but may also be in an electrical field. Such fields can separate lipids across membranes and affect the conformation of a flexible polymer, e.g. DNA. While any polymer distribution is energized, so is the distribution of all interacting polymers, small molecules, and ions by pumps and metabolism. The polymer, through its interactions is then additionally energized. Finally, since the energized polymer is attached to other energized polymers the whole system cannot relax so that the polymer has to some degree a configurational energization. Some if not all of these factors are seen in the most obvious features of biological molecules—they are integrated in an organization.

Now these interactions are dependent on the presence of elements other than those of organic chemistry. First we note that proteins are largely made from H, C, N, O, S, nucleic acids from H, C, N, O, P, and polysaccharides and lipids from H, C, O. The surfaces generated by these polymers give rise to very

different ways of binding 'inorganic' ions as well as organic molecules. Each and every one of the binding ions has an energized distribution so that all the polymers are energized differentially. The polynucleotides and polysaccharides give a narrow range of electrostatic interactions, e.g. with Na^+, K^+, Ca^{2+}, and Mg^{2+}. Membrane lipid two-dimensional surfaces come into this category too. Proteins bind a vastly richer variety of metal ions and anions. Finally, there is covalent modification which is only really different from cation binding in its kinetics. Thus all the polymers can be phosphorylated, amidated, methylated, and so on. Apart from the thermodynamic differences kinetic distinctions arise as follows and each class of polymer can be modified:

Simple cation and anion binding	Phosphorylation (esters)	Methylation
Electrostatic	(Covalent)	Covalent
Fast	Moderately slow	Very slow

Biological systems do not understand chemical bonding but they behave as if they do since evolution is a teacher *par excellence* and, following the functional value of chemical modification of polymers, we see how all biological polymers are wonderfully kinetically controlled as if it were guided by expert hands.

Before concluding, we must stress again that the chemical and physical combination of the four types of polymer has produced a network of interactive energized phases. It is essential that we do not allow analyses of individual extracted molecules or of artificial vesicles, i.e. highly localized analyses, to deceive us into thinking that molecular structure holds the key to biological activity. The structures are valid important starting points, no more. Mobile energized molecules and semiliquid phases are organized in many very different ways and need different thinking. The next Chapters 8–9 try to show the roles played by different elements and only in Chapter 21 do we attempt discussion of the integration.

Further reading

Blackburn, G. M., and Gait, M. J. (eds) (1990). *Nucleic acids in chemistry and biology.* IRL Press, Oxford University, Oxford.

Bock, G. and Harnett, S. (eds) (1989). Carbohydrate recognition in cellular function. Ciba Foundation Symposium, **145**.

Creighton, T. E. (1984). *Proteins, structures, and molecular properties.* W. H. Freeman, New York.

Darnell, J., Lodish, H., and Baltimore, D. (1990). *Molecular cell biology* (2nd edn). Scientific American Books, New York.

Doi, M. and Edwards, S. F. (1986). *The theory of polymer dynamics.* Oxford University Press, Oxford.

Evans, W. H. and Graham, J. M. (1989). *Membrane structure and function*, In Focus Series. Oxford University Press, Oxford.

Harding, S. E. and Rowe, A. J. (ed.) (1988). *Dynamic properties of biomolecular assemblies*, Royal Society of Chemistry Special Publication, No. 74. Royal Society of Chemistry, London.

Klug, A. (1983). From protein structure to biological assemblies. *Angewandte Chemie*, **22**, 565–82.

Lewin, B. (1989). *Genes IV*. Oxford University Press, Oxford.

Kotyk, A., Janáček, K., and Koryta, J. (1988). *Biophysical Chemistry of membrane functions*. J. Wiley, Chichester, U.K.

Oxford University Press, Electronic Publishing. A variety of biological molecular structures are available.

Perutz, M. F. (1978). An account of haemoglobin. *Scientific American*, **249**, 68.

Richardson, J. S. (1981). The astronomy and taxonomy of protein structure. *Advances in Protein Chemistry*, **34**, 168–339.

Sanger, W. (1983). *Principles of nucleic acid structure*. Springer Verlag, New York.

Scientific American (1985). *The molecules of life: readings from Scientific American*. W. H. Freeman, New York.

Singer, S. J. and Nicholson, G. L. (1972). The fluid mosaic model of the structure of cell membranes. *Science*, **172**, 720–31.

Vallee, B. L. and Williams, R. J. P. (1968). The entatic state of enzyme active sites. *Proceedings of the National Academy of Sciences of the United States of America*, **59**, 498–505.

Vance, J. E. (ed.) (1987). *Biochemistry of lipids and membranes*. Benjamin/Cummings, Menlo Park, California.

Williams, R. J. P. (1989). NMR studies of mobility within protein structure. *European Journal of Biochemistry*, **183**, 479–97.

PART II

The roles of individual elements in biology

8

Sodium, potassium, and chlorine: osmotic control, electrolytic equilibria, and currents

8.1	Introduction	223
8.2	Passive diffusion	225
8.3	Gated channels	230
8.4	Channel selectivity and possible constructions	230
8.5	Active transport: pumps	232
8.6	The nature of selective pumps	236
8.7	Building electrolytic circuits	236
8.8	Simple salts and the conditions of polyelectrolytes	239
8.9	Enzymes requiring potassium	240
8.10	Ion energies: summary	240
8.11	Conclusions	241

8.1 Introduction

There are large amounts of sodium, potassium, and chloride ions in sea waters but not very much in fresh waters, yet life abounds in both media and to a very good approximation it holds the internal free ion concentrations of these elements constant, not only in cells but also in the circulating fluids of higher organisms (Table 8.1). Even small deviations from normal levels in man are recognized as symptoms of malfunction or disease. (Common salt is not to be

Table 8.1 Concentration of the main free cations and anions in living cells and their environment (mmol l^{-1})

System	Na$^+$	K$^+$	Ca^{2+}	Mg^{2+}	Cl$^-$	HPO$_4^{2-}$	HCO$_3^-$
Sea water	460	10	10	52	550	<1	30
Valonia	80	400	1.5	50	50	5	(10)?
Red-blood cells	11	92	10^{-4}	2.5	50	3	(10)?
Blood plasma	160	10	2	2	100	~3	30

Group IA elements

Li	(rare)
Na	common
K	common
Rb	(rare)
Cs	(rare)

Group VII elements

F	(not uncommon)
Cl	common
Br	(rare)
I	(rare)

Elements related to Na, K, and Cl

Table 8.2 Binding constants K ($l\ mol^{-1}$) and substitution rate constants k at 25°C

$$M^+ + EDTA^{4-} \underset{k_2}{\overset{k_1}{\rightleftharpoons}} [M(EDTA)]^{3-}$$

Cation	log k_1	log k_2	log K
Li$^+$	7.7	~4.0	~3.7
Na$^+$	7.7	~5.0	~2.7
K$^+$	7.8	~6.5	~1.3
Rb$^+$	8.2	~8.0	~0.2
Cs$^+$	8.3	~8.0	~0.3

$K = k_1/k_2 =$ stability constant and will be some 10^3 times lower at pH = 7.0.

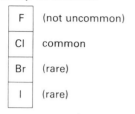

despised within biology.) The constancy of the levels can only be maintained in the biological organisms which dwell in the sea by rejecting much Na$^+$ and Cl$^-$ while accumulating K$^+$. By way of contrast all organisms in fresh waters must accumulate Na$^+$ and Cl$^-$ and they must also accumulate K$^+$. Again in multicellular organisms quite generally the cytoplasm of cells is relatively rich in K$^+$ and poor in Na$^+$ and Cl$^-$ compared with the circulating fluids. Table 8.1 shows some data for the single-cell algae *Valonia* in sea water and for red blood cells surrounded by blood plasma. Obviously, all organisms expend vast energies in keeping these internal/external gradients and they all do so in a very similar way. The balances themselves are related to the general problem of homeostasis treated in Chapter 21. There can be little doubt that these three elements, Na, K, and Cl, are and always have been essential components of life. (As a reminder, elements directly related to them in the periodic table are shown in the margin.)

In this chapter we discuss two related functions: one connected to the concentration (activity) of the same ions, which is associated with the generation of gradients and electrical potential differences, and is a determinant of the osmotic stability of cells, and the other connected to the mobility of these ions, Section 4.5, and hence to the flow of electrolytic charge in living organisms. Only these three elements have simple electrolytic functions and occur in biological systems very largely in the form of monatomic free hydrated cations, Na$^+$, K$^+$ (and note H$^+$), or anions, Cl$^-$. All three ions bind weakly to all organic molecules (see Table 8.2 which gives an example of a very strong chelating agent but relatively weak binding at pH = 7) so that they can only be retained by membrane enclosures very much as in some electrochemical cells. Functional significance in the bound state is rare, but we must remember the binding by cyclic ligands (Section 2.5) which can be used as drugs or poisons.

It is probable that the rejection of the most abundant anion, Cl$^-$, and cation, Na$^+$, was required early in the development of cells to maintain the osmotic balance necessary for cell stability. The inside of cells is of necessity rich in anionic organic molecules, for example DNA, RNA, lipids, and substrates, which implies that some inorganic materials had to be rejected to maintain osmotic balance with the sea. It was also necessary that some positive ions should neutralize the excess negatively charged groups of the biological organic molecules inside the cells since many are ionized at physiological pH, e.g. the carboxylate, phosphate, or sulphate groups of proteins, lipids, nucleic acids, or their constitutive units, as well as of any other compound present. While rejection of chloride partly compensates for this anion accumulation, residual sodium but largely potassium, and to a certain extent magnesium, in effect serve the function of neutralizing negative charge in cells.

Rejection and accumulation require movement across membranes and, although the means by which the individual ions can traverse the membrane have already been referred to in the previous chapters (summarized in Fig. 8.1), some thermodynamic consequences of the processes were not discussed. We shall look first at the connection of the thermodynamic balances to the two main types of movement: (1) passive diffusion down the concentration gradient through (gated) channels; and (2) active transport, 'pumping', against the concentration gradient coupled to an energy-yielding

Fig. 8.1 The two ways in which the simplest cations (and anions) pass through membranes. On the right is a channel which may have a gate; it allows facilitated diffusion. On the left is an ATPase pump. The enzyme is remote from the channel here and it energizes the movement across the membrane usually pumping Na^+, Ca^{2+}, H^+, or Cl^- out of the cell. There are also specific combinations of gated channels which work as exchangers, e.g. the Na^+/Ca^{2+} exchanger.

process, usually the hydrolysis of ATP. Passive diffusion is a spontaneous process driven by an increase of entropy; active transport results in a decrease of entropy and must be driven by an input of energy. Notice that the reverse of pumping is energy capture from pre-established gradients (H^+, Na^+) since it could lead to ATP formation from a cation gradient (see Chapter 5). A steady state of all three ions is maintained in all living systems and can be analysed using thermodynamic expressions but this implies that all movements are controlled.

8.2 Passive diffusion

The free-energy change, ΔG, that occurs during diffusion depends only on the difference of concentration and potentials between the compartments separated by the membrane, reaching zero at equilibrium. Here we leave aside potentials except those due to electrical fields.

For uncharged species ΔG, the free energy of transfer between two compartments of different concentrations, C, is given by

$$\Delta G = RT \ln(C_2/C_1)$$

and it will be negative for processes down the concentration gradient and positive in the reverse direction, representing in this case an energy barrier to be overcome by coupling to an energy-yielding reaction.

For charged electrolytes, however, besides the concentration gradient one must also consider the overall charges in the two compartments, and the difference in free energy between them will then depend on the electric potential difference ψ generated across the membrane

$$\Delta G = RT \ln(C_2/C_1) + z\psi\mathscr{F} \tag{8.1}$$

where z is the charge of the ion considered and \mathscr{F} is the Faraday constant.

Now, biological membranes are not equally permeable to all species due to their particular composition and organization. Among the charged species, membranes are usually impermeable to the ionic charges of the amino-acid

residues (in proteins or not) and to free or bound phosphate, relatively impermeable to sodium ions, to prevent lysis, and more permeable to the larger potassium or chloride ions (but permeability can vary greatly for different types of membranes or even for the same membrane under different conditions). The difference in permeability adjusts the distribution of diffusible ions, as described by Donnan in 1911, and causes what is called a 'Donnan equilibrium' and an electrolyte potential.

To understand this effect let us consider a 'cell' containing only potassium and chloride ions and a charged (anionic) protein, placed in an isotonic solution of potassium chloride

$$
\begin{array}{c|c}
\begin{aligned}
[K^+] &= 100 \text{ mM} \\
[Cl^-] &= 50 \text{ mM} \\
[Prot^-] &= 50 \text{ mM} \\
\text{Inside (cell)}
\end{aligned}
&
\begin{aligned}
[K^+] &= 100 \text{ mM} \\
[Cl^-] &= 100 \text{ mM} \\
\\
\text{Outside}
\end{aligned}
\end{array}
\tag{8.2}
$$

The osmotic pressure of the two sides is equal (since it depends on the number of particles per unit volume), but the concentration gradients of the charged protein and of chloride are not. Since the membrane is permeable to chloride there will be a tendency for this ion to enter the cell. Its movement cannot be compensated by the diffusion of the charged protein in the opposite direction because the membrane is impermeable to this species, whence the diffusion of chloride must be accompanied by a simultaneous movement of potassium ions in the same direction to attempt to maintain the electroneutrality of each compartment.

The final situation can be derived from eqn (8.1), applied to both K^+ and Cl^-, i.e.

$$
\Delta G_{K^+} = RT \ln \frac{[K^+]_{in}}{[K^+]_{out}} + z\mathscr{F}\psi = 0,
\tag{8.3}
$$

$$
\Delta G_{Cl^-} = RT \ln \frac{[Cl^-]_{in}}{[Cl^-]_{out}} - z\mathscr{F}\psi = 0.
\tag{8.4}
$$

Since ψ is the same in eqns (8.2) and (8.3),

$$
\frac{RT}{\mathscr{F}} \ln \frac{[K^+]_{out}}{[K^+]_{in}} = \frac{RT}{\mathscr{F}} \ln \frac{[Cl^-]_{in}}{[Cl^-]_{out}}.
\tag{8.5}
$$

It follows that $[K^+]_{in}[Cl^-]_{in} = [K^+]_{out}[Cl^-]_{out}$, where the brackets represent concentrations or, more accurately, 'activities' of the two ions. The equilibrium condition is, therefore, that of the equality inside and outside the cell of the product of the activities of potassium and chloride.

From this result and considering that $[K^+]_{in} = [Cl^-]_{in} + [Prot^-]_{in}$, one easily derives the final situation, given to the nearest whole number, as

$$
\begin{array}{l|l}
[K^+] \quad = 114 \text{ mM} & [K^+] = 86 \text{ mM} \\
[Cl^-] \quad = 64 \text{ mM} & [Cl^-] = 86 \text{ mM} \\
[Prot^-] = 50 \text{ mM} & \\
\end{array}
\tag{8.6}
$$

$$(\leftarrow H_2O)$$

Inside (cell) || Outside

At this stage, further net movement but not exchange is prevented by the difference of potential generated across the membrane, which can be calculated from eqn (8.3) as

$$
\psi = \frac{RT}{\mathscr{F}} \ln \frac{[K^+]_{out}}{[K^+]_{in}} = 0.059 \log \frac{86}{114} \approx -7 \text{ mV}. \tag{8.7}
$$

It must be said that the value obtained is only a rough indication since we have used several approximations, e.g. total concentrations instead of activities of *free* ions, and neglected volume variations and pH changes that may occur, etc. As to the volume variations the establishment of this 'Donnan' equilibrium has resulted in a net movement of particles into the 'cell' thereby raising its osmotic pressure; the inside becomes hypertonic to its environment and water will enter the cell increasing its volume (see expression (8.6)). (Remember, however, that plant cells and bacteria have a retaining outer wall to prevent such expansion.)

If in the above treatment there had been no diffusible ions outside and only protein$^-$ and K^+ ions (100 mM) initially in the cell, outward diffusion of K^+ would have left an unequilibrated K^+ potential at the membrane. However, we must include one more diffusible ion, the proton. The possible change in pH becomes clearer when considering the case of a negatively charged protein having K^+ as counter-ion on one side of a semipermeable membrane and fresh water on the other side. Naturally, the tendency will be for diffusion of K^+ out, but, since the protein cannot diffuse, the process would be prevented if it were not for the dissociation of water. This does indeed occur and H^+ tends to accumulate on the protein side of the membrane while the corresponding OH^- ions accumulate, along with K^+, on the side of the initially pure water until equilibrium is established. If there are other ions present, as in real biological systems, their effect has to be considered following similar lines of reasoning. Since Ca^{2+} and Mg^{2+} amongst the cations are much lower in concentration and have a high tendency to combine with biological ligands, the electric field potential will be mainly due to Na^+ and K^+ gradients (and to H^+ gradients in particular cases) but there are situations where the calcium potential is important.

This rather detailed description is given to show that from the very beginning of cellular enclosures there was of necessity a link between the simplest inorganic ions, Na^+, K^+, and Cl^-, the pH, and the electrolytic potentials of cells which are crucially important in living systems. Evolution has amplified these effects using energy to vastly increase the gradients by pumping (see *active* transport, Section 8.5).

There are some particular cases of passive diffusion in which the substance

or electrolyte considered does not diffuse by itself but the movement is promoted or facilitated by carriers (see Section 2.5) or by proteins in the membrane (named channels; Fig. 8.2) in a way similar to the catalysis of a reaction, i.e. the protein has no effect on the thermodynamics of the system but acts on the kinetics of diffusion by lowering the energy of activation for transfer and accelerates the rate at which diffusion occurs (Fig. 8.1). As with enzymes, channels facilitate the process in both directions, depending on the concentration gradients, and can be inhibited both competitively and non-competitively. It is important to note that the walls of the channel may be coated with water molecules (Fig. 2.11). In fact, most cells maintain an open potassium channel much of the time but closed channels, if any, for all other ions.

Facilitated diffusion is particularly common for molecules formed in one compartment but needed in another, e.g. in the transfer of sugars and amino acids across cell membranes, but notice that facilitation is very selective. Facilitated diffusion can also be used to destroy cells. The introduction of a small ring chelate carrier (Fig. 2.6) or a protein, e.g. a colicin channel (Fig. 8.3) into a membrane can generate unwanted diffusion and kill a cell. These weapons are known both to bacteria and to drug chemists.

Let us suppose that the membrane is only permeable to one ion, e.g. K^+. If we place electrodes across the cell membrane we can follow the net current

Fig. 8.2 (a) A schematic representation of the 24-helix bundle in a membrane which forms a Na^+ channel. (b) A cross-sectional view of the channel. Such views of channel proteins developed from the work of R. Henderson and N. Unwin.

Fig. 8.3 A detailed description of a protein, colicin A, which can form a channel in a membrane. Notice the multiple helices (shown in detail below) which can be inserted into a membrane leaving a hole through it (Fig. 8.2). The structure is due to the work of G. Tsernoglou.

which flows as we change the applied potential. This potassium current is zero when no potential difference is applied despite any steady-state Donnan potential. The current at any applied potential will also be zero if we close the channels. In fact, we can only observe the potential of those ions which carry current. In the example given at the beginning of this section (expression (8.2)), there could be two extreme cases where the observed current was due to the K^+ current (potassium channels open) or where it was due to Cl^- ions (chloride channels open). (Physiologists do not use the simple thermodynamic equation (8.7) for membrane potentials but a related Nernst equation which includes the permeability of the ions; see 'Further reading'.) Now manipulation of channel opening and closing for different ions has led to profound developments in evolution.

8.3 Gated channels

The channels for simple diffusion may be held closed so that, in the absence of an effector molecule which opens the 'gate' of the channel, there is no ion flow. The usual biological situation is that the gate has a receptor for an effector; on binding to the receptor the gate is opened by the binding energy which causes conformation changes. (Where these conformation changes involve movement of charges on the proteins in the membranes there is a charge drift or gating current *in* the channel without net flow *through* the channel.) Ideas on how the gated channels operate are being developed (see Section 8.5). The opening of a gate allows an influx of ions which can be used as a message either by changing the local electrical potential ψ or the local chemical concentration, C.

A gated channel can be held open or closed by a chemical, as just discussed, or by an applied voltage. The latter are called voltage-gated channels and are opened by the switching of the applied voltage.

8.4 Channel selectivity and possible constructions

Table 8.3 Relative permeabilities of the sodium and potassium channels†

Ion	Na$^+$ channel	K$^+$ channel
Na$^+$	1.00	<0.01
K$^+$	0.09	1.00
Tl$^+$	0.33	2.30
NH$_4^+$	0.16	0.13
Guanidinium$^+$	0.13	<0.01
H$^+$ (H$_3$O$^+$, H$_9$O$_4^+$)	~250.0	Low?
CH$_3$NH$_3^+$	<0.01	<0.01

† Relative permeabilities are ratios to Na$^+$ =1.0 for sodium channels and to K$^+$ =1.0 for potassium channels.

The differences in selectivity between Na$^+$ and K$^+$ channels are remarkable (Table 8.3). Note that:

1. Only Tl$^+$ approaches K$^+$ in permeability in the K$^+$ channel. No larger or smaller ion passes easily. (Tl$^+$ is a useful probe of K$^+$ channels.)

2. The K$^+$ channel excludes NH$_4^+$, which has the same radius, better than does the Na$^+$ channel.

3. Large H-bonding spherical molecules, e.g. guanidinium, pass the Na$^+$ channel.

4. No methylated compounds pass through either channel.

We can propose an explanation for these facts. Though both channels are made from protein helices (Fig. 8.3), the Na$^+$ channel made by the proteins must be the larger. This is the inverse of expectation. It would appear from the data that its low selectivity, when open, could be due to the ease with which ions move in or out through a water layer which partly fills the channel. This would account for high permeability by H$^+$ and very large H-bonding ions which could displace water as they pass. (Note that in fact little H$^+$ does pass since it is at 10^{-7} M.) By way of contrast, the K$^+$ channel is less likely to be hydrated and matches the K$^+$ radius in size. Only Tl$^+$ enters equally to potassium and selectivity is made to be remarkably high since, perhaps, it involves temporary ligation to somewhat polar protein sidechains, Tl$^+$ > K$^+$. The difference between Na$^+$ and K$^+$ channels must be seen against the background of the very great difference in solubility of their salts with hydrophobic anions such as tetraphenyl borate, where Na$^+$ > K$^+$ by 10^6. Highly structured water could generate a proton as opposed to an inorganic ion channel (see Section 2.10).

The description of a channel has been greatly aided by the study of peptides. It is clear that even a single short helical strand of polypeptide, such as gramicidin, can make a voltage-gated channel (Fig. 8.4). The details of such channels are not important here but the shape of the interior of these helices is, in that their exact shape decides the selectivity of ion movement. The interested reader should refer to the literature on the peptides gramicidin and alamethicin. The known features of non-synthetic channel proteins are consistent with a multiple set of helical strands forming a helical bundle across the membrane. Today synthetic peptides have been made which generate such helical channels. A typical composition is

$$H_3^+ N—(Leu—Ser—Ser—Leu—Leu—Ser—Leu)_3—CONH_2.$$

Notice that rates of movement of the ions, Na^+, K^+, and Cl^-, in these channels are close to the diffusion limit, i.e. $10^{10}\ s^{-1}$ for water exchange (Fig. 4.3), but that binding to such ligands as ethylenediaminetetracetate, EDTA, is not quite as fast (Table 8.2). Entry, exit, and passage through channels remains fast even if weak binding to the channel walls occurs.

Fig. 8.4 The probable structures of peptides which form (gated) channels in membranes. Examples are given in the text. The channel helices are usually formed from 20 hydrophobic amino acids.

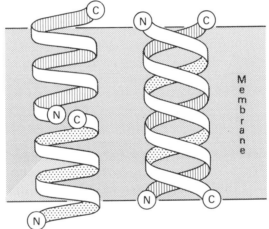

Before proceeding, it may be helpful to give a brief reminder of certain structures of cells. The above-mentioned membrane is the cytoplasmic membrane, but the considerations of ion equilibrations across it through channels when some ions or molecules are trapped inside apply to all vesicle membranes inside cells as well as to the cytoplasmic membrane. Prevention of swelling is managed partly by ion balances, as described, and partly by tension either in the outer membrane or in filamentous structures of the cell. Cells have different structures to maintain stability. Plant cells and many unicellular organisms have a strong outer wall as do mitochondria; animal cells, especially, have a network of inner fibres connected to a protein network under the cytoplasmic membrane (Section 3.5). Vesicles are controlled in shape by the filament networks in the cytoplasm. In this chapter we must note that the Na^+, K^+, and Cl^- balances are linked to the pressures in cells and thence to the mechanical tensions in various parts of an organism.

8.5 Active transport: pumps

Much larger Na^+/K^+ concentration gradients are found in biological systems than those one would expect from Donnan equilibrium considerations, and other mechanisms besides diffusion, facilitated or not, must be operative. For example, a typical K^+ concentration ratio for intra- and extracellular fluids may be of the order of 20:1 (see Table 8.1) and, since the membrane is permeable to potassium in the resting condition, potassium channel open, this gives an approximate potassium membrane potential of about $-80\,mV$ (inside negative relative to the environment). If the membrane channel is permeable to more than a single ion type, the effect of other ions must also be considered when examining observed potentials. Whereas the direction of the potassium gradient is that expected from the presence of ionized protein and other non-diffusible anions, e.g. organic phosphates inside a blood red cell, the sodium gradient is opposite from that expected and this can only happen by a process of active transport ('pumping') coupled to an energy-yielding reaction. If the energy input stops, the transport ceases and sodium ions will enter the cell slowly through appropriate holes or channels. The measured distribution of sodium and potassium inside and outside the blood red cell corresponds to a steady-state condition in which sodium ions are continually moving in by diffusion, through leaks, and pumped out by active transport across specific routes, whereas potassium ions follow the reverse process (see Fig. 8.5). In the resting state of cells it is the sodium channel which is held closed, with the potassium channel open and the pump working.

The mechanisms of generation of this steady state include an enzyme with ATPase activity, i.e. promoting the hydrolysis of ATP, an energy-yielding reaction, which is active only in the presence of Na^+ and K^+ (as well as Mg^{2+}, which acts as a cofactor; Fig. 8.1). The pathways for the ions in the pumps are formed by an association of proteins which cross the membrane, so that each kind of ion may be passed through a series of protein conformational changes to successively weaker and weaker binding sites as it traverses the membrane until being delivered (see Fig. 8.6). The probable motions within the pump are due to relative rotation/translation of helices as in Fig. 8.7 and it is these movements which provide adjustments to binding constants and to gates so as to generate flow. (Photo-halide pumps are described in Chapter 18.)

Pumping systems do not have to be electrically neutral. The Na^+/K^+ ATPase pump moves $2K^+$ in for every $3Na^+$ moved out so that the membrane potential which is established is large and directly related to the hydrolysis energy of ATP.

The establishment of concentration gradients, such as those of sodium and potassium, and, as we have seen, of H^+ (Chapters 1 and 5) provides a means by which free energy can be stored in a living cell very much as in an electrochemical cell. This energy can be made quickly available for various processes, e.g. for the transmission of impulses (messages) in nerve cells; for exchange reactions so as to accumulate or reject other chemicals, e.g. amino acids; for coupling with other chemical reactions, e.g. the 'high-energy' intermediates (ATP) for oxidative and photophosphorylation; for synthesis of new materials, e.g. in the process of replication of bacteria; and for motility

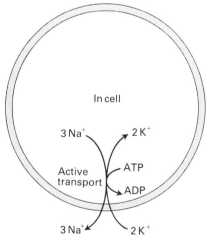

Fig. 8.5 The pumped movements of sodium ions out of and potassium ions into the cell. ATP hydrolysis is required for the movements of the ions against their gradients. The actual coupling is of two K^+ for every three Na^+ so that the pump is electrogenic, i.e. it develops charge and chemical gradients (see Fig. 8.7).

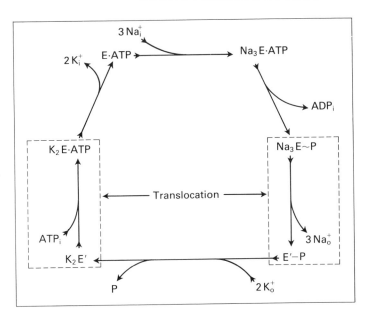

Fig. 8.6 Biochemical reaction scheme for active Na–K transport. Translocation by the enzyme, Na, K-ATPase, across the membrane corresponds also to the biochemical conformational changes of the enzyme from E (Na$^+$ inside the cell) to E' (Na$^+$ outside the cell), or vice versa. The whole process could run backwards to make ATP from large ion gradients but the process would lose energy. ATP, adenosine triphosphate; ADP, adenosine diphosphate; P, inorganic phosphate. K_i^+ and Na_i^+, K_o^+ and Na_o^+ are internal and external potassium and sodium ions, respectively. Note that here and elsewhere ATP and ADP are really $Mg^{2+} \cdot ATP$ and $Mg^{2+} \cdot ADP$.

motors. These links to simple ion gradients are of extreme importance in biology.

As stated, the potential recorded by placing electrodes across a membrane tells us which ions can flow across a membrane. Resting cells are kept highly impermeable to Na$^+$ and most other ions, e.g. Ca^{2+}, Cl$^-$, and HPO$_4^{2-}$, but they are permeable to K$^+$ in the sense that the specific K$^+$ channels are open. Remember there is no loss of K$^+$ despite the huge gradient since the potential ψ and the pumps oppose *net* flow, while allowing exchange. (There are cells of a specialized nature where the resting potential is not a potassium potential but the interested reader must consult a book on physiology to obtain insight.)

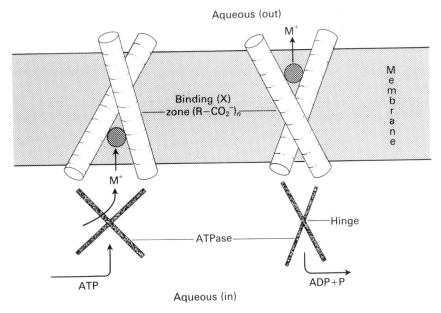

Fig. 8.7 A schematic model for an ATPase pump (or the ATP-synthase). The degradation of ATP to give an ion gradient takes place with movement of M$^+$ (H$^+$, Na$^+$, Ca^{2+}) as it flows from a low to a high concentration (see Chapter 5), left to right in the figure. There are a series of helices in the membrane forming a channel. Binding of nM$^+$ with high affinity at X, (RCO$_2^-$) (left) causes the helices to change conformation to a highly energized state (right) in which binding is much lower so that M$^+$ is released. The movement of the helices is transmitted from a groove-opening reaction which releases ADP (right). Thus energy from ATP makes the gradient. The process can be reversed to make ATP from a gradient.

Through a sudden disturbance, e.g. a light touch on the skin, the K^+ channel in a nerve cell (Fig. 8.8) can be blocked and the Na^+ channel opened when the flows of ions are as shown in Fig. 8.9. The measured Na^+ potential across the membrane is now similar but opposite in sign to the K^+ potential (Fig. 8.10). Moreover, the steady state is lost and a large net inward current of Na^+ will flow which disturbs the potential on the membrane locally.

Fig. 8.8 Motor neuron with myelinated axon.

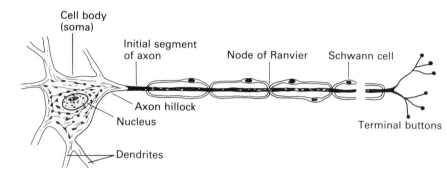

Immediately, the Na^+ channel is closed and the current reverts to a K^+ current. In nerve cells this alternation of potassium steady-state potentials with pulsed sodium potentials due to opening and closing of K^+ and Na^+ gates gives rise to a travelling wave of depolarization. This is the electrolytic message which is largely due to transverse movement of ions and local eddy currents generating a 'travelling Na^+ gate open/K^+ gate closed' message (Fig. 8.9). It is the travelling gating and ungating which is the current message. (At the end of the nerve cell the gates which are opened are those for Ca^{2+} ions or

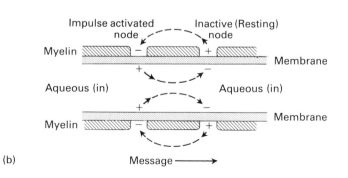

Fig. 8.9 Local current flow around an impulse in an axon, shown as a longitudinal slice. Note that current flow is represented as movement of positive charges. (a) The situation in non-myelinated nerves; (b) that in myelinated nerves.

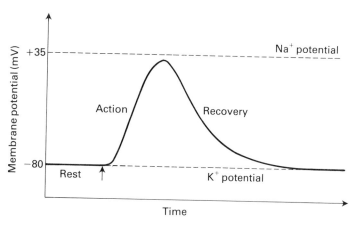

Fig. 8.10 The action potential at a nerve or muscle due to opening of Na$^+$ channels followed by their closing and the re-opening of potassium channels.

for organic chemical messengers since a new type of message is required; see Chapter 10.) This system is the basic process of nerve cells everywhere.

(While we have given a very simple account of relationships between membrane potentials and concentrations gradients in Sections 8.2 to 8.5, there are today extremely sophisticated descriptions of the relationships between membrane voltages, currents, concentration gradients, and, above all, permeabilities and their mechanisms. The interested reader is advised to read an appropriate physiology text; see 'Further reading'.)

In Fig. 8.11 we put side by side two types of circuitry. The first is man's electronic circuit. Man uses electrons in narrow zones of metal or semimetal devices surrounded by the broad insulator, air, or non-conductors such as plastics. Biology cannot make such materials and its only electronic circuits are very different (see Chapters 4 and 12). They carry very little current and are not used for message transmission at all. The major biological circuits are based on electrolytics in elongated cell protrusions where the internal conducting aqueous phase is broad and the insulator layer narrow. However, the currents are not the flow of ions along tubes as might be thought but depolarization waves along the surfaces of the elongations. Such a circuit construction is possible when, as in biology, the current-carrying medium is water and when fatty acid (or glass) membranes are the insulators. Notice that in circuits the power is along the wire of man-made electronic devices, but across all the membranes of the current-carrying devices of biology. Now these biological circuits can incorporate switches (gating systems) or condensers

Fig. 8.11 An illustration of the way in which man's electronic circuits based on metals or semiconductors are related to biological electrolytic circuits. The insulator in the latter is the membrane: the battery in the membrane is described in Chapter 5, the switch is a gated channel, and the condenser is a vesicle store (see Chapter 9). Flow of charge is of the essence of biology (see Chapter 21). S, switch; B, power supply; C, condenser; e, electron; M, metal cation.

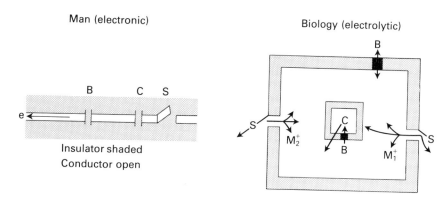

(large concentrations of ions locked in vesicles which may be suddenly released by triggering currents of ions) (Fig. 8.11). In fact, electrolytics has the same versatility as electronics. The evolution of such circuitry is seen at its best in the brain.

8.6 The nature of selective pumps

Certain points from previous chapters can now be re-introduced. In the first instance, the manner of generating ion selectivity can be analysed specifically for Na^+ against K^+ as was described in Chapter 2. The action of ion gating of pumps can also be referred to models of helix/helix motions in membranes. Driving the opening by binding of a ligand is not unlike the adjustment of the binding site of haemoglobin by allosteric effectors (Chapter 7) or movements in calmodulin (Chapter 10). A pump must introduce a one-way drive into the channel opening and closing. A simple picture is that of change from one closed configuration to another driven by ATP hydrolysis. In the movement the gating device is transferred from 'in front of' to 'behind' the ion to be pumped (Fig. 8.7). Simultaneously, the binding constant of the ion to the pump is reduced. The pump returns to its initial unloaded state via a low-energy state. This style of pumping is possible for any molecule or ion.

8.7 Building electrolytic circuits

8.7.1 The positions of pumps and channels

In a real cell the distribution of the pumps and channels for ions is not isotropic along the cell membrane so that there are zones where ions are taken in which are laterally separated from zones where ions are pumped out. This allows ion gradients to develop across rows of epithelial cells, for example (Fig. 3.12). Interestingly, it also allows current to flow via concentration gradients along a cell membrane which are laterally controlled. Another example of the positioning of the channels is that of the sodium channels of nerve cells which are found overwhelmingly at the nodes of Ranvier and not along the whole body of the cell (Fig. 8.8).

8.7.2 The ATPase pumps: some general comments

We have described the Na^+/K^+ ATPase pump for making gradients of these ions and we have noted earlier that H^+ gradients and ATP formation are related reversibly at mitochondrial and chloroplast membranes. In the next chapter we shall describe $Ca^{2+} \cdot$ ATPases, but undoubtedly there are many other ATP-dependent movements of ions and compounds. In the analysis of these pumps it has been found that there are two kinds: (1) without formation of *covalent* intermediate protein phosphorylation; and (2) with such phosphorylation (Table 8.4). There may be quite considerable similarities in

Table 8.4 Classes of ATPase membrane pumps

Class 1: no intermediate		Class 2: phosphorylation	
Mitochondrial	H^+	Cytoplasmic	H^+
Chloroplast	H^+	Cytoplasmic	Na^+/K^+
Prokaryote	H^+	Cytoplasmic	Ca^{2+}
Vacuole	H^+	Vesicular	Ca^{2+}
Vesicle	H^+	Bacterial	Mg^{2+}?
Archaebacteria	Na^+	Acantharia	Sr^{2+}?
		Desmid	Ba^{2+}?

Note. Class 2 proteins are single chains while class 1 proteins are made from multiple subunits.

all the pumps insofar as the construction of the routes through them is concerned; all appear to be based on cross-membrane helices that are mobile with respect to one another. The helical structures generate selectivity of the route. The coupling to ATP is probably through helix/helix motion (Fig. 8.7). The distinction between the two types of ATP-coupled transport lies in the ATPase and its connection to the helical channel. There is known to be close sequence homology between the pumps of class 2, i.e. Na^+ and Ca^{2+} pumps, for example. Here the last step is dephosphorylation of the protein (Fig. 8.6) and the loss of energy in this step leads to a high degree of irreversibility in the energetics. In class 1 pumps there is no phosphorylation step (see Fig. 5.12 and irreversible losses are just those associated with heat loss in any machine and may be small. The desirable differences between the two classes are obvious in that certain steps must not be readily reversed. These pumps are all part of biological electrolytic circuits.

The positioning of these pumps along membranes and the lateral currents induced cause concentration gradients which in one way or another interact with membranes and filaments. Ion gradients are not to be dissociated from considerations of cell shape. Micro-organization of electrical activity is then associated with internal and external macroscopic structure.

The connection between the shape of a cell and the positioning of pumps will be made by the interaction of the product of the pumps with the filamentous internal structure. At the same time, as the outer membrane is extended or contracted by the changed tension developed in the filaments, the curvature of the membrane changes. This change alters the positioning of the ATPase pumps in the new more curved or straightened regions of the membranes since the chemical occupancy is a component of the surface energy (see Fig. 3.6). Pumps and channels therefore develop local current circuits in association with shapes and tensions in cells and the two are interactive. It is likely that the proton and calcium pumps have a larger part to play than the Na^+/K^+ systems since Ca^{2+} and H^+ interact more strongly with the filament structures (see Chapter 12).

8.7.3 Exchangers

We have stressed the nature of pumps and channels and we saw that in some of the pumps more than one ion was involved. To keep the total system of ions under homeostatic control there are also regulating exchangers which are

probably under tight feedback control so that *relative* concentrations of ions are strictly managed. Examples are the Na^+/H^+ and Na^+/Ca^+ exchangers of mitochondria, which are specific for pairs of ions and are controlled. It is observed that some of the exchangers are used to move organic molecules across cell membranes, e.g. the uptake of amino acids. This is another part of homeostasis since it ensures that there is a relationship between organic chemistry in the cell and inorganic balances. In this section of the book we wish to keep our attention on the connections between charged particles, potentials, and currents, turning only now and then to organic ions and metabolism. The exchangers are likely to be structured like pairs of co-operative gated channels.

8.7.4 Organic anions and cations as current carriers

Just as variations in organic chemistry could lead to the development of selective coenzymes such as NADH and NADPH or different nucleotide triphosphates, so variations on inorganic chemistry lead to different message carriers. Thus Mg^{2+} and Ca^{2+} are used sometimes in place of Na^+ and K^+ and HPO_4^{2-} can replace Cl^-. This diversity of current carriers makes it possible to develop fantastically complex parallel circuits since current carriers which are chemically different (note the isolated nature of the electron in man's circuits) can be filtered by specific gated channels or pumps (Fig. 8.11). Now, there is an obvious limit to the variations on the theme of inorganic ions for making parallel processes. The variations are vastly increased by using organic ions and evolution could not fail to find these conductors. It is known that there is a huge variety of ionic messengers (and many non-ionic ones) which carry 'current'. The obvious ones use acetylcholine and adrenalin which carry single positive charges. Note that the ions are based on quarternary bases, which do not go through K^+ or Na^+ channels (Table 8.3) and cannot bind through H-bonds to proteins in a general way but flow as do the inorganic ions Na^+ and Cl^-. The number of possible parallel processes is virtually infinite. Organic reagents which act as blockers of organic current carriers are also an extremely important group of drugs.

We shall see that, when the simple process of current-carrying has to be changed to current (message)-carrying plus mechanical effect at a target, then the organic ions through their size, shape, and chemical variation can and do bind at special sites and in this respect they are an advance on Na^+ and K^+ and are related more closely to Ca^{2+} electrolytic currents. In fact, we now stray from primary messages to secondary action and the inorganic chemistry switches from that of Na^+ and K^+ to that of Ca^{2+} and Mg^{2+} while the Cl^- chemistry switches to HPO_4^{2-} chemistry. We shall see that second organic messengers, for example after acetylcholine (a cation), utilize the organic chemistry of phosphate, e.g. cyclic AMP (cAMP), inside cells.

In Chapter 22 we shall make a final attack on the problem by introducing as pollutants, drugs, medicines, etc. yet other ions, such as Li^+, Cs^+, Sr^{2+}, Ba^{2+}, Be^{2+}, and we shall allude to the huge family of organic molecules aimed at blocking channels, affecting gates and pumps, and so on, noting how they are based on sophisticated ways of thinking about ions in solution, their interactions with biopolymers, and their flow.

Acetylcholine

$CH_3CO \cdot O - CH_2 - CH_2 - N^+(CH_3)_3$

Adrenalin

$CH_3 \overset{+}{N} H_2 - CH_2 - \underset{\underset{OH}{|}}{CH} -$ (benzene ring with OH, OH)

Glutamate

$NH_3^+ - \underset{\underset{CH_2}{|}}{CH} - CO_2^-$
$\overset{|}{C}H_2$
$\overset{|}{C}O_2^-$

8.7.5 Currents of ions and morphogenic patterns: growth

The chemist understands very well the general idea of biological electrolytic currents and from the preceding sections he can see the way in which these currents are used in nerves to send messages. They are sodium/potassium currents. The physiologist is also aware of chloride and calcium currents in different tissues. There is, however, a wider issue than the general mechanism of ion-transfer in messages; there is the impact of the current carrier on the synthetic/degradative organic system. At some stage a current has to be transferred to a mechanical or chemical event if it is to serve a function. Before examining this point more closely we must look at where the circuits are. We know the topography of nerve circuits (see Section 8.7.1); we also know the microtopography of electronic currents in the membranes of chloroplasts and mitochondria (Chapter 3) but we do not know the external cell/cell or internal cell/vesicle currents at all well. In Section 6.6 local as well as general transmembrane ion flow was described. This means that there will be local as well as general lateral flow to rebuild the concentration. Now, if inlets for ions are placed along a surface at different distances from ion pumps then there will be a constant lateral current, dissipating energy, along a membrane surface both outside and inside the cell. Are such currents related to control over cell shape and do changes in the currents affect growth? There are now a number of observations which lead thought in this direction. Proton as well as metal ion currents have been observed. Is there a connection to morphogenic fields? The morphology of a living object is a dynamic state and it requires homeostatic control just as the chemical status of cellular metabolism must be controlled (see Chapter 21). Here it will not just be the inorganic ions which will be involved but a whole range of flowing organic messages.

8.8 Simple salts and the conditions of polyelectrolytes

Biological polymers are polyelectrolytes. Some, like DNA, have an extended almost linear series of charges while others, e.g. fats, give two-dimensional arrays and yet others, proteins, give tightly curved surfaces often almost spherical. These polyelectrolyte surfaces are stabilized by the surrounding ionic atmosphere due mainly to Na^+, K^+, and Cl^-. The interactions are of two kinds, one quite general and the second quite selective. The subject is complex since the polyelectrolytes can interact with one another and with a huge array of soluble organic anions and cations. For this reason we give here only references to the problems in biology associated with surface potentials of polyelectrolytes and membranes. The nature of the polymer matrix of a cell is adapted to the controlled ionic conditions (see Chapter 7).

A particularly important consideration is the effect of cations on the anionic poly-electrolytes outside the cell. The anionic groups are of two kinds. The anions of strong acids such as $(RO)_2PO_2^-$ and $ROSO_3^-$ and the anions of weaker acids, RCO_2^-. All occur in the polysaccharides of connective tissue, see for example Section 7.22. While none of these anions has a sufficient affinity for the major cation outside cells, Na^+, clusters of RCO_2^- could well bind and become cross-linked. Clusters of the other anions remain ineffective and thus they are particularly important in keeping an open external matrix.

8.9 Enzymes requiring potassium

The genuine specific requirement of enzymes for Na^+ or K^+ is limited to a few special cases although the stability of all states (conformers) of monomers and polymers in cells is connected to the ionic strength of about 0.2 M. Here we do not need to refer further to the general balance of ionic conditions in a biological aqueous medium since it is clearly true that the reactions of most biological systems have some dependence on the ionic strength which is largely either NaCl (outside) or KCl (inside). A striking case of the latter is in the halobacteria where all the biological polymers have evolved to match high salt conditions. Putting such considerations aside there are a few enzymes which have a simple stoichiometric requirement for K^+, but the present authors know of none where Na^+ is required. The examples are listed in Table 8.5. The value of the dependence is not known.

Table 8.5 Some potassium-dependent enzymes

Enzyme	Ion dependence
Pyruvate kinase	$Tl^+ > K^+ > Rb^+ > Cs^+ > Na^+ > Li^+$
Dioldehydratase	$Tl^+ > NH_4^+ > K^+ > Rb^+ > Cs^+ > Na^+ > Li^+$
Na^+/K^+ ATPase (K^+-site)	$Tl^+ > K^+ > Rb^+ > Cs^+ > Na^+ > Li^+$

NB. Tl^+ is an excellent probe (NMR and EM) for K^+ sites generally. K^+ sites are sometimes blocked by Ba^{2+}, an ion of the same size.

It is interesting to note that Tl^+ can substitute for K^+ and this substitution has been used to probe both ion channel and enzyme sites (see Table 8.3. and Table 11.12). The NH_4^+ ion does compete for some of these sites, but generally K^+ and NH_4^+ do not react at the same sites in biology. Much greater selectivity is found for the small organic cations such as acetylcholine which act as transmitters.

8.10 Ion energies: summary

While nobody denies that life as we see it is linked to DNA, RNA, and proteins, it is rare indeed to see the statement that life is linked directly to sodium, potassium, and chloride chemistry. In some highly interactive manner and as the biopolymers evolved, not before or after them, the sodium/potassium balances had to be present since all the major polymers are anions, and currents later became increasingly sophisticated. The dynamic homeostasis of these elements is now very precisely maintained and small deviations in levels of less than 1 per cent can be a symptom of a weakness in an organism. There is a deep-seated need to maintain these ion balances which interact with the states of the polymers via controls on osmotic pressure (affecting the mechanical stresses in the filaments of the cell), on electrical potentials (affecting the membranes), and on local states of polyelectrolyte

polymers both inside, including DNA, and outside cells. The temptation in the study of physiology is to look at fluctuations from the steady state (rest), thus addressing the problem of messages as the fundamental feature of Na^+/K^+ chemistry in life. It is not. The fundamental link is between two types of energized structure even at rest. One is represented by the polymerized states of organic molecules and a second by the resting concentration gradients of simple ions. The most direct way in which to see this link is through the chemical forms of energy—the bound pyrophosphate of NTP, e.g. ATP. Through NTP the whole of polymer biosynthesis is maintained and through NTP all gradients are generated. However, conversely, through ion gradients ATP can be made. This circle of interchange of energy from gradients (electrical or concentration), to weak covalent compounds (NTP), to biopolymers is the essence of life. We shall show that not only Na^+, K^+, and H^+ but also the other essential bulk and trace elements are involved. None of these conditions are static so that at all times currents flow. It is an essential feature of living systems as we know them that, in fact, current always flows. This is a part of homeostasis.

There is no sense in denying the fundamental need for metabolism of C, H, N compounds in biology for synthesis but there is equally no sense in forgetting that sustaining any such activity depends on concentration and voltage controls so that an enclosed volume is stable physically. Evolution is built on organic synthesis but also on physical chemistry so that enforced introduction of control of these three inorganic elements, Na, K, and Cl, in biology led step by step to signals concerning the charges and concentrations of ions around spherical single cells, to signals along cells, and from cell to cell in shaped organizations, to nerves and muscles, and eventually to electrolytic computers—the brain.

8.11 Conclusions

We wish to make certain bold statements to remind the reader of the significance of the chemical elements Na, K, and Cl, in life. These elements are central to:

(1) osmotic control;

(2) electrolytic balances and current, e.g. nerve action;

(3) stability of polyelectrolytes, DNA, membranes, etc;

(4) connections to chemical uptake of organic metabolites.

There has then to be an interactive system of feedback messages with many cellular processes and in higher organisms this extends to cell/cell and organ/organ communication so that points (1) to (4) concern fluids of animals which are very tightly controlled solutions of Na^+, K^+, and Cl^-. The network of signals is linked to other inorganic and organic transmitters and in Chapter 21 we shall find it is linked to growth using hormonal factors such as the glucocorticoids. This link will take us again from membrane phenomena to the heart of the cell, DNA, where K^+ is undoubtedly involved again as a

polyelectrolyte stabilizer. We suspect that the system of communication from outside to inside the cell has developed from osmotic and charge control using at first only an open K^+ channel and pumps, outwardly directed, for Na^+ and Cl^-, to the extremely sophisticated system of channels and pumps observed today in multicellular organisms, which have been further decorated by special connections to organic messengers.

Further reading

Eisenman, G., Aqvist, J., and Alvarez, O. (1991). Ion binding and channels. *Journal of the Chemical Society's Faraday Transactions* **87**, 2099–109.

Grinstein, S. (ed.) (1989). *Na$^+$/H$^+$ exchange*. CRC Press, New York.

Guyton, A. G. (1986). *Textbook of medical physiology* (7th edn). W. B. Saunders, Philadelphia.

Lamb, J. F., Ingram, C. G., Johnson, I. A., and Pitman, R. M. (1988). *Essentials of physiology* (2nd edn). Blackwell Scientific, Oxford.

Parker, M. W., Pattus, F., Tucker, A. D., and Tsernoglou, D. (1989). The structure of colicin. *Nature*, **337**, 93–6.

Stein, W. D. (1990). The Na/K ATP-ase. *Journal of Theoretical Biology*, **147**, 145–60.

Stevens, C. F. (1991). Ion channels. *Nature*, **349**, 657.

Williams, R. J. P. (1970). The biochemistry of sodium, potassium, magnesium and calcium. *Quarterly Reviews of the Chemical Society, London*, **24**, 331–60.

Williamson, J. R., Raghuraman, M. K., and Cech, T. R. (1989). Cation binding to DNA. *Cell*, **59**, 891.

9

The biological chemistry of magnesium: phosphate metabolism

9.1 Introduction 243
9.2 The spatial distribution of magnesium 244
9.3 Magnesium chemistry 245
9.4 Magnesium pumping in biology 248
9.5 Very strong binding of magnesium 249
9.6 Magnesium in walls and membranes 250
9.7 Magnesium enzymes: magnesium and phosphates 250
9.8 Magnesium and muscle cells 255
9.9 Magnesium and polynucleotides 256
9.10 Competition with polyamines 257
9.11 Polymeric equilibria: tubulins, DNA, and the cell cycle 260
9.12 Sol/gel equilibria and magnesium 261
9.13 Magnesium and lipids 261
9.14 Magnesium and chlorophyll 262
9.15 The use of manganese as a magnesium probe 265
9.16 Pollution and Mg^{2+} biochemistry 266
9.17 Lithium and magnesium 266
9.18 Conclusion 266

9.1 Introduction

Magnesium in biological chemistry is a Cinderella element: we know its hidden power and personality only indirectly since we are unable to label and follow it in a sensitive manner. Analysis *in situ* by X-ray fluorescence using the electron microscope is difficult since the element is light and present in the millimolar range; a very useful radioactive isotope is not available; there are few good spectroscopic tools (contrast calcium); even atomic absorption spectral methods need care because of the interference by phosphate with which magnesium is usually associated in biology. In these circumstances a

full description of the biological role of magnesium is hazardous. Two features stand out, however. The first is the general involvement with phosphate compounds and phosphate metabolism and the second is the binding in chlorophyll pigments. These two roles are quite different since in the first magnesium is a freely diffusible cation in aqueous solution involved not only in catalysis but in a multitude of controls, while in the second magnesium is an immobilized element in a light capture device in a membrane. We shall describe the first role before going on to chlorophyll.

Note first that the charge density on Mg^{2+}, radius 0.6 Å, is higher than that on any other freely available charged centre in biology. As a consequence, it dominates binding to negative centres of high opposite (negative) charge density, e.g. pyrophosphate. This equilibrium chemistry has been explored in detail and we give in this chapter only an outline of observations, especially those concerning competition with positive polyamine centres. In order to bring this magnesium/phosphate chemistry into a biological context we must first observe the concentration of magnesium in various compartments.

9.2 The spatial distribution of magnesium

We begin with a description of magnesium ion levels outside cells. Magnesium is plentiful in the sea (5×10^{-2} M) but is at a relatively low level in fresh water ($<10^{-4}$ M). It is seen that magnesium, like sodium, must be rejected by organisms in the sea but pumped into organisms in fresh water to maintain the free magnesium concentration in the body fluids of nearly all cells at around 10^{-3} M. However, the presence of bound magnesium outside many cells may well be essential also to maintain the integrity of their outer surfaces (see Section 9.6). Again magnesium in the outer fluids greatly affects the laying down of mineral phases. It is usual for calcite to be the major mineral of freshwater life but for aragonite to be the dominant mineral of life in sea water and we know, of course, that *in vitro* magnesium ions inhibit the growth of calcite much more strongly than the growth of aragonite. Magnesium is also a prominent minor element in bone and in calcitic shells.

Inside cells we have no clear knowledge of the distribution of magnesium other than to say that it is universally approximately 10^{-3} M. For a multicellular animal the bloodstream or extracellular fluids are also kept at about the same level (10^{-3} M) so that there is virtually no gradient of magnesium ions across the cytoplasmic membrane in a range of multicellular organisms. The function of magnesium at these membranes is then very different from that of calcium, sodium, or potassium ions. Magnesium does not act there as a current carrier or a trigger. An exception is that magnesium has an energy-dependent distribution across the thylakoid membrane where this metal ion may be sequestered in the thylakoid by phosphate and carboxylate groups at high pH. It is then quite likely that the acidification of the inside of the thylakoids on exposure to light leads to the simple displacement of bound magnesium by protons. As a consequence there is movement of the cation into the cytoplasm and magnesium acts as a slow trigger or control over cytoplasmic photosynthetic reactions of plant cells

Note, however, that the free magnesium ion concentration always remains in the range $0.5–2.0 \times 10^{-3}$ M.

The question arises next as to the magnesium levels in the Golgi and other cysternal internal vesicle compartments of higher cells. Although it is not certain, the authors believe that these compartments are relatively low in magnesium but high in calcium and manganese due to the positioning of cation pumps (Fig. 3.16). We need to look further at the comparative functions in compartments for the three divalent cations, magnesium, manganese, and calcium, after we explore the chemistry of magnesium in more detail.

9.3 Magnesium chemistry

The magnesium ion is very different from the calcium ion. The differences are in structure, thermodynamic affinity for certain classes of ligand, and reaction rates. All the differences spring from the difference in radius (Mg^{2+}, 0.6 Å; Ca^{2+}, 0.95 Å) when measured in an octahedral environment of oxygen donor atoms. The first obvious feature is in the structure of compounds and complexes since it is the case that within the A-group cations of the periodic table there is a sharp structural separation between the ions of radius less than 0.75 Å and those of larger radius (Fig. 9.1). This distinction puts a diagonal line in the periodic table, leaving Li^+, Mg^{2+}, Al^{3+}, and Sc^{3+} on one side and Na^+, Ca^{2+}, and Y^{3+} (Ln^{3+}) on the other. The smaller ions are strictly octahedral in the symmetry of their complexes with, in the particular case of Mg^{2+}, very rare different geometries. The bond distance Mg–O is very close to 2.05 ± 0.05 Å in all cases (Table 9.1). The contrast is very strong with the

Table 9.1 Some typical magnesium salt structures

Salt	Mg(II) co-ordination no.	Mg–O distances (nm)	
		Minimum	Maximum
Mg hexa-antipyrine · $(ClO_4)_2$	6	0.206 (Ca, 0.230)	0.212
$MgS_2O_3 \cdot 6\,H_2O$	6	0.205	0.212
$MgSO_4 \cdot H_2O$	6	0.204	0.209
$Mg(HPO_4) \cdot 6\,H_2O$	6	0.200	0.212
$Mg_2P_2O_7$	6	0.200	0.211
$Mg(CH_3CO_2)_2 \cdot 4\,H_2O$	6	0.200	0.210

See Tam and Williams (1985) in the 'Further reading' for sources of the data.

Fig. 9.1 A simple division of the periodic table to indicate some structural divisions associated with the diagonal relationship.

IA	IIA	IIIA	
Li	Be	B	Tetrahedral
Na	Mg	Al	Octahedral
K	Ca	Sc	
Rb	Sr	Y	
Cs	Ba	Ln	Irregular co-ordination geometry

Table 9.2 Some stability constants, log K, of magnesium, calcium, and manganese complexes at zero ionic strength and $T = 25°C$

	Log K		
Ligand	Mg^{2+}	Ca^{2+}	Mn^{2+}
Acetate	0.8	0.8	~1.0
Glycine*	~3.0	1.5	3.5
Iminodiacetate*	3.7	3.4	
EDTA*	9.1	11.0	14.1
EGTA*	5.5	11.2	12.5
4,5 Dihydroxybenzene 1,3 disulphonate*	5.9	6.0	8.6

* Smaller donor centres, N or O, assist Mg^{2+} binding relative to Ca^{2+}, e.g. unsaturated amines and phenols. Data for pH = 7 are > 10^3 fold lower.

EGTA

irregular co-ordination geometry, bond angle, bond distance, and co-ordination number (7–10) for the larger cations including Ca^{2+} (see Table 10.3). The explanation is that in the case of the smaller cations the central field of the cation dominates the co-ordination sphere while in the cases of larger ions the second and perhaps third co-ordination spheres have a more important role to play so that a compromise, irregular structure results around, say, Ca^{2+}.

The easy adjustment of the larger ions to steric crowding affects the relative thermodynamics of binding (see Chapter 2). In Table 9.2 we show that, as the number of co-ordinating groups increases and especially where these groups are weaker donors than water, but are linked together through covalent bonds, i.e. chelating ligands, then the stability constants for calcium exceed those for magnesium. Similar features are seen in the chemistry of the lanthanide cations when compared with that of Al^{3+}. Even trends within the lanthanide series follow this expectation. The value of this difference of the biological chemistry of magnesium relative to that of calcium is apparent everywhere in biology. The difference appears in large opposed heats of formation of the complexes of ions such as magnesium which are only partially offset by entropy gains. However, Mg^{2+} is not a powerful enough acid to bind effectively to such hard centres as phenolates ($pK_a = 11.0$) at pH = 7. Thus, structural and pK_a factors force the strong association of Mg^{2+} with organic polyphosphates, highly charged anions, e.g. ATP^{4-}. At the same time the stability of Mg^{2+} complexes is insufficient for it to bind to the common small anions of metabolic pathways (Table 9.3).

Table 9.3 Free substrate binding to Mg^{2+} and Ca^{2+} at pH = 7

No binding	Binding
(Simple carboxylates)$^-$	Oxalate^{2-} (?)
(Kreb's-cycle molecules)$^{2-}$	ADP^{3-} (other nucleotides)
(Amino acids)$^{+-}$	ATP^{4-} (other nucleotides)
(Mononucleotides)$^{2-}$	IP$_3^{4-}$, IP$_4^{5-}$, etc.
(Sugar phosphates)$^{2-}$	(see polynucleotides)
(Sugar sulphates)$^-$	

There is a noticeable, though weak, tendency for the small cations, including Mg^{2+}, to be able to bind to nitrogen donors and here there is again a marked contrast with Ca^{2+} (this is a matter of the polarizing power of the cation) so that the possibility arises of designing a small-cavity ligand from nitrogen donors carrying negative charge and excluding calcium on two grounds, ligand donor type and cavity size. Such a ligand is the chlorin ring. We return to the difficulties of the incorporation of Mg^{2+} into this ring in Section 9.14.2, but the competition here is no longer from Ca^{2+} but from transition metal cations such as zinc. Protein surface cavities can be formed in principle to fit Mg^{2+} from O-donor ligands but this is unusual.

The kinetic difference between calcium and magnesium, and, in fact, for all metals above as opposed to those below the line of Fig. 9.1, are due to packing more than to the total bond energy. As stated in Chapter 4, the exchange of

water is approximately 10^3 times slower from around Mg^{2+} than from around Ca^{2+} (Table 9.4); the magnesium octahedron is not easily broken. The result, again well-known, is that calcium provides the major, fast, trigger ion of biology using quite strong binding in very rapid exchange, e.g. with calmodulin. It is easy to show that magnesium cannot act as such a fast trigger ion due to the slow on-rate of its reactions. However, the very use of calcium in this way means that it had first to be excluded from many parts of intracellular biological space in order to build a triggering gradient. Presumably, this exclusion allowed Mg^{2+} to take over the major role of anion charge neutralization in the cytoplasm, e.g. with ATP^{4-}. The binding constant here does not inhibit reaction rates. In general, magnesium is not held with a binding constant greater than 10^4 in biology, but there are some notable exceptions where function is hardly likely to be related to exchange. We shall see that where magnesium is used in control reactions it is a relatively slow control.

Table 9.4 Rates of water exchange

Fast (10^{-10} to 10^{-7} s)	Medium (10^{-7} to 10^{-5} s)	Slow ($>10^{-5}$ s)
Na^+, K^+, Rb^+, Li^+ Ca^{2+}, Sr^{2+}, Mn^{2+} Ln^{3+}, Cu^{2+}, Zn^{2+}, Fe^{2+}	Mg^{2+}, Co^{2+}, Ni^{2+}	Al^{3+}, Fe^{3+}

See Section 4.5.2.

The manganese ion, Mn^{2+}, lies between magnesium and calcium in size, and can therefore be used as a spectroscopic probe for the properties of both of them, but it has a higher affinity for N-donors than either of them (Table 9.2). We return to the uses of Mn^{2+} chemistry as a probe in the study of Mg^{2+} systems later, but note here that Mn^{2+} appears to be excluded from the

Table 9.5 Some magnesium, calcium, manganese, and zinc enzymes* and their locations

Magnesium	Zinc†	Calcium	Manganese
Enzymes			
ATPases	Phospholipase C	Phospholipase A_2	Glycosylases
Kinases	Peptidases	Thermolysin	(See Chapter 14)
Enolase	Alkaline phosphatase	Saccharases	
Glutamine	(plus Mg^{2+})	(See Chapter 10)	
synthetase	(See Chapter 11)		
Compartments			
Cytoplasm	Extracellular	Extracellular	Golgi vesicle

* The metal ion is at the active site in all the examples cited. In some other enzymes the metal is probably at the active site but there is no definite evidence. Structural use of a metal away from the active site is not included in this table.
† Zinc is not so restricted to a particular compartment as other metal ions (see Chapter 11), but most of its hydrolytic enzymes operate outside cells.

cytoplasm of higher cells and to be functional in organelles and in Golgi vesicles, so that competition between the ions within biology is slight.

Using the differences in size, biological systems remove both free manganese and calcium ions to levels below 10^{-7} M while keeping the magnesium ion at about 10^{-3} M in the cytoplasm. The outward pumping of the first two ions concentrates them in reticula, e.g. sarcoplasmic and endoplasmic, and possibly in Golgi bodies, and also outside cells. Therefore, of the divalent cations magnesium dominates the cytoplasm but is of less consequence inside reticula. We must look for the differential use of the three cations with different enzyme systems placed in different compartments (Table 9.5).

9.4 Magnesium pumping in biology

We need to consider how the concentration of Mg^{2+} is maintained in cells or indeed in body fluids. To do this we have used a general way of inspecting the functions of metal ions by writing a formal equation for all metal ion biological functions

$$\text{Function} \propto K \cdot [M] \cdot \{\text{structure factors}\}.$$

We have described in Chapter 2 and in Section 9.3 the binding constants, K, and the structure factors, the latter containing kinetic effects, but we have not dealt in detail with the energized maintainance of the concentration, $[M]$. In the case of magnesium it is a general rule that, in a steady state of a metabolizing cell, the free $[Mg^{2+}]$ is close to 10^{-3} M in the cytoplasm. This statement does not refer to the condition outside the cell. In fact, it appears that, no matter what the external concentration, from about 10^{-6} M to $10^{-1.3}$ M (the sea), the internal $[Mg^{2+}]$ remains in the millimolar range in all cells. However, there are undoubtedly fluctuations when single (bacterial) cells are subjected to gross environmental changes.

Now, while higher organized systems maintain their cells in external circulating fluids of 10^{-3} M magnesium, unicellular systems cannot control the environment. As we have seen fresh water can be low in magnesium, $<10^{-4}$ M, or even 10^{-6} M, yet unicellular organisms remain able to multiply. Below this level they enter stationary phase no matter what the level of other essential nutrients. (In fact, it appears that the cells become very run down in the virtual absence of Mg^{2+}.) Studies on *E. coli* established the nature of the required magnesium pumps. Their formation requires synthesis of special proteins, as shown by genetic studies, which are inserted in the cytoplasmic membrane. What we know about these proteins suggests that they resemble the calcium pumps but why they act selectively on magnesium is unknown. We are forced to imagine that they have special acceptor 'holes' which fit small divalent cations. Interestingly, nickel, which has a similar size and stereochemistry, may well enter cells via the magnesium pump. When considering pumping we have to keep in mind too that this is a kinetic energized process, so that information about binding constants alone could be misleading. There is the possibility that weak binding, say log $K=4$, connected to an *energized* pump could lead to accumulation even from very dilute solution, 10^{-6} M, via

two steps

$$Mg + L_1 \rightleftharpoons MgL_1 \xrightarrow{\text{E}} Mg(\text{inside})$$

where L_1 is a ligand, E is a pump, probably an ATP-hydrolysing enzyme, and the last step is considered to be irreversible. Here there need not be more than 1 per cent formation of MgL_1, with $K = 10^4 \text{ M}^{-1}$ so that pumping could be effective down to external $[Mg^{2+}]$ of 10^{-6} M. Again the selectivity of the pump may depend not on the selectivity of binding for magnesium but on the recognition of the size and shape of MgL_1 in the pump, E. This type of selection is related to the difference between an enzyme substrate and an enzyme inhibitor. Both may bind equally, but only the first flows along a reaction co-ordination because of its shape and size. The selection of magnesium rather than calcium may depend on a similar principle.

Living systems in the sea must reject Mg^{2+} and their outward pumps must also use ATP but at present they are not known; however, there is some evidence for outward pumping of magnesium. The chromaffin granule, a vesicle of the adrenal gland, is surrounded by an unusual membrane which maintains in the interior of the granule a pH of 5.8, a very high concentration of adrenalin >0.1 M, and ATP at about 0.025 M. (Note that this is not $Mg \cdot ATP$ which is everywhere present in the cytoplasm.) The level of cations present is extraordinary in that there is some calcium (10 mM) but virtually no magnesium (Table 9.6). To maintain such strange gradients the chromaffin granule must have a series of exclusion mechanisms since it is known that the contents of the granule are free in solution. By some device, magnesium is excluded from being a partner of ATP, as it is in the cytoplasm, even to a much greater degree than calcium is excluded. Once again, this is extraordinary considering that the ratio of magnesium to calcium in the cytoplasm is 10^4. It would appear that this membrane is not too unlike the plasma membrane (and the Golgi membrane?) in that there are outwardly directed pumps, i.e. from the cytoplasm, for H^+, Ca^{2+}, and Mn^{2+}, while there are, if anything, inwardly directed pumps for Mg^{2+}, i.e. out of the vesicle but into the cytoplasm.

Other granules are very different. For example, in the platelet cells there are storage vesicles of pyrophosphate which has both calcium and magnesium with it as well as 5-hydroxytryptamine.

Table 9.6 The contents of chromaffin granules (mM)

Na	K	Mg	Ca	P	S	Cl
43	28	low	10	624	82	33

9.5 Very strong binding of magnesium

A very few cases of very strong binding of magnesium ions in biology are known besides the case of chlorophyll (Section 9.14). The major cases, which are well established, are in the ATP-synthetases of thylakoids and mitochondria and in the ATPases of muscle for which binding constants well in excess of 10^8 M^{-1} have been described. The function of magnesium in these enzymic systems cannot be related to turnover or exchange. Now the rigidity of the co-ordination sphere of this ion points to the possibility that it is being used to deliberately prevent motion in certain regions of organized systems. Thus, here Mg^{2+} has almost the opposite function to that of calcium and possibly zinc which, even where they form cross-links, will allow swivelling motions

due to the easy distortion of their co-ordination spheres. However, although we do not know the nature of these sites, we see clearly that biological systems can build centres which have very high Mg^{2+} binding constants. Even within the design of organic ligands for Mg^{2+}, e.g. EDTA, very few show such binding strengths at pH = 7 (Table 9.2). It is probable that the very tight unusual binding of magnesium is a property of the protein fold and its energy.

There are two other compartments where tight binding of Mg^{2+} may be present—the ribosomes and the nucleus. Little is known about them as yet.

9.6 Magnesium in walls and membranes

While examining the divisions of biological space we need to observe that much of it is occupied by protein, polysaccharide, and lipid, especially in and surrounding membranes, and much magnesium is found in these membrane and wall zones. Magnesium is used here to cross-link the carboxylated and phosphorylated anionic polymers, and shares this role with calcium ions. The binding of the two is somewhat different, of course, and so they act co-operatively in stabilizing external structures of cells. This is a very general phenomenon and is such that strong chelating agents, such as $EDTA^{4-}$, disrupt cell wall and cell–cell structures. A major binding agent in the cell walls of *E. coli* is teichoic acid (Fig. 9.2).

9.7 Magnesium enzymes: magnesium and phosphates

While magnesium is not generally associated with carboxylate-containing substrates, binding to phosphate substrates is clearly an imperative function in all cells. However, the affinity of magnesium for phosphate or phosphate monoesters is not determinant at the concentration in cytoplasm, since it does not form complexes with simple phosphate anions, HPO_4^{2-} or RPO_4^{2-}, nor does it form insoluble salts with them (Table 9.3). In order to bind to Mg^{2+} further negative charges or dipoles are necessary in the molecule which carries the phosphate. The most striking examples are pyrophosphate and other condensed polyphosphates, e.g. ATP (Table 9.7), for which the binding constants are close to 10^4. In biology these phosphates are largely Mg^{2+}

Table 9.7 Magnesium mononucleotide complexes

Mononucleotide complex	Example
Mg · adenosine triphosphate	Kinases, ATP-pumps, contraction
Mg · guanosine triphosphate	Protein synthesis, membrane responses, tubulin polymerization, elongation factors (ETu)
Mg · uridine diphosphate	Saccharide reactions and glycosylations
Mg · cytidine triphosphate	Choline synthesis (lecithins)

NB. NTP with Mg^{2+} ties together many degradative and synthetic routes with mechanical devices (see Chapter 6).

complexes, and most of the energy transfer due to NTP and the phosphate transfers of kinases depend on Mg^{2+}. Other important phosphates that could form magnesium complexes of high stability constant are the multisubstituted phosphates of sugars, for example, inositol (tri-), tetra-, penta-, and hexaphosphates. The binding to these phosphates rather than to carboxylate ligands, e.g. molecules of the citric acid cycle, arises through the high local charge density $(2-)$ on more than one terminal group.

Now, molecules such as ATP are bound into proteins to a large degree through their organic adenine centres, often leaving the phosphate chain carrying its magnesium freely exposed or weakly bound. The activation of the triphosphates for transfer of the terminal phosphate is then brought about by using an anion centre on the protein, as shown in Figs 9.3 and 9.4. The figures indicate that magnesium bound between the β- and γ-phosphates protects the ATP, bound or free in solution, from hydrolysis, while on the enzyme surface the protection is removed from the γ-phosphate by movement of magnesium to the β- and α-phosphates when it is also bound by a protein carboxylate or other groups (Fig. 9.4). Here magnesium is a locally mobile cation between phosphates of ATP or on the protein surface even though absolute rates of movement can be quite slow. Catalysis also often involves minor migration of substrate groups relative to the attacking centres of the enzymes but also a major conformational groove-closing reaction (Chapter 7), which probably involves all the helices in Fig. 9.3. The full value of magnesium in nucleotide phosphate transfer reactions of the cytoplasm can only be appreciated by reference to books containing detailed compilations of the mechanisms of the relevant enzymes (see 'Further reading'). These show that magnesium is a totally dominant cation in these biomechanical, NTP-dependent enzymic reactions and it is therefore an absolutely essential element. The enzymes involved extend from traditional kinases to the G-proteins involved in many receptors, the elongation factors, ATP-synthases (β-subunits), myosin heavy chain (muscle) ATPases, and the ras proteins (tyrosine kinases). The feeling that in this group of proteins there is a mobility device with which Mg^{2+} is involved and that this device is of extreme importance for the phosphate homeostasis is increased by the observation that single mutations in the ras proteins are associated with tumourogenesis. The mechanism of action of many of these proteins (enzymes, kinases, pumps, contractile devices, etc.) is likely to be fundamentally the same in that there is a site for NTP associated with a β-sheet domain and that this domain is linked via a helical hinge section to other domains (Figs 9.3 and 9.4 and see Fig. 7.9). The closing of the gap between domains, or its opening, can be coupled to many devices as in ATP pumps (Chapter 8). In this sense we have repeatedly used knowedge of the properties of phosphoglycerate kinases in order to appreciate the function of internal mobility in all of these magnesium-dependent biomechanical devices. The role of magnesium in this activity must not be overlooked.

The reactions of phosphate condensates are not confined to the organelles and the cytoplasm, that is, to the transfer of phosphate from the nucleotide triphosphate. There is, in the Golgi vesicles, the apparatus for sugar transfer using the reactions of UDP^{3-}. The binding constant of magnesium to the condensed pyrophosphate unit $[R–O–P–O–P]^{3-}$ is always about 10^4 so that, in the Golgi, where magnesium ion concentration may well be low, Mg^{2+} can still be bound. Manganese can substitute for magnesium and it is very difficult

Fig. 9.2 Segment of a teichoic acid which has D-alanine and N-acetyl glucosamine residues. Note the glycerol phosphate backbone.

Fig. 9.3 The essence of the structure of phosphoglycerate kinase. β-sheets are labelled A–N and helical strands are labelled I, V, VI, XII, and XIII. The inset diagram shows the bound substrates. The Mg^{2+} is close to Asp 372. (After the work of C. F. F. Blake and H. Watson.)

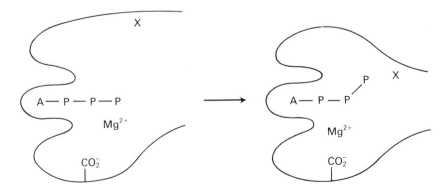

Fig. 9.4 A schematic representation of the movement of Mg^{2+} in a kinase and the groove-closing reaction that is required to bring substrate X near to the transferable phosphate at ATP.

Fig. 9.5 The structure of yeast enolase after the work of J. M. Brewer. The Mg^{2+} ion is shown as a shaded cycle. The enzyme is mainly a β-barrel with linking helices.

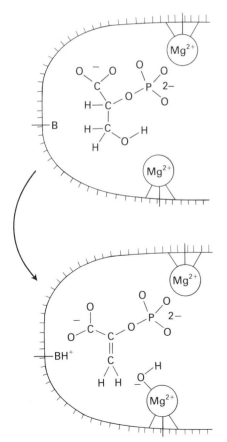

Fig. 9.6 The proposed mechanism of reaction of enolase. (After A. Mildvan.)

to know which enzymes in the Golgi compartments are magnesium-activated and which are manganese-activated.

There are many enzymes which do not involve the nucleotides and which are magnesium-requiring (Table 9.8). In some of these the substrate is a magnesium complex, e.g. isocitrate lyase, while in others the magnesium enzyme is the active species, e.g. enolase. The binding of magnesium in such enzymes is probably through two or three carboxylates and perhaps one imidazole (Table 9.9). Reference to the section on stability constants shows that such sites are selective against manganese in the cytoplasm where $[Mg^{2+}] = 1 \times 10^{-3}$ M and $[Mn^{2+}] < 10^{-8}$ M. We advise the reader to look at wall charts of metabolic pathways to see the importance of these magnesium enzymes. Notice the β-sheet constraints at the active site.

A case of the mechanism of action of magnesium in an enzyme reaction is provided by enolase. A picture of the active site is shown in Fig. 9.5 and of the possible pathway of action in Fig. 9.6. Undoubtedly some enzymes, such as enolase, use two magnesium atoms, and in the particular case of pyruvate kinase one magnesium and one potassium, so that the way in which the two (three) metal ions work together could be very complex (see Chapters 4 and 11 on zinc-catalysed acid/base reactions). In the pyruvate reaction magnesium is thought to stabilize an intermediate enolate.

This section by no means gives a complete list of the deep involvement of magnesium in enzymatic reactions. One way of recognizing this is to examine the enzymes of glycolysis (Chapter 5), glucogenesis, the pentose shunt, and of carbon dioxide assimilation (Fig. 9.7). *In almost every major metabolic path in the cytoplasm magnesium is deeply involved but it is also associated with very much polynucleotide biochemistry.*

Table 9.8 Some examples of magnesium-requiring enzymes

Enzyme	Requirement
Kinases (generally)	Phosphate transfer from ATP · Mg
Adenylate cyclase	ATP · Mg is the substrate
ATPases	ATP · Mg is the substrate
Alkaline phosphatase	Activator of the enzyme but is not essential; Mg binds the enzyme
Enolase	Magnesium is bound to the enzyme and not directly to substrate
Isocitrate lyase	Mg · isocitrate is the substrate
Methyl aspartase	Magnesium is bound to the enzyme
Ribulose bisphosphate carboxylase	Magnesium binds to intermediates and enzyme
Myosin ATPase	Mg · ATP is the substrate

Table 9.9 Magnesium-binding sites in enzymes

Enzyme	Binding site
Several kinases	ATP, one protein carboxylate
Alkaline phosphatase	Possibly three metals at active site
Xylose isomerase	Sugar, two carboxylates
Enolase	Three carboxylates
Ribulose bisphosphate carboxylase	Two carboxylates, one histidine
Glutamine synthetase	Two carboxylates, one histidine

Several magnesium-dependent enzymes have more than one metal ion binding site (see Section 4.3).
Some of the magnesium appears not as a catalyst but as a conformational control.

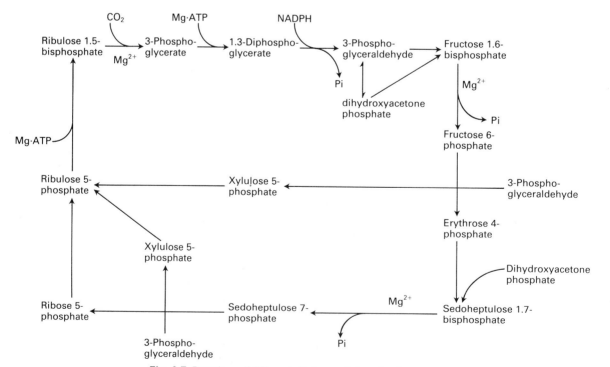

Fig. 9.7 Reactions of CO_2 assimilation and the role of magnesium.

9.8 Magnesium and muscle cells

9.8.1 Activation of tension

The tension in fibres is of extreme importance for the shapes of single cells but it is best known in the study of muscle cells of multicellular organisms. Here a model called the sliding filament model is used to explain the development of tension. The fibres are shown in Fig. 9.8. On one fibre, labelled M for myosin, there is a headgroup which can bind in different ways in de-energized and energized states to the second fibre, actin (A). The myosin contains an ATPase. The hydrolysis of $Mg \cdot ATP$ drives the head of myosin through a series of conformational changes so that the energy released causes the filaments to slide and so to reduce the overall length of the muscle fibrillar bundles (see Fig. 10.12). (The process in muscle cells is activated by calcium invasion of a cell.) For further details see books on physiology in the 'Further reading' to Chapter 8 but here notice the total dependence of the process on Mg^{2+} and that the motions in the headgroup of the myosin may be very similar to those of the ATPase pumps while the ATPase itself may have hinge movements like that of phosphoglycerate kinase.

9.8.2 Magnesium and muscle relaxation: calcium buffering

The fastest muscles require very rapid relaxation, i.e. very fast lowering of calcium ion concentrations after an activating pulse. The usual mode of

Fig. 9.8 The reactions of $Mg \cdot ATP$ which drive contraction. The sliding filament series of steps is shown in (a) with the enzyme steps in (b) (see text). AM, actomyosin; A, actin; M, myosin. There is also a tightly bound Mg^{2+} within the actomyosin complex, see Section 9.5.

relaxation of the pulse is the outward directed calcium pumps using the energy of $Mg \cdot ATP$ (see Chapter 10). A faster mode of relaxation involves the use of a soluble protein chelating agent for calcium, parvalbumin, which is designed in a special way. It has an absolute *higher* affinity for both calcium and magnesium than the trigger calcium proteins, troponin C and calmodulin. As a consequence, in the resting cell it is present as $Mg \cdot$ parvalbumin and is only able to accept calcium, from a calcium pulse, quite slowly, i.e. at the rate at which Mg^{2+} is released. The trigger proteins are metal-free before the pulse and accept calcium at the diffusion limit 10^{-9} s. Thus, despite a weaker affinity for calcium, troponin C and calmodulin take up calcium before parvalbumin since the latter only releases Mg^{2+} in some 10^{-3} s. The sequence of events for calcium binding following a pulse is then

$$Ca^{2+}(in) \rightarrow trigger\ proteins/parvalbumins/Ca^{2+}\ pump \rightarrow Ca^{2+}\ (out).$$

Parvalbumin acts relatively quickly (10^{-3} s) reducing the calcium in the cell cytoplasm before calcium rejection by the pump which has a turnover of only $100\ s^{-1}$. Very fast muscles are found to contain excess parvalbumin up to millimolar levels. If all this is a correct interpretation, the cycle of activation and relaxation of muscle is a wonderful example of the function of kinetic controls using calcium/magnesium ratios in receptor systems.

9.9 Magnesium and polynucleotides

9.9.1 Structures of DNA and RNA particles

The determination of the structure of transfer RNA (tRNA) revealed that there are different classes of magnesium associated with this folded polynucleotide. It was already known, of course, that magnesium was often associated with polynucleotides but there was no knowledge of the co-ordination chemistry. Figure 7.17 shows that the magnesium is found in one conventional binding mode and also in one particularly interesting structural trap. The fold of tRNA is in considerable part just a conventional double-stranded helix and Mg^{2+} does not bind very well locally in the short grooves of this structure (see Section 7.11) at high salt concentration though it may be retained by the total surface potential. The fold also contains a half crucifix-like structure where three strands intersect. It is here that there is an interesting binding of Mg^{2+} in a partially dehydrated state. It would appear as if the stability of the tRNA fold, and hence its functioning, is Mg^{2+}-dependent. This is a very pretty example of cross-linking of oxyanion centres by a cation. Since the Mg^{2+} will not allow rapid rearrangement of its co-ordination sphere, this half crucifix is likely to be an essential maintained part of the fold during activity. The half crucifix form does connect the various essential exterior and functional parts of the tRNA in its transferring capacity. It is probable that certain regions of other RNAs and DNAs have similar folds and a parallel need for Mg^{2+} stabilization (Table 9.10; Fig. 7.18) but the Mg^{2+} binding here could be very strong due to the greater collection of negative charges. Of course Mg^{2+} is required in all steps of transcription and translation, in large part through the involvement of nucleotide triphosphates. It is also known

Table 9.10 Magnesium requirements in polynucleotides

Species	Requirement
tRNA	At least one strong site and many weaker sites Fold stabilization Half-crucifix structures
rRNA	Stabilization of fold and association of 30S, 50S, and 70S subunits. Crucifix structures at attachment points?
mRNA	Attachment to ribosomes Crucifix structures at attachment points?
DNA	Crucifix structures
Ribozymes	Mg^{2+} acts to bind the RNA enzyme to the RNA substrate

that Mg^{2+} stabilizes the bending of DNA into particular curved structures (Section 7.11).

9.9.2 Ribozymes

The discovery of catalytic activity associated with RNA and especially its connection with splicing reactions has led to speculation that these RNA enzymes, ribozymes, could have been some of the earliest catalysts in biology. Be this as it may it is noticeable that they, like many another enzyme-handling phosphate hydrolyses, are Mg^{2+} dependent even *in vitro*. Other metal ions that are much stronger Lewis acids do not interfere. The reason could well be that these cations, such as copper, bind preferentially to the nitrogen bases of the RNA and the active metal ions are therefore $Mg^{2+} > Mn^{2+} > Ca^{2+}$. Nickel, Ni^{2+}, does work quite well so that a preferential size factor is also operative. In cells, Mg^{2+} is, of course, the only ion available for binding to the weak O-donor sites of the RNA since it is available at 10^{-3} M. It is likely that the Mg^{2+} is regulatory too so that, if it falls below some 10^{-4} M, the ribozyme is inactive just as the ribosome is inactivated. This could be important in unicellular organisms.

9.10 Competition with polyamines

In the context of RNA and DNA it is interesting to compare Mg^{2+} chemistry with that of a wide range of other cations. The binding of a series of cations ranging from simple ions such as Mg^{2+} and $[Co(NH_3)_6]^{3+}$ to complex polyamines such as $NH_3^+-[(CH_2)_2NH_2]_m^{m+}-(CH_2)_n-NH_3^+$ (Fig. 9.9) with a series of anions ranging from HPO_4^{2-} and RCO_2^- to $X-(CH_2CHX)_mCH_2-X$ (where X is carboxylate or phosphate) has been explored (see Fig. 9.10). The framework cations and anions are intended to be models for biological molecules such as spermines and histones on the one side (cationic) and DNA, carboxylated proteins, and sulphated polysaccharides on the other (anionic). Molecules studied in these model analyses have either mobile or rigid frameworks. A parallel study with encapsulating cyclic ligands and amines has

Putrescine

Fig. 9.9 The structures of some important polyamines.

Spermidine

Spermine

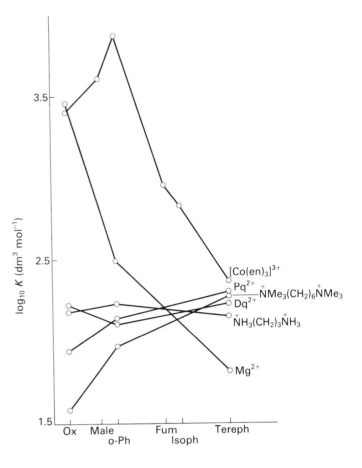

Fig. 9.10 The binding constants to dicarboxylate molecules by a variety of cations. As the negative charges are moved apart, Mg^{2+} binding falls rapidly but that of the organic amines remains steady or increases. Ox, oxalate; Male, maleate; o-Ph, orthophthalate; Fum, Fumarate; Isoph, metaphthalate; Tereph, paraphthalate; Pq^{2+}, paraquat; Dq^{2+}, diquat; en, ethylenediamine; Me, methyl.

been made but here we are more concerned with open surfaces. The findings can be stated simply using the data and figures such as Fig. 9.10.

1. Where the anionic ligand has a high local charge density, e.g. $C_2O_4^{2-}$ and $P_2O_7^{4-}$, then its binding of small cations such as Mg^{2+} is much stronger than its binding of polyamines of any kind.

2. Where the charge distribution is spread out on the anion, but not too spread out, then binding of polyamines is the stronger.

3. All binding of polyamines to the anions is weak unless the charge product (anion charge times cation charge) exceeds 12.

4. Given a large number of spread-out charges, stereochemical factors become important in charge-matching and hydrophobic interactions between frameworks (Fig. 9.11).

5. Intermediate anion charge density leads to competition.

6. In general, $-PO_4^{2-} > -CO_2^- > -SO_4^-$ in binding to all cations.

(In passing we observe that, in a sense, calcium is intermediate in character between magnesium and the polyamines since it is large, but in cells it has little impact since it is only 10^{-7} M.)

This set of observations is important in that it leads us to an understanding of the roles of frameworks of positive charge as found on labile molecules such

(a)

(b)

Fig. 9.11 The fitting of a polyamine into the groove of DNA seen so as to reveal the interaction with (a) the phosphate diester groups and (b) the stacked bases. (After the work of G. Richards.) N.B. Polyamine binding changes the structure of DNA.

as spermine, or on more specific polyamine cation surfaces such as those on the surfaces of proteins like histones and cytochrome c. These polycations bind to matching spread-out polyanion surfaces, such as the diester phosphates in the grooves of DNA and RNA double helices, or they bind to the carboxylates on the surfaces of partner proteins such as cytochrome c oxidase. At equal concentrations magnesium ions cannot compete for such surfaces in the cytoplasm of cells, while the polyamine cations do not interfere with the binding of Mg^{2+} to ATP^{4-}. We see too that in the intermediate situation, given by only one carboxylate or perhaps two on an enzyme surface and a terminal phosphate $ROPO_3^{2-}$ on a substrate, there is the ready possibility for the use of a bridging Mg^{2+} ion at the active site of hydrolysis without competition from amines. Alternatively, if amino groups of the protein side chains are suitably placed, the binding site for the anion can utilize the different selectivity of electrostatics while not involving Mg^{2+} (see 3-PG^{3-} in Fig. 9.3). It will then be the presence of special selective factors, say imidazole sidechains, and of relative charge density in proteins which will assist or prevent Mg^{2+} binding.

The preceding discussion has neglected the use of the hydrophobic framework of the polyamine cation (Fig. 9.9). Clearly, it is possible to develop part of the cationic molecule so that the interaction with a protein or DNA surface is enhanced. It is this component of organic molecules which adds a new dimension to the competition between organic cations and inorganic cations. The analysis of this competition shows that Mg^{2+} may well hold DNA, for example, in one conformation, while on synthesis of competing polyamines, say during a cell cycle, the Mg^{2+} is removed and DNA is switched both in structure and function. (NB. There is the obviously related competition between Na^+, K^+, and organic cations such as acetylcholine for protein surfaces in pumps and receptors; Chapter 8.)

Once again, notice that it is in the cytoplasm that the competition or co-operation between Mg^{2+} and organic polycations occurs and Ca^{2+} is not involved but, outside the cell, it is often calcium and amino groups which act in competition or co-operatively in enzymes and other macromolecules.

Finally, sulphated polymers do not bind so well to cations so that using the sulphate group biology can produce free anionic surfaces as in sulphated polysaccharides (Chapter 7), which therefore make up a passive extracellular matrix.

The above discussion has concerned competition between organic molecules and inorganic cations in terms of stability constants. There is a way of modulating this competition in cells through metabolism. While Mg^{2+} concentration is relatively fixed in the cell cytoplasm, that of the polyamines can vary grossly during the cell cycle. There is then a possibility that the timed relative concentration of polyamines and magnesium is an important factor in cell division.

9.11 Polymeric equilibria: tubulins, DNA, and the cell cycle

So far the account of magnesium biochemistry in this chapter has dealt with simple equilibria, where L is any ligand,

$$Mg^{2+} + L \rightleftharpoons MgL^{2+}$$

and the competition for L by Ca^{2+}, polyamines, or protons. There are further important reactions involving magnesium which are of the kind

$$nMg^{2+} + mL \rightleftharpoons Mg_nL_m^{2n+}.$$

The analysis of such reactions shows that they can be extremely co-operative ($n > 5$) when we almost have a phase change as in a precipitation reaction

$$Mg^{2+} + L^{2-} \rightleftharpoons MgL(\text{solid}).$$

Here an extremely small change in the concentration of magnesium switches the system from a one 'phase' to a two 'phase' system (see Fig. 7.28). An intermediate situation between the above is one in which there is a level of co-operativity where n is some number considerably greater than one. A plot (compare Fig. 7.27) of formation of Mg_nL_m against magnesium concentration now shows a rapid change as [Mg] increases while the system remains soluble. In this case changes in magnesium concentration do not bring about an abrupt solubility change, but only a very rapid switch in the solution, sol \rightleftharpoons gel, conditions and this could apply to DNA and RNA structures. We now turn to a different case where we treat the development of a filament as if it were like a gel reaction.

The polymerization of one of the polymers, tubulin, associated with the filamentous network of a structured cell can be pictured as in Fig. 7.14. There is involved the monomer of tubulin, L, the cation, M^{2+}, and GTP, the nucleoside triphosphate. The formation of the polymer is linked to the energy state of the cell through GTP and the concentration levels of the cation M^{2+}. It is observed that, while magnesium aids polymerization, calcium opposes it.

The balance between these cation concentrations could decide the internal structure of cells but it is likely that $[Mg]^n$ is the effective agent since calcium in cells is so low and $Mg \cdot GTP$ is dominant. We now see that relatively small changes in $[Mg]$ and $[GTP]$ during the cell cycle could lead to formation or dissolution of the tubulin filaments. Again minor changes in the environment, such as stress on the filaments leading to their dissolution, will be poised opposite the level of $[Mg]^n$. We are suggesting here that, in part, Mg^{2+} controls the *mechanics* of the cell-division cycle. As stated in Section 9.10 this control could extend to features of DNA itself and the competition with polyamines might be involved.

9.12 Sol/gel equilibria and magnesium

The polymerization of polysaccharides and of glycoproteins, all of which are negatively charged and are frequently parts of the cell exterior, controls the stability of membranes and cell–cell interactions. These cation-dependent polymers are often parts of adhesive surfaces of various kinds in different organisms. The systems lose stability in the presence of metal-binding agents such as EDTA. The suggestion is frequently made that the exact structure of these open polymer meshes is dependent on condensation between 'sol' and 'gel' states (Fig. 7.27). The switching between the two states is well known in model systems of intertwined polymers. Once again the switch sol/gel is dependent on very small changes in divalent ion concentrations, including $[Mg^{2+}]^n$ and $[Ca^{2+}]^n$. Particularly important here are the outer cross-linked membranes of Gram-negative bacteria which contain teichoic acids (Fig. 9.2).

The overall picture of these polymeric equilibria is that, differently from calcium-dependent second-messenger switches in the cytoplasm, which appear to be co-operative to only a small degree and which act very rapidly, magnesium-dependent switches are almost always phase-like changes dependent on $[Mg]^n$, where n is large, and are relatively slow.

9.13 Magnesium and lipids

The synthesis of many lipid-soluble molecules in biology is based on the transfer of the isopentenyl unit. The unit, C_5, is carried by pyrophosphate, so that there is coupled to synthesis a loss of the pyrophosphate bond energy. The pyrophosphate moiety is synthesized from ATP. In this reaction, the part played by magnesium is at several levels from the initial formation of the pyrophosphate from $Mg \cdot ATP$ to the reactions of the pyrophosphate esters in condensation reactions. These reactions are on the route to the synthesis of terpenes and cholesterol.

Another use of magnesium is in the synthesis of choline-based lipids such as phosphatidyl cholines (lecithins) where cytidine triphosphate is the energy source driving reaction.

9.14 Magnesium and chlorophyll

9.14.1 Chlorophyll binding in proteins

There are two questions of great interest in the association of magnesium with chlorophylls, as for example in Fig. 9.12, which gives the dominant molecule for light capture in biology. The first is what is magnesium for and the second is how is chlorophyll placed where it is. The requirement for magnesium rather than another metal ion in this light-capture molecule is perhaps due to the fact that an element of such low atomic number does not enhance fluorescence, and does not itself have redox properties. Small size allows other ions to fit in the chlorin cavity, but fluorescence is much greater from zinc chlorins and unwanted redox properties are to be found for Mn, Co, Fe, Ni, and Cu heterocycles. While serving no obvious chemical function, magnesium does not interfere with the photo- and free-radical reactions of reaction-centre chlorins and clearly can assist them. Moreover, the presence of magnesium in the chlorin ring gives the ring an approximate tetragonal symmetry and hence enhances the molecular extinction coefficient around 650–700 nm relative to the lower-symmetry metal-free molecule while moving the absorption maximum to longer wavelengths. Figure 9.13 illustrates these properties.

The binding of magnesium in chlorin proteins is peculiar in two ways. The chlorin is an N-donor, and the further binding to the protein is through another N-donor, from a histidine imidazole (Fig. 9.12). The co-ordination sphere is then five nitrogen atoms with the magnesium somewhat out of plane. It is difficult to believe that this arrangement could be a thermodynamically

Fig. 9.12 (a) The formula of chlorophyls a and b and their binding in proteins; (b) the binding of a protein histidine to the magnesium of chlorophyll.

Fig. 9.13 The absorption spectra of chlorophylls and of the reaction centres, P 680 (P 700). Light is harvested by a series of chlorophyll a and b molecules, Fig. 9.14, and then radiation transfer to a special chlorophyll pair (P 680 or P 700) occurs which is the reaction centre for converting light to free-radical energy, Section 9.14.2.

Fig. 9.14 The chlorophyll-containing protein for light harvesting which is in the chloroplast thylakoid membranes, after B. W. Matthews.

stable form of simple magnesium co-ordination chemistry in the presence of many free oxygen donors, e.g. from water. The chlorin/magnesium complex itself may well be stable to hydrolysis, as it is effectively an encapsulation by a ring dianion, but binding to such a site by ions such as Fe(II) or Zn(II) is so much more powerful that some selective insertion pathway has to be imagined (see Section 9.14.2). The binding to the imidazole of the protein in the authors' opinion is another encapsulation reaction but this time of the chlorophyll as a whole which is encapsulated by a helical *protein* in an '*organic* solvent' (Fig. 9.14; see Chapter 2). Effectively this is a solvent extraction reaction and no doubt it is aided by the large lipophilic tail of the chlorophyll molecule. The co-ordination chemistry is itself an interesting case of the domination of the fold energy of the protein over the ion co-ordination chemistry but the positioning in space of the complex is a property of a special protein.

Now, as well as capturing light, the special chlorophyll molecules in a reaction-centre dimer, Fig. 9.13, generate the first step of energy transduction in that at this centre the photon generates a diradical and to and from it electrons flow to give the first membrane potential (Fig. 5.3).

The chlorophyll itself has an interesting property in which it differs from porphyrins: there are no sidechain anionic carboxylate groups. It may well be that these groups that connect electron transfer reactions to proton transfer reactions in iron porphyrins are unwanted in a light-capture reaction centre, see Section 13.6.3.

9.14.2 Magnesium insertion in chlorophyll

Magnesium ions are inserted into the chlorin ring through the action of a chelatase enzyme and without the enzyme this cation could not combine with

chlorins nor could it overcome the much greater affinity of the chlorin site for the first-row transition metal ions, e.g. Zn^{2+}, Cu^{2+}, and Fe^{2+}. The requirement is for a series of reactions comparable with the reactions for the pumping of magnesium ions across membranes using ligands, L,

$$Mg^{2+} + L_1 \rightleftharpoons MgL_1 \text{ (reversible),}$$

$$MgL_1 + L_2 \longrightarrow MgL_2 + L_1 \text{ (irreversible).}$$

The reaction scheme is such that magnesium is the preferred partner of L_1 and other cations M are excluded from binding to L_2, etc. by kinetically irreversible steps. We know, for example, that ATP binds Mg^{2+} but not other metal ions in cells since the product $[M][ATP]K$, where K is a binding constant, is large for magnesium but small for most other ions. This is due to a similar stability constant K for most divalent ion binding to ATP (Table 2.2) but vast differences in $[M]$. Most Zn^{2+}, Mn^{2+}, Cu^{2+}, and Fe^{2+} ions are removed by one of three mechanisms: (1) binding to N,S ligands of protein; (2) oxidation Fe^{2+} to Fe^{3+}; and (3) pumping into vesicles (see Chapters 2 and 3). The case is very different where L is a nitrogen donor such as chlorin whence it is essential to avoid direct *reversible* binding of Mg to chlorins.

Given that there is a very small amount of ML_1 other than MgL_1 this could still be dangerous in a transfer system but for the fact that MgL_1 must have a recognizably different structure. The structure, i.e. the size and stereochemistry of a complex, e.g. $Mg \cdot ATP$, is a high degree selective to Mg^{2+}. The *recognition* of such a structure can be used in the kinetic control

$$MgL_1 + \text{chlorin} \longrightarrow \text{chlorin} \cdot Mg + L_1,$$

which is effectively irreversible due to the nature of encapsulation. We know that the enzymes for this last reaction exist but *in vitro* even they usually incorporate Zn into chlorin in the presence of vast excesses of Mg^{2+}.

In these biologically interesting chlorophyll complexes in proteins magnesium is five-co-ordinate and the magnesium atom would appear to be just too large to fit the cavity of the chlorin. The misfit is very slight since model complexes are known where magnesium sits either in the plane of a porphyrin or about 0.3 Å out of the plane. The first group of complexes is six-co-ordinate, but they have unusually long 2.25 Å axial bonds while the second group is five-co-ordinate and the axial bond length is 2.1 Å which is normal. The Mg–N bond to the porphyrin nitrogen in the ring is about 2.05 Å and it could be that the magnesium has to squeeze into such a ring making for long axial bonds (Fig. 9.12).

All the magnesium chlorophyll pigments in membranes do not absorb at exactly the same wavelength (Fig. 9.13). Just as copper is tuned to its function of electron transfer (see Section 7.8) so the absorption of dye-stuffs, here chlorophyll, is tuned to the process of energy transfer. Here the tuning consists of modifying the electronic absorption spectrum so that it moves to longer wavelengths the closer in space the molecule is to the reaction centre, a chlorophyll dimer, which has its absorption bands at the longest wavelength. The system can be written as a photon cascade $h\nu \rightarrow$ chlorophyll a \rightarrow chlorophyll b \rightarrow reaction centre chlorophyll. The control of the absorption of the dye molecule—the same organic molecule in different environments—is an entatic (strained) situation. In essence the protein has evolved to control the

relative stability of the ground and excited spectroscopic states. It can do this by minor changes in the co-ordination bond-lengths of Mg^{2+}, by adjusting the planarity of the chlorin ring, or by altering the solvation by the sidechains of proteins. A comparison of this trick for a photon pipe with the adjustments around haem to generate an electron pipe (a wire) is intriguing (see Chapter 13).

9.15 The use of manganese as a magnesium probe

The chemistries of magnesium and manganese as divalent ions are rather similar despite the difference in radii and in affinity for N-donor ligands. This similarity can be put to good use provided that the system under study has been shown to be *functionally* similar when one cation replaces the other. Manganese(II) is a good EPR reagent and a good NMR relaxation reagent. Other probes can be used; for example Cr(III) is used in Cleland's reagent, ATP(Cr), but has no activity; again the Ln(III) ions can be used but do not often show activity in Mg^{2+}-dependent systems. In those structural studies where $[Mg(H_2O)_6]^{2+}$ rather than the simple cation Mg^{2+} may be important, it is found that such cations as $[Co(NH_3)_6]^{3+}$, $[Cr(NH_3)_6]^{3+}$, and so on are useful substitutes. Particularly in the hands of Cohn and Mildvan (see 'Further reading') Mn^{2+} has been the major valuable probe of the sites of enzyme activity of magnesium (see Fig. 9.15). Most important has been the use of Mn^{2+} in probing phosphate transfer reactions. It would appear that Mg^{2+} acts as a mild Lewis acid while performing the major role of aligning the anionic reactants in the enzyme site. If anything, Mg^{2+} is a structure-based catalyst not an attacking group. Where a stronger acid is required in order to break substrate bonds then it is usual to find a zinc metallo-enzyme rather than a magnesium metal–complex. Notice that the ease of hydrolysis follows the order Anhydrides > esters > amides. Magnesium is not used to assist amide hydrolysis or synthesis (see peptidases and urease in Table 6.3).

Fig. 9.15 An illustration of the use of Mn(II) as probe for Mg(II) activity in an enzyme (after the work of M. Cohn and A. Mildvan; see Mildvan in 'Further reading'). The enzyme is pyruvate kinase. Other features of note are: (1) the use of two metal ions, here the second is potassium; (2) the necessary closed pocket; and (3) the co-ordination involving but one histidine and possibly two carboxylates Y and Z.

9.16 Pollution and Mg^{2+} biochemistry

There is little doubt that in model systems Mg^{2+} can be replaced by a variety of cations such as Mn^{2+}, Co^{2+}, Ni^{2+}, and Zn^{2+}. Sometimes Ca^{2+} will act as a replacement. The activity of the enzyme is often only moderately impaired since the main function of Mg^{2+} is to neutralize negative charge. In biological systems Mg^{2+} functions are not open to this type of replacement chemistry since the levels of all divalent cations are $> 10^4$ times lower. Accidental contamination could still be a problem and here we must concern ourselves today with the polluting elements such as Al^{3+} and possibly Be^{2+}. There is the possibility that Al^{3+} or possibly $[AlOH]^{2+}$ could occupy Mg^{2+} sites of various kinds in cells. The suggestion arises that these sites could be phosphorylated tubulins, and there are some who believe that this could cause diseases of the nervous system such as Alzheimer's disease (Sections 3.5 and 3.6), but recent evidence does not support this. The systems for uptake of Mg^{2+} are likely to be especially vulnerable to Ni^{2+} since the ion has the same simple geometry and size.

9.17 Lithium and magnesium

Figure 9.1 shows that there is a relationship in the periodic table between Li and Mg, termed the diagonal relationship, and a similar relationship between Na^+, Ca^{2+}, and Y^{3+}. The relationships are based on the fact that the elements above the line form small ions, radius less than 0.75 Å, while the ions of the elements below the line have radii greater than 0.75 Å. The result of this size difference is a markedly greater affinity by small ions for anions of high charge density, F^-, OH^-, and all phosphates $ROPO_3^{2-}$, than is shown by the larger cations and a considerable difference in kinetics also exists even in the exchange of water of hydration (Table 9.4). In biological systems we have seen how this has led to the close association of Mg^{2+} and phosphate metabolism. Lithium (beryllium and aluminium) are not essential elements, but they can interfere with biochemical processes *in vivo* by interacting with phosphate-containing molecules. One interesting possibility is the involvement of lithium with phosphate compounds. This can occur at two levels: (1) lithium could interact with the phosphate compounds of the Na^+/K^+ or Mg^{2+}/Ca^{2+} pumps; or (2) it could interact with the inositol phosphates. Lithium is, of course, a major psychotherapeutic drug, and its action may in part be based on these possibilities. Could Ln^{3+} compounds be useful here while interacting with Ca^{2+} sites?

9.18 Conclusion

The high concentration of magnesium ions in very many biological compartments is put to use in assisting the neutralization of many anionic centres in biology. This neutralization can allow binding together of anions, it

can protect anions from hydrolysis, or it can be used to assist chemical reaction. The role of magnesium in stabilizing some structures and energizing others is to be found in all the types of biopolymer, DNA, RNA, proteins, polysaccharides, and lipids. The role extends to interaction with a multitude of small molecule anions including di- and triphosphates. The movement of Mg^{2+} from compartment to compartment seems to be less important than the maintenance of a level close to 1.0 mM almost everywhere in biological systems. The whole is, of course, an energized steady state and we must be aware that through small changes in Mg^{2+} concentrations in the cell cycle magnesium could have dramatic effects dependent on $[Mg^{2+}]^n$ and the variable concentrations of polyamines.

Given the central role of phosphate in all these polymers (DNA, RNA), in metabolism, in bioenergetics, and in metabolic pathways especially inside cells (DNA, RNA, protein, polysaccharide, and lipid synthesis), it is very difficult to escape the conclusion that cells evolved with magnesium as the essential partner for anions, especially phosphates, in biological solution. Thus magnesium is and always was an element equal in importance to phosphorus in biology. Part of the problem in recognizing the role of the metal cations is that chemists and biochemists write compounds which are salts as if they were uncharged. Examples are DNA and RNA and frequently ATP, while the real species are $DNA^{n-}(M^+)_n$, $RNA^{n-}(M^+)_n$, and $ATP^{4-} \cdot Mg^{2+}$. Like phosphate and potassium the role of magnesium in living processes has evolved little since the early periods of life and it has few vital functions outside cells. Here is a striking difference from Na^+, Ca^{2+}, and Cl^- which were initially rejected from cells and so evolved in value late in evolution especially with the coming of multicellular systems.

Further reading

Brewer, J. M. (1981). The nature of enolase. *Critical Reviews of Biochemistry*, **11**, 109–54.

Halliwell, H. (1984). *Chloroplast metabolism*. Oxford University Press, Oxford.

Kitzing, E., Lilley, D. M. J., and Diekman, S. (1990). DNA-crucifix structures, *Nucleic Acids Research*, **18**, 2671–83.

MacDonald, T. L. and Martin, B. R. (1988). Aluminium and magnesium competition. *Trends in Biochemical Science*, **13**, 15–19.

McKee, V. and Rodley, G. A. (1988). The structure of magnesium/chlorophyll complexes. *Inorganica Chimica Acta*, **151**, 233–6.

Mildvan, A. S. and Grisham, C. M. (1974) Divalent cations in enzymes. *Structure and Bonding*, **20**, 1–22.

Navaratum, N., Ward, S., Fisher, C., Kuhn, N. J., Keen, J. N., and Findlay, J. B. C. (1988). Cation activation of galactosyl transferase. *European Journal of Biochemistry*, **171**, 623–9.

Sigel, H. (1990). Metal ion reactions with nucleotides. *Coordination Chemistry Reviews*, **100**, 453–540.

Tam, S.-C. and Williams, R. J. P. (1985). Electrostatic binding of anions and cations. *Structure and Bonding*, **63**, 103–51.

Wacker, W. E. C. (1980). *Magnesium and man*, pp. 1–165. Harvard University Press, Cambridge, Massachusetts.

10

Calcium: controls and triggers

10.1	Introduction	268
10.2	Free calcium ion levels	270
10.3	The calcium ion	271
10.4	Protein ligands for calcium	272
10.5	Magnesium/calcium competition	275
10.6	The resting state of calcium in cells	276
10.7	Calcium triggering: calmodulins	276
10.8	The calcium trigger proteins	278
10.9	S-100 proteins	280
10.10	Other triggering modes	281
10.11	Calcium and protein phosphorylation	283
10.12	Calcium buffering and calcium transport in cells	283
10.13	Calcium currents: movement through membranes, channels, gates, and pumps	284
10.14	Internal calcium-induced proteases	286
10.15	General remarks concerning control systems in cells	287
10.16	Extracytoplasmic calcium	288
10.17	Extracellular enzymes and calcium: digestion and blood clotting	289
10.18	Calcium proteins of biominerals	292
10.19	Calcium biominerals	293
10.20	Intra- and extracytoplasmic calcium balances	294
10.21	Summary	296

10.1 Introduction

It is almost impossible to overemphasize the role of calcium in *advanced* biological systems (Fig. 10.1). Everywhere one turns in the study of controls over metabolic pathways there is a calcium-dependent step (Table 10.1). To those unfamiliar with the deep involvement of calcium in metabolism we give just three examples of central controls: (1) in light harvesting and its coupling

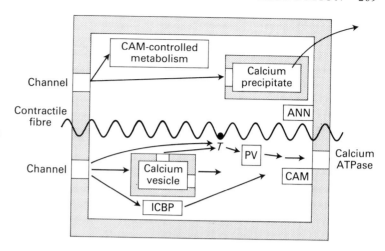

Fig. 10.1 A diagram to illustrate the intracellular activities which are controled by the calcium ion concentration. Included are calmodulin (CAM) control over metabolism, biomineral formation, tension (*T*) in filaments, and pumping (ATPase). ANN is the protein Annexin, which controls release of vesicle contents and contacts both membrane and filaments. The calcium transport is controlled by intestinal calcium binding protein, ICBP, and calcium buffering is limited by parvalbumin, PV.

to photosynthesis of dioxygen; (2) in the control of dehydrogenases in oxidative phosphorylation; and (3) in a multitude of kinase reactions. While metabolic control is one role of calcium there are others.

1. Calcium controls the mechanical stability of walls of some cells, of membranes of others, and of the tension in filaments in a whole variety of cells. Muscle contraction is stimulated by calcium as is exocytosis.

2. Calcium is intimately associated with fertilization, cell division, and hormonal activities.

3. Calcium salts, carbonates, phosphates, and oxalates, dominate biominerals.

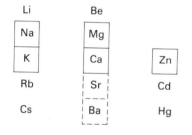

(A) Metal ions involved in homeostasis (in boxes)

The difficulty in an attempt to summarize such a wealth of activities can be seen by looking at the flood of books and conference reports on calcium in biology. The purpose of this chapter, therefore, has to be an attempt to give these detailed functional observations a rational basis at the risk of oversimplification. Only in one of the final chapters, Chapter 21, will this calcium biochemistry be pulled together with the biological chemistry of the other metals (see (A)) and non-metal elements which we believe to be involved

Table 10.1 Calcium-controlled events in cells

Activity	Controlled events or systems
Photosynthesis	Dioxygen release
Oxidative phosphorylation	Dehydrogenases
Receptor responses	(a) Nerve synapse
	(b) IP_3-linked reactions
Contractile devices	(a) Muscle triggering (actomyosin)
	(b) Cell filament controls (tubulins)
Phosphorylation	Activation of kinases
Metabolism	Numerous enzymes
Membrane/filament organization	Annexin-like proteins provide tension
Cell division (?)	S-100 proteins(?)
Cell death (?)	Internal proteases (?)

in the homeostasis of cellular activity in a very general way. It will be made very apparent that there is a special relationship between calcium, magnesium, phosphate, and zinc, especially in hydrolytic reactions of advanced cells.

10.2 Free calcium ion levels

In Chapter 1 we discussed the levels of different metal ion concentrations in sea and fresh water. There is a huge variation in calcium concentration as anybody knows who attempts to wash clothes in hard as opposed to soft water never mind in the sea. Hardness of water is closely related to its calcium ion content. Despite this variation from, say, 10^{-6} to 10^{-2} M calcium in solutions outside biology, the free calcium ion inside biology is very, very, tightly controlled both inside cells and their compartments and in extracellular fluids. The free calcium in extracellular fluids is held close to 10^{-3} M and this allows a cellular system to manipulate the precipitation of carbonates (shells), phosphates (bones), and oxalates (plant structures) in a species- and organ-dependent manner by the way in which phosphate, bicarbonate, and oxalate concentrations as well as those of calcium itself and some special biopolymers are pumped into different zones. (Here again there are numerous books on this topic.) We stress the constancy and the absolute need for constancy of free extracellular calcium levels in complex organisms since there is a very narrow concentration range between desirable calcium for mineral precipitation in one location and the slightly lower desirable calcium levels for the maintenance of extracellular structures such as connective tissue at the other. There is little room for error as we well know by observations on ageing populations who suffer from osteoporosis, calcifying arteries, and other malfunctions of calcium control.

In the cytoplasm of cells a similar very, very careful control of free calcium levels is required. Here the level is near 10^{-7} M, but the authors believe that this level is not exactly the same in different cell types and that the differences are related to the biological functions of the cells. We shall see that those calcium-binding proteins which help to buffer calcium are present in different concentrations in different cells. Despite the fact that cytoplasmic free calcium can be measured relatively accurately it is difficult to get the *very accurate* data necessary about these low concentrations. This is especially the case since cells have shapes and, although it is assumed generally when these measurements are made that there is a homogeneous cytoplasm, there is good reason to believe today that not all cells are homogeneous in cytoplasmic calcium ion concentration due to the positioning of their pumps and channels (see Chapter 3). Calcium ions are then in a steady state of flow. In passing note that, if free calcium in the cytoplasm reaches 10^{-5} M, a cell appears to commit suicide by turning on internal proteases.

Now there are further well-defined compartments in eukaryotic cells which lie behind membranes (Fig. 10.1): they are the organelles and vesicles described in Chapter 3. Figures 10.11 and 10.12 indicate how calcium is moved into many of them from the cytoplasm but we do not know the steady-state free calcium level in these microvolumes. It could be as high as 10^{-3} M

(in the sarcoplasmic recticulum). In general, vesicle space, as opposed to organelle space, is high in free calcium but different vesicles may be widely disparate.

It is against this background that we have to follow the flow of calcium in the resting state and the ability of cells to be switched in activity by allowing pulses of calcium to enter the cytoplasm from outside or from vesicular systems. The pulse varies from the activation of a reversible muscle twitch to the irreversible fertilization of an egg, which is a morphogenic step.

10.3 The calcium ion

Calcium is a large divalent cation (0.95 Å radius) and it is available, since there is plenty of it in most natural waters and its common salts are not very insoluble. It is the intermediate strength of its binding in inorganic lattices and to organic molecules, as compared with the stronger binding of cations at the other end of transition metal series, such as zinc, which makes it so useful in control of conformational changes and then cell activities. However, switches of control (triggering) must be fast as well as of intermediate strength. We have analysed not only calcium binding and structure but also calcium kinetics in Chapters 4 and 9 and hence we can summarize our chemical knowledge under three headings.

1. *Calcium binding* is selective, as a result of: (a) its ability to interact with neutral oxygen donors such as carbonyls and alcohols, an interaction largely absent from magnesium chemistry in water, together with anions, hence also defeating competition from sodium; and (b) its ability to bind to a large number of centres at once—again an ability not shared by magnesium or any of the small cations such as zinc. These two factors give calcium a selective binding chemistry based on its charge-to-size ratio (Tables 10.2 and 9.2) and, in addition, they make it an ideal cross-linking agent, as is seen in the structures of its compounds and complexes.

2. *Calcium structures* differ from those of the other available divalent element, magnesium; calcium is like sodium, potassium, strontium, barium, and the lanthanide cations in the high co-ordination numbers often observed in their compounds and complexes (Tables 10.2 and 10.3). The geometry is

Table 10.2 The properties of some cations

Property	Na^+	K^+	Mg^{2+}	Ca^{2+}	Mn^{2+}	Ln^{3+}
Size (Å)	1.00	1.33	0.60	0.95	0.75	1.05 to 0.85
Co-ordination number	6(+)	8(±)	6	7(±)	6(+)	7(+)
Bond lengths	Irregular	Irregular	Regular	Irregular	Regular	Irregular
Bond angles	Irregular	Irregular	Regular	Irregular	(Regular)	Irregular
H_2O exchange(s)	10^{-10}	10^{-10}	10^{-6}	10^{-9}	10^{-7}	10^{-7}–10^{-9}

Mn^{2+} and Ln^{3+} are useful probes for Ca^{2+} activities. All these cations bind to oxygen centres almost exclusively, with the exception of Mn^{2+}. Na^+ and K^+ bind weakly to all donors except for those in special cavities.

Table 10.3 Some typical calcium salt structures

Salt	Ca(II) co-ordination no.	Ca–O distances (nm)	
		Min.	Max.
$CaHPO_4 \cdot 2H_2O$	8	0.244	0.282
$Ca(H_2PO_4)_2 \cdot H_2O$	8	0.230	0.274
Ca 1,3-disphosphoryl-imidazole	6	0.226	0.236
	7	0.227	0.278
Ca dipicolinate · 3H₂O	8	0.236	0.257
$CaNa(H_2PO_2)_3$	6	0.231	0.233
Ca tartrate · 4H₂O	8	0.239	0.254
$Ca(C_5H_9O_7)_2 \cdot 2H_2O$	8	0.239	0.247

not that of any simple regular symmetry and is controlled as much by the second co-ordination sphere, and solvation in solution, as by the first sphere. The contrast with magnesium (or aluminium) chemistry (Table 9.1) could not be greater. (Aluminium is mentioned here because of the acid-rain problem, when aluminium may well compete for calcium sites but will not give the same structures.)

3. *Calcium kinetics* are also very different from those of magnesium in two respects. First, calcium exchanges water at close to the collisional diffusion limit of $10^{10} \, s^{-1}$ and its reaction on-rates are often diffusion-limited, whereas its off-rates are limited by binding strength. Second, within a complex the ligands on the calcium ion are *fluctional*. Magnesium (or aluminium) does not resemble calcium in either of these properties. In fact, in these respects calcium is again more like sodium and potassium and the lanthanide ions. It is the similar sizes of these ions, all around 1.0 Å in radius, which generates the similarities despite the differences in charge (Table 10.2).

With these features of calcium chemistry in mind we must turn to the properties of calcium-binding proteins, at first in general terms.

10.4 Protein ligands for calcium

The major donor groups from proteins to the calcium ion are carbonyl and carboxylate centres. These groups can be put together in many ways, but three basic and very different constructions are known. The first involves a limited set of donors close together in the amino-acid sequence where the binding of calcium does not cross-link the protein through remote parts of the sequence so as to stabilize its fold grossly (Table 10.4). This is the mode of binding in calmodulin (Fig. 10.2) and in other trigger proteins from inside cells. It is likely that in the calcium-free state these calcium-binding regions are mobile and that there is considerable electrostatic repulsion between their large number of negatively charged centres. The second mode of binding involves groups

Fig. 10.2 An outline of the domains of calmodulin. The sequence of the protein is shown where heavy circles are helical stretches. Four calcium binding hands are represented: note acidic groups D and E. The helix–loop–helix motif forms the calcium-binding 'hand' as illustrated by the concave loop between a thumb and forefinger, see Fig. 7.2.

distant in the amino-acid sequence and is largely dependent on a strongly predetermined fold which generates a rather immobile cavity for calcium. An example is phospholipase A_2 (Table 10.4 and Fig. 10.3 and 7.1). The striking differences between these two kinds of site relate to: (1) the effect of calcium on the structures; and (2) the rate of the on-reactions. The looser sequential binding site binds calcium very rapidly and adjusts the protein tertiary structure considerably (Section 10.7). The well-defined cavity type of binding site accepts calcium relatively slowly and there is virtually no conformational change. The third type of site has a longer series of anions on the surface often of a random polymer which come together through the sequence (or the fold) and amongst which the calcium can travel rapidly, for example, some phosphoproteins, the anionic glycoproteins of connective tissue, and the γ-carboxyglutamate (Gla) containing peptides of prothrombin and the bone protein osteocalcin.

Table 10.4 Calcium-binding sites in some proteins

	Sequence of ligands	Sequential?
Extracellular		
Phospholipase A$_2$	Tyr-28, Glu-30, Gly-32, Asp-49, 2H$_2$O	No
Staphylococcal nuclease	Asp-19, Asp-21, Asp-40, Thr-41, Glu-43, 1H$_2$O	No
Trypsin	Glu-70, Asn-72, Glu-80, 2H$_2$O	(Yes)
Thermolysin	Asp-138, Glu-177, Asp-185, Glu-187, Glu-190, 1H$_2$O	No
	Asp-57, Asp-59, Glu-61, 3H$_2$O	Yes
Intracellular		
Parvalbumin	Asp-51, Asp-53, Ser-55, Phe-57, Glu-59, Glu-62	Yes
	Asp-90, Asp-92, Asp-94, Lys-96, Glu-101, 1H$_2$O	Yes
Calmodulin	Asp-20, Asp-22, Asn-24, Thr-26, Thr-28, Glu-31	Yes
	Asp-56, Asp-58, Asn-60, Thr-62, Asp-64, Glu-67	Yes
	Asp-93, Asp-95, Asn-97, Tyr-99, Ser-101, Glu-104	Yes
	Asn-129, Asp-131, Asp-133, Glu-135, Asn-137, Glu-140	Yes
Intestinal calcium-binding protein	Ala-15, Glu-17, Asp-19, Gln-22, Ser-24, Glu-27	Yes
	Asp-54, Asn-56, Asp-58, Glu-60, Ser-62, Glu-65	Yes
Troponin C	Asp-27, Asp-29, Gly-31, Asp-33, Ser-35, Glu-38	Yes
	Asp-63, Asp-65, Ser-67, Thr-19, Asp-71, Glu-74	Yes
	Asp-103, Asn-105, Asp-107, Tyr-109, Asp-111, Glu(114	Yes
	Asp-139, Asn-141, Asp-143, Arg-145, Asp-147, Glu-150	Yes

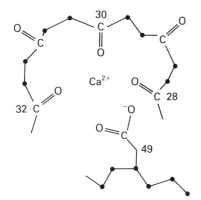

Fig. 10.3 The calcium-binding site of phospholipase A$_2$ (see Fig. 7.1). A relatively rigid hole is formed from the backbone carbonyls and the remote cross-linking carboxylates.

We can ask a series of questions about these sites. How do they avoid competition from Mg^{2+}? How is the calcium binding relayed into the structure? If magnesium should bind, would it have the same effect on the trigger? How does calcium diffuse in the cell to the sites? We must remember that the biological effect is related to calcium binding not only through binding constants, K, but also through limitations of rate based upon diffusion and the transmission of the binding through the structural change. It is therefore essential that we notice the character of the proteins involved. Finally we must ask where are the different calcium proteins within the cell and its various compartments, Chapter 3. The implications of these questions will not be addressed in this chapter and we shall return to their discussion in Chapter 21 where we describe homeostasis.

10.5 Magnesium/calcium competition

We know the structure of very few magnesium-bound proteins, but we deduced in Chapter 9 that Mg^{2+} preferred sites of high negative charge density, e.g. ATP^{4-}, co-ordination number six, and/or of small cavity size. In this context we know that Mg^{2+} ions do not bind with any great strength to such calcium sites as that in phospholipase A_2 and this we attribute to the large size as well as the low charge of the cavity. However, Mg^{2+} will bind weakly to many of the calcium calmodulin sites which are flexible and of higher charge (Fig. 10.2) with a constant close to 10^3 M^{-1}. We must suspect that, in the absence of calcium, Mg^{2+} ions (concentration near 10^{-3} M) will bind at least 10 per cent to some of these sites (and almost 50 per cent to others), since 10 per cent binding needs a binding constant of only 10^2. Even so, Mg^{2+} ions do not act so as to produce the calcium/calmodulin response since the conformational change in the calmodulin produced by Mg^{2+} is not the same as that produced by Ca^{2+}. Thus, while on the one hand magnesium may be excluded from calcium sites, on the other it may be an effective competitive inhibitor since the replacement of Ca^{2+} by Mg^{2+} does not lead to action.

A simpler example of the differences between the chemistry of Mg^{2+} and Ca^{2+} may be given by reference to the binding of the model malonate headgroup (see margin) to the two ions. The Mg^{2+} compound, $Mg(C_3H_3O_3)_2 \cdot 2H_2O$, shows simple chelation to the malonate anion, the complex is octahedral, and the bond lengths and bond angles are regular. By way of contrast, the calcium compound $Ca(C_3H_2O_4) \cdot 2H_2O$ shows a co-ordination number of above six, the geometry, bond angles, and bond lengths are irregular, and cross-linking between anions occurs. These studies should be considered in the context of the differential binding effects of Ca(II) and Mg(II), especially to the γ-carboxyglutamate (Gla)-containing proteins, such as prothrombin, since the function of calcium here includes cross-linking of the protein and a cell membrane while Mg(II) is inactive. The headgroup of Gla is malonate-like. These principal differences apply to all Ca^{2+}/Mg^{2+} competitions—for example, in calmodulins. In biology the bonding ligand is usually such that calcium and magnesium chemistries are quite distinct; hence magnesium binding does not trigger calcium proteins.

Against the background of this summary of calcium ion chemistry we shall divide the discussion of calcium function as follows. First, we describe the activity associated with calcium in the resting cell cytoplasm. This is followed by the analysis of triggering of cytoplasmic cell activities by the influx of a pulse of calcium from outside the cell. The recovery from the pulse as well as the entry of calcium requires calcium to pass through a membrane and this topic is the third in sequence. Next we look at the extracellular calcium and its steady state but we shall not concern ourselves overmuch with bones or shells (see Chapter 20 on biominerals). Finally, the calcium in vesicular systems and in organelles will be described. It must be understood that at rest the whole system is in almost perfect homeostasis so that there is always a dynamic exchange between all the calcium pools (see Chapter 21). This means that we must be aware at all times that the calcium ion is involved in cellular activities in the resting state as it is in the extracellular compartments.

Malonate headgroup of γ-carboxyglutamate

10.6 The resting state of calcium in cells

A problem that we face in the discussion of calcium biochemistry in relation to cellular activity is our lack of understanding of the cell at rest. Triggering to a new activity (see Section 10.7) is relatively well understood. At rest a cell is highly energized in all the ways described in Chapter 5 and, of course, is highly active in maintaining this state. This means not only that there is a metabolic rate at rest but that ionic gradients, units of structure, and tension in filaments are maintained in a highly regulated manner. Many of the activities depend on calcium levels and, while some new activities are produced by stimulants, many effects are just changes (up or down) of the resting actions. Two factors need particular attention. First, cells are not all alike and we would be wise to assume that their exact cytoplasmic steady-state levels of free calcium differ although they are all in the range 10^{-6}–10^{-8} M (Fig. 10.4). As far as calcium is concerned this means, for instance, that a slow posture muscle has a different free calcium concentration at rest, controlling its different metabolic rate, from that of a fast twitch muscle. Even a twofold difference in the concentration of Ca^{2+} could be sufficient to alter a cell grossly since the effect of this change will be greatly amplified if the calcium dependence is proportional to $[Ca^{2+}]^n$ where n is large. (The resting cells are constantly receiving messages, of course, which keep the cell in the rest state.) Second, cells have shapes and in Chapter 3 we have seen that the arrangement of enzymes both along membranes and internal fibrils is related to cell shape. We know that the positionings of entry channels and exit pumps for calcium are also related to cell shape, so that a calcium current flows in the cells and around them. The resting state is a state of constant calcium gradients everywhere inside and outside cells and must therefore be in constant flow. We return to this topic in Chapter 21. We describe next the activities of calcium when it is pulsed into the cytoplasm.

Fig. 10.4 The relative concentration of calcium and several other ions inside and outside cells, and the nature of the calcium ion pulse.

Metal ion concentration

10^{-8} M	10^{-6} M	10^{-4} M	10^{-2} M
In-cell calcium	←— Calcium —→ pulse	$[Mg^{2+}]$ —→	←—Extracellular calcium
Very few proteins bind calcium	Many in-cell proteins bind Ca^{2+}	Not observed for calcium	Many extracellular proteins bind Ca^{2+}
	$[Mg^{2+}]$ chromaffin granules		Na^+ and K^+ range

10.7 Calcium triggering: calmodulins

A message reaching a cell can be of two kinds—electrical, i.e. a Na^+/K^+ ion nerve message, or chemical. In the first case the wave of depolarization produced at a cell membrane can be used to open appropriately located calcium channels which had been gated previously by the voltage across the outer membrane (Fig. 10.5). In the second case the chemical, a hormone or a

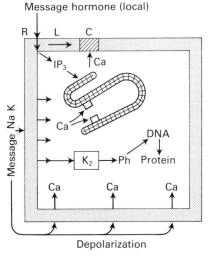

Message hormone (local)

Fig. 10.5 There are two major message modes involving calcium. In the first, calcium enters the cell through depolarization of the whole cell membrane, Na^+/K^+ message. In local triggering a chemical message at a receptor, R, activates the breakdown of phospholipid, releasing IP_3, which then releases calcium from an internal store; finally, a kinase C is activated. Other kinases, K_2, are activated in the cell. Ph, protein phosphorylation.

transmitter, binds to a receptor protein on the outer surface of a cell and, by adjusting mechanically the calcium channel attached to the receptor, it opens the channel to calcium influx (Fig. 10.5). In both cases a calcium pulse enters the cell (Fig. 10.4), and this pulse, which is short-lived due to the recovery of the voltage or the removal of the hormone or transmitter, interacts with a whole series of related calcium-binding proteins. These are called troponin C, when bound into an actomyosin contractile fibre, or calmodulins, when bound to kinases or some other enzymes, or annexins, when associated with vesicular discharges, and so on. The effect of calcium is to cause a change in cell tension or cell chemistry (Fig. 10.1) and, simultaneously, to switch on recovery especially by stimulating energy production, part of which is used to reject again the calcium via Ca^{2+} ATPase pumps.

Two further points must be made. First, the calcium pulse can be localized by the locality of the stimulation and by the speed of recovery, e.g. in nerve termini. Second, if a new pulse is repeated regularly, the whole of the cell may adapt to a quite new steady state since the resting steady-state level of calcium must be affected. A change in the rhythmic supply of calcium pulses can alter a fast muscle to a slow muscle, for example. In other words, calcium messages do not just interact with proteins but they strike to the level of RNA and DNA, and this may be due to changes in phosphorylation generated by changed steady-state resting levels of calcium acting on kinases.

The triggering process just concerns the entry of calcium through the outer membrane. A somewhat more subtle process allows the calcium to enter the cell cytoplasm from the inner vesicles (Fig. 10.5). This process requires a message to be transmitted from the outer to the inner membranes. One route for this excitation is via the release of inositol phosphates from the lipids of the outer membrane. Clearly, the process of inositol phosphate (say, IP_3) release, followed by its diffusion to an inner membrane and then calcium release, is slower than direct release of calcium but it is obvious that the space affected by the eventual release of calcium is more controlled. In some mechanisms the calcium has to backreact with the zone of the outer membrane from which the inositol phosphate was released. Here, at kinase C, it meets the lipid which released the IP_3 and which acts as an in-membrane signal. The extremely limited zone then activated, usually at the kinase (kinase C), can therefore be at the mercy of a combination of sequential events. The kinases C are phosphorylating agents for many proteins and many cellular events fall under their control. We shall return to such local effects in Chapter 21 where we shall see that other actions of calcium at phospholipases release fats in membranes, e.g. arachidonic acids which lead forward to signals via prostaglandins (see Fig. 21.16).

There is an alternative system to calcium activation based on the production of cyclic nucleotides but the two message systems are interactive through, for example, the cyclic nucleotide diesterases which are calcium-dependent, Chapter 21.

There could be a further distinction between the two kinds of message, electrical and chemical. The electrical stimulation is all over in 10^{-3} s but chemical stimulation may well be slower since there are many steps. The proteins with the fastest response times to calcium are the calmodulins (troponins) but it may be that there are somewhat more sluggish calcium-binding proteins, e.g. the calbindins, S-100 and annexin families.

10.8 The calcium trigger proteins

Table 10.5 Some internal calcium-binding proteins

Protein	Response
Calmodulin	Kinases
	Phosphatases
	cAMP diesterases
Troponin C	Actomyosin ATPase
Gelsolin	Actin organization
Calpain	Protease
Parvalbumin	Calcium buffer?
Calbindin	Calcium transport?
S-100	Cell cycle?
Annexins	Vesicle reactions and membrane attachments

NB. Most of these proteins are largely helical and undergo helix/helix motions.

The calcium trigger proteins are listed in Table 10.5 and in this paragraph, for the sake of simplicity, we shall call all these proteins calmodulins. Here troponins C are included, since they are all based on the same structural principles. The basic design of the calmodulin triggers is as follows. There are pairs of short sequences ia and ib (Fig. 10.6) which are called calcium-binding (EF) hands, placed back to back, and joined as a short two-strand β-sheet unit. This unit can be likened to the porphyrin (a sheet-like object) of haemoglobin. Leading from the unit, as in haemoglobin, are helices—two from each hand in the calcium-binding proteins (Figs 10.2 and 10.6). These helical stretches of sequence depend for their relative orientation and mutual disposition upon the exact bond angles within the β-sheet (iron porphyrin in the case of haemoglobin) and on extraneous factors which bind at the linker regions of the helices or along the helices and which, through adjusting the helices, also adjust the sheet (porphyrin). The arrangement of the helices relative to one another is therefore not fixed. The linkers or exposed helices provide the connection to other proteins and the exposed helices are the sites of binding of non-competitive drugs (Fig. 10.7) such as trifluoperazine (TFP;

Fig. 10.6 The proposed helix/helix motion in the calmodulin family of trigger proteins on binding calcium. The calcium binds at (ia) and (ib) and the sites are connected by a short β-sheet (ii). The rolling of the helices reveals hydrophobic sites for binding drugs (v) (see Fig. 10.7), and changes the structure of linking units (iv). This type of helix/helix motion may well be the general mechanical action of membrane proteins, of DNA-binding proteins, and so on. The structure is after the work of R. Kretsinger, see also Fig. 7.2.

Fig. 10.7 The binding zone of a drug, trifluoperazine (TFP), to calmodulin so as to modulate triggering.

Trifluoperazine (TFP)

see structure in margin). (Compare phosphoglycerate binding to haemoglobin.) However, given this outline of multiple helical stretches connected to several binding sites as well as to calcium hands, the choice of the binding sites for particular co-operating agents can be varied to give many combinations within calmodulin-like proteins (Table 10.5). Calcium binding itself (compare the binding of O_2 to haemoglobin) causes a small change in the geometry of the β-sheet (compare changes of iron in the haem) which twists and, in turn, rotates and moves laterally the helices (Fig. 10.8 and see Fig. 10.6). The

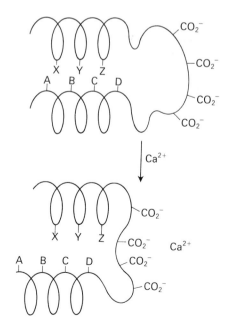

Fig. 10.8 A schematic representation of a calcium site in a subdomain (EF hand) of a calcium-binding protein. The donor groups, largely carboxylates (shown) and carbonyls, are in a sequence of a dozen amino acids. The site connects to helices which are adjusted relative to one another by the calcium binding. The rolling and the lateral motions of the helices are the essence of the triggers.

calcium binding in calmodulins therefore interacts co-operatively with any object binding to the surfaces of any other part of the protein. In this way, not only can we see how long-range drive from calcium to a change in conformation of another protein (Fig. 10.9), e.g. an enzyme such as a kinase or the troponins which control contraction, can be managed via rotation and lateral motion of helices, but we also see how the drive can be modulated by any molecule (e.g. a drug) which binds to the surface of the helices (such as TFP) or by competitive binding agents at either end of the protein. A more complex unit can be made by doubling the number of calcium-binding sequences (Fig. 10.7 and Fig. 7.2).

It is possible that the activity of many mechanical devices in biological systems is based on the principles shown in Fig. 10.8.

Fig. 10.9 A schematic representation of the sequence of binding steps where calmodulin, Cam, bound by calcium adjusts the conformation of a target protein. Two calcium levels of triggering for different targets are represented.

10.9 S-100 proteins

There is one group of calmodulin-like trigger proteins which does not act in quite the same way as those described—the so-called S-100 proteins, which were first found in brain tissue but are now known to be widely distributed. Their function is uncertain but they are believed to be associated with the cell division process. One unusual feature of the calcium binding is that one of the EF hands has very few anionic sidechains and resembles another group of proteins which are not triggers—the calbindins (Section 10.12). Another unusual feature of the proteins is that the second of the two, the S-100B domains, binds zinc with a binding constant of around 10^6 M^{-1} or greater which could be effective as a zinc-based trigger (see Chapter 11). In this case the levels of calcium and zinc may act co-operatively in cells. The more usual co-operativity is between calcium pulses and phosphorylation since the calcium pulse is often directed at kinases and phosphatases.

There are yet other series of these calmodulin-like proteins which undoubtedly enter into a huge variety of cell controls.

10.10 Other triggering modes

We have just described the very similar characteristics of calmodulin and of the calcium triggers, troponin C, of the fibrillar systems, e.g. in muscle contraction, and these activities are probably general to many cellular actomyosin filament actions. There are at least two other types of intracellular triggering which show a calcium ion dependence. They are vesicular release of chemicals by exocytosis, e.g. hormone release, and the alteration of cell shape on contact with a surface or with another cell.

10.10.1 Vesicular contents and their release: exocytosis

There is a large group of calcium-binding proteins in cells, the annexins, which does not have quite the same amino-acid sequence in the calcium-binding sites as that found in the EF-hands of calmodulins and troponins. The sequence is similar in that it is has carbonyl groups and carboxylate sidechains but it is shorter. When the proteins are isolated they have a much lower calcium-binding affinity around 10^4 M^{-1}. The proteins are thought to be bound to filamentous structures close to membranes (Fig. 10.1). It has also been observed that in the presence of anionic phospholipids the binding constant increases to around 10^6 to 10^7 M^{-1}. Clearly, only in these circumstances are they functional, since the concentration of intracellular calcium on triggering does not exceed 10^{-6} M. The function of these proteins is associated with the release of vesicular organic fast-acting transmitters or hormones, e.g. adrenalin. The following is no more than a suggestion as to how they may act.

Consider an arrangement of filaments extending to the cell membrane and which contains the negatively charged sequence of the calcium-binding protein Z near to the point of connection between proteins and membrane (Fig. 10.10) and perhaps in the membrane. Z clearly repels negatively charged phospholipids in the membrane so that there will be a distribution of these

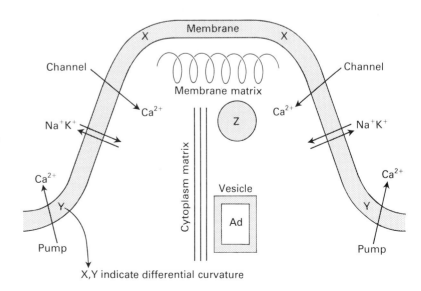

Fig. 10.10 The link between membrane, membrane-associated protein structures, calcium, and vesicle compartments. The diagram indicates that the whole is a unit of activity in that tension, curvature, positioning of pumps and channels, and movement of vesicles *all* depend on the activity of a cell in an interactive manner (see text). Cell shape is a steady-state property. Ad is adrenaline.

lipids in patches. Let us further suppose that a vesicle having neutral phospholipids in its membrane, as is known to be the case for adrenalin vesicles, is held in not too distant a relationship to this membrane and that its position is decided partly by the pressure of the filaments, pushing it toward the membrane, but partly by an inability to reach the membrane due to the presence of proteins including Z and mutual incompatibility between the neutral vesicle lipids and the charged phospholipid of the outer membrane, Fig. 10.10. Addition of calcium, i.e. a triggering pulse, will alter the disposition of all the components in that calcium ions can cross-link Z to the anionic phospholipids due to its high binding constant when both are present leaving more lowly charged lipids opposite the vesicle. The constraints on the position of the vesicle are removed and, moreover, the two membrane lipid layers, vesicle and cytoplasmic, are now compatible. Fusion follows, perhaps driven by calcium-controlled contractile factors in the filaments, and the contents of the vesicle are released. On removing the calcium by calcium pumps the former situation is restored and an empty vesicle is produced by endocytosis on the reversal of all the above steps. The vesicle can be sent back into the cell for recharging (see Chapter 3 and Fig. 3.5).

The preceding is no more than a proposal for vesicle discharge but it is given in order to show how calcium triggering could work in a rather different way from the triggering of muscle. The action is now direct cross-linking by calcium but it will generate conformational changes just the same. The calcium protein may well be designed in a way which is very similar to the design of calmodulin to allow conformation changes.

Once again it is important to observe the co-operative features of the calcium actin. In all cases the proteins have several calcium sites so function depends on $[Ca^{2+}]^n$. In the case of annexins several phospholipid molecules interact with one protein molecule. Finally there is a subtle selectivity in the construction of the helix–hand–helix unit which probably adjusts both the thermodynamics and kinetics of calcium binding.

10.10.2 Calcium, filaments, and cell shape

As we have stressed in Chapter 7 there appear to be several types of filaments in cells. There are: (1) semi-permanent multihelical structures such as actomyosins, which traverse the cell; (2) the multihelical proteins which lie under the membrane and are related to red blood cell spectrins; and (3) proteins related to properties of the nucleus as well as membranes such as tubulins. We have discussed the way in which calcium interacts with actomyosins and with filaments associated directly with the membrane. Much of this interaction can be visualized with the help of Fig. 10.10. The interaction with tubulin is less well defined and we have therefore described cation/tubulin effects under magnesium (Chapter 9) where it was stated that excess calcium opposed the polymerization of tubulin. The purpose of this paragraph is to pull together all these observations so that the dominant role of the group IIA ions, Mg^{2+} and Ca^{2+}, in the control of the shapes of cells and their mechanical properties does not pass unremarked. The construction of filaments in cells seems to be Mg^{2+}/Ca^{2+}/phosphorylation-dependent and it appears likely that constant activity leads to the shape of cells as shown in Fig. 10.10.

10.11 Calcium and protein phosphorylation

In Chapter 9 we have stressed the dependence of phosphate reactions upon the internal magnesium concentration: virtually all phosphate transfers, especially from nucleotide triphosphates, NTP, are Mg-dependent. This action is due to the association of magnesium with phosphate carrying small molecules at active sites. The role of calcium through calmodulin or calmodulin-like proteins is to switch, modulate, these enzyme actions, especially the kinases. The calcium is not at the active site and works by remote conformational control. However, it does not affect all phosphate transfers but is mainly concerned with the initiation of a chain of signals.

There is in this chain an increasingly sustained (in time) activity culminating in new protein synthesis. Earlier than this last action phosphorylation modulates the activities of all the proteins described in this chapter on calcium. Phosphorylation and dephosphorylation are intimately associated with calcium levels in an integrated system. We elaborate on the theme in Chapter 21 but the connection of calcium and phosphorylation with (1) homeostasis at rest, (2) rapid signals and relaxation, and (3) gross change in synthesis is a measure of their importance in eukaryotic cells.

10.12 Calcium buffering and calcium transport in cells

So great is the need to control calcium levels in cells that any, even moderate, excesses after stimulation must be quickly reduced so that the cell is ready for action once more. Very fast muscles, which may be called upon quickly several times a minute, contain a protein called parvalbumin (Fig. 10.1 and Fig. 3.13), which is similar to a calmodulin but has an extra non-calcium binding segment, the presence of which seems to block action in the sense that it is not known to activate any metabolism nor to bind to any construct in the cell. Since it binds both Mg^{2+} and Ca^{2+} more strongly than calmodulin does, it can act as a delayed buffer in the way described in Chapter 9.

Equally important in some cells is the redistribution of calcium within them when calcium enters locally. We pointed out in Chapter 3 that the input and output systems for ions were located in different parts of certain cells (Fig. 3.16). For example, many epithelial cells see on one side a supply of food which includes calcium ions and on the other the circulating fluids of the organism which must have a maintained Ca^{2+} concentration. It has been observed that these cells in particular have very high levels of small calcium-binding proteins

called calbindins (Fig. 10.1). (There are in fact varieties of this protein.) The location of the calbindin and is association with vitamin D activity, as well as its known binding affinity for calcium, $K_{Ca} \simeq 10^7 \text{ M}^{-1}$, suggest that its role is to transport calcium across cells from the calcium channels at one side of the epithelial cells to the calcium ATPases at the other side. This carrier transport action has been shown to be more effective than simple diffusion of Ca^{2+} ions at 10^{-7} M since calcium calbindin can be as high as 10^{-3} M. If the calbindin is correctly described, then the cellular system has gone to extraordinary lengths to manage calcium distribution within these and other cells and indeed within the organism. It is certainly the case that no other potential role of calbindins has been found. The protein is deeply associated with the maintenance of the life of the whole organism via calcium levels since its synthesis is under the control of vitamin D, which in turn is controlled by sunlight, and this vitamin is known to control too the well-being of bones through calcium levels in the blood.

10.13 Calcium currents: movement through membranes, channels, gates, and pumps

In certain cells, e.g. Purkinje cells, the Na^+/K^+ current is replaced by a calcium ion current. The full significance of these currents as well as of proton currents around cell membranes will be analysed in Chapter 21. Here we illustrate some features using the skeletal muscle cell. In Fig. 10.11 the contact zone between the sarcoplasmic reticulum vesicular system and the muscle fibres is shown. The calcium flows constantly at a very low level through the channels in the region at the tip of the vesicular systems and is buffered in this

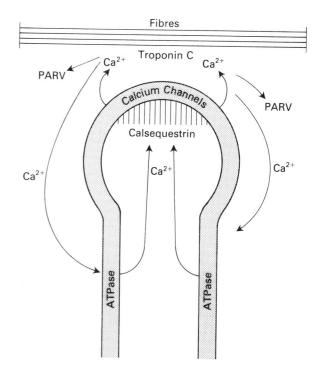

Fig. 10.11 The release system for calcium at the tip of a sarcoplasmic reticulum of a muscle cell. The calcium channels are placed very close to the contractile fibres and here calcium is stored in a calcium-binding protein, calsequestrin. The calcium is removed by parvalbumin, PARV, and then by pumps (ATPase). The whole organization has a weak current flow which can be stimulated by a nerve or chemical pulse.

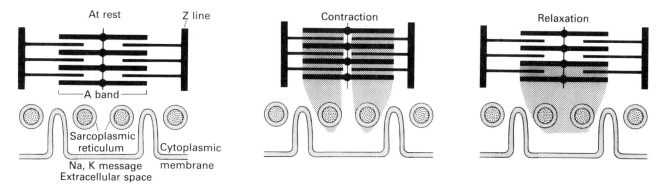

Fig. 10.12 The sliding filament model of muscle contraction. The calcium is shown as dots released from and returning to the sarcoplasmic reticulum (see Fig. 10.11). The contact between the filaments, actomyosin, which is activated by calcium (see Fig. 10.1) is via a device in the myosin head which acts mechanically by helical movements like that of the calmodulins (see reference to muscle; see also Fig. 10.16).

reticulum region by the presence of the protein calsequestrin with binding constant, $K = 10^3$ M^{-1}. Calcium levels (10^{-7} M) around the fibres are maintained by the pumps, ATPases, in a different part of the reticulum membrane. The organization is very like that in Fig. 10.10 and the whole of the activity is shape-related. A pulse from a nerve causes depolarization of the membrane (see Fig. 10.12) and release of calcium into the fibre region gives contraction and then recovery. The effect on the muscle fibres is shown in Fig. 10.12. Should the pulsing be maintained in sustained rhythmic sequences, then the system changes even its morphology since such activity changes the calcium levels in the cytoplasm and eventually affects DNA expression. This description of the muscle cell and its dependence on calcium levels may well be quite general to cells.

The movement of calcium into a cell is managed through calcium channels which may be electrically operated, voltage-gated, or opened by signal molecules which bind to receptors on the outside of membranes. Little is known about any of the voltage-gated channels for cations or indeed of the receptor-operated system except for sets of amino-acid sequences of their proteins. The sequences have been used to describe the channels as multihelical transmembrane proteins when the gating requires either charge rearrangement (voltage gates) or the realignment of the helices through a mechanical device to allow ion passage. We have described both modes of action in Chapter 7 since the principles do not differ for any ion or molecule channel. Selectivity for calcium is achieved through the operation of structural factors outlined at the beginning of the chapter. A major group of drugs is the 'calcium blockers', one of which is shown in the margin.

Calcium leaves the cell through pumping which requires the input of energy to the pump in which the calcium ion sees a high-affinity site from one side of a membrane and a low-affinity site from the other. Calcium is driven from the high- to the low-affinity state supposedly by movement of helical segments of the ATPases in the membrane. This enzyme is illustrated in Fig. 10.13 (compare the Na^+/K^+ and H^+ pumps) and the binding site for pumping is presumably in the membrane though other sites may exist. The pumping is slow, 100 times s^{-1}, hence the need for parvalbumin in fast muscles, see Section 9.8.2.

One series of dihydropyridine calcium channel blockers, e.g. nifedipine.
R, a wide variety of substituents.

Fig. 10.13 A diagram of an ATPase metal-ion protein pump here for calcium. The ATPase segment is inside the cytoplasm. It undergoes conformational changes (see Fig. 7.9) and drives rearrangement of helices in the membrane where calcium binding is adjusted (after M. Green). A simplified picture has been given in Fig. 8.7 for the general case of ion pumps.

The final movement of calcium and many other ions and molecules is through the gap junctions which join many cells keeping their metabolism in contact. The only reason for describing this junction here is that the gap junction is governed by calcium ion concentration. In the presence of 10^{-5} M calcium in any cell all gap junctions between it and neighbouring cells are closed. The cell is excommunicated and doomed.

10.14 Internal calcium-induced proteases

There is a further series of calcium-dependent proteins in cells which would appear to have calmodulin-like segments but which do not bind calcium very strongly. They are the calpains which are proteases when calcium is bound. Their function is not known but we can conceive a plausible use for what can only be a very damaging activity in a cell.

These proteins bind calcium with a binding constant of 10^5 M^{-1}. Now when the concentration of calcium entering a cell reaches 10^{-5} M a vast range of activities is switched on and a great number of otherwise dissociated units become cross-linked. The cell can be so seriously damaged that it may well be the case that to complete its degeneration is a better solution than an attempt at recovery. A switch-on of proteases activated at 10^{-5} M calcium would bring about cell death. It is known that when calcium reaches such high levels the calcium also closes the gap junctions between cells so that death of one cell does not spread to others.

In Section 10.17 we shall describe a quite different series of calcium controls on *extracellular* proteases which are stored in vesicles in cells that are not exposed to internal calcium fluxes. While we know of these intracellular proteases there is no knowledge of calcium-dependent intracellular nucleases although nucleases are required to remove damaged DNA. As we shall see some extracellular nucleases are calcium dependent.

10.15 General remarks concerning control systems in cells

In the light of Section 10.14 and the information concerning Na^+, K^+, Mg^{2+}, and Ca^{2+} ion actions let us look at the time sequences in the intra cellular control systems of biology (Fig. 10.14). We shall start from two assumptions.

1. The fastest movements possible for transfer of information are limited by diffusion of small ions or molecules. The usual number to quote is the maximum rate constant of diffusion, 10^{10} s^{-1}. An electrolytic current can not move charge faster than, say, 10–100 Å in 10^{-10} s.

2. To create an effect such as a switch of a protein to a new conformation will require some minimum of, say, 10 kcal, which means that thermally the switch is from a ground state which in the absence of the 10-kcal energy input is some 10^7 times more likely to be inactive than active. This is a reasonable value, 10^3 being too small and 10^{10} unnecessarily large. The switch is a secondary structure change. The fastest possible triggering of a control is then given by the off-rate, i.e. relaxation

$$\frac{\text{Diffusion on-rate}}{\text{Off-rate}} = \frac{10^{10}}{\text{Off-rate}} = K = 10^7$$

since it is only binding which can give energy to a chemical system. The off-rate is thus 10^3 and biological systems operate at their fastest in the 10^{-3} s range. It follows that this is the speed of the nerve with a calcium-dependent protein booster system (Fig. 10.14) of muscle contraction, of hormone release, and so on. It is also the optimal rate of protein phosphorylation controlled by calmodulin.

If a more selective response is wanted and the system can afford time then there are many alternatives, including the following.

1. We can make coming together slower so that more precise fitting is achieved, i.e. $k_{on} < 10^{10}$ s^{-1}.

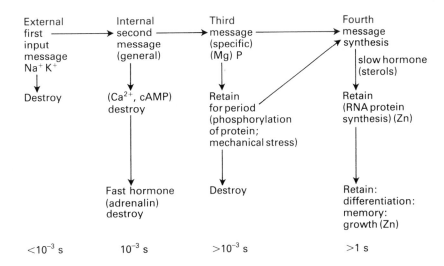

Fig. 10.14 The time sequences in biological systems and the metal ions involved.

2. We can arrange for better binding which will allow the use of rare species, most organic triggers, and a consequent slow off-rate.

In Fig. 10.15 the local concentrations of Na^+, K^+, Ca^{2+} and of the fastest organic trigger, acetylcholine, are greater than 10^{-3} M. They all act very rapidly. Molecules such as sterols are insoluble in water and difficult to transfer. They must have high binding and work as slow triggers. Metal ions such as Zn^{2+} and Fe^{2+} do not exceed 10^{-10} M in cells. They are buffered at that level. If they are to be useful in maintaining steady states then K for them has to be $>10^{10}$ when their on/off controls cannot be faster than 1 s^{-1}. They regulate and do not act as triggers (Chapter 21). We begin to see how inorganic chemistry has been put to use in extremely sophisticated ways within biology, Section 4.5.

Fig. 10.15 The left-hand cell has a Na^+/K^+ gradient. Influx of Na^+, a nerve message, causes eddy currents to flow from left to right until a Ca^{2+} channel is opened near A (acetylcholine or GABA). The calcium releases the contents of the vesicles, see Section 10.10, which now trigger the nerve message in the right-hand cell. There follows flow of Na^+ and K^+ across the membrane of this second cell.

10.16 Extracytoplasmic calcium

In a multicellular organism it is essential that all the calcium activities, i.e. in the cell cytoplasm, in vesicles, in organelles, and in the extracellular fluids, are in communication with one another since all the calcium is locked into a vast network of controls. The effect of a calcium pulse into the cytoplasm is obviously deeply connected to the non-cytoplasmic calcium concentration.

10.16.1 Vesicular calcium: exchangers

The calcium which is pumped away from the cytoplasm is not all pumped to the circulating fluids. Much is pumped into vesicles including the reticula and some is pumped into organelles. The pumps are arranged in all the respective membranes but there are also ion-exchanger systems which keep the balance between a variety of ions in different zones of the cell. These exchangers remove calcium by replacing it by Na^+ or H^+ ions, which do not act so powerfully on metabolism and are buffered by high concentrations or by binding centres.

In some reticula there are special storage proteins for calcium, e.g. calsequestrin in the sarcoplasmic reticulum, which are positioned to assist local release or stimulation (see Fig. 10.11).

10.16.2 Organelle calcium

The functioning of organelles, mitochondria and chloroplasts, is equally dependent on calcium levels as is that of the cell cytoplasm (Table 10.1). In some ways this is not surprising since these organelles may well be of independent evolutionary origin. The peculiarities are that there are differences between mitochondria from plants and animals. *In vitro* animal mitochondria can pump in calcium even to the extent of precipitating calcium salts internally although this may never occur *in vivo*. Calcium regulates some of their enzymes including some essential dehydrogenases. In plant mitochondria the uptake of calcium seems to be less. In chloroplasts the functions of calcium are not really known although it is found that this ion and chloride anions have essential roles in early events of photosynthesis especially associated with dioxygen release. It would appear that the roles of calcium are much better understood in advanced cells than in prokaryotes and organelles. Certainly the calmodulin-type proteins do not exist in prokaryotes and organelles. The significance of this finding will appear in Chapter 21.

10.16.3 Calcium in circulating fluids

In this chapter we have insisted that the level of extracellular calcium in circulating fluids is as exactingly controlled as intracellular calcium. The controls are exerted via epithelial cell pumps and through the kidney tubules. This calcium interacts directly with a variety of biological polymers and with many biological minerals which are discussed in Chapter 20. The polymers are of great variety. Some are concerned with the matrices which hold cells together—connective tissue—and these polymers include polysaccharides as well as proteins. In plant tissue the polysaccharides greatly exceed the proteins, and note that a mixture of these two forms the cell walls of unicellular organisms—both eukaryotes and prokaryotes. Calcium cross-links many of these polymers in a well-balanced sol or gel state depending on the cells under discussion (Sections 7.22 and 9.12). Binding is to carboxylate and phosphate sidechains as well as to neutral O-donor groups.

Some of the polymers involved with calcium are enzymes. Probably in the majority of cases the calcium only helps to stabilize the fold of these proteins but in several cases it assists in activation. There are two outstanding examples—digestion and blood clotting. The triggering here is reminiscent of that of calpain, the intracellular protease.

10.17 Extracellular enzymes and calcium: digestion and blood clotting

In the fluids around cells, or in specialized cells in some organisms, it is necessary to have hydrolytic enzymes to degrade polymers into small fragments so that monomers can be pumped back into cells for synthesis and energy metabolism. This applies to fats, polysaccharides, polynucleotides, and proteins. The degradation is carried out by a series of enzymes released from cells, often from vesicles where they are held in precursor or pro-enzyme forms. The need to keep enzymes in inactive states is also apparent in processes such

as blood-clotting, antigen responses, and so on (Table 10.6). Enzymes are made available from pro-enzymes by various devices including binding to the calcium ion which may or may not affect catalysis directly. In the cases of enzymes such as the protease chymotrypsin, the function of calcium is to unwind a part of the tail of the enzyme which, in the pro-enzyme form, prevents reaction. It goes almost without saying that the release of this digestive enzyme from the pancreas when food is in the stomach is due to a calcium signal to that organ. The pancreas contains vesicles which hold the pro-enzyme in a harmless condition and in a solution of low calcium.

Table 10.6 Extracellular calcium proteins

Protein	Binding centre
EGF-domains of proteins C, S, Z, and VII, IX, and X (blood-clotting)	(Hydroxy-aspartate) or two aspartates
Osteocalcin (bone) Prothrombin	γ-carboxy-glutamate plus aspartate or glutamate
Complement C_{1r} and C_{1s}	EGF-like?
Digestive enzymes	Many different cases
Connective tissue	Many examples

NB. Most of these proteins are cross-linked by –S–S– bridges and many contain β-sheet structures. Viral coat proteins often contain calcium.

Now while this mechanism of altering the conformation of pro-enzymes is used in many digestive enzymes and also, in a somewhat different manner, in the blood-clotting cascade discussed later in this section, calcium has a different influence on some nucleases and phospholipases. The case of phospholipase A_2 has been particularly well studied.

Phospholipase A_2 has the fold illustrated in Fig. 7.1. The enzyme is without activity in the absence of calcium ions and in their presence it can digest phospholipids in aggregated states. The site of calcium binding is a hole in the protein surface close to the known attacking groups for hydrolysis. The function of calcium is presumed to be to hold the phosphate group of the substrate in a suitable orientation for attack by the active site histidine. The calcium is held in a completely different manner from that seen in calmodulin. The binding groups are mostly neutral oxygen donors held on the rigid framework of the protein (Fig. 10.3) and, to limit mobility, the protein has both a β-sheet construct and helices cross-linked by seven –S–S– bridges (Fig. 7.1). A more rigid structure could hardly have been constructed, again in marked contrast with the loose assembly of helices of calmodulin.

Enzymes of the same kind as phospholipases and external proteases are found in all digestive juices and in toxins, e.g. snake toxins. Such enzymes are also needed in fertilization. Much faster peptidases are derived from very different enzymes (see Chapter 11).

In blood-clotting there is again a new calcium trigger device (Figs 10.16 and 10.17). Here the extracellular calcium has a dual role with the membrane phospholipids rather like that with annexins in cells. Once injury to the organism has occurred, the calcium activates a set of proteases, through unfolding a series of multidomain proteins, and simultaneously binds domains

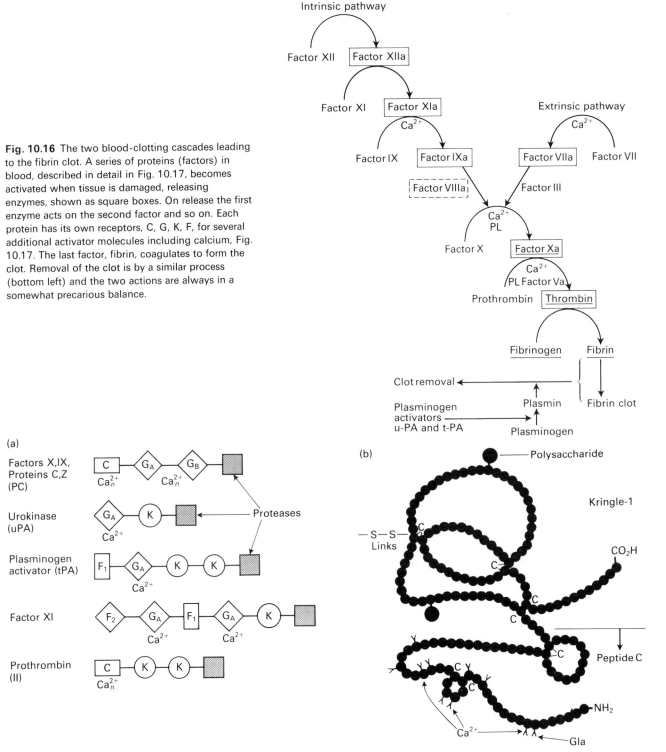

Fig. 10.16 The two blood-clotting cascades leading to the fibrin clot. A series of proteins (factors) in blood, described in detail in Fig. 10.17, becomes activated when tissue is damaged, releasing enzymes, shown as square boxes. On release the first enzyme acts on the second factor and so on. Each protein has its own receptors, C, G, K, F, for several additional activator molecules including calcium, Fig. 10.17. The last factor, fibrin, coagulates to form the clot. Removal of the clot is by a similar process (bottom left) and the two actions are always in a somewhat precarious balance.

Fig. 10.17 (a) The complexity of the domain structures of some proteins in the clotting and clot-removal sequence of Fig. 10.16. The protein domains, peptide C, growth factor G, kringle K, and finger, F, control the proteases and in turn binding factors, such as calcium and phosphatidylserine (PS), control the domain protease interaction. (b) The sequence of the calcium-binding peptide, C, and the first kringle, K, of prothrombin. The Gla units (Section 10.7) are indicated. The —S—S— bridges in (b) are shown by C.

locally to cell fragments and membranes, represented by PL (phospholipids) in Fig. 10.16. The effects are brought about by a very open binding site for calcium composed of a specially synthesized amino acid, γ-carboxyglutamic acid (Gla) (Table 7.2 and Fig. 10.17). By having several Gla units close together in sequence the effect of calcium binding is to change a very stretched out but random polymer segment into a highly organized one while leaving calcium ions (several) in part exposed so that they can bind phospholipid or other anionic headgroups of the damaged tissue. The multidomain protease, prothrombin, for example, also readjusts its domain/domain interactions so that the hydrolysis necessary to make the blood clot can occur. The role of calcium is very like that in osteocalcin where Gla groups are also involved but where the exposed calcium can bind to a bone mineral rather than a damaged cell surface. The rates of these extracellular changes are all much slower than that of the intracellular calmodulin. Figure 10.17 includes several proteins from the blood-clotting system and Table 10.6 shows two from the complement components of the immune system.

10.18 Calcium proteins of biominerals

Finally, calcium proteins are deeply involved with the laying down of biominerals—bones, shells, and teeth (Chapter 20). The proteins involved here are heavily glycosylated and phosphorylated in many cases but the exact function of these proteins is not known. What is known is that they have virtually random-coil structures in the absence of calcium. A typical small protein of this kind is osteocalcin which carries Gla (γ-carboxyglutamic acid centres.) The fold is shown in Fig. 10.18. There is a surface from which calcium ions exchange easily and presumably move readily over the great number of

Fig. 10.18 The outline structure of osteocalcin. All the Gla units with malonate headgroups, formula I in Section 10.7 (shown as G) are to one side so that the folded protein binds many calcium atoms in an almost planar array. A is aspartate. (After work of A. Atkinson.)

anionic centres. This protein binds directly to bone but there are other anionic surfaces amongst the bone and teeth proteins which do not bind directly and may act to control the diffusion of calcium or as a reservoir of calcium somewhat like calsequestrin. Bone is of course a huge extracellular calcium buffer. It is thought that the soluble proteins such as osteocalcin and casein (in milk) inhibit precipitation while the matrix calcium-binding proteins initiate crystallization.

10.19 Calcium biominerals

There is clearly a relationship in evolution between the development of mineralization and the development of organisms. An obvious point is that load-bearing structures are most easily devised in a growing tissue system using minerals. The hard organic exoskeletons of insects are disadvantageous in that they restrict growth. The connection between minerals and growth will be left to Chapters 20 and 21, however, since they concern shape. Here we must look on the mineral deposits from the point of view of solution chemistry. The minerals, especially calcium and iron salts, are in tight steady-state relationship with the extracellular (calcium) or intracellular (iron) free ion levels through the solubility products (see Chapters 1 and 2).

The insolubility of calcium salts arises from its doubly charged state, Ca^{2+}, and its large size which more nearly matches that of inorganic and organic anions. As a general rule, given equal concentrations of Mg^{2+} and Ca^{2+}, the calcium salt will precipitate with all common highly charged anions, e.g. SO_4^{2-}, HPO_4^{2-}, CO_3^{2-}, and organic anions such as dicarboxylates. The common biological minerals, with silica, are calcium carbonate, phosphate, and oxalate. The solubility products are such that at about pH $= 7$ (lower for oxalate) the salts precipitate with a calcium concentration of about 5×10^{-3} M and a given concentration of any of the anions HCO_3^-, HPO_4^{2-}, or oxalate^{2-} in the range of $1\text{--}10 \times 10^{-3}$ M. This statement is made with little precision since the precipitation reactions are sensitive to pH, temperature, pressure, general salt concentrations, and specific ion effects from the presence of especially Mg^{2+}, F^-, OH^-, and so on. Thus we have to relate the relationship of precipitation to the particular conditions in an organism in an environment, e.g. shellfish in the sea. However, it is very clear that the precipitation is very carefully controlled. In this chapter the implication is that at the site of precipitation the solubility product is only just exceeded. The uptake of calcium and the related anions in the case of HPO_4^{2-} and oxalate^{2-} must be extremely carefully balanced with the deposition. There can be but one conclusion: the calcium diet is controlled such that the concentration level of this ion is held by feedback controls at the level of the solubility product of the precipitated salt

$$[Ca^{2+}][X^{2-}] \simeq \text{solubility product.}$$

From the point of view of inorganic chemistry this statement is extremely peculiar since it implies very considerable risks from unwanted deposition and very strong limitations on the flexibility of much other chemistry. On the other

hand such a balanced chemostat in calcium and its partner anion (and presumably pH) has great biological advantages in that it sets the conditions in a very firm manner for the evolution of organic reactions and structures. The highest forms of life we know have evolved further and further toward an extraordinary set of conditions which generate almost complete dominance over the organism by fixed levels of free calcium. This is not an accident but an inevitable result of a search for the best and most sensitive devices for controlled growth (Chapters 20 and 21), for controlled response, and for interaction with external events. The fact that the control is associated with calcium is dependent upon the availability of this element and the solubility of its salts and it is very difficult to see an alternative solution for a general external chemostat given the abundances of the elements (Chapter 1). The choice of partner anion seems to have evolved (with fluctuations) toward carbonate and then to phosphate. In the first case the external environment is largely responsible for the HCO_3^- level or the CO_2 level so that the organism is bound into the environmental chemostat. In the case of calcium phosphates (bone) there is a required interactive control between free calcium and free phosphate in the external fluids independent from the environment. The latter are then the most finely tuned fluids in biology. The discovery of bone by evolution was undoubtedly an irreplaceable benefit for the development of organisms. We elaborate on the nature of this material in Chapter 20.

Two other features of this chemistry must not escape notice. First the dominant precipitates, $CaCO_3$ and $Ca_2(OH)PO_4$, contain the elements carbon and phosphorus, respectively, with all this holds for the organic chemistry of life. Second, both precipitates buffer the pH so that there is feedback between all possible pairings of Ca^{2+} to HPO_4^{2-} or HCO_3^- and to pH. We stress that solubility products are sensitive to temperature and pressure and, if the chemostat is to be maintained, then it must be a physicostat as well. While pressure control rests largely on contact with the environment, temperature control internally is imperative and is found in higher animals with bones. Surely it is this chemostat feature which allows a very great sensitivity to have evolved in biological computers, brains, which so depend on inorganic chemistry.

On looking again through this chapter the reader is asked to keep in mind the fundamental importance of the insoluble calcium salts to the whole of the network of calcium controls, which were described first.

10.20 Intra- and extracytoplasmic calcium balances

We now wish to bring into focus further features of calcium controls both of homeostasis and of the linking between the different calcium pools. We begin with a description of a common feature, the sensitivity of these controls.

10.20.1 Multiple sites for calcium

We have concentrated attention on the individual sites of calcium binding but in many proteins several atoms of calcium bind co-operatively. Examples are troponin C, calmodulins, osteocalcin, and prothrombin. As explained in

Chapter 2, multiple binding, which is co-operative, gives an added intensity to effects of change of calcium similar to dioxygen binding to haemoglobin. The extent of co-operativity increases from the monomeric sites to multico-operative sites of crystals. The great co-operativity in crystals is the subject of Chapters 20 on biominerals and 21 on homeostasis.

The use of multiple binding sites on the surfaces of the extracellular calcium proteins is somewhat strange since several of the sites must be saturated. Their binding constants exceed 10^3 in the presence of 10^{-3} M calcium. Good examples are provided by the Gla-containing proteins of the blood-clotting system and by osteocalcin. A possible explanation is that it is only on binding the last one or two out of some four or five calcium ions that the active form of the protein is created. We may then write

$$\text{Protein} + n\text{Ca}^{2+} \rightleftharpoons [\text{Protein} (\text{Ca}^{2+})_n] \xrightarrow{m\text{Ca}^{2+}} [\text{Protein} (\text{Ca}_{n+m})]$$

unfolded *fold*(1) *fold*(2)

$$K = > 10^3 \text{ M}^{-1} \qquad K < 10^3 \text{ M}^{-1}$$

In fold (2) all of the calcium atoms match a surface of either a phospholipid membrane or a crystal. This mode of triggering is very dependent on free calcium and clearly resembles a very sensitive buffer.

10.20.2 Cell shape

The steady-state, dynamic shape of a cell depends on the interaction of the outer membranes with both the *internal* matrix and the *external* connective tissue and its tension. Calcium and magnesium ions are deeply involved with both and they require metabolic energy constantly to maintain gradients. We can therefore anticipate that the future will show that there is a close involvement in the maintenance of a steady state of flow, now inside and outside cells, of these two ions and of sodium and potassium currents with cell shape. Of course, these flows are monitored by hormonal (and morphogen?) organic chemicals too, but in the context of this book we can be forgiven for thinking that the complicated organic hormone/inorganic messenger relation-ships of present-day life are latecomers in evolution compared with the initial system using filaments made from proteins and glycoproteins and inorganic ions only.

10.20.3 Cell–cell interaction

In the development of multicellular systems the connective tissue guides the interaction between cells. As explained in Chapter 7 it is vital that this matrix is controlled and, by the very nature of the anionic polymers involved, one of the controls is the level of free calcium. A very simple scheme for the role this metal ion could play is shown in Fig. 10.19.

10.20.4 Cell morphogenesis

Most of the description of calcium biochemistry concentrates upon triggering. We have insisted that the steady state of free calcium at rest is of equal

Fig. 10.19 The role of calcium in cell/cell interactions shown in schematic form.

importance in homeostasis. The homeostatic condition can only be maintained if any pulse of calcium entering a cell is followed rapidly by calcium expulsion and provided that no calcium-dependent message affects the cell's DNA and its translation activity. The opposite is required in morphogenesis. The expression of DNA through to proteins must be altered. We do not know how this is achieved but it has to be the case that in some way the level of calcium following the initial pulse is sustained after, say, fertilization. It has been suggested that the calcium is released from vesicular stores in the egg. Does it then interact directly with DNA or is this process dependent on phosphorylation? How are external as well as internal proteins changed in concert and how is morphogenesis connected to these changes?

10.21 Summary

The purpose of this chapter is to give the reader an impression of the multiple roles of calcium inside and outside cells. Although it was easy to describe triggering we have insisted that this is not usually a switch from zero to 100 per cent activity but a change from a basal resting cell state to a somewhat different one. In most cases the cell must recover from the triggered state since the calcium is a tool for maintaining the integrated *resting* activity of the organism. In a few cases the change of calcium may be permanent as in fertilization, differentiation, and cancer induction. The cause of the change may be massive invasion of the cell (fertilization), maintained switches in hormonal or electrical activity at the cell membrane, or mutation but all of them are associated with new calcium resting levels. An attempt at a more detailed description of the integrated calcium circuits will be made in Chapter 21. Should the maintenance of calcium levels fail, then the cell will become damaged (Chapter 22) and this may affect a whole organ or organism.

Now, although it may not have been noticed, we have concentrated attention in this chapter on advanced eukaryotic cells and especially on cells of multicellular systems. The accompanying intracellular proteins, e.g. calmodulins, and the chemically modified extracellular proteins containing

γ-carboxyglutamic acid and hydroxyproline developed late in evolution. They do not appear to be present in prokaryotes which use calcium almost exclusively in extracellular protection, outer membranes or coats, and in assisting digestion by stabilizing extracellular enzymes. Calcium is rejected by calcium pumps but is not greatly used as a signal. The strong indication is that the overwhelming use of this element developed at the time dioxygen entered the atmosphere in quantity. It was then necessary to integrate its new functions into the total activity of the organism (see Chapter 21). Just as with the glucocorticoid controls of Na^+ and K^+ there is a sterol hormone, vitamin D, which links the levels of calcium to the activities of DNA. [N.B. The sterol hormones are synthesized by steps which require dioxygen and iron.] Interestingly, both sets of sterols also connect to zinc, which takes us to the next chapter.

We shall see in the last chapters of the book that while the levels of free calcium are related to hormone activities so is the growth and nature of bone. The temptation is to think in the following evolutionary way. At first biological cells rejected calcium along with sodium and chloride while externally the calcium stabilized walls. With the development of more controlled external polymers calcium precipitates became an integrated protective part of the outside of the cell, and allowed stability to initial multicellular assembly. The need arose for cell–cell communication and what could provide a better message system than the expelled ions Na^+, Cl^-, and Ca^{2+}, the last as the trigger. As sophistication developed so it was necessary to have more complex signalling back and forth from organs to link cell assemblies in steady-state growth to DNA regulation. The calcium signals were therefore linked by feedback mechanism to the host of transmitters and hormones formed from organic chemicals which so confuse us today. As we stated at the beginning of this chapter, in advanced organisms calcium seems to be involved everywhere. Table 10.7 serves as a reminder of some features we have mentioned.

Table 10.7 Some biological functions of calcium

Function	System or effect
Structural unit	Bone, shells, teeth, cell walls
Triggering	Muscle, synapse, hormone release
Control	Morphogenesis, fertilization, sporulation
Catalysis	Control of digestive enzymes (proteases), blood clotting
Current flow	Heart, barnacle muscle, general around cells(?)
Cell lysis	Protease trigger
Sensors (gravity)	Otoliths
Cross-linking	Calcium proteins, cell fusion

Further reading

Allen, T. J. A., Noble, D., and Reuter, H. (ed.) (1989). *Sodium–calcium exchange.* Oxford University Press, Oxford.

Anghileri, L. J. (ed.) (1990). *The role of calcium in biological systems*, Vols 1–5. CRC Press, Boca Raton, Florida.

Axelrod, J. (1990). Phospholipase A_2. *Biochemical Society Transactions*, **18**, 503–7.

Berridge, M. J. and Irvine, R. F. (1989). Inositol phosphates and cell signalling. *Nature*, **341**, 197–205.

Calcium-binding proteins in health and disease (1980–90). A regular series of volumes with different publishers.

Cohen, P. and Klee, C. B. (ed.) (1988). *Calmodulin*. Elsevier, Amsterdam.

Goldbeter, A., Dupont, G., and Berridge, M. J. (1990). Calcium oscillations and protein phosphorylation. *Proceedings of the National Academy of Sciences USA*, **87**, 1461–5.

Heizmann, C. W. and Hunziker, W. (1991). Intracellular calcium-binding proteins. *Trends in the Biochemical Sciences*, **16**, 98–103.

Huber, R., Schneider, M., Mayr, I., Romisch, J., and Paquer, E.-P. (1990). Annexin calcium-binding sites. *FEBS Letters*, **275**, 15–21.

Kater, S. B., Mattson, M. P., Cohen, C., and Connor, J. (1988). Calcium regulation of the neural growth cone. *Trends in Neurological Science*, **II**, 315–21.

Petrobon, D., Di Virgilio, F., and Pozzan, T. (1990). Calcium homeostasis. *European Journal of Biochemistry*, **193**, 599–622.

Squire, J. M. (ed.) (1990). *Molecular mechanism of muscle contraction*. Macmillan, London.

Williams, R. J. P. (1970). The biochemistry of sodium, potassium, magnesium and calcium. *Quarterly Reviews of the Chemical Society, London*, **24**, 331–65.

General reference (1991). *Cell Calcium*, **12**, 61–268, is devoted to calcium flow.

11

Zinc: Lewis acid catalysis and regulation

11.1	Introduction to Lewis acids	299
11.2	Zinc in biological space	302
11.3	Availability and concentration of free Zn^{2+} ions	303
11.4	Types of protein associated with zinc	303
11.5	Zinc exchange rates	307
11.6	The number and selectivity of ligands to zinc	307
11.7	Zinc as a catalytic group in enzymes	308
11.8	The use of zinc in biological space and time	312
11.9	Regulatory roles of zinc	312
11.10	Zinc and peptide hormones	314
11.11	Summary of intracellular zinc: is zinc a master hormone?	314
11.12	Extracellular zinc	315
11.13	Conclusion	317
	Note. Probes of metal sites by isomorphous substitution	318

11.1 Introduction to Lewis acids

At the beginning of each chapter on a given metal in biology we should have in mind the question as to why it has been selected for its particular tasks. In the present case the question itself has two parts: (1) why zinc rather than another metal ion; and (2) why use a metal ion at all since zinc is only an acid catalyst comparable to the proton. As we stressed in Chapter 6, there are certain advantageous properties of metal ions both in their binding strengths and their rates of exchange of ligands which make them a potential source of chemically and hence biologically interesting selective properties. Zinc, Zn^{2+}, is the case in point here. Its most obvious distinction is its highly concentrated charge, since it is a small ion (radius 0.65 Å). In themselves, however, an electrostatic charge of two and small size give quite modest binding to anions such as carboxylate and phosphate even where several anions occur together (Table 2.2). This is due to the competition from water of hydration, see

Divalent ions

Trivalent ions
Lewis acids available
to biological systems

Table 11.1 Why zinc is so valuable in biology relative to other metals

Availability > nickel, cadmium, iron, copper, etc.
Strongly retained > manganese, iron(II), magnesium
Fast ligand exchange > nickel, magnesium
No redox reaction; contrast copper, iron, manganese
Flexible co-ordination geometry > nickel, magnesium
Good Lewis acid (only copper amongst divalent ions is better)
Supplies hard base, hydroxide (only Cu(II), Fe(III), Mn(III) are better)
Common in cells and may well be a link between many controls and regulation

Chapter 2. Furthermore, the electrostatic binding by zinc is a property shared almost equally with Mg^{2+}, and to a lesser degree Ca^{2+}, as well as with other metal cations such as Cu^{2+} and Ni^{2+}, but such charge concentration is not known amongst organic ions. Thus, although this electrostatic part of its chemistry is a feature of certain metal ions, it is not peculiar to zinc. The second characteristic of metal ions is their high affinity for electrons as shown by the energy required to get from the gas atom M to M^{2+}, i.e. their ionization potentials (Fig. 2.5). The ionization potential is a rough guide to Lewis acid strength and here zinc differs from Mg^{2+} and Ca^{2+} as well as, quite obviously, from organic cations, since its electron affinity is much higher. While it is therefore a strong *Lewis* acid, see the margin, it remains similar to copper and nickel in this respect and in all model properties described so far (see Fig. 2.8). All these three ions can then bind strongly to donors such as thiolates and amines, which is a property of strong Lewis acids not shown by magnesium or calcium. Zinc does differ from copper and nickel in yet other respects, however; for example, it does not show variable valence (oxidation state change) and could therefore be preferred in some biological compartments for this reason alone since redox activity introduces the risk of free radical reactions. On the other hand Zn^{2+} is still quite like Cd^{2+}, and Cd^{2+} is known to be able to substitute for Zn^{2+} in many zinc enzymes. It might have been thought that biology would have made some use of cadmium, but cadmium is a relatively rare element and zinc is much more available (Chapter 1). We see, therefore, that zinc is not quite unique, though it is highly selective as a cation; it carries a special set of structural and thermodynamic properties (Table 11.1) as well as availability, and these are further enhanced in value when rates of reaction are considered.

Reactions require reorganization of atoms. Fast reactions will be more easily carried out by metal ions which, while polarizing groups, can take up and release molecules from their co-ordination spheres rapidly (Chapter 4). We can write K, the binding constant, as a ratio of rate constants

$$K = \frac{k_{on}}{k_{off}}.$$

The values for both k_{on} and k_{off} for zinc reactions usually greatly exceed those of nickel (see Fig. 4.3) while K is roughly the same. Moreover, the rearrangement of groups on the surface of zinc without release, i.e. fluctional reactions, is usually fast unlike those of nickel. The differences arise from the fact that in a field of ligands the zinc ion is not effectively polarized so that its ligand field stabilization energy is zero (Section 2.5). As a result, zinc, but not nickel, passes readily from one symmetry of its surrounds to another without exchange and exchanges ligands quickly. Such factors favour a higher catalytic activity since they allow rapid atom flow including substrate binding, changes in the bound intermediates, and product release, to, on, or from the metal ion, in enzymes. If we use I_2, the ionization potential to the doubly charged state, i.e. the polarizing power, multiplied by k_{on} (i.e. the rate constant for water release from the co-ordination sphere) as a guide to acid catalysis, zinc looks like a better catalyst than almost any other divalent cation readily available to biology except copper, which is objectionable on redox grounds. It may even be that, if we examine more highly charged relatively available cations such as Al^{3+}, Fe^{3+}, and Co^{3+}, zinc (Zn^{2+}) would still turn out to be

the best catalyst in neutral aqueous media since in this medium the higher valent ions are hydrolysed and lose their functional strength by binding to hydroxide ions. The higher valent ions also exchange water much more slowly and are less fluctional.

The need for zinc in catalytic sites of enzymes may not even then be thought to be obvious since many enzymes which hydrolyse esters or amides, the major substrates of zinc enzymes, do not require metals. As already mentioned, Zn^{2+} has highly localized charge and high electron affinity and is therefore a very effective attacking group in a sense quite different from that of any organic centre or even bound protons (NB pH = 7.0). Looking at the substrates it attacks (in enzymes) we see first that they are often small molecules such as H_2O, CO_2, and C_2H_5OH, and that another major target is amino- or carboxyl-*terminal* peptide links generally. A clear possibility therefore is that zinc enzymes can be used to attack substrates in a more or less chemically non-selective but physically constrained site. Enzymes without such powerful attacking centres as zinc can attack only large substrates (with very high selectivity) by using many points of attachment. The active site groups are here relatively poor attacking groups *per se*, and attack in these non-metal enzymes is based on very well regulated orientation and strain in the substrate generated by the precision of binding. Zinc enzymes would then be reserved: (1) for attack on small substrates which cannot be handled by organic groups in a multipoint attachment; and (2) for less specific but very rapid attack when they could work even with relatively weak binding of substrate. Lack of selectivity, which would usefully go with powerful attacking groups, is of obvious advantage in digestion, for example. One case is the zinc nucleases. Speed (also at the price of selectivity) could also be advantageous in hormone turnover, and we notice that zinc is particularly associated with peptide hormone release from propeptides and peptide hormone destruction.

Metal ions need not be used just in catalysis since they can be triggers of systems or play a structural role. We know that triggering, like catalysis, requires good binding with selective recognition, but also reasonably rapid on and off reactions since there is a need to stimulate a biological system quickly but to relax it quickly too. The role of calcium in catalysis and triggering activity was described in the previous chapter. Amongst the divalent cations zinc, like calcium, has the required properties for a trigger or control ion since it binds well and exchanges rapidly (Section 11.5). Naturally, it is used in different circumstances from those in which calcium is used, as the calcium ion binds to O-donor ligands only but zinc binds largely to S- and N-donors in biological systems. Moreover, zinc is at a much lower concentration than calcium. Since zinc binds much more strongly than calcium its exchange is that much slower; hence it is not so much a trigger as a *regulator* (or a hormone?). The similarities (and differences) between zinc and calcium lead to further parallel (not identical) activities in biology where both metal ions are of extreme importance in structural cross-linking but once again different chemical specificity is involved. It may be that the fluctional properties of both these metal ions are useful in all these functions.

Finally, since the zinc ion is a dissociable but indestructible unit and can therefore be used as a homeostatic control between many reaction systems, it could be employed simultaneously as a catalyst and as a control. We must ask if there is some hidden cross-talk between different uses of this metal ion.

11.2 Zinc in biological space

Fig. 11.1 The variety of roles played by zinc inside and outside cells. P, protein; MT, metallothionein.

Fig. 11.2 The compartments of pollen and of sperm showing the uneven distribution of zinc and calcium. Note the localization of the ATP(ase), E_1, to drive the cell movement.

Before making statements about particular functions a few general remarks about the three roles of zinc (acid catalyst, control ion, and structural ion) and their spatial distribution in cellular systems are worth careful note. At the risk of overgeneralization we make these statements.

1. Zinc is the most common catalytic metal ion in the cytoplasm of cells but is less common than some group IA and IIA metal ions. Iron is more common in membranes, manganese in some vesicles or organelles, and copper largely outside the cytoplasm and outside cells. Nickel, cobalt, and all other metal ions are not common anywhere in eukaryotes.

2. Zinc enzymes in cells are involved in a great variety of enzyme reactions in the cytoplasm.

3. Zinc digestive enzymes are active outside cells or in vesicles.

4. Zinc is rarely intimately associated with a membrane.

5. The structural role of zinc extends from filamentous (keratin-like structures) to the organization of chromosomes, largely inside cells.

6. The use of zinc in DNA regulatory functions is confined to eukaryotes.

The peculiarities of the chemistry of zinc, summarized in Table 11.1, are then put to particular use in particular parts of biological space (Fig. 11.1), which differ greatly from those used by iron or copper, for example. Particularly fascinating examples are given by the male gamete (Fig. 11.2) and the male animal reproductive tract which are very high in zinc. We turn first to the more chemical problems.

11.3 Availability and concentration of free Zn^{2+} ions

Table 11.2 Logarithmic concentration* of elements in sea water (mol dm^{-3})

Na	−0.3	Ca	−2.0	Ni	−8.1
Mg	−1.3	Sc	−11.0	Cu	−8.4
Al	−7.7	Ti	−8.0	Zn	−8.1
Si	−4.0	V	−7.5	Mo	−7.0
P	−5.7	Cr	−8.4	Sr	−5.0
S	−1.6	Mn	−9.3	Cd	−9.0
Cl	−0.3	Fe	−9.0		
K	−2.0	Co	−10.7		

Data from Bruland, K. W. (1983). Trace elements in sea water. In *Chemical oceanography*, Vol. 8 (ed. J. P. Riley and R. Chester), Chapter 45. Academic Press, London.
* These are not the free ion levels but the total concentrations (see Chapter 1).

Biological systems are energy-consuming. Energy economy is therefore important and this implies that biology will use available elements rather than those which have similar properties but are much less available (Table 11.2). Zinc is relatively available even in sea water and is amongst the most available of the trace elements. It is more available than copper, iron, cobalt, or cadmium. Considering its potential value, it is not surprising that it is relatively common in biological systems, about as common as iron, but there is an additional feature. As we shall see, zinc has assumed great value in biology and here and there it is quite highly concentrated. The relative concentration of free zinc ions within biological cells varies from about 10^{-9} M in the cytoplasm of many cells to about 10^{-3} M in some vesicles. (Calcium has a similar localized concentration in cytoplasm and vesicles.) Clearly, a mechanism must be available for moving the element into selected compartments by pumping and the elevated concentrations then allow triggering as with calcium. These pumps, which to date are only predicted to exist, require selective binding like all the other activities associated with zinc described in Section 11.2. In extracellular fluids, zinc may well be pumped down to a lower level than the intracellular value 10^{-9} M but, since it can be released from vesicles, the level may be much higher locally for a short time. In fact, zinc concentrations are high in a variety of vesicles of nerve terminals of the brain, and, in particular, in 'mossy fibres' of the hippocampus. It is not known why zinc should be present there, but one interesting possibility is that its release under stimulus assists the production of 'long-term' memory in the form of growth of the nerve. We consider another possibility in Section 11.10. A great deal of work is required to define fully the way in which zinc is distributed. A similar comment concerns the transport and circulation of zinc in the blood of higher organisms where an albumin appears to be involved but is poorly described.

We turn next to the types of chemical neighbours with which zinc is associated in biology and we start with proteins since a great deal is now known about zinc proteins from crystallographic and chemical studies.

11.4 Types of protein associated with zinc

In Chapter 7 we described three very different extreme types of protein: (1) the extended β-pleated sheet, (2) sets of α-helices; and (3) almost random structures, in relation to their relative mobility and therefore their functional potential (Table 11.3). There are, of course, a variety of proteins which have combinations of β-sheets and α-helices in them. Inspection suggests that, even in these proteins, the $\beta\alpha$ alternating structures are relatively rigid, for example, in domains, while where there are patches of structure containing only helices these patches are internally mobile. We observed in Chapter 10 that the trigger proteins, e.g. those which bind calcium, are almost totally helical and mobile and that this is also true of the *allosteric* oxygen-carrier proteins which

Table 11.3 Major fold characteristics of some zinc proteins

Helical	β-sheet	Mixed*	Random
Insulin, Zn†	Carbonic anhydrase, Zn	Carboxypeptidase,† Zn	Metallothionein‡, Cu, Zn, Cd
S-100, Zn, Ca	Superoxide dismutase†, Zn, Cu	Aspartate transcarbamylase, Zn	σ-factor IIIA(?), Zn
Phospholipase C†, Zn		Alcohol dehydrogenase, Zn	
		β-lactamase, Zn	

* The active site is always closely associated with the β-sheet.
† Cross-linked by the metal ion or –S–S– bridges.
‡ Folds only on binding metal ions.

use iron or copper. On the other hand, the electron transfer proteins which contain copper, the blue proteins, are based on immobile β-sheets (Section 7.8). Of course chemical cross-linking also reduces mobility. Cytochrome c, another electron transfer protein, although it is helical, is cross-linked three times and is also of low internal mobility where it is cross-linked. Such electron transfer proteins do not require gross mobility and, in fact, that would be disadvantageous (see Section 4.7). Turning our attention to enzymes generally, it is usually true that active sites are held in proteins close to or on β-sheet-like structures and we may suppose that the relative rigidity of these structures helps to control specificity of binding and the directionality of attack by acid/base or redox centres. The stability of the fold also prevents exchange of metal ions from such a protein. We look first at zinc enzymes in this context.

Crystal structures are available for several *enzymes* which require zinc at the active site (Table 11.3). It is immediately clear that with one or two exceptions they are based on extensive β-sheets and we presume that it is the strength of the fold which generates the peculiar and particular active-site geometries of the metal ion (Fig. 11.3) since these proteins fold in the absence of the zinc. We have described the imposition of a special geometry at the metal by the strength of the fold as an entatic (strained) condition of the metal geometry (Section 7.8) and we shall return to this feature in Section 11.7.3 since probes allow the metal binding in the protein to be inspected. An enzyme in which zinc is not at the active site is superoxide dismutase, another β-sheet enzyme, where copper is the active metal ion. Here both the binding of the zinc in a β-sheet fold and the β-sheet itself now stabilize the unusual entatic, rhombic geometry around the copper ion, whereas the zinc geometry is a conventional tetrahedron. Zinc is but an additional stabilizing factor and, when we look at copper biochemistry, we shall see why it was necessary to build such a stable structure (Chapter 15).

Fig. 11.3 The immediate donor atoms in some zinc binding sites where S is a thiolate of cysteine and N is an imidazole donor; contrast calcium and magnesium. The co-ordination geometry is generated by the relatively rigid protein.

Now the active-site attacking groups of enzymes, zinc in the present case, must retain some considerable local mobility since, unlike the active sites of electron transfer proteins, considerable relaxation of the local structure allows for more rapid on/off reactions of substrates and products and for the co-ordinates of the whole system, attacking groups plus substrates, intermediates and products, to *flow* along the reaction pathway easily. The zinc substrate binding on one of its sides only and to sidechains of proteins, all of which are flexible since they have β–CH_2– sections to their sidechains, is very different from the constrained environments of metal ions in Fe_4S_4, haem, vitamin B_{12}, or F-430. The flexibility at the open-sided zinc enzyme site will be illustrated by an example later.

Turning to other kinds of enzymes to which zinc is bound but not at the active site there are crystallographic data for a few (Fig. 11.3). In these proteins the metal acts as a cross-link between thiolate groups, usually four, or perhaps a combination of thiolates and histidines to make a total of four ligands. This is a double or knotted cross-link and is found in intracellular proteins, e.g. alcohol dehydrogenase. The cross-link could be, in part, a replacement for the –S–S– or calcium bridges of extracellular proteins since inside a cell most –S–S–bridges are unstable to reduction and calcium cross-links are unstable since in cells there is very little calcium. In this case zinc binding (cross-linking) controls the amount of the protein in the cell by stabilizing it in bound form so that it is not hydrolysed (Section 11.9). In the cross-links in alcohol dehydrogenase and aspartate transcarbamylase, zinc has a geometry which is now that of a conventional tetrahedron.

So far, all the zinc binding we have examined is strong with binding constants $\geqslant 10^9$ M^{-1}. In insulin complexes, where the zinc and insulin are retained in a *vesicle*, the complexes are weak and the concentration of free zinc must be high. The zinc is cross-linked to this *helical* protein in two different but conventional ways, one each in two different forms of the protein through histidine sidechains. In the two forms the helices are slightly differently disposed relative to one another and the zinc site geometries are closely related to an octahedron. There is a remote link here to calcium- and zinc-binding to calmodulins (see the protein S-100, Section 11.12) where the metal ions bind to and modulate the properties of a helical protein (see Fig. 10.6). The possible reason for the presence of zinc in insulin vesicles will be analysed in Section 11.9, but we do not know why insulin should have a secondary structure which is inherently mobile but a cross-linked interior. Notice particularly that conventional zinc binding geometry, α-helical structure, and non-enzymic function occur together. There is an exception to this generalization in that phospholipase C is a helical protein. However, it is strongly cross-linked by *several* zinc ions and mobility must again be constrained.

Metallothionein is, in the absence of zinc, a random coil protein and only on binding zinc, cadmium, or copper does it form a series of turns so as to hold the metal ions in two types of cluster of M_4 and M_3 units, respectively, formed by regular tetrahedral, thiolate, donors. The probable reason for having such clusters of metal ions in this protein will be discussed in Chapter 21 on homeostasis. The structure of metallothionein bound by different metals, zinc, copper, and cadmium, is not the same but the essential core structure is fixed. While it is reasonable to assume that all the other proteins described so far are

Table 11.4 Zinc finger proteins

Protein	Function
Many DNA promotor region binders	Receptor for hormones*, etc.
Aspartate transcarbamylase	Nucleotide base synthesis
Kinase C(?)	Critical signalling(?)

* Sterols, thyroxime, retinoic acid, etc.

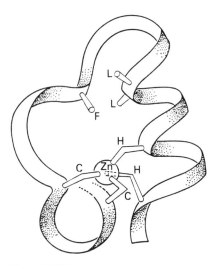

Fig. 11.4 The structure of one kind of zinc finger as determined by NMR methods. Here certain amino-acid residues are shown: L, leucine; F, phenylalanine; H, histidine; and C, cysteine. (After R. E. Klevit and P. E. Wright.)

very highly zinc-selective, this is not true of metallothionein which is in effect a scavenger of heavy metal ions and the protein structure clearly adapts itself to the particular metal ion. There is no question of an entatic state here. The value of such a common scavenger will be examined again especially in the discussions of the interactive role of zinc and copper in biology, and the protection of biology from pollution caused by some metal ions (Chapter 22).

We have deliberately left until last a further extremely important group of zinc proteins which have no enzymatic activity, the zinc finger proteins (Table 11.4), which have only been found in eukaryotes. Some of these zinc proteins bind to DNA promoter sites, but others are part of critical enzymes and appear to have a structural (or control?) role. Their full structures are not known, but the structure of the common repeated zinc finger sequence has been determined by NMR methods (Fig. 11.4). The zinc is in repeating loops of protein of some 30 amino acids (Fig. 11.5). In the protein (free from DNA) the fingers are not likely to be well structured and the zinc, even in the bound state, is rather exposed. Once again these non-enzymic proteins differ markedly in structure from the zinc enzymes. Unlike the enzymes, in which one or at most two or three zinc ions are involved, the zinc fingers can occur in multiples of up to 30. We return to their probable function in Section 11.9.

As stressed in previous chapters, it is important to observe whether the ligand centres of the protein for metal binding occur close together in the sequence or not. The distribution of these two different constructs between calcium trigger and calcium enzyme sites was very clear in Table 10.4. The situation is much the same for zinc. The DNA binding zinc finger is composed of a run of residues closely in sequence and we must expect that zinc removal will be fast and alters the fold at least locally, as is required of a control. The connections of the metal in enzymes, by way of contrast, are usually to residues at large distances apart in the sequence and so are related to total fold stability. We shall see that, much as for calcium, the two different types of protein allow very different zinc exchange rates.

It is known that there are other zinc DNA-binding proteins which are not closely related to zinc fingers and it will be of great interest to find their structural and dynamic features.

Fig. 11.5 An example of a usual sequence around zinc sites in two zinc fingers. The circled amino acids are conserved and are either ligands or part of the hydrophobic core.

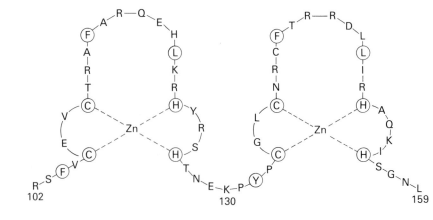

11.5 Zinc exchange rates

Table 11.5 Exchange rates of zinc proteins

Protein	Ligands	pH	Exchange rate $t_{1/2}$(h)
Carboxypeptidase	N_2O_3	8,0	24
Alkaline phosphatase			
Site A	N_2O_3	8,0	84
Site B	N_1O_3	8,0	24
Gene 32 protein	S_3N	7,8	12
Gal4	S_4	7,8	1–2
	S_4	7,8	1
	S_4	7,8	0.75
Metallothionein	S_4	7,0	~1

This table is from unpublished work of Professor J. E. Coleman, Yale University, New Haven, Connecticut, USA, and is presented with his permission. The exchange rates were measured in the presence of cadmium ions. We have also had helpful discussions with Professor J. Kaegi of the University of Zurich, Switzerland.

The exchange rates of metal ions from protein complexes are not an intrinsic feature of their metal ion binding strengths but are dependent on protein fold energies; this has already been observed for calcium. In the case of zinc it is well known that many zinc enzymes can be separated from cells without significant loss of the metal. This could be due to the presence of at least 10^{-10} M free zinc in the solutions used in extraction, but it is also found that zinc, incorporated in a β-sheet/helix structure, does not exchange over days (Table 11.5). The zinc in the zinc fingers and metallothionein exchanges over a period of the order of 10 h; thus, here zinc has the same exchange time as any strongly bound reagent in biology such as a sterol hormone. Exchange rates of this order are suitable as regulators over periods of the order of days, but cannot be functional in rapid triggering. We return to this topic in Chapter 21.

Now that we have examined the general fold of zinc proteins we must return to the sites themselves and their activities.

11.6 The number and selectivity of ligands to zinc

The proteins in Table 11.3 nearly all bind zinc through at least three groups, two of which are imidazole or cysteine (Fig. 11.3). In some cases, four such groups are used when the zinc acts as a cross-link and is blocked from interaction with any small-molecule substrate. In the enzymes, zinc can have as few as two imidazoles when it is also bound to a carboxylate, e.g. in carboxypeptidase, or it can have three such donors, e.g. in carbonic anhydrase. The fourth (and fifth) positions are then occupied by water molecules which become involved in the catalytic act, or, at least, are readily replaced allowing access of substrate to the zinc. The zinc site is then organized in several ways: (1) type of ligands; (2) angular distribution of ligands around the zinc; (3) bond distances; (4) leaving groups; and (5) the next nearest neighbours and surrounds of the metal. The last term includes the whole of the active site groove in which the zinc sits. We should be able to rationalize all the peculiarities of these different parts of the zinc sites in terms of the reactions which they carry out. An example, carbonic anhydrase, is given in Section 11.7.1.

Against the background of the preceding structural and chemical character-ization of zinc sites we must examine the sites of other metal ions in proteins in an attempt to appreciate selectivity. We observed in Chapter 2 that, if we used a naïve approach based just on stability constant data and free competition amongst cations, it was very difficult to see how the selection of zinc in a given site is so specific—an estimate suggests that zinc sites contain less than 1 per cent of any competing element. We repeat that most competing metal ions are excluded on the basis of low accumulation (Ni), of lower affinity for N(S) co-ordination (Mg), by a dislike for tetrahedral co-ordination (Cu), and,

finally, by a deliberate segregation of other elements (Cu, Mn) away from the cytoplasm or the special vesicles into which zinc is pumped. Only a combination of such factors can have made it possible for biological systems to select zinc into the above sites. For example, other proteins use the same four groups (four RS^-) to bind Fe (in rubredoxin). Again, the site of zinc in an almost octahedral hole bound only by histidines in glyoxalase is a unique situation for zinc in an enzyme so far, but a very parallel site binds iron in the photosynthetic reaction centre complex and possibly in lipoxygenases. To understand this competition we must know the redox state of competing ions, their concentration locally, and so on before the effects of all the factors mentioned in Chapters 2, 3, and 5 can be appreciated. All such matters are considerations of effective stability constants, Section 2.4.

11.7 Zinc as a catalytic group in enzymes

11.7.1 Zinc as a Lewis acid

A zinc ion is a good electron acceptor, or Lewis acid, in inorganic and organic chemistry. As such, it polarizes groups to which it binds and either increases the attacking power of a bound base or increases the probability of attack on the bound group by acting as an acid. An example of the first case is bound water which becomes an effective bound OH^- at some pH, usually above 9.0 for model complexes,

$$Zn^{2+}(H_2O) \longrightarrow Zn^{2+}(OH^-) + H^+,$$

which means that zinc as an acid can generate a very effective localized hydroxide ion (base) which can attack other molecules. This is the proposed function of zinc in carbonic anhydrase where the OH^- attacks CO_2,

$$O=C=O + OH^- \longrightarrow HCO_3^-,$$

and we next take carbonic anhydrase as an example of a typical zinc enzyme.

Water is a substrate in this reaction and in carbonic anhydrase the zinc ligands, all neutral imidazoles, and the protein structure are 'designed' to make the zinc of $Zn(H_2O)$ as acid as possible so that the ionization $H_2O \rightarrow OH^-$ occurs at pH = 7. In addition, the movement of the substrate atoms along the path of intermediates is aided by the rapid known switching of co-ordination geometry of the zinc between four and five co-ordination which reduces barriers between steps (Fig. 11.6). Furthermore, the groove which leads in the enzyme from the surface to the active site is designed as a proton conductor by the incorporation along it of several imidazoles. It is the combination of the geometry, bond angle and bond length, the ligand groups, the site and mobility, and the immediate surrounds, effectively the solvent, of the zinc complex, together with the more remote proton conduction groups which makes the enzyme so powerful. The overall structure is typical of an enzyme in that it is largely a β-sheet (Fig. 11.7). We begin to see why a very large molecule might be required to carry out even such a simple reaction with speed and selectivity.

An example of a second use of zinc is in a direct hydrolysis of zinc ester or

Fig. 11.6 A proposal for the pathway of formation of bicarbonate from carbon dioxide using carbonic anhydrase as catalyst. The function of a base, B, is not clear.

amide substrate complex where the zinc acts as an attacking acid on this second substrate while making water attack it as well (see Section 4.3).

$$Zn^{2+} \leftarrow O = C \begin{smallmatrix} R \\ \\ OEt \end{smallmatrix} \quad + \quad OH^- \longrightarrow \left[Zn^{2+} \longrightarrow {}^-O - \overset{R}{\underset{OH}{C}} - OEt \right] \overset{H^+}{\longrightarrow} Zn^{2+} + EtOH + RCOO^-$$

No clear-cut example of this reaction path can be given but it is possible that such a path is present in carboxypeptidases. (Note again how firmly the zinc is bound in these enzymes.)

Chemists have clearly developed models of some of these metal ion catalysed hydrolyses and group transfers using inert trivalent Co(III) metal ion centres, (A). These centres substitute very slowly and therefore allow

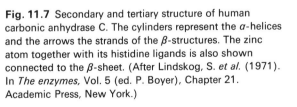

(A)

Substrates = ⟵ NH₂CH₂CO·OCH₃
 ⟵ NH₂CH₂CO·NH·CH₃

Fig. 11.7 Secondary and tertiary structure of human carbonic anhydrase C. The cylinders represent the α-helices and the arrows the strands of the β-structures. The zinc atom together with its histidine ligands is also shown connected to the β-sheet. (After Lindskog, S. *et al.* (1971). In *The enzymes,* Vol. 5 (ed. P. Boyer), Chapter 21. Academic Press, New York.)

analysis of single steps of reaction while they show rates for these single steps of the same order of magnitude as those of the enzymes themselves. Of course, in this chemistry there is no flow of substrates since the products cannot leave the cobalt(III). Now, the rate of steps in these model reactions can be enhanced by the presence of more than one metal ion in the synthesized complexes. When one turns to the zinc enzymes handling phosphate esters and some other substrates, one observes that, indeed, in several of them there is more than one metal ion as the model chemistry suggests; the usual metal ions are the divalent cations Mg^{2+} and Zn^{2+} (Fig. 11.8). Examples of this type of enzyme are given in Table 11.6. The question arises as to how these divalent ions can be as effective as the trivalent cobalt(III) even in the single steps of reaction. Consider their different ability to generate a bound hydroxide ion as earlier in this section. The difference in hydrolysis constants for $M(III)(H_2O)_6$ and $M(II)(H_2O)_6$ is greater than a factor of 10^6 for cobalt(III) against $Zn(II)$ or about 10^{10} for cobalt(III) against magnesium(II) so it is probable that the magnesium ion in an enzyme will act only to orient the phosphate-containing substrate. The zinc is then the enzyme attacking group and we know from the studies of carbonic anhydrase that this metal in an enzyme active site bound to three imidazoles is specially activated to $Zn-OH^-$ at a pH not far from 7.0. The hydrolysis constant of zinc in the enzyme is some 10^3 greater than for free zinc, but, clearly, the enzyme site provides still extra catalytic factors. It is intriguing that here model chemistry is giving much pause for thought about the exact mechanism of enzymatic metal ion catalysis. The discussion applies

Table 11.6 Examples of zinc enzymes using several metal ions

Enzyme	Metal ions
Phospholipase C	Three closely related zinc ions
Alkaline phosphatase	Two zinc and one magnesium ions
5' Nucleotidase	Two zinc ions

○ M^{2+}

Fig. 11.8 The structure of alkaline phosphatase showing the grouping together of three metal ions in the active site. Notice the β-sheet. (After J. Coleman and H. W. Wyckoff.)

equally to those *acid* phosphatases which contain not zinc or magnesium but two atoms of either iron or manganese.

11.7.2 Zinc and redox reactions

Zinc is also found in several enzymes which are involved in redox reactions. A typical example is alcohol dehydrogenase

$$NAD^+ + C_2H_5OH \longrightarrow NADH + CH_3CHO + H^+.$$

The metal, as a Lewis acid, assists in the activation of the alcohol to make hydride removal easier. The acceptor in the enzymes is usually NAD (NADP). A particularly important example of these redox reactions is an enzyme in the shikimate pathway where the hydride removal step uses the zinc-dependent enzyme dehydroquinate synthase. The importance of the pathway is that it leads to the synthesis of the three aromatic amino acids, phenylalanine, tyrosine, and tryptophan. The enzyme, like several other zinc enzymes, is in the cytoplasm. The question arises as to whether the zinc has more than a catalytic role in this particular enzyme. The parallel with the use of zinc in nucleic acid synthesis, that is in aspartate transcarbamylase, is clear in that zinc is again linked to some of the most fundamental steps in biological synthesis of major polymers. Finally, zinc appears in the enzyme laevulinic dehydratase, which controls porphyrin synthesis and hence a vast range of redox reactions. As for magnesium and calcium, the involvement of zinc in so many different pathways leads us to ask if it has a very extensive role in controls in biology.

11.7.3 The entatic state and probes

In all the above enzymes zinc can be replaced *in vitro* by other divalent metal ions (Fig. 4.2). Outstandingly, cobalt(II) is the best replacement for activity despite the fact that the expected order of Lewis acid catalytic power is

$$Cu > Zn \le Ni > Co > Fe > Mn.$$

(Note that this switching of metal ions at the active site of an enzyme is a parallel inorganic study to site-specific mutagenesis of organic protein sidechains). The implication is clear: the enzyme site is often designed for a site symmetry which is not octahedral and both direct structure determination and spectroscopic probe analysis show this. In several cases the site is of irregular symmetry, four or five co-ordinate, which is readily attained by Zn^{2+} or Co^{2+}. As stated in Section 11.4, the enzyme imposes a geometry on the attacking divalent metal ion to optimize attacking strength and minimize barriers for reaction. Similar comments have been made earlier concerning electron-transfer enzymes. The name 'entatic state' has been coined to draw attention to the power of the evolved protein fold to optimize 'design' around groups even though this introduces local metal ion strain and, therefore, some loss of stability. The general approach is not too different from Pauling's notion concerning the 'design' of binding cavities for substrates where he pointed out that a binding 'design' which forced the substrate to a structure matching a transition state, another strained state, was advantageous. (NB Note the different problem where in helical proteins the strain forces motion in the helices by *rack* action.)

11.8 The use of zinc in biological space and time

Table 11.7 Zinc and filamentous structures

Filamentous structure	Role of zinc
Collagens	Zinc collagenases
Proteoglycans	Zinc stromelysin
Denatured collagens	Zinc gelatinase
(Chromosomal proteins)	Zinc structural role?
Keratins	Zinc cross-links

Today it is difficult to know where all the zinc inside cellular systems is located. The stress of biological studies on chemical change and therefore on enzymes has left the structural and mechanical parts of cellular systems much in the background. However, cells are highly internally structured in dynamic organizations; from DNA outwards to ribosomes and to membranes there are vast numbers of filamentous constructions in cells and additional ones outside them. The role of zinc in these structures, though suspected from analysis, is as yet unknown. Methods are needed for the full analysis of the zinc associated with DNA and RNA, not now just with the known function in some polymerases but with filaments and here we can only mention such polymers (Table 11.7). We will return to the description of zinc and external filamentous structures when we have looked at other complementary zinc functions but note immediately that much connective tissue is degraded by zinc enzymes.

11.9 Regulatory roles of zinc

Zinc is probably a major regulatory ion in the metabolism of cells; the 'feel' for this role comes from gross biological considerations though their molecular nature begins to appear, e.g. in zinc fingers (in association with appropriate properties, structural, dynamic, and location, while bound to DNA) and in the link to amino-acid, nucleotide, and haem syntheses.

As in the case for regulatory calcium, the regulatory role of zinc can be made to depend on a high power of the free zinc ion concentration by the binding of several metal ions to one protein. This type of co-operativity is familiar in haemoglobins and calmodulins and possibly in insulin, but more obviously in the transcription σ-factors, e.g. IIIA (see Table 11.3), where as many as 10 zinc atoms may occur (Fig. 11.5). The equilibrium reaction must now be written in a series of steps up to n where maximum co-operativity is n

$$A(Zn^{2+})_n \rightleftharpoons B + nZn^{2+}.$$

Alternatively, each zinc complex $A(Zn^{2+})$, $A(Zn^{2+})_2$, $A(Zn^{2+})_3$, etc. could have a separate role to play so that the gearing of reactions in a cell could be regulated by zinc concentration in a smooth fashion. It is the authors' opinion that these control roles of zinc, in which the zinc bound states can be active or inhibitory in regulation, will be found in many more cases than we know at present, especially associated with fertilization and growth. As stated in Section 11.4, an extensive group of zinc proteins is now known which bind to promotor regions of DNA (Table 11.4). These zinc finger proteins are connected to hormones deeply associated with developmental process, e.g. thyroxine and many steroid hormones. Lack of zinc has long been associated with various problems of growth and of healing of damaged tissues. In any integrated view of the role of zinc it is essential to keep in mind the possible connection between a variety of enzyme activities (Table 11.8) and control of the expression system via DNA.

Table 11.8 Zinc enzymes in synthesis

Synthetic step	Zinc enzymes
RNA	RNA polymerase
DNA	Reverse transcriptase
Viral synthesis	Terminal dNT transferase
Transfer RNA	tRNA synthetase
Essential amino acids	Dehydroquinate synthase
Essential nucleotides	Aspartate transcarbamylase

The high dependence on Zn^{2+} concentration can also be of value in homeostasis provided the zinc ion is effectively 'buffered' (Chapter 21); here we note again the clusters of metallothionein. As described in Chapter 21, the most sensitive homeostatic mechanism is based on a high power, n, of the free metal ion concentration, here $[Zn^{2+}]^n$. However, we have to note that the same protein seemingly maintains the homeostasis of copper (and cadmium). Let us suppose that the random coil protein, metallothionein, folds differently according to the metals to which it binds; there can then be two sets of receptors for copper and zinc metallothioneins as well as one for the apoprotein on DNA, but this means that the metabolisms of the two metals have become inextricably linked. Why might this be of value? We return to this point in Chapter 2.1.

There is an alternative way, different from direct regulation, in which zinc could act as a control. As we know from Chapter 7, protein folding is sometimes dependent on factors other than the sequence, e.g. various types of cross-linking. The examples we have given are metallothionein, cytochrome c, and osteocalcin. It could well be that all zinc structural units seen as zinc 'fingers' *in enzymes* fall into this class. This would mean that zinc controlled the amounts of protein produced as determined by the binding constant of an initially bound state before the fold was completed

Here, the final state, Fold(2), is shown as irreversible. Such a control is thought to operate within the zinc (finger) domain of aspartate transcarbamylase. Interestingly, while this enzyme controls the synthesis of bases for DNA production, in turn, its rate of reaction is modulated by nucleotide concentration. Kinase C is another critical protein to which this control may well apply.

The excitement over the discovery of zinc in transcriptase factors and its involvement in the core proteins of chromosomes must not hide the importance of the discovery of zinc in reverse transcriptase and in RNA polymerase (Table 11.8). Once again notice that these enzymes have high rates of reaction but may not be very discriminating with regard to the nucleotide bases. As stated in Section 11.1, zinc is most valuable where speed is most important, but its aggression as a Lewis acid implies that care must be taken in the selection of targets or else a general attack must be anticipated on peptide, phosphodiester, or polysaccharide bonds. These observations suggest that we should look more closely at some of the major targets of zinc hydrolases and their locations. While we make these observations we must be aware that the nucleotide and amino-acid polymerases must be of high fidelity and that this is managed by units other than the zinc enzymes in the ribosomes. Where the zinc (or calcium) enzymes act in digestion as proteases, however, there is little requirement for discrimination in general hydrolytic attack.

11.10 Zinc and peptide hormones

A feature of the zinc concentration is its localization in particular organs. As we have seen, parts of the male reproductive systems of both plants and animals are well known as being zones of high zinc concentration, but there are areas of the brain, especially the hippocampus, which have high zinc. Why?

Now there is a requirement for the activity of many peptide triggers and hormones in this part of the brain such that both their release and their removal must be very rapid. Consider the usual reaction sequence for a peptide hormone, where E is an enzyme,

$$\text{Stored precursor} \xrightarrow{E_1} \text{Hormone (transmitter)} \xrightarrow{E_2} \text{Local hydrolysis of hormone (relaxation).}$$

The familiar parallel is with stored acetylcholine or adrenalin and *local* rapid degradation after action by esterases and amine oxidases (Cu-proteins). The rapidity of catalysis by E_1 and E_2 and their location are equally important. Zinc enzymes are at their best in such situations and the known distribution of zinc in certain parts of the brain may well be linked to this function. Certainly the zinc enzyme peptidase 24.11 and its close relatives are associated directly with the destruction of enkephalins and other peptide hormones (Table 11.9). Is zinc release from nerve terminals a part of this system?

We conclude that there must be a huge range of hormones which have a connection with zinc biological action and we stress that these activities arose after the advent of molecular dioxygen in the atmosphere.

Table 11.9 Zinc proteins related to peptide hormonal action

Hormone	Mode of association with hormones
Insulin	Zinc associated with hormone storage
Angiotensin	Zinc in antiotensin-converting enzyme
Enkephalin	Zinc in enzyme enkephalinase
Several peptides	General hydrolysis by endopeptidase 24.11
Neurotensin	Degradation by zinc-dependent enzyme
Other peptides	Amino and carboxypeptidases (Zn) at synapses

11.11 Summary of intracellular zinc: is zinc a master hormone?

Inspection of zinc activities—

(1) in transcription factors, zinc fingers;

(2) in the core proteins of DNA(?), i.e. the division processes;

(3) in RNA synthetase and reverse transcriptase;

(4) in many cytoplasmic enzymes, some extremely critical for the synthesis even of the monomers of proteins, DNA, and RNA;

(5) in many enzymes for the degradation of external peptides (hormones) and structures (collagen, see Section 11.12.1);

(6) in its storage in critical zones of the body, e.g. the male reproductive tract—

leads one to consider the possibility that the internal homeostasis of zinc (by metallothionein) is finely tuned so that zinc is one of the dominant *cytoplasmic* growth hormones. Rapid cell division will then depend upon the presence of stored zinc and its transfer. This suggestion that zinc is effectively a hormone is no more than an hypothesis, but the importance of zinc, which equals iron in concentration in living systems, must not be missed because it is so hard to follow by analysis. We shall see again the all-embracing role of this metal when we turn to extracellular activities closely connected to growth. Perhaps in early evolution and still in prokaryotes iron and manganese played the main role of communicating with DNA (Chapters 12 and 21), but in the evolution of the eukaryotes this role has been taken over by zinc in the cytoplasm. The importance of zinc then increased again as development of complex multicellular systems evolved which were dependent on hormones such as peptides, thyroxine and sterols. Is all this chemistry linked to the increased concentration of dioxygen some one billion years ago? As stressed earlier, zinc is not redox-active and yet has particular value in organisms developed after a dioxygen atmosphere was generated. Is this change due to the relative solubilities of oxides and sulphides, Fig. 1.12?

11.12 Extracellular zinc

11.12.1 The export of zinc enzymes: digestion

Extracellular enzymes can be generated in two ways: first, by direct export of protein across the outer membrane when, if they are zinc enzymes, they find and fold around zinc in the extracellular fluids; second, by synthesis in vesicles. Here the protein is secreted into a space where zinc is also secreted. Both modes are common in zinc biochemistry and we should ask why both are used.

It is particularly noticeable that zinc proteases are stored in vesicles in such systems as the pancreas (digestive carboxy- and aminopeptidases) and in venoms (Table 11.10). The proteases associated with hormone and transmitter release from peptides are frequently zinc-dependent too, as are the peptidases 24.11 for their destruction. Thus, in nutritional studies zinc deficiency can show up in many ways associated with the variety of modes of activity, e.g. of zinc proteases such as of collagen, transmitters, hormones, and

Table 11.10 Zinc in degradation processes

Degradative process	Zinc enzyme
Pancreatic juice action	Carboxypeptidase
Venom haemorrhagic action	Zn proteases
Extracellular digestion (yeast)	Zn aminopeptidase
Extracellular digestion (bacteria)	Thermolysin
Breakdown of DNA and RNA	3'Nucleotidase
Peptide hormones	Peptidases 24.11
Connective tissue degradation	See Table 11.7

digestive enzymes. It must not be forgotten that zinc too is in other extracellular hydrolytic enzymes, e.g. phospholipases, nucleases, and saccharases, all active in digestion. (In passing observe the β-sheet structures of these enzymes; Fig. 11.9.)

Fig. 11.9 The fold of carboxypeptidase showing the buried metal ion connected to a β-sheet. (After W. Lipscombe.)

Given present-day knowledge we may like to think that the earliest use of zinc in evolution was in fact as a *general internal acid catalyst* to assist synthesis and as *general external acid catalyst* to assist digestion. In a reducing sulphide atmosphere it could do both better than any other available cation. Subsequently, and while the digestive function has remained, the external proteases have developed as part of the enzymic mechanism for the degradative management of the *extracellular* matrix (Table 11.7; see also Chapter 21) and of the controlled release and hydrolysis of hormonal peptides. Furthermore, in the cell, zinc has become associated with multiple regulatory systems associated with growth—in one sense—an obvious necessary further connection. Clearly, the postulated development of the biological chemistry of zinc parallels very closely the proposed evolution of the functions of calcium. Do they connect at the S-100 proteins?

11.12.2 Cross-linking and hardening of extracellular matrices

There is growing evidence that zinc is used to cross-link phenolic materials in the strengthening of teeth of insects and lower animals. These teeth or mandibles are made from chitins. The presence of high zinc in the teeth has not

been definitely associated with phenolate sidechains, but this remains highly probable. The movement of the zinc down hollow central tubes of the teeth toward the cutting edges is thought to occur by the transfer of vesicles high in this metal.

11.12.3 Zinc in solid-state devices

Zinc also occurs in some simple crystalline materials within biology. The most obvious are the deposits of zinc phosphates in what may be storage vesicles of such animals as snails. The real purpose of the vesicles is unknown, but they could act as depositories of unwanted excess zinc until it is rejected by exocytosis. Another certainly useful zinc deposit is crystalline zinc cysteinate in the tapetum of night-seeing animals. The crystals act as reflectors giving added sensitivity to the eye. It seems likely that these crystalline deposits were also laid down in vesicles.

11.13 Conclusion

This chapter demonstrates the diversity of uses of zinc as illustrated in Fig. 11.1. It is not quite true to say that zinc has as many uses as sulphur, but the time has come to look at the biology of zinc in a way similar to that in which we look at sulphur. This may stop the search for too many particular symptoms of zinc excesses or deficiency states since such states may well generate a general impairment of many functions.

The 'Further reading' for this chapter, which introduces the principles behind the use of zinc in biochemistry, is deliberately short. The chemistry of zinc is well described in modern textbooks of inorganic chemistry (e.g. Bertini *et al.* 1986). The bio-inorganic chemistry has been described in detail in three review books by experts in particular topics (Sigel 1983; Spiro 1983; Mills 1988). We see our purpose as pulling all these lines together so as to provide a guide starting from the chemistry and biochemistry of the element. Naturally, this has had to involve some speculative comment, but the deep involvement of zinc in biology is gradually becoming apparent.

In a later chapter on homeostasis an attempt will be made to bring the chemistry of this metal into a picture of a network of interactive communication systems in the cell, many of which use elements such as Ca^{2+}, Fe^{2+}, Mg^{2+}, HPO_4^{2-}, as well as zinc (Table 11.11).

Table 11.11 Communication networks

Simple current carriers	Triggers (second messengers)	Hormones
Na^+, K^+	Ca^{2+} (Mg^{2+})	Zn^{2+} ?
Acetylcholine	cAMP (cGMP)	Steroids

Other control elements are not usually classified in the above manner, e.g. Fe^{2+}, Mn^{2+}, ATP, ADP, since they serve many other functions.

Note. Probes of metal sites by isomorphous substitution

Table 11.12 Some probe (isomorphous) substitutions

Natural metal	Probe substitution
K^+	Tl^+, Rb^+
Mg^{2+}	Mn^{2+}, $Cr^{3+}[M(NH_3)_6]^{3+}$ where M = Co, Cr
Ca^{2+}	Mn^{2+}, Lanthanides (M^{3+})
Zn^{2+}	Co^{2+}, (Mn^{2+})

It has proved particularly easy to substitute zinc in protein (enzyme) sites by other metal ions. In particular, Co(II), Mn(II), and Cd(II) have proved rewarding in exchange studies. The detailed spectroscopy of the active sites of the zinc enzymes made possible by these substitutions and the kinetic analysis that has been carried out with them has revealed the very subtle nature of the structure and dynamics of active sites. This work, which precedes site-specific mutagenesis of organic groups but is parallel to it and more amenable to interpretation, is often overlooked by biochemists who do not follow the intricacies of the analysis available through the spectroscopy of metal ions. Their potential as reporter groups is of another order from that provided by organic centres. In this book little of this work is recorded but many deductions depend on the depth of work given in volumes devoted to this type of analysis. Table 11.12 gives some possibilities not only for zinc but also for many other metal ion substitutions.

Further reading

Berg, J. (1986). Potential metal-binding domains in nucleic binding proteins. *Science*, **252**, 485–93.

Bertini, I., Luchinat, C., Maret, W., and Zeppezauer, M. (ed.) (1986). *Progress in inorganic biochemistry and biophysics*, Vol. 1. *Zinc enzymes*. Birkhauser Verlag, Basel.

Buckingham, D. A. (1976). Hydrolysis by metal hydroxide complexes. In *Biological aspects of inorganic chemistry*, Vol. 37 (ed. A. W. Addison, W. R. Cullen, D. Dolphin, and B. R. James), pp. 141–97. Wiley Interscience, New York.

Dixon, N. E. and Sargeson, A. M. (1983). Metal ion roles in metalloenzymes. In *Zinc enzymes* (ed. T. G. Spiro). Wiley-Interscience, New York.

Kägi, J. H. R. and Kojima, Y. (ed.) (1988). Metallothionein II. *Experientia*, **Suppl. 52**.

Klug, A. and Rhodes, D. (1987). Zinc fingers. *Trends in Biochemical Science*, **12**, 464–9.

Kuby, S. A. (ed.) (1991). *A study of enzymes*, Vols. I and II. CRC Press, Boca Raton, USA.

Mills, C. F. (ed.) (1988). *Zinc in human biology*, *Human nutrition reviews*. Springer Verlag, Berlin.

Pan, T. and Coleman, J. E. (1990). Zinc cluster regulatory proteins. *Proceedings of the National Academy of Sciences USA*, 2077–81.

Sigel, H. (ed.) (1983). *Metal ions in biological systems*, Vol. 15. *Zinc and its role in biology and nutrition*. Marcel Dekker, New York.

Spiro, T. G. (ed.) (1983). *Metal ions in biology*, Vol. 5. *Zinc enzymes*. Wiley Interscience, New York.

Vallee, B. L. and Auld, D. S. (1990). Zinc enzymes and protein. *Biochemistry*, **29**, 5647–59.

Vallee, B. L. and Williams, R. J. P. (1968). Probes for metal ions in enzymes. *Chemistry in Britain*, **4**, 397–406.

12

Non-haem iron: redox reactions and controls

12.1	General introduction to transition metals in biology	319
12.2	Introduction to iron biological chemistry	322
12.3	Iron uptake	323
12.4	The non-haem iron proteins	326
12.5	The iron/sulphur proteins	326
12.6	Fe/S centres active as enzymes	329
12.7	Location of Fe/S proteins	330
12.8	Why are there Fe_nS_n clusters in biology?	331
12.9	The organization and synthesis of ferredoxins	332
12.10	The Fe–O–Fe cluster	333
12.11	Mononuclear non-haem/non-Fe/S iron and oxidative enzymes	335
12.12	Iron and secondary metabolism	337
12.13	Binding ligands in non-haem/non-Fe/S proteins	338
12.14	Iron as an acid catalyst	338
12.15	Summary of non-haem iron functions	339
12.16	Iron regulation and homeostasis	339
12.17	Iron buffering	340
	Note. The discovery of Fe/S centres	341

12.1 General introduction to transition metals in biology

In the introductory chapters we stressed that the different mineral elements had different biological functions and that these were based on a selection corresponding to potential for activity. This is the course evolution had to find, but much thought is necessary to see the complications which inevitably arise since many elements are capable of more than one function (see Chapter 6). The problems of understanding have already increased as we have moved from the chemistries of Na^+/K^+ to those of Mg^{2+} (note phosphate

(A) Transition metals

chemistry and chlorophyll), to those of calcium (note fast resting state controls and triggering), to those of zinc (note catalysis and slower control). However, all these activities relate to an increasing Lewis acid strength and a decreasing availability from group IA to group IIA elements and then zinc. Moreover, in evolution we have little reason to suppose that the relative availability of these elements changed grossly from the anaerobic to the aerobic eras even from one biological niche to another. When we leave this group of elements and turn to the transition metals (A) we meet quite new problems since: (1) the elements have increased functional capacity in that while they remain good Lewis acids they can be both π-electron donors and redox-active; (2) their availability has undergone remarkable changes during evolution (Chapter 1); (3) biological systems have synthesized selected organic partners for a considerable fraction of each of the elements, e.g. Fe (haems), Co (corrin, B_{12}), Ni (F-430), and Mo (a pterin), (remember Mg (chlorin)), while leaving much of the element in a freer form; and (4) functional properties, e.g. redox potentials, are now very sensitive to ligand co-ordination, including its geometry. There is then complexity to be met which is made even less simple by biological diversity during evolution. We have seen that advanced multicellular organisms developed special systems of control based on calcium and zinc with the proteins calmodulin and zinc fingers, respectively, and there arose a major difference between prokaryotes and eukaryotes. This distinction, although not so immediately obvious in the biological chemistry of transition metals, is equally important for them; for example, the redox chemistry of eukaryotes is largely removed into compartments separated from the cytoplasm. There are further new problems for us to appreciate, however, since the chemistry of the transition metal ions is not just specialized according to different functions, or between prokaryotes and eukaryotes, but has evolved differently in different higher organisms. There was a primitive separation between organisms heavily dependent on iron, nickel (especially), and cobalt, the archaebacteria, and other prokaryotes where the dominance of iron is more obvious. It is also clear that copper is of little significance in most of these organisms relative to its multitude of roles in multicellular eukaryotes while in these eukaryotes the role of nickel and cobalt is further diminished. Finally, the dependence on vanadium is highly specific amongst eukaryotes. Element role in evolution is obviously specialized and has clearly changed with biological niche and with time for the transition elements as well as for zinc and calcium. Understanding rests here as elsewhere upon an analysis of the proteins associated with the redox-active metal ions. In Tables 12.1 and 12.2 we divide them into two extreme groups which separate α-helical, more mobile structures, from β-sheet or cross-linked, immobile structures (see Chapter 7) for reasons which will become apparent.

Given this multitude of problems we have decided to start the description of the redox role of the elements with the one transition metal element certainly common to all life—iron. As we stressed in Chapter 1, there is no way life could have existed without the introduction of a redox-active metal to carry out functions such as the activation of small molecules for redox chemical change and for the transport of electrons. It is these functions which allow the essential energy-capture reactions related to redox processes to be connected to the equally essential acid/base functions connected to polymerization of phosphate in ATP. The connection is made at the non-metal level via the

Table 12.1 Less rigid structures in redox proteins

Redox protein	Structural feature	Function
Cytochromes c	Helical (a) two thioether bridges	Electron transfer
Cytochromes c'	Helical (a) one thioether bridge	Allosteric change on binding NO/CO
Cytochromes b (membrane)	Helical (a)	Electron transfer, proton pumping/coupling
Cytochrome oxidases (b or a, a_3)	Helical (a)	Electron transfer O_2 reduction proton pump/coupling
Cytochrome P-450	Helical (a) small β-sheet	Hydroxylation
Iron binding protein of ribonucleotide reductase	Helical (a)	Changes conformation on redox reaction
O_2-carrier proteins	Helical (a)	Allosteric

chemistry of H^+, X—H, and H^- (see Chapters 1 and 5), but only through the assistance of metal-ion redox catalysis using overwhelmingly iron, at first in simple complexes and then in haem.

Starting from iron we shall progress to the biological chemistries of manganese, copper, and molybdenum plus vanadium and tungsten, and then to those of cobalt and nickel. We shall be in a position, at this stage, to look at

Table 12.2 Highly immobile tertiary structures in redox proteins

Redox protein	Structural feature	Function
Blue copper proteins	One β-sheet domain	Electron transfer
Copper oxidases	Three β-sheet domains	O_2 oxidation to free radicals
Peroxidases		
HRP	–S–S– cross-linked helices	H_2O_2 free radical reaction
CcP	β-sheet domain plus helices	
Cytochrome b_5	β-sheet domain plus helices	Electron transfer
Superoxide dismutases		
Cu/Zn	β-barrel	Superoxide dismutation
Mn/Fe	$a\beta$-structure	
Flavodoxin	β-sheet	Electron and hydride transfer
Ferredoxins	β-sheet	Electron transfer
Rubredoxin	β-sheet	Electron transfer

the comparative roles of non-metals. Here we shall find that, corresponding to the functional division of metals $Na^+/K^+:Mg^{2+}/Ca^{2+}:Zn^{2+}$ from the redox-active transition metals, there are the divisions of the heavier non-metals Cl^-, silica, and phosphate (RPO_4^{2-}) from S and Se, where we again deal with the acid/base chemistry of the first three (Chapter 18), separately from redox chemistry of the second two (Chapter 19). This leaves aside the major 'organic chemistry' of metabolism, synthesis and degradation, based on H, C, N, and O, which is well described in standard biological texts and will receive only passing reference here. The chemistry of biological minerals in which the metals and non-metals combine has a chapter of its own

(Chapter 20). At this stage we shall be left with the way in which all this chemistry is brought together in some harmonious fashion to form an active organism. Chapter 21 on homeostasis speculates on this problem, before we describe the effect of man's activities upon the elements in life.

12.2 Introduction to iron biological chemistry

The involvement of iron in biology is complicated, perhaps more complicated than that of any element other than the more common H, C, N, and O. It is not just involved in oxidation–reduction catalysis and bioenergetics but is connected to many control systems and to several acid–base reactions (Table 12.3). In an effort to simplify the description of its biochemistry we shall divide the material into:

(1) uptake and storage of iron;

(2) iron/sulphur proteins;

(3) additional iron proteins using simple co-ordinating ligands;

(4) feedback controls from uptake to enzymes;

(5) haem proteins (which will be described separately in Chapter 13 since they are so widely distributed and important).

Table 12.3 Some iron proteins and enzymes: localization and function

Protein or enzyme	Localization	Function
Haemoglobin (Fe)	Cytoplasm	Transport of O_2
Transferrin (Fe, Mn)	Outside cells (plasm)	Transport of Fe and Mn(?)
Ferritin (Fe)	Cytoplasm	Storage of Fe
Rubredoxin, ferredoxin (Fe)	Cytoplasm (prokaryotes) and membranes	Electron transfer
Hydroxylases (Fe)	Cytoplasm and membranes	Hydroxylation
Peroxidases (Fe)	Vesicles and outside cells	Utilization of H_2O_2
Acid phosphatase (Fe, Mn)	Vesicles and outside cells	Hydrolysis of phosphate groups
Superoxide dismutases	Mitochondria (prokaryotes)	Dismutation of O_2 (Fe, Mn)

As a consequence of this division, largely in terms of iron co-ordination chemistry, functions of similar kinds but not always in similar locations will appear under different headings. For example, electron transfer chains involve (2) and (5), oxidases can be found from (3) and (5) as can dioxygen carriers, while controls (4) are linked to (2) and (3) via free iron but to (5) via haem iron. To remind the reader of the way in which some of the different iron proteins are involved in one organization, here energy capture, we show in Fig. 12.1 the essence of the electron transfer chain of mitochondria.

The first problem we meet in the biological chemistry of iron is its low

Redox potential (mV)

Fig. 12.1 The energy-transducing respiratory chain of mitochondria. Note the Fe/S proteins and their redox potentials. See also Photosystem I and the reaction centre complex and its Fe/S proteins.

availability today, generated by the upsurge of dioxygen (see Chapter 1). Due to this fact all living organisms in aerobic zones had to develop complicated uptake and control mechanisms to remain viable since in the presence of dioxygen at pH = 7 iron is precipitated as $Fe(OH)_3$.

12.3 Iron uptake

The basic problem of iron incorporation is shown in Fig. 12.2. There are a multitude of steps in which iron is moved from place to place in a living organism. Starting with prokaryotes, the simplest involves the use of small chelating agents which are synthesized by many prokaryotes and used to scavenge the external media for iron (Fig. 12.3). The design of these reagents is such that they are soluble in water when not bound by iron, but on binding

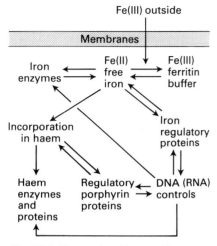

Fig. 12.2 The uptake of iron and its subsequent controlled distribution and metabolism.

they turn inside out and become able to pass through lipid membranes. In more complex systems the iron chelate is taken up by a receptor and channelled to or through the inner membrane. A very similar final transfer mechanism exists for the uptake of cobalt in vitamin B_{12}. Parallel mechanisms are available in the membranes of mitochondria, but here the iron is taken up as its citrate from the cytoplasm. (In plants iron may also pass up the stem associated with citrate.) The iron is released to the cytoplasm of the prokaryote by a variety of methods, including reduction to iron(II) and destruction of the carrier. The metal can be kept as iron complexed with protein, and is then distributed to many compartments even to the extent that some iron oxidase proteins are extruded, e.g. from soil bacteria giving oxidases to break down lignins. Alternatively, iron can be incorporated into porphyrin (Fig. 12.2) and used as discussed in Chapter 13. The excess iron is held in a number of storage forms the best known of which is ferritin, but the name ferritin covers a number of protein/iron stores which differ considerably.

The uptake of iron is complicated further in higher organisms where

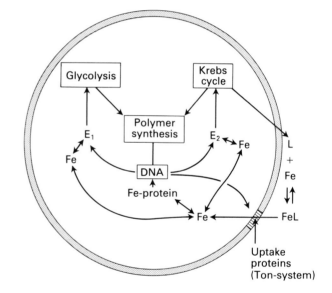

Fig. 12.3 A schematic representation of iron metabolism in bacteria such as *Escherichia coli.* Note the feedback loops. E, enzyme; L, Ferrioxamine B; R_1, H; R_2, CH_3, shown below.

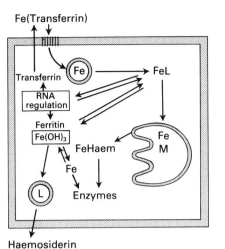

Fe(Transferrin)

Haemosiderin

Fig. 12.4 A scheme corresponding to that of Fig. 12.3 for the uptake of iron into an advanced eukaryote cell. Note: (1) the feedback control; (2) the role of the mitochondria, M; (3) the danger of excess iron leading to haemosiderin deposits. FeL = Fe citrate or Fe protein.

L = ligated degraded Fe complex.

proteins handle the metal in body fluids (Fig. 12.4). The proteins called transferrins (Fig. 12.5) are recognized, when bound to iron, by special receptors on the outer membrane of cells and are then engulfed in minimicrosomes (vesicles) (Fig. 12.4). The vesicles transport the transferrins to a lysosome where the metal is released. Part of the mechanism depends on the fact that carbonate is bound initially with the iron to transferrin at pH = 7 and the acidity of the lysosome destroys this complex. The iron probably as free Fe(II), is passed into the cytoplasm where at the more alkaline pH it forms a series of non-haem iron proteins and Fe(III) complexes with citrate or with other carriers, and some of it is subsequently pumped into the mitochondria or chloroplasts. In these organelles it can form complexes with a variety of proteins or it can be handled in another capture chelatase converting the iron to haem iron. The haem is incorporated into a variety of proteins, not only in mitochrondria and chloroplasts, but finds its way to a number of vesicles, e.g. peroxisomes and cell membranes, or to cellular dioxygen carrier proteins in circulating fluids. We return to the intercommunication network between the various haem and non-haem proteins after we have described each of the classes of proteins.

In the early days of evolution the problem of iron uptake may not have existed since in the anaerobic sulphide-rich media there could well have been much Fe(II) and small clusters of Fe(II)/Fe(III) sulphides. The uptake principle could then have been simple incorporation of the iron, possibly as sulphide complexes, and biology would have had direct access to both electron

(a)

(b)

Fig. 12.5 (a) The outline fold of transferrin, a hinged protein. (b) Details of the binding site for iron; the open side interacts with bicarbonate. (After E. N. Baker and T. Blundell.)

transfer and hydrogenation catalysis. Iron/thiolate complexes may have been central to control also, see Chapter 21. We need not speculate further except to note that, to this day, the incorporation of iron sulphide complexes is still an essential feature of all biological systems. The way in which these sulphides are formed in the presence of dioxygen is something of a mystery.

12.4 The non-haem iron proteins

The series of these proteins include very distinctive classes. The most prominant are the iron sulphur proteins which are taken to include mononuclear rubredoxins, and the various multinuclear ferredoxins, 2Fe/2S, 3Fe/4S, 4Fe/4S, and some higher clusters, in which all iron atoms are bound either to thiolate or sulphur atoms with only one or two exceptions (Fig. 12.6). The exceptional cases where one co-ordinating atom is not sulphur will be discussed after the main group has been described. The second well known series is based on the oxo-bridged unit, Fe–O–Fe, first seen in haemerythrin, the dioxygen carrier of certain worms. The final group is a set of mononuclear iron proteins where the ligands are nitrogen and oxygen donors. The functions of these three classes of proteins are very different.

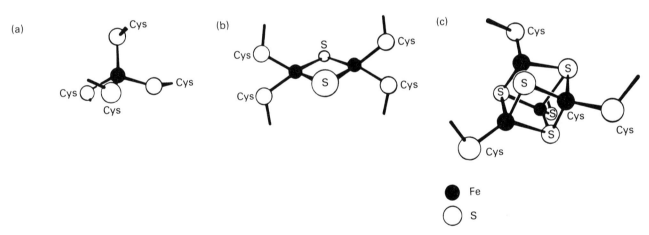

Fig. 12.6 The co-ordination in iron/sulphur proteins: (a) rubredoxin; (b) Fe_2S_2 ferredoxins; (c) Fe_4S_4 ferredoxins. The Fe_3S_4 unit is like that of Fe_4S_4 with one Fe removed. ●, Fe; ○, sulphur—those marked 'S' are sulphide, S^{2-}, while those marked 'cys' are cysteinate sulphur.

12.5 The iron/sulphur proteins

The structures of several of these small proteins are known (Fig. 12.7). To a first approximation in the vast majority of them each iron is in a tetrahedron of RS^- alone (rubredoxins) or together with S^{2-} donors (ferredoxins). The fold of these small proteins is based on β-strands forming small β-sheets (Fig. 12.7); the iron co-ordination sphere is usually saturated and is held in a stable

Plant ferredoxin

Peptococcus aerogenes
Ferredoxin

Clostridium pasteurianum
Rubredoxin

Chromatium vinosum
HiPIP

● Fe ○ S ⊙ Cys–S

Fig. 12.7 The folds of some typical iron sulphur proteins. Note the β-sheet constructions. (After the work of many authors including L. Jensen, K. Fukuyama, and R. Stroud.)

condition by this fold. The cluster is held in a protective groove in the protein (see Section 7.7). The exact stereochemistry even in rubredoxins (some four classes of this type of protein are now known) is clearly dependent on the exact protein fold. The proteins are variously soluble in water or membrane phases and in this stable form in either phase they appear overwhelmingly to have a simple single function: they are part of electron-transfer wires (Fig. 12.1). (Exceptions are described later.) The redox switch of high-spin Fe(II) to high-spin Fe(III) involves only a small change in bond length due to the fact that it occurs in an interactive sulphide cluster. There is little relaxation energy change (see Section 4.7) so the sites are ideal for function. Conformational strain induced by the protein fold, which is undoubtedly observed in some of the Fe/S proteins, is not necessary for the basic function, but may be used to control redox potentials as may the general surrounds of the iron centre.

For the most part the iron/sulphur proteins have very negative redox potentials (Table 12.4) while covering the range from 0.0 to -0.5 V. Their frequent occurrence in the early steps of oxidation of organic molecules, e.g. NADH (i.e. bound H_2) and succinate (Fig. 12.1) suggests that they belong to

Table 12.4 Midpoint potentials* of examples of different iron–sulphur cluster types

Cluster	Protein	E_m (mV)
[Fe]	*Clostridium pasteurianum*: rubredoxin	−58
[Fe]	*Clostridium thermoaceticum*:	
	Rubredoxin 1	−27
	Rubredoxin 2	20
[Fe]	*Desulphovibrio gigas*: desulphoredoxin	−35
[2Fe–2S]	Spinach ferredoxin	−420
[2Fe–2S]	*Spirulina maxima*: ferredoxin II	−310
[2Fe–2S]	*Nostoc* strain MAC: ferredoxin II	−455
[2Fe–2S]	Adrenal ferredoxin	−270
[3Fe–4S]	*Azotobacter vinelandii*: ferredoxin I	−424
[3Fe–4S]	*Desulphovibrio gigas*: ferredoxin II	−130
[4Fe–4S]$^{1+:2+}$	*Clostridium pasteurianum*: 8Fe ferredoxin	−403
[4Fe–4S]$^{1+:2+}$	*Bacillus stearothermophilus*	−280
[4Fe–4S]$^{1+:2+}$	*Desulphovibrio gigas*: ferredoxin I	−455
[4Fe–4S]$^{2+:3+}$(HIPIP)	*Chromatium vinosum*	356
[4Fe–4S]$^{2+:3+}$(HIPIP)	*Paracocuss* sp	282

Clusters in complex electron-transfer systems

Cluster	Protein	E_m (mV)
[2Fe–2S]	Xanthine oxidase (bovine milk)	−300
[2Fe–2S]	Rieske centre	
	Bovine heart	260
	Spinach chloroplasts	290
[4Fe–4S]	Photosystem I (spinach chloroplasts)	
	Centre A	−550
	Centre B	−590
	Centre X	−705
	Succinate dehydrogenase (bovine heart mitochondrial complex II)	
[2Fe–2S]	Centre 1	0
	Centre 2	−260
[4Fe–4S] or [3Fe–4S]	Centre 3	70

* Potentials are versus the standard hydrogen electrode.

the most primitive parts of energy-capture machinery (see Chapter 5). At an even earlier stage in evolution they may have been involved in direct hydrogen transfer reactions from dihydrogen and in photosystems. This function brings us to the exceptional Fe/S proteins which have substrates other than the electron, e.g. dihydrogen (see Section 12.6).

As mentioned some of the Fe/S clusters as well as some rubredoxins are rather different from the majority. Two exceptional features are found (Table 12.5). The first is the high redox potential of the cluster in one protein which is part of the mitochondrial and chloroplast redox chain. The protein is called HIPIP (high potential iron protein). This protein is on the oxidizing side of the quinone oxidases of energy capture devices and is therefore at a higher potential than the membrane cytochrome *b* (Fig. 12.1). One of the ligands here may well be a histidine nitrogen. It is not easy to think of any role of such a protein which could not be performed by a haem protein, but see

Table 12.5 Unusual Fe/S centres in electron transfer system

Protein	Fe/S	Unusual features
Nitrogenase	P–Fe$_4$S$_4$	Spin states
High potential iron protein	Fe$_4$S$_4$	Nitrogen co-ordination
Aconitase*	Fe$_4$S$_4$	Loses Fe, binds ligands
D. gigas protein	Fe$_4$S$_4$	Loses Fe
Sulphite (nitrite) reductase	Fe$_4$S$_4$	Fe/S unit bound to haem
Hydrogenase	H–Fe$_4$S$_4$	Unusual EPR, binds H$_2$?

* At least six dehydratases are known.

below. The second group of unusual Fe/S proteins are enzymes as opposed to electron-transfer proteins. The functions are now much more complicated since the acid/base properties of the iron atoms become important. The iron must be to some degree exposed, and some of it can dissociate making a potential link to controls.

12.6 Fe/S centres active as enzymes

It has become clear recently that the variety of Fe/S centres which bind to substrates rather than transfer electrons is quite considerable. Table 12.5 refers to some of these, but many more are similar dehydratases. At the same time physical studies indicate that these centres are quite different from those that are simply involved in electron transfer. As in many other cases the physical studies involve electron paramagnetic resonance (EPR), Mössbauer, extended X-ray absorption fine structure (EXAFS), magnetic circular dichroism (MCD), and absorption spectroscopies. The peculiarity of the sites rests not just in the type of protein ligation to the iron atoms but also in the site geometry, in the number of iron atoms in a cluster, and in the exposure of at least one iron atom to exchange with iron in the medium. The types of co-ordination vary and include one bound histidine and one bound carboxylate from the protein. When the substrate binds, the single exposed iron of, for example, an Fe$_4$S$_4$ cluster, can become six co-ordinate and the centre probably generates a series of trapped free radical intermediates. Two examples stand out—dihydrogen-reacting enzymes and aconitase, for which there is a structure available. In aconitase the iron centre acts directly on the substrate, aconitic acid (Fig. 12.8), which is involved in the citric acid cycle (Fig. 12.4 and Fig. 5.9), allowing the creation of a short-lived free radical. The reactive cluster is Fe$_4$S$_4$, which is able to lose an iron atom, but the consequently formed Fe$_3$S$_4$ cluster is not active. The suggestion follows that the switch from active to inactive forms of the enzyme by loss of iron is a control on the citric acid cycle related to iron availability. Other enzymes in this class are the ferredoxins III from *Clostridia*, which are thought to be associated with reactions of dihydrogen and which, due to exchange of iron, control these reactions. It is probable that the association/dissociation of one iron atom of a cluster, possibly like that of HIPIP, is involved in the ferroxidan bacteria which utilize the redox reaction of Fe^{3+} of the environment as an

Fig. 12.8 The proposed active site of aconitase showing substrate bound to one iron. (After H. Beinert.)

energy source. In conclusion it would appear that the effect of the protein fold on the Fe/S centres in these enzymes is much stronger than in the simple electron transfer proteins, so that the idea of stressed states with easy transition between intermediates, as in zinc and copper enzymes, becomes plausible (see entatic state).

In passing, we note that the association of redox centres with free radical chemistry and especially with rearrangement of molecular skeletons mooted here for the Fe/S enzymes is common to vitamin B_{12}, some copper enzymes, and the Fe–O–Fe centres described in Section 12.10.

The observation of freely dissociable iron from clusters and its possible connection with control also leads forward to a general discussion of control of cellular metabolism based on the levels of iron in a cell (see Chapter 21).

12.7 Location of Fe/S proteins

These proteins are rarely found in extracellular space. Thus, in bacteria the proteins are usually in the membrane or in the cytoplasm and less often in the periplasmic space and this is also true for the organelles, mitochondria and chloroplasts. It would appear to be the specific function of copper (blue) proteins and cytochromes c and not of Fe/S proteins to carry out electron transfer in the external aqueous phases especially at the high redox potential of aerobic life. The Fe/S proteins are not very resistant to oxidation by dioxygen. Exceptionally, however, there are a few Fe/S complex proteins in the extracellular fluids of higher organisms. Amongst these proteins are the xanthine (aldehyde) oxidases and it is not clear why such a centre has been used when apparently a more stable group could have been found. In early evolution very little electron transfer chemistry could have been necessary outside a cell, but this changed with development concomitant with the advent of dioxygen (see Chapters 15 and 17).

The membrane Fe/S proteins are associated with photosystem I, with very low potential reactions of dihydrogen (hydrogenases) and with reactions of dinitrogen (nitrogenases) which also release dihydrogen. In Chapters 3 to 5 we referred to the NADH and NADPH coenzymes as forms of bound H_2. The

oxidation of these coenzymes is directly connected to membrane Fe/S proteins or to flavins, which themselves transfer electrons to Fe/S at a potental of around -0.5 V. Very many of these systems are membrane bound.

12.8 Why are there Fe_nS_n clusters in biology?

The first question is 'Why do such clusters form at all?' General chemistry leaves us with a good understanding of the selective uptake of Fe(III) into these highly anionic sulphide sites (Section 2.5.7). It is also reasonable that in a reducing (H_2S) atmosphere inorganic clusters of Fe(III) with Fe(II) and sulphides will form. The studies of Holm (see Further reading) have been very revealing as to the nature of this non-biological chemistry which resembles anaerobic biology. The problem for the chemist is to appreciate how such species can be formed today in the interior of cells which are exposed to dioxygen. Indeed, it is not always possible even to extract the ferredoxins without damage from dioxygen. This is a general problem, which concerns also sulphide clusters containing other metals, e.g. Mo, V, Ni.

If we believe that biological systems have developed in an understandable evolutionary sense, the question must be asked also as to why Fe_nS_n clusters are found in virtually all biological systems. This question can be put at two levels; first, what is the particular value of clusters; and, second, if they have such a value why was an Fe_nS_n cluster chosen rather than some other metal cluster?

Now, apart from functioning as an electron transfer agent, the cluster of M_nS_n, where M is now any transition metal of variable valence, must have come into ready equilibrium with two redox reactions if it was to become biologically useful. The first is the reaction with protons and dihydrogen (the dihydrogen could be bound as NADH)

$$M_nS_n^{p+} + H_2 \longrightarrow M_nS_n^{(p+-2)} + 2H^+,$$

and the second is the reaction with sulphide

$$M_nS_n^{p+} + S^{2-} \longrightarrow M_nS_n^{(p+-2)} + S\downarrow.$$

These two reactions are connected through H_2S formation. Given a suitable organization and energy source they could be utilized taking hydride (H_2) originally from a carbon source and in place of the dioxygen used today in mitochondrial reactions there would be a sulphur acceptor when the reaction system would run between -0.5 and 0.0 V. It may well have been a primitive necessity to carry out these reactions and it may also have been that the catalysis of the reactions was present before life. The question could then be answered as to why Fe_nS_n was chosen in particular in early evolution for this function. It may well be that of all such M_nS_n clusters, assuming clusters are the best catalysts for the above reaction sequence, the iron complexes were the most available. Notice that iron sulphides readily give persulphides, pyrites, which is a step on the way to polymerized sulphur. In the earliest forms of life the inorganic reaction system $Fe/S/H_2$ may have formed the basis of energy, life-giving processes themselves.

We note again that there are no other M_nS_n clusters in biology. There are a few other biologically active small clusters of iron (Fe_2O) and manganese (Mn_2O and possibly Mn_4O_4 in the dioxygen carriers and oxidases), but none are based on sulphur chemistry. While there are no other M_nS_n clusters, mixed clusters based on Fe_nS_n units, e.g. Fe_6MoS_8 or Fe_6VS_8, exist in nitrogenases. We discussed the chemistry which allows these clusters to exist in Chapter 3. These leaves us with an unsolved problem—the synthesis of Fe/S clusters in aerobes.

12.9 The organization and synthesis of ferredoxins

The ferredoxins introduce a problem in metal/protein chemistry which we have put aside so far: the fact that they are incorporated into organized assemblies. The self-assembly is here essential to function since it controls spatial ordering and the electron transfer distance (Chapter 4). The selective, almost specific, recognition of surfaces by proteins is often a problem since the surface residues are very mobile and it must be remembered that some mobility in the organization may be essential for function (Chapter 7). Moreover, the combination of several of protein units must be within a *particular* membrane. Here the location of protein synthesis may be important since several of the Fe/S proteins of the electron transfer chains of mitochondria and thylakoids are coded in mitochondrial and chloroplast DNA and not in cytoplasmic DNA. In fact, there is at least one mitochondrially coded peptide or protein in each of the three major particles of the cytochrome chain and one in the mitochondrial ATP synthetase (see Fig. 12.1). In the development of a complex cell it may be important to restrict the site of synthesis of a given protein since this may decide the membrane into which it and then other proteins are incorporated.

There is a further interest in the synthesis of Fe/S proteins. When an Fe/S protein is made there does not appear to be much excess in the cell of either iron, in an Fe/S cluster or not, of sulphide or of its apoprotein. Here we return to the remarkable selectivity of combination of metals and proteins which is greatly enhanced if the concentration of specific partners, chosen metal or metal complex and chosen protein, is matched by feedback control to DNA (see Chapter 2). The implication is that the synthesis, and perhaps the destruction, of the proteins is related to the free Fe or Fe complex concentration and that the uptake of Fe or Fe complex is related to the availability of protein for metal incorporation.

A disturbing feature of this selective uptake, mentioned in the previous chapter, is that the zinc finger sites look exactly like the Fe(III) sites of rubredoxins (and ferredoxins), yet the two must not be confused. Now, there are no zinc fingers in prokaryotes, and in eukaryotes much Fe/S chemistry is confined to mitochondria and chloroplasts. The suggestion is clear that the coming of dioxygen brought about a radical reorganization in the ways in which Fe/S proteins were positioned and the way in which $(RS^-)_4$ co-ordination chemistry was used. Possibly at this time Fe_2O clusters appeared also. As we discussed in Chapter 2, the selection in these proteins

may not be at the level of the donor atoms in the co-ordination sphere, but through selection of cavity size and of second and third co-ordination sphere sidechains due to the fold of the protein, and of pumping into compartments.

12.10 The Fe–O–Fe cluster

It was thought at one time that the Fe_2O cluster was confined to certain dioxygen carriers, the haemerythrins. This seemingly primitive carrier is pictured in Fig. 12.9. It is assumed to bind dioxygen as the peroxide anion O_2^{2-}, since the Fe(II) of the deoxygenated state is converted to the oxo-bridge Fe(III) condition on uptake of O_2. The protein itself has a four multihelical bundle fold. The expectation then is that such multisubunit haemerythrins can be co-operative since the oxidation state of the iron would be expected to adjust the relative disposition of the helices as there is a radius contraction of >0.1 Å from ongoing high-spin Fe(II) to high-spin Fe(III). In fact, some haemerythrins are co-operative, but the co-operativity is not as high as in haemoglobins. Examination of the structure shows that the Fe–O–Fe unit cross-links the helices and must constrain motion to some degree. Even so the proteins illustrate again the idea that multihelical proteins are mechanically adaptable in contrast with the Fe_nS_n clusters in β-sheet proteins (Tables 12.1 and 12.2).

The Fe–O–Fe centre is easily recognized by the peculiar high intensity of the spin-forbidden d–d absorption bands in the visible region of the spectrum as well as by use of Mössbauer spectroscopy. It was these peculiarities which led scientists to describe the iron ribonucleotide reductases, the second to be found in the Fe–O–Fe series of proteins, in terms of this bridged oxo structure (Fig. 12.10). However, this protein is an enzyme and contains an even more unusual feature—a tyrosine free radical which remains stable within the protein all through extraction procedures. It would seem that the purpose of the iron centre is to assist in the formation of this radical. The starting point for such an assertion is that the reaction

$$\text{Ribose} \longrightarrow \text{deoxyribose}$$

is a reaction requiring first a hydrogen atom abstraction, a radical

Fig. 12.9 The structures of the haemerythrin-active site. (After L. Jensen.)

Fig. 12.10 The structure of the active site of ribonucleotide reductase (a). The protein is helical and its structure is shown at (b). (After Norland *et al.*, see Further reading.)

intermediate. The two-electron reduction then goes via effective OH˙ abstraction and removal. (See the similar redox and rearrangement reactions in coenzyme B_{12} catalysis in Chapter 16 and note that the enzyme with coenzyme B_{12} must have come earlier in evolution and perhaps is being replaced by the iron-dependent ribonucleotide reductase.) Thus, it is believed that the tyrosine radical acts as a hydrogen-abstracting agent prior to reduction of the ribose. Notice the extreme example of an unusual energy state of a group, here tyrosine, in a protein, which parallels the formation of entatic states of metal ions (see Table 7.10).

There is another very powerful and unusual enzyme in this Fe–O–Fe series—methane oxidase. There are no known cofactors in this enzyme other than the Fe–O–Fe centre. Once again we notice that the reaction is probably a hydrogen atom abstraction

$$CH_4 \longrightarrow CH_3^{\cdot} + H^{\cdot}.$$

(The easiest way in which to break a C–H bond at a saturated carbon is the radical path; Section 4.12.) The suggestion then arises that the reaction requires a free radical in the protein, when the initial step parallels that of the cytochrome P-450 reactions which include ω-hydroxylation of long chain fats (see Chapter 13) and the reactions of certain other monoxidases, e.g. copper galactose oxidase. There is the distinct possibility that the P-450 catalysts are a recent development of great value especially in eukaryotes, though they occur in prokaryotes, while the non-haem radical catalysts are usually found only in prokaryotes and vesicles of eukaryotes.

A very instructive feature of the Fe–O–Fe enzymes is that the domain which includes the Fe–O–Fe unit is helical and presumably adjusts conformation as this domain and the tyrosine free radical are formed. Here as in other helical 'enzymes' a required order of reaction is found which is compatible with a series of conformation changes (Section 7.4).

12.11 Mononuclear non-haem/non-Fe/S iron and oxidative enzymes

The examples of this class of enzyme include those which act outside cells, e.g. those expressed by soil bacteria, and do not involve a cycling of oxidation states. It is the substrate which is attacked by molecular oxygen. The function of the iron is to activate the substrate, S, for ease of O_2 attack

$$Fe(III)S \longrightarrow [Fe(II)S^{\cdot}] + O_2 \longrightarrow S-O_2 + Fe(III).$$

where $[Fe(II)S^{\cdot}]$ represents a partial charge-transfer complex. A typical reaction is the addition of dioxygen across an activated benzene ring. The enzyme is maintained in the Fe(III) state so that there is no loss of iron.

The iron monoxidases in cells which contain a single iron atom usually go through the Fe(II), Fe(III), and higher states such as FeO(IV). The mechanism of oxidation is due to insertion of oxygen atoms derived from dioxygen. A typical case is the reaction of the penicillin ring cyclase, reaction A, and many such enzymes are found in prokaryotes, but there are

Isopenicillin N
synthase

Epimerase

Ring expansion
-hydroxylase

Penicillin

Reaction scheme A

O_2 +

Prolyl in a protein

α-keto-
glutarate

Reducing
agent

Hydroxyprolyl in a protein

+succinate
+CO_2

Reaction scheme B

parallels with eukaryote enzymes such as the connective tissue lysyl and prolyl hydroxylases, reaction B, used in collagen maturation; see also the synthesis of hormones and prostaglandins in Fig. 12.11. Another interesting feature is that the oxidation is linked to the use of α-ketoglutarate as a hydrogen donor. This substrate connects the metabolism back to the citric acid cycle and we are reminded of the presence of isocitrate in the enzyme nitrogenase, Chapter 17.

Once again the fact that the iron in the resting state of the enzyme, Fe(II), may (freely) dissociate connects the activity of the enzymes of this kind in prokaryotes to the levels of free iron in the cell according to the equilibrium

$$Fe \cdot E \rightleftharpoons E + Fe^{2+}.$$

Active Inactive Control

Now, while it is clearly permissible for the lower organisms to have such enzymes with freely dissociable iron in their cytoplasm (see Table 12.5), this is no longer agreeable in the eukaryotes and especially in multicellular organisms. Free Fe(II) and Mn(II) or, for that matter, Cu(I), Co(II) and many other free transition metal ions, easily activate molecular dioxygen as

$$O_2 \longrightarrow O_2^{\cdot} \longrightarrow H_2O_2 \longrightarrow OH^{\cdot}(\longrightarrow H_2O).$$

The three central compounds in this series are powerful oxidizing agents and are known to be mutagenic and carcinogenic; hence the risk of adventitious

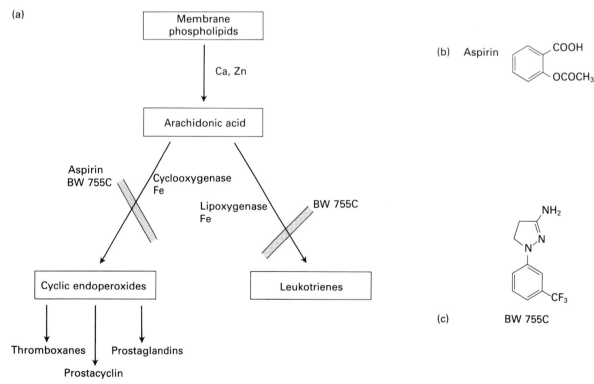

Fig. 12.11 (a) The pathway of prostaglandin synthesis from arachidonic acid (see reaction scheme B), produced from membrane phospholipids. Prostaglandin production can be blocked by (b) aspirin and (c) BW 755C, which inhibit the cyclo-oxygenase system. BW 755C also inhibits lipoxygenase and blocks the production of leukotrienes.

dioxygen reactions had to be contained. It is a general observation that the reactions of the above intermediate steps of reduction of dioxygen are handled with great precautions in eukaryotes.

1. Many of the reactions involving iron are in organelles or vesicles, e.g. mitochondria, peroxisomes, reticula, or vesicles.

2. Many of the reactions outside cells use copper proteins which do not release copper easily.

3. Iron enzymes are held in the Fe(III) state so as to protect them from loss of iron even in vesicles.

4. The iron oxidases in the cytoplasm are often haem proteins which do not release iron, e.g. cytochrome P-450, and/or are on membranes or in microsomal fractions.

5. Some of the enzymes have more than one metal to ensure that intermediates between O_2 and H_2O are not released, e.g. cytochrome oxidase.

[An interesting enzyme which is conveniently placed in this group is the superoxide dismutase of prokaryotes, mitochondria, and chloroplasts; this enzyme is replaced by a copper/zinc superoxide dismutase in the cytoplasm of all eukaryotes. Here we meet another Fe(III) enzyme which has a very closely related Mn(III) form. The character of all three superoxide dismutases (see Chapter 14 on copper proteins) is that the redox potential of the M(I)/M(II), Cu, or M(II)/M(III), Fe, or Mn, is placed approximately between those of the $O_2/O_2^{\cdot-}$ and the $O_2^{\cdot-}/H_2O_2$ couples. The superoxide dismutases are a clear example of proteins in which there is a switch away from a risk-carrying iron (or manganese) enzyme (prokaryotes) to a safe copper enzyme (eukaryotes)].

12.12 Iron and secondary metabolism

Table 12.6 Some non-haem iron oxidases

Enzyme	Characteristics
Lipoxygenases	Oxidation of arachidonic acid
Prostaglandin oxidases	Cyclization and epoxidation
Penicillin cyclases	Cyclization (also cephalosporins)
Lysyl oxidases	Precursor of collagen cross-link
Prolyl oxidases	Precursor of collagen cross-link

NB. The iron is bound by N,O-donor ligands and may well be in a six co-ordinate site.

The substrates of many of the above enzymes and their products are of considerable interest in secondary metabolism (Table 12.6 and schemes A and B in Fig. 12.10), which produces compounds outside the major required metabolic paths. These secondary metabolites are usually made by oxidative reaction so that either hydroxyl or halogen atoms are introduced into organic molecules. Halogenation will be described in Chapter 19. Both haem and non-haem enzymes are active in halogenation and hydroxylation. In prokaryotes the enzymes seem to be merely a part of the catalytic armory for the production or removal of offensive chemicals. An obvious example is penicillin, but many halogenated compounds and sterols fall in this class. In multicellular eukaryotes it appears that evolution has adapted these compounds so that they become part of the messenger or hormone systems. Another use is in the synthesis of the progenitors of the cross-linking agents of extracellular matrices such as hydroxyproline and hydroxylysine.

There is a peculiarity about the production of some hydroxylated compounds especially in plants. The wide range of alkaloids, which appear to have little value today, may well have a defense role, and perhaps they are waiting to become messengers, or hormones.

12.13 Binding ligands in non-haem/non-Fe/S proteins

As we have seen there is a large series of iron enzymes (non-haem/non-Fe/S) which cover a variety of functions. The binding site, whether in the Fe–O–Fe or the single Fe series of proteins, is to either several histidines or histidine and carboxylate ligands and the geometry approaches octahedral (Figs 12.9 and 12.10 and Table 12.7). The disposition of the ligands and probably the bond lengths are such that these sites do not favour either zinc or copper relative to other protein (N/S) centres of lower co-ordination number which remove these other metal ions by binding them more strongly (see Chapter 2). The iron sites do not form very stable complexes, especially with the reduced metal state, so that exchange takes place, certainly in prokaryotes

$$Fe(II) \cdot E \rightleftharpoons Fe(II) + E.$$

This allows a control over reaction systems which is dependent on the free Fe(II) concentration. We shall describe the way in which this concentration connects to iron-controlled reactions generally in Chapter 21. The secondary structures of some of the proteins in this group, e.g. superoxide dismutase, are not built upon simple helices, as is haemerythrin, but include considerable amounts of β-sheet as is common amongst enzymes generally.

Table 12.7 Co-ordination in non-haem, non-Fe/S proteins

Lactoferrin (transferrin)	Octahedral	2Tyr, His, Asp, CO_3^{2-} (2)
Superoxide dismutase	'Octahedral'	3His, Asp, $(H_2O)_2$
Haemerythrin	Octahedral (I)	3 His, $2CO_2^-$, O^{2-}
	(II)	2His, $2CO_2^-$, O^{2-}, H_2O
Ribonucleotide reductase	Octahedral	His, 2/3Glu, O^{2-}, H_2O
Photoreaction centre Fe(II)	Octahedral	4His, $2H_2O$?

NB. The iron centres of enzymes such as penicillin cyclase, prolyl oxidase, and prostaglandin synthetase are likely to be similar to the above.

12.14 Iron as an acid catalyst

The requirements for catalysts in relatively acidic media (pH $<$ 5) which prevent or reduce the stability of the binding of divalent ions to proteins may well be the cause of the development of a few M(III)-dependent hydrolytic enzymes in the acidic (pH $=$ 4 to 5) zones of some lysosomes (Fig. 12.4). The major proteins are either Fe(III)- or Mn(III)-dependent. As in a number of other enzymes which require Fe(III) it would appear that Mn(III) is an acceptable substitute. Some of the enzymes have a pair of metal ions. The advantages of a site with more than one metal ion in acid–base catalysis have been discussed in Section 4.3, but there the metal ions were quite far apart in the enzyme. Perhaps the use of the M.O.M. cluster is just to stabilize the M(III) state.

12.15 Summary of non-haem iron functions

We see in this chapter that non-haem iron is often involved in redox chemistry and that much of this chemistry involves free radicals. The free radical may be just an electron-transfer intermediate, but it could be part of the ground state of an enzyme or it might be involved in the reactions of substrates. (Free radical reactions will be apparent again in cobalt and copper biological chemistry.) Some of these activities are listed in Table 12.7 where we include the mononuclear iron site of the photoreaction centre. (This site is of uncertain function, but its structure is known to be like the others in that iron is in a rough octahedron co-ordination sphere including four histidines. The site in lipoxygenases, Table 12.6, may be similar). Undoubtedly a critical feature of all this iron chemistry is the compartment it is in and we shall keep this in mind, especially in Chapter 21. The locality of Fe/S proteins in association with low redox potential chemistry (H_2, NADH, H_2S, aconitase) differs clearly from that of high redox potential chemistry (O_2) where the Fe is bound by imidazole and carboxylate. Haem chemistry will highlight this point again (see Chapter 13). A further essential point is the *control* over iron uptake and use.

12.16 Iron regulation and homeostasis

Because the role of iron is so central to redox metabolism it is probable that there is no life without it today. In some theories of early evolution (see Section 12.8) iron takes part in the catalysis of extremely primitive reactions which were in existence before they were incorporated into living systems. At first the iron/sulphur/hydrogen reactions may have 'buffered' life's chemistry in tune with the environment. Be this as it may, today iron is required absolutely. In its absence oxidative systems must be shut off so that essential polymers and monomers are preserved while scavenger molecules are sent out to find this metal. The implication is clear. Iron concentration must be connected to DNA-regulated synthesis of its own scavenging chelating agents and to a considerable number of other pathways. Starting from this premise the iron (Fe) uptake (U) regulatory (R) protein (FUR) was looked for and found by J. B. Nielands (see Lorenzo *et al.* 1987 in 'Further reading'). It binds both iron and DNA and it is presumed that the way in which it alters its own conformation and that of DNA when iron is bound is such as to prevent the synthesis of enzymes for the production of scavenger chelates (siderophores and ferroxamins).

Looking at the synthesis of intermediates either for scavenging for iron or for the synthesis of porphyrin (Fig. 12.12), we see that they are derived from the citric acid cycle. In this cycle there are key enzymes which can be controlled by the levels of free iron since the iron dissociates from them with some ease. An example in the citric acid cycle is aconitase, and in the pathway from the product of glycolysis, pyruvate, to the citric acid cycle, it is pyruvate decarboxylase. Both enzymes use Fe_4S_4 clusters. Of course, there is also indirect control of haem synthesis since the insertion of iron in porphyrins uses a chelatase enzyme.

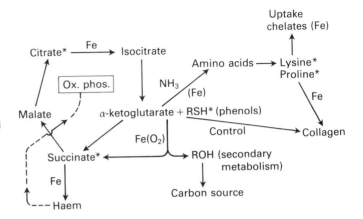

Fig. 12.12 The link between the citrate cycle and iron controls. It is suggested that iron levels are essential for switching between the different metabolic paths required for uptake and energy generation and for balancing synthesis and degradation.

Apart from control over basic metabolism common to prokaryotes and eukaryotes, much of which is locked into organelles in the eukaryotes, there are a variety of secondary metabolic pathways which depend on hydroxylation. In fact, most secondary metabolites have an oxidative step requiring an iron enzyme. Any list of such products would include a wide range of hormones including sterols, thyroxine, and adrenalin. Lack of iron is critical to the levels of all these compounds not only at the cellular but at the multicellular level. Distribution, as well as uptake and incorporation, is essential.

Finally, iron enzymes are linked not only to the metabolism inside cells but also to extracellular products such as hydroxyproline for connective tissue. Connective tissue is based on *extracellular* collagen but the requirement for 2-oxoglutarate as well as iron in the synthesis of hydroxyproline, a part of collagen, once again links control back to the intracellular citric acid cycle.

Given the complexities of control and its link with several elements as well as iron we shall leave further discussion until Chapter 21.

12.17 Iron buffering

When describing the biological chemistry of sodium, potassium, calcium, and zinc we observed different ways of keeping ion concentrations constant. The first three ions are managed, at least in part, by pumping mechanisms with feedback controls which operate against leaks. In the case of calcium and zinc there are, in addition, binding proteins which bind more than one metal ion and so form buffering agents sensitive to $[M]^n$ where n may be greater than three. Calcium is also buffered by precipitates such as bone and shell. Iron buffering is managed by a small precipitated phase in a protein capsule, ferritin (Fig. 12.13), which is a four-helical-bundle protein (12 copies) sometimes containing haem *in vivo* and which very probably is an allosteric protein. The protein is not the same in bacteria and eukaryotes and neither is the precipitate. We turn to the complexities of such forms of storage in Chapter 20 on biominerals and in Chapter 21 on the homeostasis of cells. As stated in Section 12.8 in evolution Fe/S clusters probably controlled (buffered) cells on the early sulphide-rich earth.

Fig. 12.13 (a) Schematic representation of a ferritin subunit. Four long helices, A, B, C, and D, form a bundle with the shorter fifth helix, E, lying at an acute angle to the bundle. There is only one short region of intermolecular anti-parallel β-sheet in the loop L. Some of these proteins, perhaps all, bind haem (Chapter 13). The way they pack together is shown in (b). (After P. Harrison.)

(a)

(b)

Note. The discovery of Fe/S centres

As a side-issue it is interesting to review briefly how the Fe/S centres were discovered, see Gibson *et al.* in Further reading. The spectroscopy of the Fe/S proteins, especially their EPR and Mössbauer spectra, generated a very interesting discussion in the 1960–70 period. This book is not concerned with such analyses but here they provide an interesting insight into the difficulty faced by chemists in the examination of biological molecules. When the EPR signals were first observed physical scientists refused to believe that these proteins contained iron. It was thought and firmly stated that no metal centre with five or more d electrons could have an EPR signal with $g < 2.0$. There are very good theoretical arguments for this if the iron is in an isolated site and no iron models had shown such signals. Proof of the origin of the signal as being due to a coupling of spins on *two* iron atoms was provided when two groups of scientists were brought together by one of us (R. J. P. W.) to analyse the EPR spectra. When a satisfactory theory of the spectra based on spin-coupling between two or more iron atoms had been established the biologists believed that the condition of the iron atoms was unusual. Again dispute arose in which chemists who could synthesize related small-molecule clusters argued that there is nothing particularly unusual about them. As the number of Fe/S clusters increased, however (see Table 12.4), it became clear that there was a wide variety of them with different EPR signals and very different redox potentials. Slowly it has become obvious that clusters of this kind are held in

particular energy (spectroscopic) states by the fold of the protein to which they are attached. The general position is that which is maintained throughout this book: each metal site in a protein can only be understood thoroughly with reference to the total protein which holds it and gives it its function. Parallels, but not copies, can be studied in small molecule models.

Further reading

Baker, E. N. *et al.* (1987). Structure of lactoferrin. *Proceedings of the National Academy of Sciences of the United States of America*, **84**, 1769–73.

Beinert, H. and Kennedy, M. C. (1989). Iron–sulphur clusters. *European Journal of Biochemistry*, **186**, 5–15.

Gibson, J. F., Hall, D. O., Thornley, J. H. M., and Whatley, F. R. (1966). The nature of the iron–sulphur clusters in proteins. *Proceedings of the National Academy of Sciences of the United States of America*, **56**, 987–1003.

Green, J., Trageser, M., Six, S., Unden, G., and Guest, J. R. (1991). Iron-binding transcriptional regulator, FNR. *Proceedings of the Royal Society of Biological Science*, **244**, 137–44.

Harrison, P. M., Andrews, S. C., Ford, G. C., Smith, J. M. A., Treffry, A., and White, J. L. (1987). Ferritin: an iron sequestering molecule from microbes to man. In *Iron transport in microbes, plants and animals* (ed. G. Winkelman, D. Van der Helm, and J. B. Nielands), pp. 445–75. VCH Verlagsgesellschaft, Weinheim.

Holm, R. (1977). Model complexes and iron/sulphur proteins. In *Biological aspects of inorganic chemistry* (ed. A. W. Addison, W. R. Cullen, D. Dolphin, and B. R. James), pp. 71–112. Wiley Interscience, New York.

Jacobs, A. and Worwood, M. (ed.) (1980). *Iron in biochemistry and medicine.* Academic Press, London.

Knoff, D. B. (1988). Photosystem I. *Trends in the Biochemical Sciences*, **13**, 460–1.

Lorenzo, V. de, Wee, S., Herrero, M., and Nielands, J. B. (1987). Operator sequences of the aerobactin operon of plasmid Col V-K30 binding the ferric uptake regulation (fur) repressor. *J. Bacteriol.* **169**, 2624–30.

Mann, J. (1987). *Secondary metabolism.* Oxford University Press, Oxford.

Moura, I., Tavares, P., Moura, J. J. G., Ravi, N., Huynh, B. H., Liu, M.-Y., and Le Gall, J. (1990). Unusual non-haem iron centres. *Journal of Biological Chemistry*, **265**, 21596–602.

Norland, B., Sjöberg, V. B. M., and Eklund, H. (1990). The structure of the iron-containing subunit of ribonucleotide reductase. *Nature*, **345**, 593–6.

Ponka, P., Schulman, H. M., and Woodworth, R. C. (ed.) (1990). *Iron transport and storage.* CRC Press, Boca Raton, Florida.

Rouault, T. A., Stout, C. D., Kaptain, S., Hartford, J. B., and Klausner, R. D. (1991). Iron-regulated RNA-binding protein and aconitase. *Cell*, **64**, 881–3.

San Pietro, A. (ed.) (1965). *Non-haem iron proteins.* Antioch Press, Yellow Springs, Ohio. This book gives a wonderful insight into the beginnings of this field.

Wachterschäuser, D. (1988). Origin of life and iron sulphides. *Microbiology Reviews*, **52**, 452–86.

Williams, R. J. P. (1982). Iron and manganese in controls. *Federation of European Biochemical Societies Letters*, **140**, 3–10.

Williams, R. J. P. (1989). Electron transfer in biology. *Biochemistry International*, **18**, 475–99.

13

Haem iron: coupled redox reactions

13.1	Iron in porphyrins	343
13.2	Properties of isolated haem units	344
13.3	Classification of haem proteins by iron properties	347
13.4	Classification of haem proteins by secondary structure	348
13.5	Where are haem proteins in cells?	352
13.6	Haem protein functions I: electron transfer	353
13.7	The surfaces of haem proteins	357
13.8	Haem-protein functions II: storage and tranport	358
13.9	Haem-protein functions III: oxidases and dioxygenases	361
13.10	Substrates of haem enzymes and secondary metabolism	365
13.11	Haem in controls	366
13.12	The synthesis of haem	367
13.13	Summary of haem-iron functions	367
13.14	Control of cell activity and iron	368

13.1 Iron in porphyrins

Our knowledge of the biochemistry of iron porphyrins is probably more extensive than that of any other metal-factor unit: literally hundreds of proteins from different sources have been examined. The functional value of this cofactor is immediately seen by the diversity of its uses within biology (Table 13.1) and by the amounts of its proteins in a wide variety of cell types. The availability of the proteins has led to very detailed studies of them by chemists and biologists, but this book is not concerned with the detailed physical chemistry of the haem or its proteins. We shall concentrate here on appreciation of the functional value of haem proteins and perhaps the best place to start is with an examination of some features of the isolated haem units.

Table 13.1 Modes of incorporation of iron–protoporphyrin in different proteins to generate different functions in different cells, organelles, and membranes

Metalloprotein*	Iron co-ordination number	Electronic states†	Oxidation states
O₂ carrier			
Haemoglobin	5	HS/LS (O₂)	II
Myoglobin	5	HS/LS (O₂)	II
NO binding			
G protein	5 ?	HS/LS (NO)	II
CO binding			
Cytochrome c	5	HS/LS (CO)	II–III
O₂ activator			
Oxidases	5 ?	HS/LS (O₂)	II–VI
Cytochrome a₃	5 ?	HS/LS (O₂)	II–VI
P-450	5/6	HS/LS (O₂)	II–VI
Peroxidases (O₂)	5	HS	II
H₂O₂ activator			
Peroxidases (H₂O₂)	5	HS	III–V
Catalases (H₂O₂)	5	HS	III–V
Electron transfer			
Cytochromes a	6	LS	II–III
Cytochromes b	6	LS	II–III
Cytochromes c	6	LS	II–III

* All these proteins are helical to a large degree.
† HS, high-spin; LS, low-spin.

13.2 Properties of isolated haem units

To a first approximation the porphyrin ring is a di-anionic chelating agent which provides a relatively rigid hole for a cation, close to but smaller than the diameter of high-spin Fe^{2+} or Fe^{3+}, with four pyrrole nitrogens as its ligands (Fig. 13.1(a), (b)). As stated in Chapter 2, the fact that iron and not some other metal is almost invariably associated with the porphyrin rings in biology is not because of the thermodynamic bias of metal iron equilibrium constants: starting from high-spin aqueous ions the binding constants to porphyrin must follow the Irving–Williams series of Fig. 2.8. It is known that of all the metals in this series only Fe^{2+} is incorporated *in vivo* but all the other elements have been put into the ring by chemical syntheses. There are closely related rings in biology which do not contain iron but do contain magnesium, cobalt, or nickel, Fig. 2.12. Iron is associated with porphyrin by a biologically evolved kinetic pathway—an energized synthesis—just as cobalt is associated with corrin, magnesium with chlorin, and nickel with F-430, to the exclusion of the other metal ions in each case. The synthesis of all these rings has a common route via the uroporphyrins and we must look later at the precursors in the synthetic pathway seeking for any feedback controls of interest. The synthesis of iron porphyrin is, of course, partly controlled by free iron levels at the

(a)

(b)

Fig. 13.1 (a) The axial ligands, methionine and histidine, to the iron porphyrin in cytochrome *c*. (b) The formulae of protoporphyrin bound here through thioether links as in cytochrome *c*.

insertion step in mitochondria. The very fact that evolution has produced this compound in such quantity speaks to its functional value as a chemical.

The iron porphyrin complex has some immediately interesting physical and chemical properties where it differs from simple iron.

1. It is limited in water solubility and, especially after screening of the iron by two extra axial ligands (Fig. 13.1(a)), readily enters hydrophobic solvents.

2. It is easily oxidized by molecular dioxygen to higher iron oxidation states; then there follows a slow attack on the ring structure. There is, in fact, an enzyme, haem dioxygenase, which assists degradation of free haem to eliminate it from those parts of an organism where it would be a dangerous oxidation catalyst.

3. The organic substituents around the outside of the ring are not fixed (contrast vitamin B_{12}) and a variety of porphyrins are chosen for different functions.

4. The high-spin iron in the ring, whether in the Fe(II) or the Fe(III) state, is quite close in energy to the low-spin state, and indeed to intermediate spin states (Fig. 13.2), and this spin-state equilibrium balance is one of the delicate controls which the porphyrin ring together with the axial ligands to the iron and the proteins hold over the metal. The low-spin state of iron has special chemical properties.

5. While the ligand, porphyrin, is obviously a good four-centred σ-donor, it is a negatively charged $>N^-$ donor from two centres, it is also an effective π-acceptor. The energy level diagram of the iron–porphyrin unit makes it clear that the metal and the ligand are a strongly co-operative unit. In the low-spin state Fe(II) is made into a good π-donor, i.e. a π-base, which strongly assists binding of O_2, CO, and NO.

6. Porphyrin compounds, like chlorins and phaeophytins (Chapter 9), can be oxidized to useful free radicals. The metal and the ligand oxidation states can be treated separately to some degree depending, of course, on the symmetry of the orbitals in which the unpaired electron(s) sit. NB High-spin iron is always a weak (multi) free radical.

7. Some of the ring substituents are reactive; for example, the vinyl groups can suffer addition to give thioethers, e.g. cytochromes *c* (Fig. 13.1(b)), and various oxidations or reductions are possible to give haems *a* and rare iron chlorins.

8. The hydrophobic nature of the porphyrin surface makes for strong interaction with hydrophobic small molecules as well as with hydrophobic protein surfaces.

9. Iron does not dissociate from haem to any degree so that haem iron is almost a new element distinct from iron.

10. The redox potentials of the iron porphyrins are readily adjusted by the proteins (see Fig. 13.3).

11. Haem can act as a cell regulator quite separately from free iron.

Given the complexity behind these properties it pays us to simplify the description of haem binding in proteins.

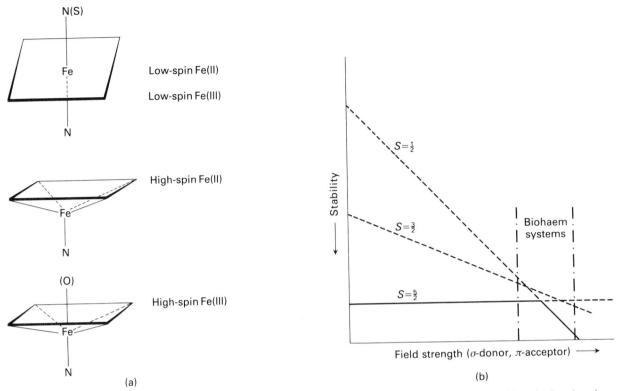

Fig. 13.2 (a) A generalized picture of the variety of high- and low-spin iron porphyrin structures. Note that bond lengths, bond angles, and the conformation of porphyrin are involved. (b) The manner in which the stabilities of the spin states (S) of Fe(III) haem varies with the field strength of the fifth and sixth ligands. A similar diagram can be drawn up for the Fe(II) ion.

Fig. 13.3 The redox ranges of haem proteins and a comparison with the redox potentials of other metal centres. The haem group in all cases is in a cytochrome. Note Fe(IV). Rb, rubredoxin; HIPIP, high potential iron protein.

13.3 Classification of haem proteins by iron properties

Table 13.1 classifies the haem proteins by function. (A different approach is to classify them by the chemical bonding.) In the table, in addition to functions, we give spin and oxidation states for the metallo proteins and also classify them as containing five- or six-co-ordinate iron. The next valuable descriptive feature is the ligation from the protein sidechains. If there are two axial ligands making a rough octahedron around the iron the two donors are almost invariably one imidazole (histidine) plus either a second imidazole or a thioether of methionine but cases of lysine, carboxylate, or phenolate binding are known. In the methionine complex the haem is usually covalently bound through thioether links via its vinyl groups to the protein. This construction is a feature of cytochromes *c* (Figs 13.1 and 13.4), but there are variations of this structural pattern with but one thioether link, cytochromes *c'*. Such is the complexity of this cytochrome *c* group of proteins, most if not all of which are extracellular, that there are books which deal with them alone (see 'Further reading'). The next group of proteins, with two ligand histidines, can have the haem bound covalently as above or, more usually, not bound to the protein. The two series of proteins correspond roughly to a further group of cytochromes *c*, that is, those covalently linked into the protein, and the cytochromes *b*. Again there are occasional variations from these general statements since it is known that a further few cytochromes with two axially co-ordinated protein sidechains are bound by a lysine $-NH_2$ group and a

Fig. 13.4 An outline, somewhat impressionistic, picture of cytochrome *c*. The haem and the ligands are in the centre and the haem is covalently linked to the sequence. Note the many helices. The arrow shows a hypothetical protein-coupled track for protons due to known H-bond breaks at Gly-41, Thr 78, Met 80, Tyr 67, Lys 78, and a re-adjustment within helices 60–69 and 70–77 with redox state changes (see Fig. 13.8).

histidine, cytochrome f (see also ferritins in Fig. 12.13). Finally, within this pattern are cytochromes which are based on oxidized ring sidechains, cytochromes a, and which have two histidines co-ordinated. Almost all the cytochromes a, b, and c are in low-spin states. We turn to their protein structures later, but the haem in them is usually well buried in a hydrophobic slot formed by several strands, often of helices (Fig. 13.4). To a first approximation they are electron transfer proteins and the noticeable feature of them is that they cover a redox potential range Fe(II)/Fe(III), from -0.3 to $+0.4$ V (Fig. 13.3), providing a redox range reaching to much higher potentials than the Fe/S proteins. It may well be that the development of the higher potential proteins followed the coming of dioxygen during evolution.

The second group of haem proteins are five-co-ordinate in the Fe(II) state and may or may not have one water molecule in the sixth position when in the Fe(III) state (Table 13.1). The iron is largely high-spin in both conditions and is usually bound by histidine as its fifth ligand but there are a few cases of thiolate, phenolate, or even carboxylate binding. Usually the proteins do not bind covalently to the haem, but the haem lies in a hydrophobic pocket. There are variations in the haem substituents in, for example, spiro-haem of sulphite oxidase and haem a_3 of cytochrome oxidase. The outstanding feature of these proteins is that they are 'designed' to transport or bind dioxygen (sometimes NO also) or to activate such molecules as hydrogen peroxide, superoxide, or dioxygen rather than to be electron transfer proteins. Their names give their functions, for example, dioxygenases and dioxygen transfer proteins use the sixth co-ordination position as a binding centre for dioxygen. In the course of many of their reactions as oxidases the iron cycles through oxidation states (II), (III), and (IV) and the porphyrins often pass through a radical state. The iron also cycles through changes of spin state and, with these changes and changes of oxidation state, the porphyrin ring alters its ruffling. Then, unlike most of the first group of haem proteins, the redox and binding reactions are obviously coupled to changes in the haem and its ligands, and to some degree can be linked to co-operative changes in the protein. The degree of coupling is an essential part of different functions. Coupling is frequently associated with allosteric effects, e.g. in the uptake of dioxygen by haemoglobin. We shall need to examine the structures of the associated proteins in order to appreciate why and where coupling between haem and protein occurs since this is one of the most fascinating features of haem biochemistry.

13.4 Classification of haem proteins by secondary structure

We have now divided the haem proteins generally by function and by structural co-ordination chemistry, but it is the protein that gives each of them its specific functional character. Of course, this applies to the second substrate they bind, i.e. in addition to O_2 or H_2O_2, but first and importantly we need to look at the protein fold. It appears that, quite unlike Fe_n/S_n proteins where the active site is usually multiply cross-linked to thiolate sidechains of the protein and the strands of the protein form β-sheets, i.e. cross-linked H-bonded structures, the essence of haem proteins is a loose hold on the iron through

perhaps only one co-ordination group, a relatively loose hold even on the porphyrins, and a secondary structure around the haem of helices or loose strands. Helices, unless they are cross-linked by –S–S– bonds or joined to β-sheets, are intrinsically mobile relative to one another (see Chapter 7). It would therefore seem that helices plus haem form an excellent structural motif for the development of many *mechanical* functions. Our next step in this analysis is to see in which haem proteins the helices are cross-linked and in which they are not, so that we can know where evolution has clamped the haem to prevent co-operative motion of haem and protein during reactions. It is the development of co-operative interaction between iron and many proteins through the use of the porphyrin frame that has led us to use *coupled reactions* in the title for this chapter.

Table 13.2 classifies the secondary structures of haem proteins. The first group includes the *c*-type cytochromes which, as already stated, are structures which are covalently linked to the protein. The proteins themselves do not fold without the haem but, once the fold is completed and the thioether cross-links have formed, the cytochromes *c* are the most stable and have the least internally mobile sidechains of all proteins studied in depth (Fig. 7.7). They are invariably simple electron transfer proteins in which electron gain or loss is accompanied by little coupling to changes in protein conformation (but see Section 13.6.3). The haem cannot exchange from these proteins and very generally they are placed outside cells where dissociation would be a problem.

The other two groups are also based, at least in part, on helices around the haem. There is one group in which the haem is wrapped around by helices, usually four, but where these helices are linked to a β-sheet (Fig. 13.5). These proteins, cytochromes b_5 for example, are simple uncoupled electron transfer proteins but operating in a somewhat different way, in a lower redox potential range, and in a different cell compartment from cytochromes *c*. The structure of the third group is similar around the haem, but there is no β-sheet section. These proteins may also take part in electron transfer and in binding of and in the reaction with small molecules such as dioxygen but their reaction, even electron uptake or release, is connected to a protein conformation change which may be linked to energy capture devices or to the relay of information. Some of these proteins are found in the energy-capture membranes (Fig. 13.6). The protein of cytochrome a_3 is of this kind too but the haem is different and it has a high redox potential.

The group also includes the dioxygen carrier proteins; here the helical structure (Fig. 7.3) is more complicated than in the cytochromes *b* and the haem is open-sided. The helical structure leaves a narrow groove to the iron.

Table 13.2 Secondary structures of haem proteins

Secondary structures	Examples
Helical but cross-linked by sulphur links	Cytochrome *c,c'*, peroxidase
Helical plus β-sheet	Cytochrome *c* peroxidase, cytochrome b_5, catalase
Helical	Membrane cytochromes *b, a, a_3, o*, cytochrome P-450 (largely), haemoglobin, myoglobin

NB. Note the difference from the Fe/S proteins where the Fe/S is locked in a β-sheet. Haem is usually surrounded by a more mobile framework of α-helices or loose strands.

Fig. 13.5 An outline structure of cytochrome b_5. There is a β-sheet holding the helices which in turn hold the haem in a relatively rigid protein with two histidine, H, ligands. (After F. S. Matthews.)

Fig. 13.6 The critical section of the cytochromes b of the energy-capture membranes. It is considered that the haem units are held as shown by histidines protected by helices of which two are shown. The numbers are of the sequence. Note the buried propionate carboxylates (P) and compare cytochrome c while contrasting cytochrome b_5.

The co-operative helix/helix movements linked above to electron transfer are also similarly linked here, but more importantly they are linked to dioxygen uptake and release by several haem groups in a multihaem protein, leading to the allosteric equilibria of haemoglobins. Notice that this linkage is much reduced in the dioxygen storage protein, myoglobin.

Some members of the first and second groups of proteins are also based on five-co-ordinate haem, but now either the helices are cross-linked by –S–S– bridges, as in peroxidases and oxidases, some of which use superoxide, or they are attached to a β-sheet, as in cytochrome c peroxidase. These proteins have second substrates as well as dioxygen or hydroperoxide, do not have mobile folds, are usually found outside cells or in vesicles, and are freely diffusing molecules but uncoupled systems. The distinction between coupled and uncoupled proteins is strongly stressed in Table 13.1 and 13.2.

A further point of interest here is the degree to which the haem is buried. In electron transfer proteins the haem lies just buried so that the iron is about 10 Å from the surface. In proteins which pick up dioxygen or hydrogen peroxide and where a small organic substrate has to be oxidized, e.g. peroxidases, the haem is perhaps somewhat deeper in the pocket and in these the substrate usually sits in a groove close to the entrance to the haem. In catalases, which only destroy H_2O_2 and must not undergo electron transfer reactions, the haem is buried some 20 Å into the interior. In all proteins the haem is displaced away from the centre of the protein. Once again, we see how the protein conditions the metal ion both directly in its co-ordination geometry and indirectly at greater distances so that functional value is optimal. When the haem is buried, either its propionate sidechains are buried in the protein (cytochrome c and membrane cytochromes) or they are exposed to water (cytochrome b_5 and haemoglobin). The significance of these arrangements will become clear later.

Finally, there is a group of haem proteins in which the helical structure dominates but where the haem co-ordination to the protein fluctuates in the reaction cycle. In one subsection of haem proteins, the iron ligands are a thiolate from a cysteine sidechain and one water. These are the cytochrome P-450 enzymes (see Chapter 7). They are known to be largely based on helices but to have some short β-strands and a small β-sheet not too far from the haem. The second substrates are bound here and are very inert molecules, often simple hydrocarbons. There is a compulsory order of addition of substrates, and reaction pathway, which demands a co-operative series of protein rearrangements (Section 7.5). In both these latter respects the enzymes are very different from simple peroxidases and oxidases and resemble cytochrome a_3, although they differ from cytochrome a_3 in their location in a cell. Cytochrome a_3, which is in a membrane, probably uses a similar mechanical action of helices in order to act as a proton pump. The haem protein which binds NO and modulates guanidine cyclase enzymes is likely to be of this kind too.

Whereas we have seen that the haem does not dissociate from the cross-linked proteins and enzymes, such as cytochrome c and peroxidases, dissociation from several of the other proteins, e.g. cytochromes b_5, is much easier. As a consequence it is possible for free haem to be a control molecule acting much like free iron, Section 13.11.

13.5 Where are haem proteins in cells?

The low-levels of haem proteins in some cells and even organs is reflected in the fact that they are almost white (contrast red, haemoglobin-containing cells with white blood cells) but the distribution is often due to the different content of compartments in cells. The conclusion is obvious that cells with a high oxidative capacity, which have lots of mitochondria, or cells which carry oxygen will appear darkly coloured. In plant tissue a parallel contrast exists between green, chlorophyll-containing tissues, and white tissues, e.g. between leaves and roots. Since different haem proteins are widely distributed in compartments it is easier to start this description of the topographical distribution of haem proteins with a statement of the compartments in which there is no red colour. Haem proteins are very rare outside cells of higher organisms, i.e. in circulating fluids. They are also virtually absent from the nucleus, from lysosomes, from the Golgi apparatus, and from the interior of reticula. The compartments where they are present in greatest amounts are the membranes of aerobic prokaryotes and their periplasmic space, the membranes of organelles, and the aqueous phase of peroxisomes (Table 13.3). In

Table 13.3 Haem protein organization in eukaryotes

Haem protein	Compartment
Haemoglobin	Specialized red blood cells
Myoglobin	Cytoplasm of red muscles
Peroxidases	Peroxisomes
Catalases*	Peroxisomes
Cytochromes (c)†	Mitochondrial membrane or periplasm
Cytochrome P-450*	Microsomal membrane
Cytochrome b_5	Microsomal membrane
Cytochromes a, b	Mitochondrial or thylakoid membranes

* Haem enzyme that, in prokaryotes, is free in the cytoplasm.
† Haem enzyme that, in prokaryotes, is on the outside of membranes.

prokaryotes and in mitochondrial and chloroplast membranes they form the basic electron transfer functions at high redox potential (cytochromes a, c_1, f, and some cytochromes b; Table 13.1). This function extends to the periplasmic space of prokaryotes and organelles where there are several cytochrome c proteins. There are also some cytochrome b electron transfer proteins anchored to the cytoplasmic membrane. In strict contrast, the haem catalases and peroxidases in eukaryotic cells are in peroxisomes while in prokaryotes they are again in the periplasmic space. The haem oxidases may be associated with a variety of membranes in eukaryotes and prokaryotes. For example, the P-450 cytochromes belong to the eukaryote reticula membrane, but these same enzymes are in the cytoplasm of prokaryotes. The low level of haem proteins in the cytoplasm of eukaryotes is most striking and yet the cell with the most haem in it, the red blood cell, has all its haem protein, haemoglobin, in the cytoplasm. Now this is a most peculiar cell since in many higher species it has lost its nucleus. It is almost as if haem protein reactions and the nucleus

were best kept separate. Perhaps we should understand this as a further removal of eukaryotic DNA from any danger due to side-reactions of an oxidative kind. Haemoglobin is, in principle, only a dioxygen carrier, but there is constant risk of some superoxide production together with Fe(III) haemoglobin. The blue appearance of blood under some conditions is due to this oxidation and the haemoglobin is protected from oxidative reactions in many red blood cells by the presence of a strong soluble reductase using cytochrome b_5 as an electron carrier.

Finally, we should not forget the possible restriction of location based on synthesis. The cytochrome b membrane protein and one subunit of cytochrome oxidase are synthesized via mitochondrial DNA. We can only ask if it is this localization of synthesis which keeps these two proteins within the mitochondrial membrane? It will be interesting to see which proteins are synthesized by the chloroplast DNA. We have also noted that during the course of some haem-protein syntheses the protein itself or the haem and the protein are cross-linked. These proteins are external to cell or organelle cytoplasm or membranes.

13.6 Haem protein functions I: electron transfer

13.6.1 Simple electron-transfer proteins

We now turn to the more detailed examination of the functions of haem enzymes. The most obvious function consists of electron transfer reactions often at redox potentials higher than 0.0 V (Fig. 13.3 and Table 13.4). The least complicated examples are the cytochromes c of the periplasmic space of bacteria and organelles, the cytochromes c_1 and f on the outer surface of the same membranes, and perhaps the cytochrome a of cytochrome oxidase. In each case electron transfer is not complicated to a marked functional degree by coupling of this reaction to other changes. The structures of these electron transfer proteins are governed by the features listed in Tables 13.1, 13.2, and 13.4, which apply equally to the simple electron transfer proteins based on Fe/S and Cu (see Chapters 12 and 15 and the discussion of the theory of electron transfer in Chapter 4).

Table 13.4 Effects of proteins on haem electron-transfer sites

Control over structure of co-ordination sphere affects the relaxation energy (entatic states)
Control over structure, first- and second-co-ordination spheres, and solvent exposure, affects redox potential
Degree of burial affects the distance of electron transfer
Degree of burial protects from adventitious small-molecule reaction
Surface charge and surface structure affect the recognition between proteins
Internal mobility affects relaxation energy
Internal mobility affects long-range coupling possibilities whether this mobility refers to single proteins, pairs of proteins, or particles
Coupling can be to energized proton diffusion paths (pumps) in membranes

13.6.2 Multihaem electron-transfer proteins

There are a number of proteins which contain more than one haem (Fig. 13.7). We leave to one side here the membrane haem proteins cytochrome b and cytochrome aa_3, which provide a path for electron transfer across or into membranes, since these proteins are coupled and are part of energy-transducing devices, as described in Section 13.6.3. In some sulphur bacteria there are water-soluble multihaem cytochromes, c_3, c_7, and so on which contain from three to eight haem units. They are usually found in the periplasm of the bacteria. They are of unknown function but one hypothesis is that they are electron storage devices rather like condensers in man's electronic circuits. They then hold n electrons ready to be released on contact with a reactant. These c-type proteins have a low redox potential and can act between the very low potentials of the particle I and particle II of mitochondria, that is between NADH or even H_2 (-0.5 V) and quinones (0.0 V). They could scavenge for electrons much as cytochrome c scavenges between particles II and IV. Another possibility is that, through small conformational changes, they have a control function to which we turn next.

Fig. 13.7 The outline structure of a cytochrome c_3. (After R. Haser.)

13.6.3 Electron transfer/proton transfer coupling: models

A major consideration in bioenergetics is the coupling of electron and proton flow in the mitochondrial and chloroplast membranes. There are electron-transfer haem proteins involved, cytochromes aa_3 and cytochromes b, and it is commonly thought that the redox reactions and some binding reactions, e.g. of O_2, are also coupled to proton movements by the readjustments within their helical frameworks. In this way ideas on proton transfer in energy coupling are obviously linked to ion pumping by reversible ATPases (see Chapter 8). Unfortunately, the details of structures of membrane proteins are

extremely difficult to obtain and working models developed from smaller proteins are the most we can devise as yet. One such working model which we describe next follows from very detailed studies of cytochrome c (Fig. 13.4), which shows minor coupling of redox reaction to the protein matrix and H-bond rearrangements. We give this example in detail to illustrate the subtle nature of bio-energy coupling (see also bacteriorhodopsin, Section 5.8).

As explained above, cytochrome c has the following structural features and redox state dependent changes (Figs 13.4. and 13.8(a), (b)).

(a)

(b)

Fig. 13.8 (a) A crystallographic study showing the major changes in structure of cytochrome c on redox reaction which lie between residue 40 and 80. Note the internal water molecule and its movement (after T. Takano and R. E. Dickerson). The heavier lines are for the reduced state. The grey arrow is the line of H-bond breaks. (b) A diagram using crystallographic studies showing main region of changes seen by NMR (in solution) of the structure of cytochrome c on change of redox state. Taken with Fig. 13.8(a) the suggestion is that the electrostatic energy change from Fe^{2+} to Fe^{3+} causes a shift of the propionates and a series of hydrogen bond rearrangements (arrow) occurs in the region shown (see Fig. 13.4).

1. A haem iron bound to a methionine sulphur (Met-80) and a histidine (His-18) where the numbers refer to the position in the sequence.

2. Two long helices from residues 1–12 and 87–102. The first helix is rigidly fixed in space by groups immediately following it, cysteine 14 and 17 which form covalent thioether bridges to the vinyl group of the protoporphyrin haem. The second helix is locked into this first helix.

3. Following residue 18 there are a series of turns on one side of the haem running to residue 38 which is an arginine holding the buried propionic acid sidechain of the haem (Fig. 13.4). Effectively all these parts of the protein are relatively rigid, i.e. from residues 1 to 38 and from 87 to 102.

4. The sequence from 38 on contains a set of rather poorly defined helices from residues 50 to 55, 60 to 69, and 70 to 76 before it runs along the haem through Met-80 to the terminal helix, 87–102.

5. The haem propionates are buried in the protein and associated with an internal hydrogen-bond network which even includes water molecules (contrast haemoglobin and cytochrome b_5).

It is the set of loose helices and the hydrogen-bond network between residues 38 and 86 which is of interest here. On the electron-transfer redox reaction, Fe(II) to Fe(III), there is not just some alteration of the ligand, Met-80, and the strand from residues 79 to 85 but there is a rearrangement of the hydrogen bond network in which some three or four H-bonds are broken and some made (see Fig. 13.8 for details). At the same time the accessibility of this region of the molecule to protons from the solvent is considerably changed. Thus we observe on a small scale features which are demanded in a proton/redox pump: (1) the redox-coupled alteration of proton-binding accompanying helix readjustment; and (2) a change of access of protons to these regions of the protein which is equivalent to a gating and is essential for redox-controlled proton flow. (In Chapter 4 proton flow is described in terms of rotational diffusion requiring the making and breaking of H-bonds); (3) the presence of anionic proton binding groups deep in the membrane, here we note particularly the propionates, which take part in the H-bond network and which are responsive to the redox state change of the iron.

13.6.4 The membrane electron transfer proteins: protein coupling

A particularly intriguing feature in the possible connection between redox reactions and proton energization and gating (pumping) is the potential value of these *buried* haem propionate sidechains in membrane-bound proteins. The protonation and deprotonation of such carboxylic acid centres as well as their connection into H-bond networks are adjusted by the redox and binding states of the iron in several cytochromes (see above) much as the reactions of haemoglobin are coupled to protonation states elsewhere. It may well be that these or other buried carboxylate groups associated with redox reactions (e.g. with haem or quinones (Fig. 13.9) or of light-absorbing centres) or other energizing reactions, such as those of ATP-dependent pumps, are essential parts of the energization of protons in membranes and that, through changes in helical structures, movements of these protons are gated. Thus electrostatic energies of buried charges could become extremely valuable in coupling. The

Fig. 13.9 The electron-transfer events in the cytochrome *b* of particle III of mitochondria and thylakoids (Fig. 11.1), are coupled to the uptake and releases of protons by the quinone Q at its redox sites. The coupling probably uses the buried propionate sidechains of the haem and other acid anion centres of the proteins. QH_2, carries the protons over the longest distance.

importance of the haem unit and the association with helical structures appears to have a very deep significance. Figure 13.9 shows the structure of the *b* type membrane proteins which have buried propionates and quinone (Q) coenzymes which are often close to buried carboxylate protein side-chains, that change protonation state with redox reactions. These structures clearly have the potential to act as pumps or gates coupled to electron flow. We return to this topic after examining the uptake of small molecules by haem proteins.

13.7 The surfaces of haem proteins

We interrupt the discussion of haem-protein function with an important diversion about their surfaces. The haem proteins are of three different kinds with respect to organized multiprotein systems:

(1) isolated proteins, catalase and peroxidase;

(2) carrier proteins which bind to several sites—cytochromes *c*;

(3) proteins bound in complexes, e.g. cytochromes *a* and *b* of mitochondrial complexes III and IV. These proteins are at best only laterally mobile in their membranes.

It is useful to discuss these observations here since amongst electron-transfer proteins the difference between these three classes greatly affects functional value. The description of their surfaces, which generates site location and the degree of restriction on diffusion, has to be analysed in terms of the charged and hydrophobic groups on the surfaces and the matching of surfaces to one another (see Chapter 7).

The simplest position was discussed in Chapter 5 where we analysed electron transfer in the reaction centre of photosystems and noted that the organization was tightly maintained (see also Figs 13.9 and 13.15). Now at the other extreme there are proteins such as cytochrome *c* which need to diffuse between many partners. They scavenge the intermembrane spaces for electrons and must be able to recognize several target proteins, and to bind to them and to dissociate from them rapidly. As explained in Chapter 7, the best

Fig. 13.10 The binding regions of cytochrome *c* due to the presence of positively charged sidechains are shown in black. There is a low level of selectivity of binding. Negatively charged sidechains are cross-hatched; hydrophobic groups are stippled.

device for the first group is a lock-and-key or a dye-and-mould fit and here hydrophobic interactions may well be very valuable. The best device for the second is a hand-in-glove matching using electrostatic surfaces and a more mobile underlying frame may also be of value. Perhaps these requirements explain the charged surface (Fig. 13.10) and the internal mobility of one side of cytochrome c near its binding surface. It is an adjustable protein. Finally, peroxidases have even more hydrophilic surfaces due to glycosylation so that they are totally protected from binding (see Chapter 7).

13.8 Haem-protein functions II: storage and transport

13.8.1 Dioxygen: storage, transport, and signalling

The study of haemoglobins and myoglobins has been as intensive as that of almost any other proteins. The following features are outstanding.

1. Dioxygen binding alters the haem iron spin states, the ruffling of the haem plane, and the iron–ligand bonds (Fig. 13.11(a)).

2. The changes in the iron porphyrin and its ligands on dioxygen binding are relayed to changes in each protein unit to different degrees. Internally there are sidechain and mainchain helix adjustments (Fig. 13.11(a), (b)).

3. In multi-subunit haemoglobins, see Fig. 7.3, the conformational changes are transmitted from one subunit to another (Fig. 13.11(c)) giving allosteric, co-operative, dioxygen-uptake curves.

4. The binding of protons and several small molecules adjusts the binding constants and the co-operativity through alterations of the helical structures.

5. Amino-acid changes are species or even family specific, of course, and give a wide range of affinity and co-operativity differences leading frequently to advantages in particular biological niches but also to diseased conditions, e.g. sickle cells.

The analysis of haemoglobins has thus led to major development in our knowledge, but for details the reader must go to specialist books on this protein (see 'Further reading').

Finally, there is a control link via a dioxygen-binding protein, probably another haem protein, to the synthesis of the globin proteins themselves. The body adjusts such synthesis even at different altitudes. Thus, dioxygen is regulatory through its binding to haem in proteins other than haemoglobin.

There are other haem proteins of unknown function which behave in a somewhat similar way on binding to ligands. In particular, the cytochromes c' of bacteria are allosteric, four-helical bundle, haem proteins which alter conformation on binding CO, NO, or CN^-. It is not known if they act so as to control metabolism in the presence of such poisons. There is the distinct possibility that control was the primary function of haem proteins in evolution before their use in electron transfer, which could always have been done at low potential by Fe/S proteins, or before their handling of reactions of dioxygen. The value of signalling the presence of poisons, e.g. NO, CO, or even O_2

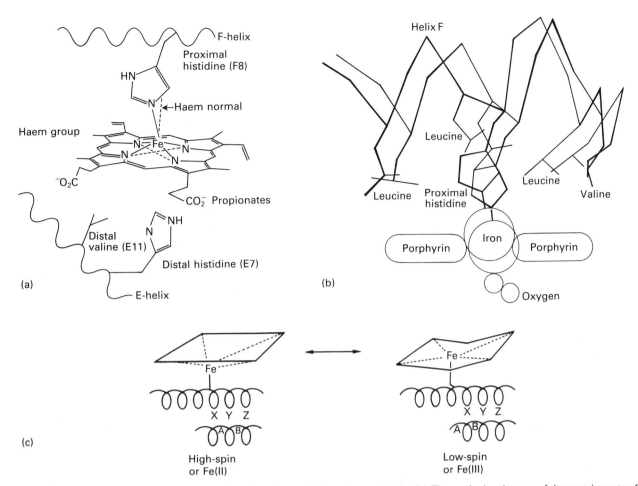

Fig. 13.11 (a) A detailed picture of the O_2-binding sites of myoglobin or haemoglobin. (b) The particular changes of the protein parts of haemoglobin on binding O_2. (After Pertuz, Baldwin, and Clothia.) (c) A sketch of the way in which helices slide (schematic) on change of spin state with O_2 binding in haemoglobin. (Note the relative positions of AB and XYZ.) This causes allosteric co-operativity.

cannot be overestimated and in evolution the neutralization of environmental poisons may have always been the first step before making use of them. An example is provided perhaps by nitric oxide.

13.8.2 Nitric oxide and haem receptors

Nitric oxide, NO, is an intermediate in the reduction of nitrate or nitrite to ammonia, and it has usually been considered as a dangerous poison. Surprisingly, NO is also found to be produced by oxidation of L-arginine *in vivo*. The NO acts in fact as a endogenous trigger of smooth muscle relaxation although it is produced in nearby endothelial cells. The generation of NO requires an oxidase which is probably a pterin-dependent enzyme similar to the enzymes for the hydroxylation of phenylalanine, tyrosine, and tryptophan in the cytoplasm of cells. The effect of NO on relaxation is uncertain but is thought to result through the stimulation of the GTP→cGMP reaction where cGMP is the muscle-relaxing agent. The enzyme for this cyclization may well

be controlled, here by haem, when the reaction,

NO + high-spin bound haem ⟶ low-spin bound haem · NO,

is an allosteric trigger reminiscent of haemoglobin and cytochrome *c'*, as explained above in Section 13.8.1 (Fig. 13.12).

Fig. 13.12 The *a*-helical fold of cytochrome *c'*, shown as a dimer. The iron is five-co-ordinate. (After P. Weber.)

13.8.3 Cytochrome *c* and cyanide

Once again we can use cytochrome *c* to probe the effects of ligand binding to haem proteins just as it was used to follow the effects of redox state change. Cyanide is a powerful poison for O_2-binding centres much as is CO. Cyanide

Fig. 13.13 The changes in cytochrome *c* (a) binding in the native state when (b) the ligand, 80, is replaced by the ligand, CN⁻. The main movement involved extends over very few amino acids from 78 to 85 but there are several H-bond rearrangements. This type of movement is typical of proteins in binding reactions where the main frame is not adjusted. However, here the protein is opened to proton motion in a quite new way (see Fig. 13.8) and such proton movements give the possibility of generating controlled proton wires in membranes.

binding to cytochrome *c* requires: (1) a groove-opening reaction; (2) a ligand displacement; and, therefore, (3) a protein rearrangement. The changes are similar to the changes with redox state of cytochrome *c* and in haemoglobin. In the cytochrome the reaction, which has quite a high activation energy, involves several hydrogen-bond breaks, as described for the redox state change, and minor adjustments in the whole sector of the protein from residues 40 to 85 in the sequence (see Fig. 13.13), but most atoms move little more than 1.0 Å except, of course, for the haem ligand at 80.

13.9 Haem-protein functions III: oxidases and dioxygenases

There are in fact relatively few *simple* haem oxidases or dioxygenases and many of these are of bacterial origin (Table 13.5). Where possible, oxidation of organic molecules is carried out by complex oxidases starting from hydrogen abstraction so as to keep the energy of the reaction with dioxygen controlled in the cytochrome chain of mitochondria. Of course, this is not a possible route for the oxidation of very resistant molecules such as simple aliphatics, e.g. alkyl chains, terpenes, and aromatics like benzene and naphthalene compounds. Here very aggressive attack is needed—see cytochrome P-450 in Section 13.9.3. There are also hydroxylated compounds from which energy could be abstracted during oxidative degradation, but, where the compound is overwhelmingly important, we find that both cytochrome P-450 and peroxidases are used to prepare them. Finally, oxidation is required before the export of compounds and is carried out in vesicles, or is required outside the cells to assist in food-supply or to neutralize poisons, hormones, etc. (see Fig. 13.14). The haem enzymes normally work only in vesicles.

Table 13.5 Haem monoxygenases and dioxygenases

Haem protein	Functions
Monoxygenases	
Cytochrome P-450	Hydroxylation of a large range of inert organic molecules
Dioxygenases	
Aromatic ring oxygenase	Bacterial ring cleavage reactions
Haem oxygenase	Removal of haem outside cells?
Indole amine dioxygenase (tryptophan pyrrolase)	Attack on viruses? Uses superoxide?

We tackle the different oxidases in the following order:

(1) cytochrome oxidase so as to link immediately to electron transfer proteins;

(2) cytochrome P-450;

(3) peroxidases and catalases.

Fig. 13.14 Reactions of dioxygen which are generated by haem catalysts, summarizing the utilization of haem enzymes in organic chemical reactions.

13.9.1 Cytochrome oxidase

This is a membrane protein with four metal ions which are thought to be essential, but it may well be that there is also extra copper, one zinc, and one magnesium which control the fold. The four active metal ions are two copper and two haem iron atoms. One copper, Cu_A, and one haem, haem a, are largely involved in electron transfer functions, while one copper, Cu_B, and one haem, haem a_3, carry the O_2 molecules through its four steps of reduction to $2H_2O$. While so doing they neutralize four protons and also pump four protons across the membrane in which the enzyme sits (Fig. 13.15(a)). The states of the metal ions have been the subject of extraordinarily intensive study, but, while we may start from a state $Cu_B(I) Fe_{a_3}^{2+}$, the interactions with dioxygen are so very intimate that they are hard to follow and there is variation depending on small factors, e.g. pH (see Fig. 13.15(b)). The function of the unit is to capture the energy between the redox potential of O_2 ($+0.8$ V) and that of cytochrome c ($+0.25$ V) as a proton gradient. This is probably achieved, in parallel with the reactions of cytochrome P-450 (see Section 7.4), by a cyclic series of conformational changes of helices here roughly arranged perpendicular to the membrane. This description, flimsy as it is, is sufficient for the purpose of this chapter since it gives the essence of the value of haem Fe with its flexibility inside proteins in contrast to the rather rigid Fe_4S_4 units. The obvious parallels between this description and that of the mechanical movements of dioxygen carrier proteins should not be missed. An hypothesis as to the helical motions that could be involved in proton pumping and gating has been given in Section 8.5 while describing ATP/proton pumps.

The proton movements in cytochrome oxidase are shown in outline in Fig.

4H⁺

Aqueous (out)

Cyt c

Cu_A

a

Fe

M
e
m
b
r
a
n
e

e

a₃

Fe O₂

Cu_B

2H₂O

Aqueous (in) 4H⁺ 4H⁺

(a)

$$Fe^{2+}:Cu^+$$

$$O_2$$

$$Fe^{.3+} \, O_2^{2-} \cdot Cu^{2+} \qquad Fe^{3+}:Cu^+$$

$$e, 2H^+ \rightarrow \qquad \qquad e$$

$$FeO(IV):Cu^{2+}(OH_2) \qquad Fe^{3+}:Cu^{2+}$$

$$e \qquad \qquad \rightarrow 2H_2O$$

$$Fe^{.3+}OH^-:Cu^{2+}OH^- \qquad 2H^+$$

(b)

Fig. 13.15 (a) The proposed outline structure of cytochrome oxidase (after M. Wikström). The enzyme links O_2 reduction to water with proton gradient generation. (b) The intermediate steps of the reaction of cytochrome oxidase.

13.15(b). It is here that the detailed knowledge of features of cytochrome c and its redox reactions becomes of value. In Fig. 13.8 we showed the way in which H-bond breaking was linked to the changes in the states of the iron. This can happen either on ligand binding (O_2, CN^-) or on electron transfer. It is likely that in the case of cytochrome oxidase all changes of condition of the iron will be coupled to protein movements and H-bond breaking. It is dangerous to break down proton movements too closely into association with particular reaction steps.

Before leaving the description of cytochrome oxidase it is worthwhile to draw attention to certain parallels in the descriptions of other haem enzymes and a wide variety of enzymes which use mechanical devices. First, in Chapter 9 we described the kinases in terms of hinge-bending using α-helical zones and movements of domains to limit reactions. We shall see a very similar description applies to the P-450 cytochromes. Second, we analysed the coupling of the ATP hydrolysis to pumping of ions including protons, or the reverse in ATP-synthases. The mechanism was one where energy was applied to sets of α-helices in membranes so that the location of (anionic) groups would be relocated by applied energy gating and/or pumping of ions (Fig. 8.6). The same description has just been used for cytochrome oxidase where gating and pumping of protons is required. (Note that here the proton pumping is secondary to the stoichiometric reaction of protons and dioxygen.) Now in each of these cases there are parallels with the α-helical motions of calcium triggers (Chapter 10) which can also be connected to kinases, and with dioxygen carrier proteins (Chapters 12 and 13). While the individual states of the metal ions in all these proteins may be unlike the states normally observed in inorganic chemistry, i.e. entatic states, it is the case that the activity is connected to mechanical *rack* action which involves protein motions and energization of α-helices and to further energizing of metal ion states. This rack action must not be confused with the static entatic states of metal ions generated by binding of metal ions to rigid protein frameworks as in conventional enzymes and electron transfer proteins based on β-sheet or cross-linked domains (see Chapters 7 and 15).

In the description of non-haem iron enzymes we observed that there were a number of oxidases. Similar oxidases are not so common in eukaryote as opposed to prokaryote cells. It may be the case that in evolution haem is taking over these oxygenase functions. It remains the case that all such activity is largely confined to compartments outside the cytoplasm.

13.9.2 Cytochrome P-450

We have described the character of compulsory-order enzyme reactions in Chapter 4. In P-450 oxidases the sequence of reactions is: (1) binding of organic substrate RH; (2) iron reduction to high-spin Fe(II); (3) O_2-binding to low-spin Fe(II) or as superoxide bound to low-spin Fe(III); (4) reduction of $Fe(II) \cdot O_2$ to FeO; and (5) attack of O on RH to give ROH, probably using radical intermediates (Fig. 13.16). There are a series of conformational switches which generate this pathway. They are allowed by the protein which is very largely helical (see Section 7.4). The P-450 enzymes are concerned with animal hormone metabolism as well as detoxification (Fig. 13.14).

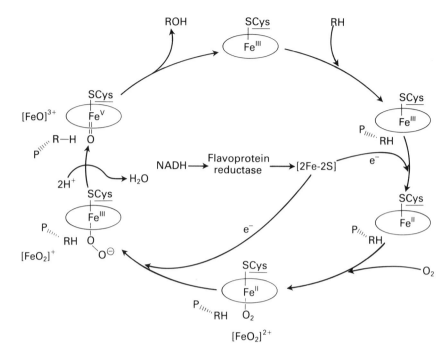

Fig. 13.16 The pathway of the reactions of cytochrome P-450.

13.9.3 Peroxidases and catalases

The peroxidases are intimately concerned too with plant hormone metabolism, e.g. the removal of indole acetic acid, and with the production of the animal hormone thyroxine. In the plants they also help lignification and destruction of lignins in protection and growth. In effect, they provide ways of producing free radical polymerization and depolymerization outside the cell (compare copper oxidases such as laccases and collagen cross-linking enzymes). The reactions are often of low selectivity and both phenols and indoles are attacked. The reaction involves long-range electron transfer from the substrate to the oxidized iron porphyrin (Fig. 13.17), but there is no

Fig. 13.17 The active site of horseradish peroxidase is buried in the middle of the protein and is accessible only to small molecules such as H_2O_2, O_2, and CN^-. The second substrate binds near the surface. Note the –S–S– bridges and the glycosylation, Ch. (After Y. Morita.)

Fig. 13.18 The intermediates in peroxidase reactions.

suggestion of coupling to protein conformational changes since the proteins are heavily cross-linked. An outline reaction mechanism is given in Fig. 13.18.

The catalases, like the peroxidases, are in vesicles—in this case the peroxisomes—since many of the oxidative reactions in these enclosed zones generate the dangerous chemical H_2O_2. The structure of catalase is such that the haem is hidden in a deep pocket presumably to prevent the escape of even more reactive intermediates such as OH^{\cdot} and O_2^-. As a consequence, catalases are only effective against small primary peroxides and, of course, they can only act in aqueous solutions. Cells have two other methods of disposing of peroxides—using vitamin E in membranes and a selenium enzyme in the cytoplasm (see Chapter 19).

13.10 Substrates of haem enzymes and secondary metabolism

Table 13.6 Haem enzymes and hormones

Haem enzyme	Hormones (synthesis or removal)
Cytochrome P-450	Sterols Ecolysome (plants)?
Peroxidase	Thyroxine Indole acetate (plants)
Globin-like (?)	Nitric oxide

The oxidative chemistry of haem iron and indeed of iron generally is closely linked to secondary metabolism, i.e. not the major synthetic and degradative routes of biology but side-products. The products of this metabolism early in their evolution may have been merely useful poisons, e.g. hydroxylated saturated ring compounds and halogen compounds. Some are listed in Table 13.6. Further development would require that protection from these poisons evolve; hence, the same metabolites probably came to be destroyed by similar mechanisms to those that were used to produce them. Much activity of unicellular systems especially in connection with oxidative secondary metabolism appears to be summarized in this way (see Section 12.12).

In multicellular organisms messengers were required. These messengers are hormones and they look very like the poisons of an earlier evolutionary period. In particular enzymes such as haem oxidases and peroxidases are involved both in syntheses and degradation of hormones. This is a new role for iron quite separate from electron transfer and the primary metabolism of C/H/O compounds to give energy. What we lack at the moment is a connection from the production and destruction of hormones to their use in the cell machinery (see Chapter 21). It appears to be the case that the hormones have to be present in an organism in well-controlled steady-state, quantities. Their concentrations must be under feed-back control and we shall ask later to what extent this feed-back involves levels of metals such as haem iron.

13.11 Haem in controls

Haem is turned over quite rapidly in cells. It is synthesized through insertion of iron into protoporphyrin by ferrochelatase which is one of the enzymes which binds Fe(II) in moderately fast exchange. Excess protohaem is destroyed by haem oxygenase. There is then a steady state in haem which is related to the level of free iron only in part.

The haem binding to some proteins is made irreversible by cross-linking to the protein, e.g. in cytochromes c,c', and f. These proteins are possibly all extracellular. In cells, the haem proteins are based on hydrophobic binding in pockets, e.g. in cytochromes b, myoglobin, haemoglobin, and cytochrome P-450. (Presumably cytochrome b-type proteins are the most primitive since they are found in all bioenergetic chains with redox potentials around 0.0 V). The binding constants for the free haem are estimated to be around 10^8 M^{-1} giving a free haem concentration in the cytoplasm of 10^{-8} M. Thus, haem in its dissociative reactions is not very different from Mn^{2+}, Fe^{2+}, and even Zn^{2+}. It is this that allows it to be used in regulation.

Now haem also communicates with the genes which control synthesis of the proteins which bind it and with those which make protoporphyrin. In particular, the early steps of porphyrin synthesis are switched off by excess haem and the haem oxygenase is switched on. The protein syntheses switched on by haem include that of cytochrome c, some subunits of cytochrome oxidase, and of cytochrome P-450.

The feature which is most striking is that, while the enzymes for porphyrin synthesis are synthesized via the cytoplasmic DNA, haem itself is synthesized in mitochondria (under Fe control) and then returns to the cytoplasmic nucleus to act as a regulator of proteins needed by the mitochondria. The haem organic chemistry is then independent of the free Fe(II) level once this level is adequate. The complicated linking of the requirements of mitochondria with those of the cell are not linked via Fe but by a haem system. Now the regulatory protein for haem synthesis has been analysed and found to contain zinc fingers. Moreover, in the δ-amino-laevulinic acid dehydratase enzyme, required in protoporphyrin synthesis, zinc is also required. Both the pathways and the protein syntheses associated with haem obviously have an iron dependence, but they also have a potential zinc regulation. Why?

Haem proteins may also act at other levels in controls. It is known that haem regulates the expression of ferritin and transferrin proteins at the transcriptional (RNA) level. Interestingly, bacterial ferritin is a haem-binding protein (the ferritin haem linkage is quite peculiar being formed in bacteria from two methionine ligands) and it is known that the haem redox reactions are important in iron release from bacterial ferritin. The release of iron from eukaryotic ferritin may also depend on haem binding to the protein. If this is the case, it must dissociate from the protein readily, as it does from b-type cytochromes, myoglobin, and haemoglobin, and all these proteins are then linked to a haem pool as well as to iron release from ferritin. The controlled release of haem from ferritin may depend on the iron load of the ferritin since this helical bundle protein has a stability (and even a redox potential) which is strongly dependent on the load of iron.

13.12 The synthesis of haem

The control over the concentrations of haem in a cell does not just rest with the level of iron and the synthesis of a protein since the porphyrin itself has to be prepared. This problem exists for the ring systems of corrin (Co), F-430 (Ni), and for chlorin (Mg). We are not concerned here with the niceties of the stereochemical problems in making such a ring but, first, with the nature of the precursor δ-amino laevulinic acid. This molecule comes directly from a combination of glycine and succinate. The enzyme δ-amino laevulinic acid synthetase is regulated by haem itself so that excess is not made. Notice that succinate and glycine (from acetate) are linked to the citric acid cycle so that switching off this cycle in mitochondria stops haem synthesis as well as iron scavenging (Chapter 12). As said in Section 13.11 the synthesis of haem from iron and porphyrins occurs in the mitochondria, indicating again that very much of the iron chemistry in biology is locked into the redox activity of primitive (prokaryote) cells or cell compartments. Strangely vitamin B_{12} is also important in the synthesis of haem.

The next step in synthesis requires δ-amino laevulinic acid dehydratase. Given the role zinc plays in advanced cells (Chapter 11), it is intriguing that this is a zinc enzyme. On another topic, note that one of the major effects of lead poisoning is that it inhibits this step of porphyrin synthesis, possibly by competition with zinc.

13.13 Summary of haem-iron functions

It is worth putting some of the functional value of haem proteins together:

1. The porphyrin traps the iron so that there is no equilibrium with free iron.
2. The porphyrin unit, like the chlorin unit, fits together very well with helical proteins. The hydrophobic character of the porphyrin means that it does not exchange easily.
3. Helices together with porphyrin generate a co-operative coupled unit suitable for mechanical energy transfer. This is expressed in allosteric systems, in preferred order reactions, and in oxidative phosphorylation couplings.
4. The porphyrin unit assists long-range electron transfer by developing low-spin states of iron so that relaxation energies are small (Chapter 4).
5. The iron in the porphyrin complex can be left with one open position for small substrates. The low-spin Fe(II) then acts as a π-donor.
6. The iron and the porphyrin readily yield higher oxidation states of very high redox potential (Table 13.7).

The fruit of this activity is a very large part of secondary metabolism and protection as well as energy capture.

Table 13.7 One-electron redox potentials of organic molecules (pH=7)

Redox potential range (V) versus hydrogen electrode at pH=7	Organic molecules	Oxygen species and iron oxidation states
> +1.0	Hydrocarbons, benzene	H_2O_2, OH·, Fe(V)
+0.5 to +1.0	Phenols, anilines, indoles	O_2, FeO(IV) H_2O_2
+0.0 to +0.5	Some sugars	O_2^-, Fe(III)
−0.5 to 0.0	Flavins (quinones), NADH	Some Fe(III)

13.14 Control of cell activity and iron

As we close this chapter on haem iron it is necessary to carry forward an impression of its functions and the relationship of these functions to the activities of other elements. The basic concept is that given in Chapter 6 where it was pointed out that there are two very different classes of reaction: (1) acid/base, as characterized by condensation and hydrolysis; and (2) redox, as characterized by electron transfer. These classes are inextricably interwoven in order that life's chemistry can work. Being so interwoven, of necessity the two must be linked by the same controls. We have seen in Chapters 10–12 that the free ion concentrations of Mg^{2+}, Ca^{2+}, and Zn^{2+} were three controls and regulators over acid/base reactions and in Chapter 18 we shall increase this list to include several phosphate compounds. (NB. These are very basic controls and we are not searching for the level of specificity inherent in local organic hormone and transmitter activity.) The control of redox activity is by iron levels, Fe^{2+}, and in prokaryotes and organelles this may dominate Mg^{2+} and Zn^{2+} control while Ca^{2+} control is absent. In higher organisms, the control over redox activity appears to have been removed to the organelles where probably it is shared between Fe^{2+}, free and in haem, and Mn^{2+}. Clearly, an exchange of information must pass between the organelles and the cytoplasm of the eukaryotes. Here we must remember that the oxidative energy of the organelles is the source of chemical energy for the acid/base steps (ATP). There is then a complicated connection between redox and acid/base chemistry, a subject that will be discussed in more detail in Chapter 21.

Further reading

Antonini, E. and Brunori, M. (1971). *Hemoglobin and myoglobin and their reactions with ligands*. North-Holland, Amsterdam.

Dolphin, D. H. (ed.) (1978). *The porphyrins*, Vols 1–7, Academic Press, New York.

Moore, G. R. and Pettigrew, G. W. (1987). *Cytochrome c: biological aspects*. Springer Verlag, Berlin.

Moore, G. R. and Pettigrew, G. W. (1990). *Cytochrome c: structural and physico-chemical aspects*. Springer Verlag, Berlin.

Poulos, T. L. (1985). Heme enzyme crystal structures. In *Advances in inorganic biochemistry*, Vol. 7, 1–36 (ed. G. L. Eichorn and L. G. Marzilli). Elsevier, Amsterdam.

Ruckpaul, K. and Rein, H. (ed.) (1990). *Frontiers in biotransformation: cytochrome P-450*, Vols 1–7. Taylor and Francis, London.

Wainio, W. W. (1970). *The mammalian mitochondrial cytochrome chain*. Academic Press, New York.

Wikström, M. and Babcock, G. T. (1990). Cytochrome oxidase: catalytic intermediates. *Nature*, **348**, 16–18.

Williams, R. J. P. (1991). Uncoupled and coupled electron transfer reactions. *Biochimica et Biophysica Acta*, **1058**, 71–4.

14

Manganese: dioxygen evolution and glycosylation

14.1	Introduction	370
14.2	Manganese chemistry	372
14.3	Monomeric Mn(III) and Mn(IV) chemistry	372
14.4	The biological chemistry of manganese	379
14.5	Manganese control systems	380
14.6	The production of dioxygen	381
14.7	Manganese and dioxygen metabolism	382
14.8	Manganese precipitates	383
14.9	The evolution of manganese functions	385
14.10	Summary	386

14.1 Introduction

The first transition metal series

The importance of manganese in living systems, though known to be very considerable, remains poorly explored in many areas. The manganese biochemistry which presently attracts greatest interest is that involved in O_2 release by photosynthesis system II. There is another well known enzyme which contains manganese and also evolves O_2, which is the superoxide dismutase of prokaryotes and of the organelles of eukaryotes. Besides these two cases, manganese biochemistry largely results from the association of the metal with a few enzymes and proteins, some of which are listed in Table 14.1. Looking at this list there would appear to be no general rhyme or reason for the way in which manganese is used, despite the fact that it has a particularly unique chemistry. In this chapter we summarize this chemistry before we turn to an attempted rationalization of the uses of the element in comparison with the uses of other inorganic elements of somewhat similar chemical potential. The first functional value of manganese is as a Lewis acid (compare magnesium, calcium, and zinc (M^{2+}), and the second, in higher oxidation states, is as an oxidation catalyst (compare iron and copper). It will be shown that compartmental separation of manganese is very closely connected to its particular functional value in all cases.

Table 14.1 Manganese enzymes and proteins

Protein	Valence	Comments on binding site	Function
Concanavalin A	II	Glu, Asp, Asp, His, H_2O, H_2O	Saccharide binding (extracellular)
Galactosyl transferase	II	Mn specific *in vivo*	Transfer of galactosyl group (Golgi)
Arginase	II	4 Mn atoms per enzyme molecule	Hydrolyses, arginine to ornithine and urea (mitochondria)
Glycosylaminase	II	S–H group near active site	Hydrolyses, guanidino-acetate to glycine and urea (mitochondria)
Phosphoenolpyruvate carboxykinase	II	Binds through O. (Fe competes not Mg)	C–C carboxylyase (mitochondria)
Pyruvate carboxylase	II	4 Mn atoms per enzyme molecule	C–C carboxylating synthetase (mitochondria)
Proline dipeptidase	II	O (and S) ligands proposed	Hydrolyses C-terminal proline dipeptides (ubiquitous)
Isopropylmalate synthase	II	S–H group near site	C–C oxo-acid lyase (mitochondria)
Mn-ribonucleotide reductase	II	Probably N, O ligands (see iron enzyme)	Reductive elimination of 2-hydroxyl from ribonucleotide (prokaryote)
Mn-dioxygenase	II	Low-symmetry Mn(II) centre	Catalyses extradiol cleavage of catechols (prokaryote)
Acid phosphatase	III	His residues essential	Hydrolyses orthophosphoric monoester (prokaryote)
Mn-catalase	III	—	Redox enzyme; acts on H_2O_2 (prokaryote)
Mn-peroxidase (ligninase)	II/III	—	Redox enzyme as above, degrades lignin (extracellular)
Mn-superoxide dismutase	III	His, His, Asp, His, H_2O	Redox enzyme; acts on superoxide radical (mitochondria, chloroplasts, prokaryotes)
Mn P_2O_7	III	O-ligands	Acts in place of SOD in lactobacillus
Photosystem II	II/IV	Assumed typical N/O donor site	Dioxygen release in plants (chloroplasts)
Imidazole glycerol phosphate dehydratase	II?		Histidine synthesis
Glutamate synthetase	II	N/O donor site	Glutamine synthesis (mitochondria)
Isocitrate dehydrogenase	II?		Citric acid cycle (mitochondria)

14.2 Manganese chemistry

14.2.1 Oxidation states

The most obvious features of the chemistry of manganese in simple aqueous solution are the stability of Mn^{2+} in acid solution and the greater stability of MnO_2 in alkaline solution and in the presence of oxygen. This chemistry is brought out in an oxidation state diagram (Fig. 1.7) which shows that oxidation states higher than Mn(IV), i.e. than MnO_2, are extremely powerful oxidizing agents in all conditions and are therefore unlikely to be of much concern in biological systems. We shall deal here only with the chemistry of Mn(II), Mn(III), and Mn(IV).

In detail, the oxidation state diagram (Fig. 14.1) shows how different is the chemistry of manganese from that of, for example, iron and molybdenum. It also differs greatly from that of the other major element among oxidation catalysts, i.e. copper. In each case there are three obvious oxidation states to discuss: II, III, and IV for Fe and Mn; I, II, and III for Cu; and IV, V, and VI for Mo. In the cases of Fe and Cu it is the middle oxidation state, Fe(III) and Cu(II), respectively, which is the most stable under mild oxidizing conditions and at neutral pH, while in the case of manganese the middle oxidation state, Mn(III), readily disproportionates to Mn(II) and Mn(IV); molybdenum differs again in that all its three states are of comparable stability. It is these differences in redox potential which evolution has used to the utmost advantage, given the additional features of the different chemistries of the most common oxidation states, Mo(VI), Mn(II), Fe(II) and Cu(I), in solution at pH = 7 and at a redox potential of about 0.0 V. It is clear to all chemists that Cu(I) is useful in similar ways to Fe(II), e.g. both bind O_2 and CO, but both Mn and Mo in all their oxidation states are very different indeed. We turn next to a description of the chemistry of the individual Mn oxidation states before looking at their biochemistries.

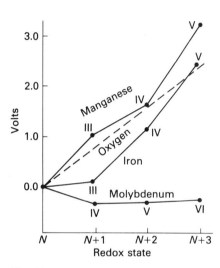

Fig. 14.1 An oxidation state diagram for Mo, Fe, and Mn, where N is oxidation state II for Mn and Fe and III for Mo, versus the normal hydrogen potential ($= -0.42$ V) at pH = 7. The slope of the lines joining two points gives their redox potential. Note the instability of Mn(III) and Mn(V).

14.2.2 Structures of Mn(II) complexes with organic ligands

When we describe the chemistry of a divalent ion, the foremost observation is the ionic radius. The radius of Mn(II), 0.75 Å, lies between that of Mg(II) and Ca(II). The half-filled, $3d^5$, shell of the ion makes it behave as a spherically polarizable ion, with no crystal-field stabilization energy; hence we expect that in the structure of its complexes it will behave in a manner between those of Mg and Ca. This is generally true; Mn(II) complexes are usually six-co-ordinate and octahedral, but there are some examples of irregular seven-co-ordinate complexes, e.g. with ethylenediaminetetracetate, EDTA, while Mg(II) complexes are almost invariably six-co-ordinate and octahedral and Ca(II) complexes are usually of irregular co-ordination with seven or more ligands. The ease of transition between the six- and seven-co-ordinate states presumably accounts for the lability of Mn(II) complexes compared with those of Mg(II) and in this respect Mn(II) resembles Ca(II). We must not be surprised then when we find that in some biochemical systems Mn(II) can substitute for Ca(II), while in others it can substitute for Mg(II) (isomorphous replacement). This substitution makes Mn(II) a useful probe of the activities of the other two ions since it is easy to follow the Mn(II) ($3d^5$) ion using either its

EPR signal or its enhancement effect on the relaxation times of the NMR signals of nuclei in molecules in its vicinity. Mn(II) can therefore be used to map biological space and has been much used in this way in many studies of enzymes, e.g. metallo-enzymes of Zn(II), Mg(II), and Ca(II), and in NMR imaging.

14.2.3 Manganese(II) equilibria and biochemistry

The main difference between Mn(II) and Mg(II) chemistry lies in the somewhat higher affinity of the manganese ion for nitrogen or possibly sulphur ligands (Fig. 2.8). If we consider first the affinity for purely oxygen donors, then Mn(II) parallels Mg(II) with but small differences, e.g. with carboxylate or phosphate donors, and this is also true of the solubility of its salts with oxyanions. As is well known, both of these cations have a somewhat higher affinity for small simple anions than has Ca, but Ca is noticeably more stable in association with a ligand carrying a larger number of donor groups (compare EGTA with EDTA in Fig. 14.2). The difference arises through the larger size of Ca(II) which also permits it to form relatively stronger lattices and so gives more insoluble salts. The following related facts can now be put together.

1. The concentration of Mg(II) in biological systems is usually about 10^{-3} M in all compartments with the possible exception of some vesicles; that of Ca(II) is 10^{-3} M in extracellular fluids but 10^{-7} M inside cells, while the

Fig. 14.2 The stability constants, log K, of some complexes of the elements in their divalent M(II) states compared with Mn(II). The symbols refer to ethylenediamine, EN; triethylenetetramine, TRIEN; ethylenediamine diacetate, ENDA; ethylenediamine tetracetate, EDTA; and ethylene-glycoldiamine tetracetate, EGTA. Note how the increase in the number of O-donor groups stabilizes Mn(II) and Ca(II) relative to other elements.

Mn(II) concentration is close to or less than 10^{-8} M almost everywhere inside or outside cells. (Especially in eukaryotic cells cytoplasm Mn(II) concentration may be much lower than 10^{-8} M, but could be higher than this in vesicles.)

2. Of the three cations Ca(II) forms the most insoluble salts.

3. Mg(II) and Mn(II) have roughly equal binding strengths to simple ligands based on oxygen donors, e.g. polyphosphates, and Ca binds only somewhat more weakly.

4. Calcium alone binds relatively much more strongly to larger numbers of donor centres, especially those of lower charge, e.g. in EGTA and in calmodulins.

It is then easy to understand the following biological observations concerning purely the O-donor centre chemistry of the three cations.

1. The insoluble salts of biology are largely Ca salts—in animal bones and shells and in plants, especially calcium oxalates; Mn is found only in a few solid systems, mostly in excretion nodules (see Section 14.8).

2. Small-molecule complex ions in biology are found in the cytoplasm of cells as Mg(II) complexes, e.g. Mg(II)·ATP, i.e. a phosphate complex.

3. Calcium can be and is used as the major signalling trigger ion of biology; this value arises from its large concentration gradient across the cell membranes.

It is seen that manganese is generally excluded from simple oxygen-donor chemistry; its low availability relative to magnesium and calcium (see Section 14.4) makes it less valuable in biological chemistry than either. The role of manganese, like that of other trace elements, is then in special situations, with specially designed protein ligands which usually have one or more nitrogen donor groups.

Now, when we turn to those more favourable circumstances for Mn(II) as opposed to Mg(II) and Ca(II), i.e. binding to nitrogen donors, there is a different competition to be faced by the Mn(II) ion, which is that from the other trace elements, especially Zn(II) and Cu(II) (see Fig. 2.8). Simple inspection shows that these two elements dominate the competition for the binding to N/S donors through the absence of Ni(II) and Co(II), and due to the very low availability of Fe(II) which is easily oxidized to Fe(III). (Biological systems make precipitates of Fe(III) in ferritin, goethite, and magnetite so the concentration of Fe(II) in biology remains very low due to its removal on oxidation.) The strength of the competition faced by Mn(II) from Zn(II) and Cu(II) can be seen in the slopes of the Irving–Williams series of stability constants (Figs 2.8 and 14.2). Clearly, if there is a limited supply of centres based only on nitrogen/sulphur donors, they will be overwhelmingly occupied by Zn or Cu; this expectation is borne out in a number of proteins. The divalent Mn(II) is left to be sequestered by some donor system of intermediate strength between that of all O-donors (Mg, Ca) and all N/S donors (Zn, Cu). Even here we must look for situations peculiarly suited to Mn(II) and its somewhat larger size since Zn(II), especially, is relatively available. As already mentioned, Mn(II) has the peculiarity that it is not only an ion of intermediate electron acceptor strength but it also has an

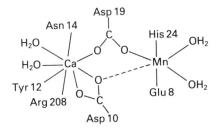

Fig. 14.3 The formula of the binding site of concanavalin A showing the difference between a Ca site with three carbonyl ligands, from Tyr 12, Asn 14, and Arg 208, and a Mn site with one nitrogen ligand, apart from carboxylates.

intermediate size between Ca(II) and Mg(II) while Zn(II) and Cu(II) are of the same radius as Mg(II) (Table 14.2). The study of complex ion formation constants has shown that framework ligands can be designed very selectively for ions of a given size so that a multidonor ligand made up from O-donors and N-donors, giving a total of six or at most seven donors and such that is not able to collapse around very small ions, is then ideal for Mn(II) once Zn(II) and Cu(II) free ion concentrations have been suitably reduced by N/S donors. Illustration of the small difference in stability constants between Mn(II) and Zn(II) complexes with ligands of this N/O type and which occupy considerable space around a cation is given in Fig. 14.2 (EGTA). We must not be surprised then at the co-ordination chemistry of Mn(II) as seen in biology. Figure 14.3 gives an example.

Table 14.2 Radii and ionization potentials (*IP*)* of some elements including manganese

	Ca	Mg	Mn	Fe	Co	Ni	Cu	Zn
Radius M(II) (Å)	0.99	0.65	0.78	0.72	0.69	0.65	0.65	0.65
IP M→M(II) (eV)	11.87	15.04	15.64	16.18	17.06	18.17	20.29	17.96

* The ionization potential, $M \rightarrow M^{2+}$, is the same as the electron affinity of the reverse reaction $M^{2+} \rightarrow M$ and is a good measure of the acceptor power of the ion, M^{2+}. Similar tables can be given for M(III), etc. The order of 'softness' is related to *IP* (see Table 2.8).

14.2.4 The kinetics of Mn(II) complexes

As stated in Chapter 4, Mn(II) exchanges water and ligands rapidly; this has some advantages in controls since, given the relatively weak binding constants of the metal in proteins, it is readily separated from a protein partner, but it also means that it is lost from proteins during many extraction procedures. Generally, Mn(II) does not form metalloproteins but exchangeable metal complexes with proteins, much as Mg(II) and Ca(II) do, in marked contrast with Cu(II) and, to a lesser degree, with Zn(II). Here Fe(II), which can also exchange quickly and forms relatively weak complexes, is similar to Mn(II). The lability of many Mg(II) enzymes and the fact that Mn(II) can replace Mg(II) at the active site has led to the unlikely suggestion that some Mg(II) enzymes are, in fact, Mn(II) enzymes. (In the cytoplasm of cells [Mg] exceeds that of [Mn] by at least a factor of 10^4, so here Mg(II) is presumably the active cation, but in the Golgi the situation may well be rather different.)

Before passing on, the reader should note that we have so far looked at Mn(II) as a competitor ion with Mg(II), Ca(II), Ni(II), Cu(II), Zn(II), and so on, for ligands. Biological systems are of great compartmental complexity, so that isolation of particular biological spaces behind membranes can be used to separate the functions of one ion from others (see Chapter 3). We have already mentioned functional separation of Ca, which is pumped out of all cells allowing its use as a trigger. Consequently we shall see that, while the above logic is well-founded on thermodynamic principles, kinetic control (pumping) can affect some of the conclusions. The evidence available suggests that Mn(II) is pumped into vesicles where its competitive strength is enhanced, e.g. in the Golgi apparatus. We turn first to the chemistry of manganese in higher oxidation states.

14.3 Monomeric Mn(III) and Mn(IV) chemistry

14.3.1 Structures

The Mn(III) ion ($3d^4$) forms complexes with a tetragonal symmetry having four short and two longer bonds in the general case and in this respect it resembles Cu(II). It is the only trivalent ion of biological consequence which falls in this structural class and therefore it has the potential to be introduced into selected sites in proteins; in this respect it differs from Fe(III) which, being a $3d^5$ cation, enters octahedral or tetrahedral sites readily (see also Mn(II)). These trivalent ions are small, of course, with radii of about 0.6 Å, and the exchange of simple ligands, e.g. H_2O, is relatively fast as for most Jahn–Teller ions, e.g. Cu(II).

The Mn(IV) ion ($3d^3$) resembles the Cr^{3+} ion in its electron configuration and hence it should seek octahedral co-ordination; however, the very strong polarizing power of M^{4+} ions leads to close association in water with one oxygen dianion, giving the oxocation MnO^{2+} whose preferred sterochemistry is that of a very distorted octahedron. The inorganic chemist will note the parallels with VO^{2+} (Chapter 17).

14.3.2 Thermodynamics

The high charge on the M(III) ions leads them to bind strongly to anions, but their small size leads them to prefer the smallest of anions such as RO^- and F^- rather than RS^-; this is true of Mn(III) and Fe(III). In addition, RS^- is open to oxidation and Mn(III) is a very powerful oxidizing agent. In this chapter we shall assume that high-valent Mn chemistry does not involve sulphur co-ordination; there are as yet no known exceptions in biology. There is, therefore, some difference from Fe(III) chemistry, which is probably due to the oxidizing power of Mn(III), but we shall stress the overall similarities between these two elements at first. A problem with M(III) ions at pH = 7 is hydrolysis, which grossly limits the free ion concentrations and, especially in the case of biological structures, it is often unclear if a bound 'water' molecule is in fact an hydroxide anion, or even if the metal ion forms a $[M-O-M]^{4+}$ complex using two M(III) ions (see Chapter 2).

Mn(IV) shows this hydrolysis to a much more marked degree and it exists generally as MnO^{2+}. The chemistry of this oxidation state in water is not well known, but we can use the chemistry of VO^{2+} and UO_2^{2+} as a guide. Table 17.2 shows that MO^{2+} complexes behave in the stability of their co-ordination compounds very like M^{3+} complexes, i.e. while one co-ordination position is totally screened by the oxo group the others act as if the residual electrostatic positive charge on the cation were three not two. It has then a strong affinity for small anions, e.g. F^- and RO^-; it too readily forms bridged hydroxy/oxo complexes and once again solubility of free ions is limited.

The relative stability of the Mn(III) complexes when compared with Mn(II) and Mn(IV) complexes is the factor which controls the redox potentials relative to the *aqueous* couple values (Fig. 14.1). It is generally true that the Mn(III) ions form much more stable complexes than the Mn(II) ions with

almost every kind of ligand. On the other hand, since little difference is expected between the stability constants of the complexes of MnO^{2+} ($Mn(IV)$) and $Mn(III)$, since MnO^{2+} behaves like an M^{3+} cation, there will remain a tendency for $Mn(III)$ to disproportionate to $Mn(II)$ and $Mn(IV)$ species in multinuclear assemblies, no matter what the co-ordination centre. Here Mn is very different from Fe where chemistry drives $Fe(II)$ plus $Fe(IV)$ towards $2Fe(III)$ even in complex ions, i.e. in the opposite direction; in a similar manner, $Cu(III)$ and $Cu(I)$ give the stable intermediate state $Cu(II)$. We can only expect to see stable $Mn(III)$ compounds in biology or elsewhere in isolated mononuclear units (see superoxide dismutase, Section 14.7, where it is bound in a rather unexpected manner. The use of manganese to make dioxygen and of iron and copper to reduce it becomes clearer.

14.3.3 Kinetics of Mn(III) and Mn(IV) reactions

The kinetic aspects of primary interest in the $Mn(III)$ and $Mn(IV)$ states are:

1. The increased stability of complexes relative to $Mn(II)$ leads to slower dissociation and there is not much loss of ions of this kind from metalloenzyme sites. $Mn(III)$ ions, therefore, separate with their proteins.

2. The relatively fast loss or exchange of OH or O from these species is of primary concern since these reactions can be one-electron (OH^{\bullet}) or two-electron transfers (O); these reactions, which concern O_2, H_2O_2, and O_2^- and their oxidation chemistry, have been described in detail in Sections 13.9 and 13.10.

The energies of the oxidation states is very important in controlling the reaction cycles of manganese.

14.3.4 Manganese cluster chemistry

In oxidation states III or IV, Mn, like $Fe(III)$ and $Cr(III)$, readily forms clusters related to simple hydroxide or oxide chemistry. The three elements can also form clusters in mixed oxidation states. The exact formulation of many such complexes is now known from X-ray diffraction studies; it is found that oxo groups can bridge two or three Mn atoms in many types of clusters (Fig. 14.4), including cubane and adamantane forms. Interest in the Mn clusters focuses, of course, on the O_2-generating proteins of chloroplasts. However, in these biological clusters the special geometry enforced by the protein may well lead to peculiarities, as is well known in Fe_4S_4, Fe_3S_4, and Fe_2S_2 clusters in proteins. (Note that some Fe_nO_n clusters are known in biology, e.g. in ferritin and haemerythrin-like proteins.)

Throughout this account of Mn chemistry there is always the running thought of the parallel chemistry of Fe. How can biology select for Mn in the higher as well as the lower oxidation states and then in clusters? It would appear that the answer must be through the preparation of *all* the Mn complexes through the $Mn(II)$ state, since $Mn(II)$ and $Fe(II)$ are far from equally available and not so similar as $Mn(III)$ and $Fe(III)$. In fact, it is suspected that $Mn(III)$ and $Fe(III)$ are so similar in binding and kinetics that some biological systems hardly distinguish between them, e.g. in superoxide

Fig. 14.4 Two complexes made by Christou and coworkers (see 'Further reading') to model the manganese mixed-valence cluster used in dioxygen formation. (a) $[Mn_4O_3Cl_6(acetate)_3(imidazole)]^{2-}$ and (b) $[Mn_4O_2(acetate)_7(bipyridyl)_2]^+$. (c) Some other proposed structures of the manganese cluster involved in dioxygen release in biology, see Fig. 14.6.

dismutases, in acid phosphatases, in ribonucleotide reductases, and in the transfer of both metal ions across membranes using transferrin and its receptors. It is an important peculiarity of the cluster biochemistry of Fe and Mn that this confusion does not arise. There can be little doubt that it is the disproportionation of Mn(III) that removes manganese from such clusters as Mn(II) and allows the build-up of oxo-species MnO(IV).

The structures in Fig. 14.4 and those of the Fe_nS_n clusters in Fig. 12.6 illustrate some similarity but we must remember there is a difference of one volt in redox potential and of 10^9 years in their evolution in biology.

14.4 The biological chemistry of manganese

From the preceding considerations the biochemistry of Mn ought to be dominated by: (1) the chemistry of Mn(II), which should be intermediate between that of Mg(II) and Ca(II) and Zn(II); and (2) the powerful oxidizing action of the stable higher oxidation state, Mn(IV). This expectation is only partially fulfilled and we shall have to try to understand in addition why considerable Mn(III) chemistry is observed in biology.

14.4.1 The occurrence and availability to biology of manganese

Manganese is not a rare element and is about as common as Fe or Zn (Fig. 1.5). At pH = 7 many of its salts are quite soluble so that at equilibrium the free Mn(II) concentration is relatively high in biological systems (10^{-8} M), even in the presence of oxygen and manganese oxide precipitates. For the same reason, in the very early periods of life, Mn(II) must have always been available, since manganese sulphides are the most soluble of all transition metal sulphides from Mn(II) to Zn(II) (Fig. 1.12). Availability of Mn(II) at about 10^{-8} M is not a problem.

14.4.2 The uptake of manganese

Assuming a reasonable concentration of this trace element as Mn(II) in the sea or in soil solution, then, initially, Mn(II) will be taken up in a similar manner to Ca(II) and Mg(II) and it is undoubtedly transferred in this oxidation state in the more reducing environment of biological systems, to various organs, cells, and vesicles. It is known to be concentrated in plant cell vacuoles and in chloroplasts, for example. This uneven distribution means, of course , that it must have an energized, pumped distribution.

In the bloodstream of animals the Mn(II) can be oxidized on binding to transferrin and then, as Mn(III)·transferrin, it may be transported into their cells, only to revert to Mn(II) on release. It is then again pumped into animal cell vesicles and organelles.

14.4.3 Pumping of manganese gradients

As stressed in Section 2.4, ions are frequently handled by pumping mechanisms of a considerable degree of selectivity. Without such specialized pumps we could not understand the differential distribution of metals within biological species, organs, cells, organelles, and vacuoles. Only if we accept such kinetic control of free ion distribution can we systematize not only the special compartments but also the particular way in which chemical partners of the metal ions are 'selected' in biology. Manganese is pumped out of the cytoplasm of all eukaryotic cells (Fig. 14.5) and into a large variety of organelle and reticula (vesicle) compartments (Table 14.3). The pump is, in all probability, very similar to that for calcium (Section 10.13). The transfer operation of pumps is not based on thermodynamic competition for sites since it will depend upon selected recognition of the stereochemistry of the

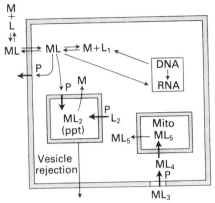

Fig. 14.5 A general description of the distribution of the elements in cellular systems. Metal ions enter on carriers ML or ML_3, often using pumps, P, and can also be rejected. Alternatively, the ML complex can be transferred to a vesicle where it may precipitate as ML_2, or it can be processed as ML_4 and placed in an organelle ML_5. Ligands or ML can interact directly with DNA. (L and P are proteins.) All of the steps are possible for manganese complexes. Mito, mitochondria.

Table 14.3 Manganese-containing vesicles

Vesicle	Known Mn protein
Lysosome (acid)	Acid phosphatase (Mn(III))
Chloroplast	Superoxide dismutase, enzymes
Mitochondria	Superoxide dismutase, enzymes
Golgi apparatus	Glycosyl-transferases
Thylakoid (acid)	Dioxygen generation
Vacuole (acid)	(Mn(II) is free)

See also Table 14.1.

compound to be transferred by the accepting compound in the pump (see Chapter 3). We remind the reader that, since the process is coupled to energy, it can be a one-way selection. Consider again that

$$\text{All M bind to } L_1 \text{ to give } ML_1,$$

$$\text{but only } M_1 \text{ allows } M_1L_1 \xrightarrow{P_1} M_1L_2 \text{ (transfer).}$$

This means that the binding of L_1 is thermodynamically controlled, but the transfer of a particular M_1 to L_2 from ML_1 is the only M transfer allowed since the protein, P_1, recognizes the shape of M_1L_1 only. The principles are exactly the same as those discussed in the placing of magnesium in chlorophyll.

14.4.4 The distribution of Mn in biological systems

Mn is known to be a mutagenic agent and it is clear that, since its binding to proteins is relatively weak, it could not be allowed in high concentration in the cytoplasm of eukaryotic cells. The problem of mutation, so serious a threat to eukaryotic DNA, is not such a problem for prokaryotes since mutation can be helpful in generating variation of such species. Mitochondria and chloroplasts are organelles which can be looked upon as descended from prokaryotic cells. It is noteworthy that these organelles, as well as the cytoplasm of prokaryotes, accumulate Mn to a much greater degree than does the cytoplasm of eukaryotes. The eukaryotic cell also pumps manganese into almost any vesicular partition of the cell rather than leave it in the cytoplasm. Some of the vesicles known to have high manganese are listed in Table 14.3; in such vesicles Mn(II) does not face such strong competition from Mg(II), which is retained in the cytoplasm of all cells. The enzymes then formed are discussed below. Particular examples are some Mn-dependent peroxidases as shown in Table 14.1.

Manganese poisoning may well arise when an excess of Mn is forced into a vesicle where it has damaging effects; for example, chromaffin granules of animals, into which Ca is deposited, can receive considerable Mn (if it is made available) which is then menacing since it can affect the adrenalin of the vesicles through oxidation reactions. To some degree Mn is also rejected to the outside of cells, but this rejection is much less notable than that of Cu or Ca; it could well be just a detoxifying mechanism. We next look at the value of this energized pumping of Mn for control of certain cellular activities in local compartments.

14.5 Manganese control systems

There is now the clear knowledge that DNA expression is controlled in part by Zn proteins, the so-called zinc fingers, especially in eukaryotic cells (see Chapter 11). The ways in which Fe (and Mn) is built into biological systems and compartments led to the earlier prediciton that, especially for the prokaryote systems, there would be regulatory genes controlling the relationship between uptake and growth. The so-called Fe (F) uptake (U)

regulating (R) protein, FUR, was looked for and found by Nielands (see Chapter 12). The similarities between the chemistry of Mn(II) and Fe(II) lead us to suppose that the control systems connected to Mn could be related to the FUR system. The basis for the idea that such a system exists is the link between Mn and photosynthesis, and the handling of sugar moieties so produced in some biological compartments. The complexity of control over such complicated interactive systems demands gene regulation. However, as with Zn so with Mn, the full ramifications of the use of this metal have not been uncovered. Zinc, we can say, is in some way like a general growth factor in eukaryotic systems; magnesium is intimately linked to phosphate transfer. Can we pin-point a central biochemical function for Mn from which we can seek its regulatory systems or is its use too generalized? Mn is connected closely to: (1) organelle functions, e.g. in thylakoids and mitochondria (see Table 14.1); and (2) the handling of glycosylation, which is a peculiarity of the Golgi apparatus. It would appear that in plants in particular we must seek a manganese pump protein which is regulated by the same gene systems as either that for the synthesis of photosystem II (dioxygen production) or that for the production of glycosylases. Table 14.1 includes some other enzymes which could be involved in metabolic control at the organelle level, mitochondria or chloroplast. Particularly interesting in the context of the comments in Section 12.16, on iron control, is the requirement for manganese in some of the enzymes of amino-acid synthesis and some in the citric acid cycle; the picture is then

where manganese controls the production of (E), the protein of manganese enzymes, but manganese excess switches off Mn(II) uptake, and manganese deficiency switches it on via free E. Very recently, it would appear that such a regulatory system has been found controlling the production of manganese peroxidases (ligninase) in *Phanerochaete chrysosporium*.

14.6 The production of dioxygen

The description of the active site for the generation of dioxygen from water is limited by the absence of a structure; what is thought to occur is shown in a scheme in Fig. 14.6. There are four manganese atoms involved and there are major influences of calcium and chloride, which must be looked upon as control functions. The exact way in which the four manganese atoms work together is not resolved, but the step(s) in the reaction, one electron oxidation at a time, can be readily followed by a variety of spectroscopic techniques.

Fig. 14.6 A chemical scheme for O_2 production at a four-manganese cluster. The S-states are the five observable spectroscopic conditions. Note the involvement of Ca^{2+} and Cl^-.

(There are many models described in the book by Graham *et al.* (1988; see 'Further reading') and see Fig. 14.4.)

Without a structure we cannot describe the manganese binding in any detail but it is likely to be similar to that in haemerythrin (Fig. 12.9), which binds O_2^{2-} as does Fe(III), while Mn(II) binds O_2 much less well; the O_2 release is effectively irreversible. The need for Ca^{2+} and Cl^- suggests that the whole unit $(Mn)_4$ is open to conformational change and control and it is tempting to expect a helical bundle like that of haemerythrin. The linking between Ca^{2+}, Cl^-, and Mn^{2+}, all of which are exported from the cytoplasm, is a very intriguing aspect of homeostasis.

A further control feature may depend on the stability of the Mn(II) binding; there is the distinct possibility that Mn(II) can dissociate from the protein complex and in this dissociation could lie a control by manganese availability over the use of light energy. Photosystem II generation of O_2 is secondary to the production of reduced CO_2 and of ATP in biological growth and it is therefore useful to direct energy away from O_2 release.

14.7 Manganese and dioxygen metabolism

As shown in Table 14.1 and referred to in Section 14.3.4, Mn is found in certain peroxidases, catalases, and superoxide dismutases (Fig. 14.7). The peculiarity then is that Mn, like Fe, is useful in all the reactions of oxygen except, unlike Fe, the direct use of dioxygen itself. The Mn peroxidases and catalases are found in a few extracellular or vesicular systems. The first is said to be associated with the strengthening of cross-linking in lignins—a function which can also be served by the haem peroxidases and Cu oxidases released from vesicles. It is probable that the haem peroxidases catalyse oxidation of the free Mn(II) to Mn(III) and that Mn(III) is the real oxidant for lignin polymerization through free radical reactions *outside* cells.

Fig. 14.7 A view of manganese superoxide dismutase showing the α-helices and β-sheet and the octahedral Mn(III) co-ordination. (After C. C. F. Blake and M Ludwig.)

Now, while we can explain why Mn and not Fe is used in dioxygen production and why Fe or Cu, but not Mn, are used in dioxygen transport, in utilization for energy production, and in many oxygenases, this leaves no reason for the use of Mn(III) in either catalases, superoxide dismutases, ribonucleotide reductases, or in peroxidases rather than Fe(III). This problem comes again in the choice of Mn(III) as well as Fe(III) in acid phosphatases (see also the comment on Mn(III) transferrin in Section 14.4.2). The simplest explanation is the higher availability of Mn(II), especially outside cells and in the vesicles which hold these enzymes. Once incorporated into a protein site of N/O donor ligands the metal is then oxidized but disproportionation is prevented by isolation of the Mn(III) in a monomeric complex. We are suggesting that the preference for Mn(III) over Fe(III) may be almost accidental in some enzymes. At the same time manganese catalase, which has two manganese atoms, could be a helpful enzyme in the understanding of dioxygen release systems.

14.8 Manganese precipitates

It has been observed that in certain plant tissues, e.g. the base of the stinging hair of the common nettle, there is a cyclic appearance during development of the hair cell of a very large deposit of manganese (Fig. 14.8 and Table 14.4). There is also excess phosphate present and we wonder if this manganese system is being used to protect oxalate while Ca·oxalate is laid down in the

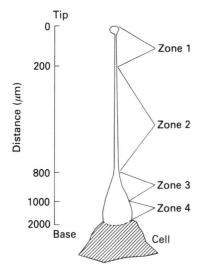

Fig. 14.8 Diagramatic representation of a stinging hair of a simple nettle showing various zones. Elemental maps of immature and mature stinging hairs, Emergence I and II, were produced using proton-induced X-ray emission combined with a scanning proton microprobe. Elements determined are Ca, Si, P, S, K, and Mn, in the zones of the figure as given in Table 14.4.

Table 14.4 Analytical data extracted from analyses performed at different points along the length of each hair (Fig. 14.8)

Distance from tip (μm)	Concentration (μg/cm²)					
	Ca	Si	P	S	K	Mn
Emergence I						
At tip	14	171	6	0.3	20	—
100	82	139	14	0.9	58	—
200	91	56	11	1.7	128	—
300	121	27	9	0.9	86	—
400	127	14	7	0.9	67	—
600	149	17	6	1.4	87	—
800	167	20	6	0.8	57	—
1000	242	84	10	0.9	34	—
1200 (at base)	110	21	60	3	124	42
Emergence II						
At tip	8	190	2	0.05	1	—
100	206	117	2	0.6	5	—
200	248	71	3	0.8	6	—
300	252	49	4	0.5	6.5	—
400	294	21	2.5	0.8	7	—
600	304	16	1.2	0.4	5	—
800	285	17	2.5	0.5	5	—
1000	246	62	3	0.7	5	—
1200 (at base)	82	5	2	0.7	30	—

Emergence I is an earlier stage of development than emergence II.

hair. The manganese would then be removing the products of the glyoxalate cycle, such as $O_2^{.-}$ and H_2O_2 which are dangerous for oxalate; this cycle also occurs in peroxisomes where Mn is in high concentration. The store, which is removed at maturity, is possibly $Mn_2P_2O_7$, a known mineral in a vesicle of a lactobacillus that is thought to act as a superoxide dismutase in just this way.

Manganese-depositing bacteria have been known for many years; specifically *Gallionella*, *Sphaerotilius*, *Leptothrix*, and *Chonothrix* have been shown to deposit manganese oxide extracellularly. Extensive work has been carried out on control of the deposition process (allied with the extracellular deposition of Fe hydroxide) and on the physiological aspects of the deposits in these systems. It can be that up-and-down movement in the water column, from oxic to anoxic zones, allows the organisms to capture energy using a cycle of changes

$$\text{Oxic} \quad Mn^{2+} \xrightarrow{\text{O}_2} MnO_2$$
$$\text{Anoxic} \quad Mn^{2+} \xleftarrow{\text{H}_2\text{S etc.}} MnO_2 \quad \text{Gravity}$$

Furthermore, intracellular granules of manganese salts have been observed within the common garden snail; these granules frequently contain a wide variety of other environmentally available metals and therefore have been interpreted as part of a cellular detoxification process.

14.9 The evolution of manganese functions

This section is purely speculative. Manganese, in the sulphide solutions that may have dominated early life, was relatively more available than any other transition metal ion since it does not have a very insoluble sulphide but see vanadium, Chapter 17. On the other hand it is a considerably better Lewis acid than either magnesium or calcium which were the two other divalent ions of easy availability. We might well concede then that an early use for Mn(II) would be as a useful acid catalyst for easy reactions such as those of phosphatases. In such reactions we know that two metal ions are better than one, e.g. as seen in many acid and alkali phosphatases to this day. Thus manganese (II) clusters may have evolved.

At some stage life developed the ability to use light linked to the discharge of electrons so as to generate reducing equivalents, but it is simultaneously necessary to have a source of reducing equivalents on the opposite side of the membrane in which light capture occurs (Fig. 5.1). In the presence of sulphide the reaction $H_2S \rightarrow S\downarrow + 2H^+ + 2e$ is an ideal source, but it requires metal ions to catalyse the reaction (see Section 17.11). The preferred metal ion must go readily through the cycle.

$$2M^{2+} + h\nu \longrightarrow 2M^{3+} + 2e \quad \text{and then} \quad 2e \longrightarrow \text{reduction system}$$

$$2M^{3+} + H_2S \longrightarrow 2M^{2+} + S\downarrow + 2H^+ \qquad 2H^+ \longrightarrow \text{gradient energy (ATP) production.}$$

Now manganese is peculiarly suitable for such a reaction as Mn^{3+} sulphides are open to ready reduction with sulphur release and metal dissociation. The reaction is best carried out in a vesicle so as to trap the sulphur produced which can then be rejected. Note that the vesicle would also be acid. It would pay to put manganese in the vesicles and to couple the whole reaction to ATP production (see Chapter 5).

Given that this chemistry is all speculative, the further consideration that, given the very high redox potential of Mn(III), the above system would come to generate dioxygen from water using a Mn(II) dimer may be acceptable. Moreover, dioxygen rejects itself by diffusion and the greater proton gradient generated in the acid vesicle can be used to make more ATP (see Chapter 5).

As a consequence of a separation of the dioxygen-generating system into plants the question arises as to whether manganese has a higher level of control in plants than in animals. The link between photogeneration of dioxygen and the incorporation of reducing equivalents in carbon assimilation followed by the extensive use of more polysaccharides by plants is suggestive but as yet we know little more than could be used to conclude that there may well be a general network of links in metabolism in plants based on free Mn^{2+}. Especially intriguing are the further connections to deoxyribonucleotides (Table 14.1) and amino acids via the shikimic acid pathway and we notice the ability of Mn(III) to act as a replacement for Fe(III) in several other important reactions. Thus manganese biochemistry underwent dramatic changes with dioxygen introduction and we note how much of this new

chemistry is put outside the cytoplasm into vesicles. It is possible that manganese importance in gene control, if it exists, is only to be found in multicellular plants. One very important reaction is the oxidation of lignin, by white rot mycetes, which utilizes free Mn^{2+} ions. It has been shown that the production of the enzyme is probably controlled by a manganese protein, i.e. it is regulated by free Mn^{2+} in the cell.

Whether this sequence of events is true is not of concern here. We must observe that some such sequence must have been part of evolution. This particular story explains why we today find manganese in (acidic) vesicles and organelles and, being there, it is not surprising that its biological control chemistry is linked to that of calcium and chloride especially in the chloroplast.

14.10 Summary

The biochemistry of manganese in all oxidation states is dominated by transfer to selected compartments and by its association with N/O donor systems. These allow the use of Mn(II), and even Mn(III), as weak acid catalysts and the cycling of oxidation states (II), (III), and (IV) in reactions of especially H_2O, O_2, H_2O_2, and O_2^-. The very special association of manganese in the higher oxidation states is with the release of O_2 from within chloroplasts. Mn(II) appears to have an additional association with glycosylation in Golgi-derived vesicles and Mn(III) is found in mitochondria and chloroplasts in such enzymes as superoxide dismutase as well as in lysosomes in acid phosphatases. Finally, there is the suspicion that there are regulatory proteins which bind manganese. The connection of all this special manganese biological chemistry with that of other elements will be described in Chapter 21.

Further reading

Archibald, F. S. and Fridovich, I. (1982). The scavenging of superoxide by manganous complexes. *Archives of Biochemistry and Biophysics*, **214**, 452–63.

Christou, G. and Vincent, J. B. (1987). A molecular 'double-pivot' mechanism for water oxidation. *Inorganica Chimica Acta*, **136**, L42–L43.

Ghiorse, W. C. (1984). Biology of Fe- and Mn-depositing bacteria. *Annual Reviews of Biochemistry*, **38**, 515–50.

Graham, R. D., Hannam, R. J., and Uren, N. C. (ed.) (1988). *Manganese in soils and plants*. Kluwer Academic Publishers, Dordrecht.

Hughes, N. P., Perry, C. C., Williams, R. J. P., Watt, F., and Grime, G. W. (1988). A scanning proton microprobe study of stinging emergences from the leaf to the common stinging nettle *Urtica dioica L. Nuclear Instruments and Methods*, **B30**, 383–7.

Khangulor, S. V., Barynin, V. V., Voerodskaya, N. V., and Grebenko, A. I. (1990). Binuclear cluster in manganese catalase. *Biochimica et Biophysica Acta*, **1020**, 305–10.

Kuhn, N. J., Ward, S., and Leong, W. S. (1991). Manganese in Golgi vesicle enzymes. *European Journal of Biochemistry*, **195**, 243–50.

Mildvan, A. S. and Cohn, M. (1970). Aspects of enzyme mechanisms studied by nuclear spin relaxation induced by paramagnetic probes. *Advances in Enzymology and Related Areas of Molecular Biology*, **33**, 1.

Phillips, C. S. G. and Williams, R. J. P. (1966). Oxidation state diagrams. In *Inorganic chemistry*, pp. 169–73. Oxford University Press, Oxford.

Reeke, G. N., Becher, J. W., and Edelman, G. M. (1978). Changes in the three-dimensional structure of concanavalin A upon demetallization. *Proceedings of the National Academy of Science of the United States of America*, **75**, 2286–90.

Sillen, L. G. and Martell, A. E. (1964). *Stability constants of metal ion complexes*, Chemical Society Special Publication, no. 17. Chemical Society, London.

Stallings, W. C., Pattridge, K. A., Strong, R. K., and Ludwig, M. L. (1985). The structure of Mn superoxide dismutase from *Thermus thermophilus* HB8 at 2.4 Å resolution, *Journal of Biological Chemistry*, **30**, 16424–32.

Williams, R. J. P. (1982). Free manganese and iron cations can act as intracellular cell controls. *Federation of European Biochemical Societies Letters*, **140**, 1–10.

15

Copper: extracytoplasmic oxidases and matrix formation

15.1	Introduction	388
15.2	Copper and electron transfer	389
15.3	Copper and dioxygen	391
15.4	Copper enzymes and nitrogen oxides	394
15.5	Superoxide dismutase	395
15.6	The transport and homeostasis of copper	396
15.7	The overall functions of copper	397

15.1 Introduction

There are five distinctive characteristics of the biochemistry of copper, four of which are chemical:

1. Cu(II) is the most effective available divalent ion for binding to organic molecules (Fig. 2.8). Since this is in fair measure due to its electron affinity, it is also the best attacking acid available at around pH = 7, but it is rarely used as such in biology.

2. Cu(I) is the most effective available monovalent ion for binding to organic molecules. Again this is due to its high electron affinity.

3. The almost equal tendency of Cu(I) and Cu(II) to form complexes with many organic ligands, notwithstanding their very different sizes and stereochemistries, makes the Cu(I)/Cu(II) couple particularly useful and adaptable in the range of redox potentials from about +0.2 to +0.8 mV, i.e. on the oxidizing side of the biological scale (Fig. 13.3).

4. Cu(I) can act as a π-donor and can bind such ligands as O_2 and CO.

5. The fifth factor, which is non-chemical, is that only one or two copper proteins are found in the cytoplasm of (eukaryotic) cells.

The Ni, Cu, and Zn subgroups of the periodic table

A most striking aspect is the strong correlation of copper enzymes and their proteins with the reactions of small non-metal compounds which only became available in quantity when dioxygen entered the atmosphere (Table 15.1). We then may make the conjecture that biological systems did not use copper

Table 15.1 Substrates for copper enzymes

Enzymes	Substrate
Simple oxidases	O_2 plus phenols, amines, ascorbate, ferrous ions, etc.
Cytochrome oxidase	O_2 plus cytochromes c, a, and a_3
Nitrogen oxide enzymes	NO, NO_2^-, N_2O + oxidizing or reducing agents
Specialist oxidases	O_2 + proteins or polysaccharides for cross-linking

extensively before the advent of an oxidizing atmosphere based on dioxygen. Now, this event coincides with the development towards multicellular organisms and we shall therefore seek a special role for copper in that development.

15.2 Copper and electron transfer

Perhaps the most common use of copper is in electron transfer, especially associated with oxidative enzymes and energy capture. These electron transfer systems are made of small 'blue' proteins that ferry electrons to reaction centres and some of them exist as freely mobile small proteins. None of the blue copper proteins are found in the cytoplasm, though some are bound to the external surface of cytoplasmic membranes. In Chapter 7 it was shown that the construction of their copper(II) sites is entatic, i.e. with a somewhat distorted tetrahedral geometry and held in that geometry by a β-barrel fold. The site geometry (Fig. 15.1) is easily recognized as one that has evolved as a design to assist electron transfer. This 'designed' centre is maintained in a

Table 15.3 Some standard redox potentials of copper proteins at pH = 7.0

Protein	Redox potential (mV)
Laccase	
Type I (blue)	+785
Type II	+500
Type III	+400(?)
Azurin	
Type I (blue)	+330
Plastocyanin	
Type I (blue)	+370
Tyrosinase	
Type III(?)	+370
Haemocyanin	> +800(?)
Dopamine monoxygenase	+310

Note that the redox potentials are higher than that of the copper aqueous ions (+170 mV) which means that Cu(I) is bound with greater strength than Cu(II).

Table 15.2 Some copper proteins and enzymes: localization and function

Protein or enzyme	Localization	Function
Cytochrome oxidase	External face of mitochondrial membrane	Reduction of O_2 to H_2O
Laccase, Tyrosinase	Extracellular	Oxidation of phenols (reduction of O_2 to H_2O)
Phenol oxidase	Extracellular	
Caeruloplasmin	Extracellular (blood plasma)	Oxidation of Fe(II) to Fe(III)? (reduction of O_2 to H_2O)
Haemocyanin	Extracellular (blood plasma)	Transport of O_2
Lysine oxidase	Extracellular	'Cross-linking' of collagen (reduction of O_2)
Ascorbate oxidase	Extracellular	Oxidation of ascorbate (reduction of O_2 to H_2O)
Galactose oxidase	Extracellular	Oxidation of primary alcohols to aldehydes in sugars (reduction of O_2 to H_2O_2)
Amine oxidase	Extracellular	Removal of hormones, e.g. adrenalin
Blue proteins	Membranes (high potential); thylakoid vesicles	Electron transfer
Superoxide dismutase	*Cytosol*	Superoxide dismutation (eukaryotes)

number of oxidases, apart from in energy transduction devices where it is necessary to pass electrons from a substrate to the dioxygen-binding centre (Fig. 15.1). These various cases are illustrated in Table 15.2 where we also note the localization of these copper proteins. Some of these reactions use high redox potentials which require that the site and its geometry are designed to give both fast reactions and high potentials (Table 15.3). The distortion of the copper site toward tetrahedral geometry has just this effect since it stabilizes Cu(I) relatively to Cu(II); in fact, it would appear, from all our knowledge of structure and function of the blue copper sites, that their proteins are rigidly designed as a match for Cu(I) and a mismatch for Cu(II). In Chapter 7 it was pointed out that, upon removal of copper from the protein, the structure of the protein and the co-ordination site remained intact. Thus the very robust character of the protein allows it, like cytochrome c, to be used as a freely mobile scavenger of electrons. It is very probable that the selection of copper in certain sites is based on the binding of Cu(I) which is the only singly charged cation which is a good Lewis acid and is available to biology.

Fig. 15.1 The active site of plastocyanin and azurin (after the works of H. Freeman and L. Jensen). (a) The β-barrel construction, the ligands, and the substrate binding sites at A and B for two different proteins. (b) Details of the active site.

15.3 Copper and dioxygen

15.3.1 Haemocyanin

There is a complex *extracellular* copper protein in many arthropods which carries dioxygen in the blood. This protein, haemocyanin, has two closely associated copper (I) atoms which bind dioxygen reversibly to give a purple $[Cu(II)]_2 \cdot O_2^{2-}$ complex. The intermediate is a $Cu(I) \cdot O_2$ π-complex. The ligands to the copper are three histidines and the $Cu(I)$ ion is stabilized in roughly tetrahedral geometry. The region around the copper atom is here based on protein helices (Fig. 15.2), which are adjustable so that dioxygen binding is co-operative as in iron-containing haemoglobin or some haemerythrins. It is interesting to compare the larger and smaller ions in the three series of Table 15.4. The use of helices to generate co-operativity via the ion size contraction on dioxygen binding is common to all these dioxygen carriers.

The haemocyanin protein itself is an extracellular protein which appears to be synthesized in a special way restricted to some arthropods (but note that arthropods are the largest group in the entire animal kingdom, 85 per cent of all animal species). Clearly, a large intake of copper is required and the special synthesis is in the hepatopancreas. The protein is put out into the circulating bloodstream as a large aggregate of monomers and in this it is quite unlike haemoglobin and haemerythrin which are found in cells. Unlike them, too, haemocyanin does not require a reductase to be kept in the $Cu(I)$ state since $Cu(II)$ haemocyanin is very difficult to make and, quite possibly, haemocyanin has a redox potential which prevents auto-oxidation. Since haemocyanin is extracellular, it can be stabilized further by binding with calcium ions, which cross-link the protein structure.

15.3.2 A comment on copper of cytochrome oxidase

Cytochrome oxidase is another protein (enzyme) which binds to dioxygen but now it uses the dioxygen in reactions. It has two copper atoms which are widely separated in space and, when we examine them, we observe that one copper, Cu_A, looks somewhat like the blue copper discussed in Section 15.2 and is presumably for electron transfer, while the other, Cu_B, is directly associated with a haem group, cytochrome a_3, at the dioxygen uptake site (Fig. 13.15). Although it is not proven, the dioxygen reaction site appears to be a combination of one iron of the haemoglobin kind, here the cytochrome a_3, and one copper, Cu_B, of the haemocyanin kind, bound by three histidines, and the surrounding protein is probably a typical helical bundle membrane protein (Fig. 13.15). As we have seen in several other biological devices, this α-helical construction is a very useful one for converting chemical energy to conformational change. In these coupled electron transfer centres we suppose that the coupling goes from the reduction of dioxygen to the mechanical movement of helices to the transfer of protons uphill across the membrane; hence the system is a proton pump. Note that this part of the energy capture of the mitochondrial and bacterial membranes is the only part which depends on copper. Similarly, the only part of the photocapture organization of the thylakoid in chloroplasts which depends on copper is in photosystem II,

Fig. 15.2 The dioxygen binding site of haemocyanin drawn schematically to call attention to the a-helical nature of the local structure (see Huber (1989) in Further reading.)

Table 15.4 A comparison of the three dioxygen carrier proteins

Larger ions	Smaller ions	O_2 transport protein
Cu(I)	Cu(II)	Haemocyanin
Fe(II) (high-spin)	Fe(II) (low-spin)	Haemoglobin
Fe(II) (high-spin)	Fe(III) (high-spin)	Haemerythrin

which leads to the production of dioxygen; in these thylakoids only a simple soluble copper blue protein is involved. In all cases the copper system is external even to the cytoplasm of the organelle or prokaryote in that it is either in a vesicle (a thylakoid), or locked in a membrane on the periplasmic surface. The potentials of all these copper proteins are around or above $+0.30$ V and, obviously, these copper-utilizing systems developed late in evolution. Some reactions of NO can also be coupled to energy capture in particular bacteria and they may use a similar system.

Recently, it has been found that there is quite a variety of membrane oxidases which resemble cytochrome oxidase. The analysis of these oxidases, all of which may be proton pumps, has shown that Cu_B is present in them all but that its partner, cytochrome a, may be substituted by other cytochromes. The second copper, Cu_A, and, indeed, a third copper are not essential and cytochrome a may also be replaced by a cytochrome b. The details of cytochrome oxidase and the way in which it could act as a proton pump have been described in detail in Chapter 13.

15.3.3 Copper and extracellular oxidases

The major extracellular oxidases to which this paragraph refers are given in Table 15.2. They usually consist of a combination of 'blue' copper (Type I), pairs of copper atoms (Type III), and extra atoms of a type of copper (Type II) associated with the two Type III copper atoms (Fig. 15.3). These enzymes have an organic substrate site which is thought to be bound some 10 Å from the Type I site, so that there is the electron transfer chain

$$\text{Substrate} \xrightarrow{\ e\ } \text{Type I} \xrightarrow{\ e\ } \{\text{Type II} + (\text{Type III})_2\} \xrightarrow{\ e\ } O_2.$$

(It is possible, at the high redox potentials used, that electron transfer is through some of the peptide bond constructs and not through space as discussed in Chapter 4). The substrate leaves as a free radical. The dioxygen remains bound and is reduced in steps to $2H_2O$, i.e. the oxidases are designed to retain $O_2, O_2^{\cdot-}, O_2^{2-}$ (i.e. H_2O+O), and $O^{\cdot-}$ before giving the second H_2O. The structure of one such oxidase is known (Fig. 15.3), and it is possible to imagine that the three copper atoms (two Type III and one Type II) can act as an interactive system. The reaction steps of the metal ion oxidation states are not very different from that of cytochrome oxidase or, in the reverse direction, from that of the release of dioxygen from the $(Mn)_4$ centre of photosystem II (Chapter 14). The role of copper in these oxidases can be contrasted, however, with that in cytochrome oxidase described in Section 15.3.2. Here we suppose that any pair of the three coppers could be involved in the initial uptake of O_2 since the inhibitor azide (N_3^-), which can be compared with $O_2^{\cdot-}$ or O_2^{2-}, is known to bridge between the type II copper and the type III copper atoms.

Although the co-ordination chemistry at the active site for the non-blue copper is somewhat like that in haemocyanin, in that histidines are involved in the binding of copper, in one striking respect the oxidases differ absolutely from the O_2-carriers and cytochrome oxidase. They are built from several β-sheet domains, like most enzymes (Table 15.5), and they are then constrained so that they do not undergo conformational changes. These oxidases are enzymes, not mechanical transduction devices, and this allows

Fig. 15.3 The scheme of binding of many copper oxidases where each domain is a β-sheet, often a β-barrel. Contrast cytochrome oxidase (Chapter 13). (See Huber (1989) in Further reading.)

Table 15.5 Secondary structures of copper enzymes

Enzymes	Secondary structure
Electron transfer proteins	Single domain β-barrel
Ascorbate oxidase	Three β-sheet domains
Galactose oxidase	Three β-sheet domains
Nitrite reductase	Two β-sheet domains
Cytochrome oxidase	Cu_A domain, β-sheet; Cu_B domain, helical (?)
Haemocyanin	a-helical at active site

Fig. 15.4 The active site of galactose oxidase (see Ito *et al.* 1990 in the 'Further reading').

them to have more selective surfaces. Some copper oxidases are extremely selective.

There are one or two copper oxidases which do not fall into this group as judged by both structures, as far as they are known, and spectroscopic data. One example is galactose oxidase (Fig. 15.4), which has but one copper with a tyrosine ligand that becomes a tyrosine radical during the reaction cycle. The reason for the use of a radical in many sugar reactions is not clear but we have seen this use of a tyrosine radical before, namely in the photoreaction-centre chemistry (Chapter 5) and we shall meet it again in riboside reductases and in some reactions and rearrangements using either Fe–O–Fe or vitamin B_{12} centres (cobalt).

A striking feature of this copper *enzyme* which we shall notice again in Section 15.5.2 and which has been apparent in the description of the magnesium, calcium, zinc, non-haem iron, manganese *enzymes*, and the previous copper oxidases is the presence of a β-sheet secondary structure at or near to the active site. This structure is not present in calcium trigger proteins nor at haem active sites which are α-helical.

15.3.4 Substrates of copper oxidases

The small substrates for copper enzymes are listed in Table 15.1. Leaving aside now the dioxygen or superoxide reaction sites, the enzymes oxidize ascorbate, phenols, some amines, perhaps ferrous ion, and some sugars. The sites of the blue proteins are shown in Fig. 15.1. The role of the 'blue' copper in some of the reactions is simply that of accepting a single electron to generate an organic free radical which can diffuse in the aqueous media outside cells. We can imagine that the objective is protection since the oxidation of phenols or indoles in this way leads to free radical polymerization observed in skins of insects and of plants, i.e. in keratinization of insect cuticle and browning, lignification of cut plant surfaces. Again, the production of ascorbate free radicals may be best seen as the necessary introduction of a feeble free radical in extracellular fluids to scavenge for dangerous free radicals. In this same sense, caeruloplasmin may remove Fe(II) from the blood, which could otherwise generate free radicals from dioxygen. Amine oxidases, also referred to in Table 15.2, are important since random oxidation would produce nitrosamines and other dangerously reactive entities; the amine oxidases help to remove excess transmitters or hormones such as adrenaline.

Reactions with larger substrates of special copper enzymes include the final

cross-linking oxidation of collagen and elastin. Collagen is produced in the cytoplasm, but processed in (Golgi) vesicles by iron oxidases before being released to the extracellular medium; in this medium it is cross-linked by a copper oxidase. The degree of cross-linking is controlled in different parts of an organism, so that the mechanical properties of much connective tissue are dependent on copper; for this reason, copper deficiencies introduce connective tissue weaknesses. We return to the interactive role of copper and zinc enzymes in the management of connective tissue in Chapter 21.

15.3.5 Copper proteins and coenzyme PQQ

Some copper enzymes require a tricarboxylated quinone, coenzyme PQQ (see structure in margin), or a similar dihydroxyaromatic, compound. Known examples are plasma amine oxidase, diamine or histamine oxidase, choline dehydrogenase, dopamine β-hydroxylase, and dopa decarboxylase. This coenzyme PQQ, which is to be found with the copper enzymes outside cells or inside vesicles, goes reversibly to an orthoquinone. There is some suggestion of a general link between this coenzyme, ascorbate, and copper in the control of connective tissue (collagen formation) through the requirement for one or all of these cofactors in lysyl and prolyl hydroxylation.

15.4 Copper enzymes and nitrogen oxides

Amongst chemicals which are today quite readily available due to the dioxygen in the air are nitrate and its reduction products, NO_2^-, NO, and N_2O. In a general way these are dangerous chemicals for multicellular organisms, e.g. nitrates must not be high in drinking water, but they are possible sources of energy (see Fig. 1.8). In fact the denitrifying bacteria utilize this possibility and some copper enzymes assist in the reduction of the lower oxides, but not of nitrate, which has to be handled first in an atom transfer

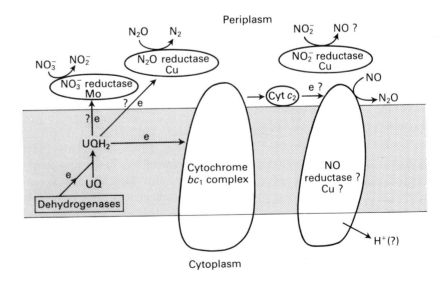

Fig. 15.5 A proposal for the involvement of the stages of nitrate reduction by cells. Nitrite reductase is a molybdenum enzyme while nitrate and nitrous oxide reductases use copper or haem iron. The reactions are largely in the periplasmic space.

reaction by molybdenum (see Chapter 17). The reactions of the lower oxides of nitrogen, like the reactions of N_2 itself, are largely confined to bacteria. Their metallo-enzymes, not all of which are copper enzymes, are usually either in the periplasmic space or in the membranes of the bacteria (Fig. 15.5). We shall draw attention to parallel molybdenum enzymes in Chapter 17.

15.5 Superoxide dismutase

15.5.1 The location

We have already stated that many copper proteins are outside eukaryotic cells and in the periplasmic space of prokaryotes. Others are in vesicles (Table 15.2). We view the contents of a vesicle as being close to extracellular space since vesicles do not contain reductive reaction pathways and, frequently, they equilibrate their internal fluid with extracellular fluids by endocytosis. As stated in Table 15.2, we find copper blue proteins in the inner solution of thylakoids which are vesicular systems of chloroplasts. We also find copper in the chromaffin granule vesicles of the adrenal gland where a copper enzyme is involved in the last step of oxidation to produce adrenalin. Thus copper, like iron, can act both in synthetic and oxidative clearance steps associated with important hormones. There is, however, an exception to this general placing of copper since virtually all eukaryotic cells have one copper protein in the cytoplasm, superoxide dismutase.

15.5.2 The reactions of superoxide dismutases

The involvement of copper in superoxide dismutase in the cytoplasm could be thought of as an anomaly. How can this be explained, especially when the prokaryotes do not have such an enzyme to remove superoxide? It is known that bacteria, algae, mitochondria, and chloroplasts manage to eliminate superoxide with iron or manganese enzymes. Thus, if we are to maintain the point that copper entered biology at a late stage, probably when dioxygen accumulated in the atmosphere, we must ask why it was necessary to supersede the prokaryote Mn and Fe superoxide dismutases? We can well believe that the first superoxide dismutases in the bacteria would be iron or manganese enzymes since their synthesis would utilize available ions. The reason for dispensing with these enzymes late in evolution could well be that these Mn or Fe enzymes are not very stable and readily lose Fe(II) or Mn(II). They would, therefore, be dangerous to DNA in the cell cytoplasm of eukaryotes and especially multicellular organisms. Since both Cu(I) and Cu(II) bind very strongly, this enzyme will not give a measurable concentration of free copper in the cytoplasm. In this light, copper superoxide dismutase is seen to provide a structure for copper with an extremely low (and slow) degree of dissociation. The structure shown in Fig. 15.6 is clerly designed in this way; the copper superoxide dismutase is again a strong β-barrel and, apart from binding copper via *four* histidine sidechains, the enzyme active-site residues are cross-linked into other parts of the structure by zinc bridges. It does appear as if a special dismutase was required in the cytoplasm, especially of multicellular organisms, which would protect DNA from O_2^- and its

(a)

(b)

Fig. 15.6 (a) The structure of Cu,Zn superoxide dismutase showing the β-sheet construction. (b) Detail of the metal-binding sites in Cu,Zn superoxide dismutase. H = histidine. (After J. S. Richardson.)

products but at the same time would not allow a high free metal ion concentration.

15.6 The transport and homeostasis of copper

15.6.1 The binding proteins

High levels of copper in living systems do not appear frequently since copper minerals are rather insoluble. Deficiency is more frequent due to the fact that copper can be removed by precipitation as, for example, Cu_2MoS_4, in the gut of many grazing animals. The deficiency leads to connective tissue weakness, as one might expect (see Section 15.3.4). Once copper is digested, its further uptake is controlled by an albumin circulating in the blood which presumably delivers the copper to the liver; the binding site utilizes four nitrogen donor centres, one of which is an ionized peptide. The liver is the source of many of the oxidases circulating in blood. These proteins are exported presumably through the Golgi and it may well be that copper is incorporated directly in the vesicles; if excess copper is delivered, then there are various powerful chelating peptides and proteins for its removal.

As stated in Section 15.5.2, copper is maintained at very low levels in the cytoplasm of cells; the excess of copper is handled by special ligands, metallothioneins (MT), which bind seven Cu(I) atoms in a thiolate complex and are common in animals, and phytochelatins, which occur mostly in plants. The exact structure of these compounds is not known, but it is worth noting that they are different from those of the zinc and cadmium compounds, especially so in the case of metallothioneins; this would allow bound copper to be recognized as Cu·MT and exocytosed through vesicles, while zinc metallothionein would remain as a zinc buffer in the cytoplasm. Metallothionein synthesis is strongly regulated and it is likely that it is always just in excess of the free Zn, Cu, and any Cd in a cell. Homeostasis is built here around feedback protein synthesis of metallothionein for these three metal ions. Note that the binding constants of metallothionein for copper is very high so that free [Cu] does not exceed 10^{-15} M in cells.

15.6.2 Copper exchange rates

There are three reasons for supposing that, for the most part, there is little signalling due to copper in cells. First, copper exchange rates are usually very slow due to the strength of Cu(I) and Cu(II) binding to ligands. A binding constant at pH = 7 of 10^{15} is to be expected in many of the proteins we have described. Second, copper is largely extracellular and it is difficult to imagine how these copper proteins would provide feedback messages to the DNA. Third, copper is usually bound deeply inside proteins and these proteins are made from β-sheets; hence exchange is inhibited, both in the forward and reverse directions, by the strength of the protein fold.

At least one protein is very different—metallothionein. It is intracellular and has a random structure in the absence of metal ions. The exchange of copper from this protein is much more rapid. Finally, there is reason to suppose that

copper exchange from dopamine β-monoxygenase is rapid. We refer to this protein again in Chapter 21.

15.6.3 Regulatory copper proteins

In certain organisms stimulation of metallothionein synthesis by copper is observed. It is by copper (I) since silver (I) has the same effect, and leads to the appearance of a copper metallothionein protein in vesicles, i.e. outside the cytoplasm. Stimulation of metallothionein by zinc also generates metallothionein but it remains in the cytoplasm as the zinc protein. The protein which in yeast stimulates metallothionein to form and bind copper is known. It is a sulphydryl-containing protein called ACE-1; this protein binds several Cu(I) atoms in clusters. (The zinc protein has not been identified yet.) Here the free copper is controlled by metallothionein synthesis and regulated by the copper (I) complex of ACE-1 to keep it below 10^{-15} M in the cytoplasm. It is likely that the zinc control protein is also based on sulphydryl-donors, but it may be a zinc finger protein rather than being a cluster protein.

15.7 The overall functions of copper

We now put to one side the use of copper in simple electron transfer at high potentials >0.4 V as is found in energy-transducing systems; we have offered an explanation of the design of the proteins in this case. Clearly, except for this function, copper is used protectively and in the cross-linking final steps of synthesis of extracellular filaments and tissues. Many copper oxidases help to produce the extracellular matrices of living eukaryotic organisms. The examples go from the chitins, the cuticles of insects, to the collagens of many species and to many resins and lacquers released by plants. In plants, the copper oxidases, again free radical oxidases, are used in a further protective manner when the tissues are damaged, producing phenol oxidation (and in man the formation of melanin), 'browning' reactions which are the beginnings of protective films of plastic; this 'browning' is a protection against further damage even from light. However, protection by copper goes further in that both in the cell cytoplasm (superoxide dismutase) and on the outside of cells (ascorbate oxidase) two copper enzymes either scavenge free radicals or perhaps produce *weak* free radicals to remove strong ones. Copper oxidases also remove excesses of transmitters—amines—but are used to produce adrenalin, a signalling substance for alerting an animal to danger.

The curiously intriguing feature is the way in which the reactions are separated from parallel oxidative functions controlled by iron (haem) or manganese enzymes. The explanation that copper was not used until the coincidental appearance of substantial quantities of dioxygen and the arrival of multicellular organisms looks to be extremely probable. The scheme of events would be:

(1) early life in the presence of extremely insoluble copper sulphides (Fig. 1.11); no copper proteins;

(2) increasing amounts of dioxygen in the atmosphere;

(3) increase in concentration of copper(II) resulting from the oxidation of copper sulphides, and the coincidental loss of iron as $Fe(OH)_3$ (see Fig. 1.11);

(4) use of copper in such energy-capture systems as cytochrome oxidase;

(5) increase in free radical attack on the extracellular matrix of unicellular organisms;

(6) evolutionary development of cross-linked connective tissue, e.g. collagen and chitins via free radical polymerization (leading to multicellular organisms).

The further use of copper in multicellular organisms could then follow:

(7) protective coats of insects, etc.;

(8) internal protection both in cells and in the bloodstream from adventitious free radicals;

(9) the development of hormonal messengers, e.g. adrenalin.

The complexity of the uses of copper is shown in Fig. 15.7. If all this is true, copper became the new essential element for multicellular development in life. It is curious to note that this development probably arose from the introduction into the environment of an extremely 'poisonous' substance (dioxygen) followed by another one, copper, and that adaptation to their presence led to such striking evolution.

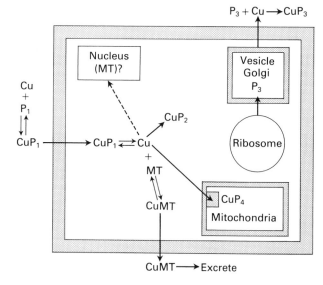

Fig. 15.7 A general scheme for the handling of copper in cells and their compartments. MT is metallothionein.

However, a multicellular organism has to grow, so that connective tissue must be continuously broken down as well as generated. Zinc enzymes degrade collagens; hence there is the obvious necessity for the zinc and copper enzymes to work together in the construction of the multicellular system. Such a dynamic homeostasis (see Chapter 21) is quite possibly made secure by the utilization of co-ordinated control over both copper and zinc by metallothio-

nein. Note that excess zinc in humans damages copper metabolism and pregnant animals including women *should not* be given zinc supplements.

Further reading

Adman, E. T. and Jensen, L. (1981). Azurins. *Israel Journal of Chemistry*, **21**, 8–12.

Agrawal, O. P. and Scheller, K. (1990). Tyrosinase and the chitin cuticle. *Zeitschrift fur Naturforschung*, **45**, 129–31.

Ciba Foundation Symposium (1980). (Eds. D. Evered and G. Lawrenson.) *Biological roles of copper*. Ciba Foundation Symposium, No. 79. Excerpta Medica, Amsterdam.

Ferguson, S. J. (1987). Denitrification. *Trends in Biochemical Science*, **12**, 354–8.

Gaykema, W. P. J., Volbeda, A., and Hol, W. G. J. (1986). Structure of haemocyanin. *Journal of Molecular Biology*, **187**, 255–75.

Guss, J. M. and Freeman, H. (1983). The blue copper electron transfer proteins. *Journal of Molecular Biology*, **169**, 521–63.

Hu, S., Furst, P., and Hamer, D. (1990). DNA and copper binding functions of ACE 1. *The New Biologist*, **2**, 544–55.

Huber, R. (1989). Copper clusters in oxidases, *Angewandte Chemie (International Edition)*, **28**, 848–69.

Ito, N., Phillips, S. E. V., McPherson, M. J., Keen, J. N., Stevens, C., Yadav, K. D. S., and Knowles, P. F. (1991). Galactose oxidase. *Nature*, **350**, 87–90.

Lontie, R. (ed.) (1984). *Copper proteins and copper enzymes*, Vol. 1. CRC Press, Boca Raton, Florida.

Messerschmidt, A. and Huber, R. (1990). The blue oxidases. *European Journal of Biochemistry*, **187**, 341–52.

Peters, T. (1985). Phytochelatins. *Advances in Protein Chemistry*, **37**, 161–245.

Rauser, W. E. (1990). Phytochelatins. *Annual Reviews of Biochemistry*, **59**, 61–86.

Richards, M. P. (1989). Copper regulatory proteins. *Journal of Nutrition*, **11**, 1062–70.

Sigel, H. (ed.) (1981). *Metal ions in biological systems*. Vol. 12, *Copper*, and Vol. 13, *Copper proteins*. Dekker, New York.

16

Nickel and cobalt: remnants of early life?

16.1	Introduction	400
16.2	The chemistry of cobalt and nickel	402
16.3	Hydrogenases	405
16.4	The reactions of vitamin B_{12}	406
16.5	The organometallic chemistry of life	408
16.6	Urease: nickel in vacuoles	409
16.7	Nickel uptake	410
16.8	Conclusion	410

16.1 Introduction

Fig. 16.1 The structure of the nickel-containing coenzyme, F-430.

We start this chapter with the remark that the functions of these two elements appear to belong to very primitive systems indeed. In some sense it looks as if the elements were at their most useful when metabolism was based on such chemicals as CH_4 and H_2, but that after the advent of dioxygen their value diminished. In higher organisms the activity of nickel is apparently confined today to one curious enzyme, urease, although the symbiotic anaerobic bacteria of these higher organisms still use nickel in some dihydrogen reactions. Free anaerobic bacteria, especially methanogens, have also kept the nickel hydrogenases and other nickel enzymes, but the methanogens belong to the special class of bacteria, the archaebacteria, most of which have been relegated to anaerobic niches within the geochemical world. They are thought to be of primitive ancestry and also use uniquely special cofactors including the F-430 ring chelate of nickel (Fig. 16.1) as well as a special sulphydryl coenzyme (see Chapter 19).

The well recognized activity of cobalt is confined to functions of vitamin B_{12} (Fig. 16.2). All other cases of cobalt-activated systems have been identified in a somewhat unsatisfactory way by adding back cobalt to extracted systems and we shall not describe them further (but see Chapter 11 on zinc enzymes). The use of vitamin B_{12} in biology may also be on the way out in an evolutionary sense, since the transfer of methyl groups and the rearrangement reactions by radical mechanisms, which depend on B_{12} in some organisms, can be done by

Fig. 16.2 The structure of the methylated coenzyme from vitamin B_{12} which contains cobalt. The B_{12} coenzyme for many other reactions has a cobalt–carbon bond to the $5'–CH_2^-$ of adenosyl, see inset.

methionyl CoA or iron- and manganese-utilizing enzymes, respectively, in other organisms. These iron and manganese enzymes are dependent on dioxygen. Vitamin B_{12} is also required in archaebacteria and this association of nickel and cobalt with early anaerobic organisms suggests immediately that organometallic chemistry may have had a considerable importance in early life forms.

Whereas in haem biochemistry there are a variety of chemical modifications of the haem itself, through such reactions as formation of thioethers or oxidation, the rings of vitamin B_{12} and, as far as we know, that of F-430 are fixed. Variation is introduced in the B_{12} series by changes in the nature of the bound base, benzimidazole, but the metal, cobalt, is not bound by sidechains of proteins, as is the case for the metals of haem and chlorophyll. The situation is less clear in the case of the nickel of F-430. It is quite possible that the

diversity of haem chemistry is of recent evolutionary origin, following the advent of dioxygen, e.g. cytochromes *a* and sirohaems, and that it was relatively simpler in the distant past (see Chapter 13).

Good reasons for seeking to eliminate nickel and cobalt from biology today are that neither seems to have a very special chemistry in acid/base catalysis which could not be replaced by zinc, which has chemical advantages (Chapter 11), and their redox functions in a dioxygen-dominated atmosphere would seem to be replaceable by the more accessible iron, copper, or manganese. The source of carbon today is virtually confined to CO_2, so that O_2 must be lost in synthesis of carbon compounds, in direct contrast with the necessary loss of H_2 when CH_4 was a carbon source. Again, cobalt is a relatively rare element and today nickel is not easily available from silicate rocks, e.g. soils, since, like Co(III), it is highly stabilized in the octahedral holes of silicate lattices. However, nickel is available from the sea and we must suppose that it is unwanted by biology since its exchange kinetics are too slow (Chapter 4) and it could be a threat to many Mg^{2+} reactions since both Ni^{2+} and Mg^{2+} have very similar structural demands.

16.2 The chemistry of cobalt and nickel

Iron, cobalt, and nickel have two features in common which separate them from earlier and later elements in the periodic table. All three elements give a variety of oxidation states and a variety of spin states and all have more than five 3d electrons in their lower oxidation states, i.e. they are electron-rich. While stability of oxidation state (III) falls along the series Fe to Cu, the stability of oxidation state (I) increases, but oxidation state (II) is dominant through all this series of elements. Through a combination of increasing electron affinity and decreasing size in any one oxidation state the balance of bound ligands shifts from oxygen donors to nitrogen donors, while sulphur donors retain a reasonable and increasing binding power relative to oxygen donors throughout the series (Chapter 2). All these factors make the availability of electrons high from Fe^{2+} to Cu^+. This general picture is complicated by the ligand field stabilization energies of the different metals in different oxidation states (Chapter 2) which allow stability to particular electronic states of the different ions to appear and also connect particular stereochemistries with these states. Table 16.1 lists some features. The elements nickel and cobalt can clearly be described as special, in that they are not only generally electron-rich, especially in lower oxidation states and in low-spin states, but in that some of their 3d electrons are forced into exposed σ- (or π-) orbitals in these low-spin states by the preferred symmetry of their complexes. In a conventional tetragonal field it is usual for a d^8 low-spin ion to have a very exposed doubly occupied d_{z^2} orbital, so that it is a donor rather like RS^-, i.e. it is a very strong Lewis base. The tetragonal low-spin d^7 ion is a reactive free radical, able to give or take an odd electron, also in the exposed d_{z^2} orbital, to or from other molecules. In this respect d^7, low-spin ions are to be compared with σ-organic free radicals. This exposure of d electrons in σ-orbitals is only possible in low-spin states after d^6, Fe(II) in the first

Mn Fe	Co Ni (Cu)	(Cu) Zn
3d shell reactivity in π-orbitals	3d shell reactivity in σ-orbitals	3d shell reactivity almost removed
O_2, NO	H_2, CH_3^-	

Table 16.1 Occurrence of ligand field states

		High-spin		Low-spin	
		Octahedral	Tetrahedral	Octahedral	Tetragonal
Iron(II)	d^6	Usual	Less common(S)*	Quite stable(N)*	Rare
Iron(III)	d^5	Usual	Common(S)	Less stable(N)	Rare
Cobalt(I)	d^8	?	?	Rare	Usual
Cobalt(II)	d^7	Common	Common	Rare	Usual
Cobalt(III)	d^6	(Rare)	Rare	Usual	Rare
Nickel(I)	d^9	(Tetragonal)	Rare	Rare	Usual
Nickel(II)	d^8	Dominant	Rare	Rare	Usual
Nickel(III)	d^7	Common(?)	Common(?)	Rare	Usual
Copper(II)	d^9	(Tetragonal)	Rare	—	—

* S and N refer to the nature of the donor atoms.

transition metal series, since the division of orbitals in an octahedral field protects at most six 3d electrons in the lower set of orbitals of Fig. 4.8. These orbitals have π- and not σ-character. We see that low-spin nickel and cobalt ions have a very special potential reactivity associated with occupied σ-orbitals as well as having π-donor capability and lower symmetry (tetragonal) than is usually seen for low-spin d^6, Fe(II). The lower symmetry and co-ordination number expose an active site (Figs 16.1 and 16.2).

At this moment it pays us to observe how these simple considerations have been optimized in the biochemical world. First, tetragonal ligands have been developed from porphyrin for Fe, Co, and Ni based on ring structures. The evolved rings are not the same for the three metal ions, Section 3.12. For iron, the porphyrin ring is highly conjugated and symmetrical to protect to a marked degree *all* the out-of-plane 3d electrons of Fe(II) (six 3d) or Fe(III) (five 3d) in π-orbitals. We have already described how the fifth and sixth co-ordinating positions, which are often filled by protein sidechains, can be used to modulate the properties of the iron in these porphyrin complexes (Chapter 12). Where one site is open, as in oxidases, the switch to low-spin d^6 gives a high affinity to binding of dioxygen. For cobalt, the ligand ring size is smaller (Fig. 16.2) giving a smaller hole to stabilize Co(III) by σ-bonding, and now in B_{12} the remaining fifth and sixth co-ordination positions are not from the protein. This makes the coenzymes from vitamin B_{12} into apparently stable octahedral low-spin d^6 complexes. However, two other factors have changed from iron porphyrin chemistry: (1) in the structure, the fifth co-ordination position is occupied by a non-protein base, usually a benzimidazole, and (2) the sixth co-ordination position is occupied by a saturated σ-carbon donor to make a metal–carbon bond. The unit becomes a transferable coenzyme and B_{12} derivatives are thought of as vitamins. The chemistry with a carbon–metal bond is quite unusual for elements normally forming rather polar bonds in water and, in effect, it provides a relatively weak bond which can break in two ways:

$$Co^{III} - CR_3^- \longrightarrow Co(I) + {}^+CR_3,$$

$$Co^{III} - CR_3^- \longrightarrow Co(II) + {}^\cdot CR_3.$$

Hence, cobalt–carbon chemistry in low-spin complexes is associated with easy development of carbonium ion and carbon-radical intermediates involving

saturated carbon chemistry and tetragonal d^8 or d^7 oxidation states. The combination of cobalt with the B_{12} ring system and carbon chemistry is a masterpiece of synthetic judgement produced by evolution.

The ligand for nickel is F-430, which is no longer a planar ring due to its heavier saturation (Fig. 16.1). This ligand can be used to provide open space mainly to one side of the nickel ion. In this complex the low-spin d^8 ion (Ni(II); compare Co(I)) can accept a CH_3^+ group to form a Ni–CH$_3$ bond. If we look upon this state as a Ni(IV)–CH$_3^-$ compound, a $3d^6$ complex, we see that it can go on to form an octahedral unit. It is now well suited to combining with CO in the absence of dioxygen, much as is the open-sided form of Fe(II) porphyrin ($3d^6$) in haemoglobin. A reaction typical of organometallic chemistry could then follow because this metal ion has two *adjacent* sites not blocked by the ring chelate, in contrast to porphyrin chemistry. CH_3CO^+ is a leaving group

$$
\begin{array}{ccc}
H_3C^- \ \ CO & & CH_3 \\
\ \ \ \backslash \ / & \longrightarrow & | \\
 & & CO^+ \\
Ni(IV) & & | \\
 & & Ni(II)
\end{array}
\qquad (16.1)
$$

for general synthesis. (The exact pathway may not perhaps be of this kind since we know that in methanogens the first carbon to enter must come from CO_2 and it must be reduced to CO before it can be activated in the organometallic synthesis (16.1); Fig. 16.3.) Insofar as this chemistry is correctly described, it is possible to see how both cobalt and nickel systems were delicately developed to *activate very simple C_1 carbon compounds*. C_1 syntheses are now unusual and most biochemistry starts from C_2, except for photosynthesis where the C_1 unit is incorporated into sugars directly. A further use of the nickel in the F-430 ring system appears to be as a desulphurization catalyst.

In conclusion, we can see how evolution has prepared ring systems so that special chemistry could be carried out with very particular and admirably useful metal ions in special spin states and geometries (Table 16.1). The

Fig. 16.3 The reaction pathway of carbon fragments in archaebacteria. x, y, and z are catalytic sites on a framework; THF, tetrahydrofolate. Note the involvements of Ni and Co (B_{12}). After the work of Wolfe and Thauer.)

elements selected, Ni and Co, have special value opposite σ-bond organometallic chemistry, i.e. M–H and M–CH$_2$R. The switch to high redox potential reactions that dioxygen brought about removed the absolute need for these low potential reactions which vitamin B$_{12}$ and F-430 achieved. We have described the reactions of haem proteins in Chapter 13, but much of that chemistry came later in historical time, and we note particularly the significance of the ring propionate in energy capture (Chapter 12). While noting the sophistication of this small-molecule chemistry, we must also note that it developed within proteins. Co-operativity in this development is demanded by evolutionary pressure so that the special reactivities at the metal centres, already highly developed by the ring ligand, were further enhanced by proteins. The imposed structures generate 'entatic' conditions in the co-ordination sphere, which are illustrated in Section 16.4. Again, selective sites for substrates evolved, which do not rely on binding to the rings. We look first at some further nickel enzymes which were required to activate another early reactant, dihydrogen, and which are still used by anaerobes.

16.3 Hydrogenases

The ability to react with dihydrogen, H$_2$, or to produce dihydrogen from water is well known in chemistry. The reactions are catalysed by a variety of compounds containing at least the following metals which are available to biology: Fe, Co, Ni, Cu. It is also easy today for man to make hydrogenation catalysts from rarer metals, such as Ir, Rh, Pt, and Ag. Clearly, the best metal ions are from the electron-rich ends of the transition metal series in the periodic table, but if we assume that copper was not available to biology in its earliest period and that cobalt was more of a rare element than iron or nickel, then these last two elements, as sulphur complexes or clusters, could well have been the preferred catalysts in biology for these reasons alone. Since the reactions are redox reactions and involve the divalent state, a considerable advantage from the use of nickel is that it dissociates less readily from a protein. The easiest reaction to envisage for the activation or release of a molecule of dihydrogen is hetereogeneous dissociation, which can be written reversibly

$$M^+ - X^- + H_2 \rightleftharpoons M^+ H^- + H^+ X^-. \tag{16.2}$$

This is the way the reaction of copper or silver acetate with dihydrogen is written and the mechanism involves the loss of one ligand from the co-ordination sphere. Now, it is known that the hydrogenases which involve iron alone are derived from somewhat unusual Fe$_4$S$_4$ clusters. It is known too (Chapter 12) that other iron clusters which are reactive and have one iron different from the other three in the enzymes lose one iron atom relatively easily. It could well be that an iron enzyme involving divalent iron and a ligand that is removed in the course of reaction is always open to loss of metal. No matter what the value and disadvantages of iron, however, the dominant system which evolved for hydrogenases uses the more strongly bound, non-exchanging nickel. The *chemistry* is written as

where removing H^+ may require a protein base, written as X in (16.2). The comparison with rhodium and iridium organometallic catalysts is worth examination for the sake of comparison with man's catalysts. Some other special reactions of nickel enzymes are included in Table 16.2 for comparison.

Table 16.2 Nickel-containing enzymes: localization and function

Enzyme	Localization	Function
Urease	Extracellular (Vesicle)	Hydrolysis of urea
CO-dehydrogenase	Membrane	Catalysis of the reaction $4H_2 + 2CO_2 = CH_3COOH + 2H_2O$ (acetogenic bacteria)
Factor F-430	Membrane	Catalysis of the reaction $4H_2 + CO_2 = CH_4 + 2H_2O$ (methanogenic bacteria)
Bacterial hydrogenases (several)	Membrane	Catalysis of the reaction $H_2 = 2H^+ + 2e$

NB. Nickel enzymes may well be involved in desulphurization too.

16.4 The reactions of vitamin B_{12}

Table 16.3 lists some of the enzymes which contain vitamin B_{12}, either in the simple form or as a coenzyme. There are four groups of important reactions:

(1) the reduction of ribose to deoxyribose;

(2) the rearrangement of diols and similar molecules;

(3) the rearrangement of malonyl to succinyl;

(4) the transfer of methyl groups.

An immediate observation is that all these reactions require the approach of quite big substrates to the cobalt since even in methyl transfer the transfer is to methionyl CoA. One side of the vitamin B_{12} ring system must be very open during reaction and we know that this is the side of the carbon σ-bond (Fig. 16.2). The benzimidazole base remains attached. The protein groove containing the cobalt vitamin must also be able to open very considerably as in cytochrome P-450 but not in haemoglobin (Fig. 16.4). Such a system could only remain secure if it did not react with dioxygen or if it was protected from dioxygen. It is known that vitamin B_{12} Co(II) and Co(I) complexes separated from enzymes are vulnerable to dioxygen and we conclude that the B_{12} enzymes must react today by preferred order mechanisms, i.e. in the order:

Table 16.3 Some examples of enzymes requiring B$_{12}$ coenzymes: localization and function

Enzyme	Localization (in bacteria)	Function
Methionine synthetase	Cytosol	Conversion of homocysteine into methionine
Methylmalonyl CoA-mutase	Cytosol	Conversion of methylmalonyl-Co-A into succinyl-CoA
Glutamate mutase	Cytosol	Conversion of L-glutamate into L-threo-β-methylaspartate
Diol dehydratase	Cytosol	Conversion of diols into aldehydes
Lysine mutase	Cytosol	Conversion of D-lysine into 2,5-diaminohexanoic acid
Ethanolamine deaminase	Cytosol	Cleavage of amino group in $NH_2CH_2CH_2OH \rightarrow CH_3CHO + NH_3$
Ribonucleotide reductase	Cytosol	Reduction of ribonucleotides to the corresponding deoxyribonucleotides

Fig. 16.4 A plausible reaction pathway for the diol-dehydratase reaction catalysed by coenzyme B$_{12}$.

blocked centre; binding of substrate to open the centre; reaction; close centre as products leave.

The exact nature of the cobalt intermediate has been established as Co(II) in reactions (1), (2), and (3), but is Co(I) in reaction (4). The use of a radical pathway follows from the homolytic splitting of Co(III) bond to the carbanion of adenosine (Fig. 16.2). The diradical formed has as its primary reaction centre the carbon radical of a sugar, which allows H-atom transfer to and from substrate while rearrangement occurs (Fig. 16.4). The mechanism has been thoroughly analysed and the Co–C bond breaking in the enzyme requires a small amount of energy (10 kcal) to be compared with 20 kcal in free solution. Such is the power of the protein.

In methyl transfer it is Co(I) which is the acceptor/donor of CH_3^+. This is a curious reaction since the redox potentials of cobalt in the vitamin indicate that Co(I) is extremely difficult to obtain in water, but there appears to be a starter reaction involving Co(II) in the enzyme reaction sequence. Interestingly, Co(I) in these complexes does not form a Co–H bond with H^+ until a pH of around zero (contrast Ni and Fe hydrogenases). The discrimination in favour of CH_3^+ against H^+ may be due to the very large size of the d_{z^2} orbital.

Of very special concern is the reduction of ribose to desoxyribose which is considered to go via the radical pathway. There is a certain bewilderment in the thought that there was an early requirement for a metal complex as complicated as vitamin B_{12} in such a central reaction; perhaps we can accept that the earliest life was based on RNA, but it looks almost impossible to conceive how a switch to DNA took place if this switch needed vitamin B_{12}. It is probable then that vitamin B_{12} had evolved much earlier to assist organometallic chemistry. The still later change for this reaction to an iron or manganese enzyme dependence of seemingly greater simplicity (Chapter 11) clearly requires a very high redox potential, i.e. of dioxygen, to generate tyrosine free radicals. Cobalt allows free radical chemistry at very low redox potentials. Something of extreme importance for the understanding of the early development of living systems is still missing but there is the suggestion that cobalt in vitamin B_{12} may be on the way out today.

16.5 The organometallic chemistry of life

Before we discuss the relevance of organometallic chemistry to early life we must examine the nature of the probable ligands. This organometallic chemistry could not be that of non-aqueous phases using such ligands as phosphines or cyclopentadienes which are so common in man's organometallic chemistry. The obvious electron-rich ligands in water are compounds of sulphur, sulphide, and thiolates. Today organometallic chemists avoid this chemistry advisedly since it is very reactive; biology could not avoid it, but it could protect this chemistry in proteins. Now, as well as providing water-soluble electron-rich compounds, small colloidal sulphide particles, M_nS_n, and anions such as MoS_4^{2-}, the early earth provided large sources of H_2S and various sulphide surfaces all far out of equilibrium in the sea. The source of the unstable minerals and chemicals was and still is volcanoes and ocean vents.

The reactions and the development of these surfaces and particles as non-living *autocatalytic systems* in reactions with CO_2, H_2, CH_4, and so on, could have been the point at which life began. It could then have generated just random polymer production, at first associated only with templates of sulphide or clay particles; the next step to ordered polymer production we all would recognize as genuinely living organometallic chemistry. Much research on the reaction of sulphide complexes, colloids, and surfaces would appear to be a priority to an understanding of these possibilities. Some residual chemistry of this kind could rest in the hydrogenases (Ni), nitrogenases (Mo, V, Fe), and ferredoxins (Fe). Note that the sulphide chemistry could have been energized from the very nature of its sources (see 'Further reading').

16.6 Urease: nickel in vacuoles

The disappearance of nickel chemistry from higher organisms can be accounted for partly by the strength with which nickel is held within oxide rocks in octahedral holes and partly by the kinetic risk which it generated since the style of redox chemistry it promotes is very vulnerable to dioxygen. There is quite curiously residual nickel in urease, where it acts in the animal and plant kingdoms as an acid catalyst in the particular pathway of decomposition of urea so as to generate ammonia. The reaction is

$$H_2N \cdot CO \cdot NH_2 \xrightarrow{H_2O} H_2N \cdot CO_2^- \cdot NH_4^+ \xrightarrow{H_3O^+} 2NH_4^+ + HCO_3^-.$$

Two nickel ions are used in the enzyme in some concerted fashion. We must suppose that some special advantage accrues from the use of nickel rather than zinc, but it is hard to state what that advantage is. Notice that the nickel is in a vesicle and, generally, nickel taken up by plants is deposited in vacuoles. Modern organisms reject nickel, but at the same time the urea cycle is very important for them, requiring nickel to keep ammonia balance.

This urea cycle carries out the net reaction

$$NH_4^+ + HCO_3^- + 2H_2O + aspartate + 3\,ATP$$
$$\downarrow$$
$$Urea + fumarate + AMP + P \cdot P + 2P + 2ADP$$

We are not concerned in this book with the intermediate steps but we observe that it is one route for the elimination of excess nitrogen (NH_4^+) as urea. Excess ammonia is very poisonous and different species have different modes for its elimination. Man, amongst many land animals, uses the urea cycle in this way. Now urease itself does the reverse in that it generates eventually ammonia from urea. While this could be advantageous where nitrogen is in short supply, e.g. in bacteria, it seems pointless in higher animals. The mechanism of hydrolysis via ammonium carbamate may be pointing to a quite different function for the enzyme but it is not known. Finally we note again that the use of this enzyme with two *nickel* atoms at the active site is indicative of a very special use of nickel bound to thiolates. What can it be and why is zinc not used? Why is it locked away in vesicles?

16.7 Nickel uptake

The uptake of nickel into bacteria and plants must be mediated by some energized system. There is the strong suggestion that the pump is, in fact, the magnesium pump. The hydrate of nickel (or of cobalt) is very similar in all respects to that of magnesium. Of course, this coincidence has its dangers since magnesium is absolutely essential and nickel must be selectively removed, e.g. by plants into vacuoles. Risks to the health of humans from nickel poisoning will be discussed in Chapter 22. The way in which nickel and cobalt are so selectively incorporated in their special ring systems is not understood, but we have described potential procedures in Chapter 2.

16.8 Conclusion

The environment in which life began could be called electron-rich; it had an atmosphere of reducing gases and much sulphide was available at the surface. Biology, in such an atmosphere, needed special catalysts to handle compounds such as CH_4, H_2, and H_2S. Cobalt and nickel centres are very valuable in these circumstances since they have reactive sulphides and their complexes could develop to give σ-donors in low-spin, open-sided systems. The advent of dioxygen changed the environment to one which was relatively electron-poor, containing O_2, N_2, and CO_2. This change demanded a switch in catalysts to metals such as iron, copper, and manganese. Such is the depth of involvement of the trace elements in life. In the 'hidden-world' of archaebacteria, major catalysts remain Ni- and Co-based (F-430 and corrin, for example), but today in higher organisms these catalysts are lost or trivial in extent when compared with those of iron and copper oxidases. In large part nickel is now rejected, cobalt as a simple ion is not even taken up and became part of a vitamin. Where nickel is used in higher organisms it is in vesicles.

Further reading

Andrews, R. K., Blakeley, R. L., and Zerner, B. (1984). Biochemistry of urease. *Advances in Inorganic Biochemistry*, **6**, 245–83.

Dolphin, D. (ed.) (1982). B_{12}. Vols I and II. John Wiley, New York.

Golding, B. T. (1990). The B_{12} mystery. *Chemistry in Britain*, 950–4.

Pratt, J. M. (1972). *Inorganic chemistry of vitamin B_{12}*. Academic Press, New York.

Thauer, R. K., Diekert, G., and Schönheit, P. (1980). Biological role of nickel. *Trends in Biochemical Science*, **5**, 304–6.

Walsh, C. T. and Orme-Johnson, W. H. (1987). Nickel enzymes. *Biochemistry*, **26**, 4901–6.

Wolfe, R. S. (1985). Unusual coenzymes of methanogenesis. *Trends in Biochemical Science*, **10**, 396–9.

17

Molybdenum, vanadium, and tungsten

17.1	Molybdenum: introduction	411
17.2	The availability of molybdenum	412
17.3	The molybdenum enzymes: a first overview	413
17.4	Physical methods for enzyme and model studies	415
17.5	Relevant model chemistry of molybdenum	416
17.6	Rates of reaction	422
17.7	Biological chemistry of molybdenum: general considerations	423
17.8	Molybdenum: conclusions	427
17.9	Vanadium chemistry and biochemistry	427
17.10	Tungsten biological chemistry	431
17.11	Molybdenum, vanadium, and tungsten in evolution	432

17.1 Molybdenum: introduction

Group V	Group VI
(P)	(S)
V	(Cr)
Nb	Mo
Ta	W

Molybdenum, vanadium, tungsten and related elements

In this chapter we shall approach the biological chemistry of a metal, molybdenum, which has some features of non-metal chemistry, in a different way from that used to describe the biological chemistry of the metals of the previous chapters. Here we shall place a stress on model structural chemistry and considerations of equilibria in solution in order to illustrate the ways in which inorganic chemists frequently work so as to understand metalloproteins. Their objective usually is to make low-molecular-weight compounds which reflect the composition and the properties of the naturally occurring molecules often before these are known in full. Molybdenum biological chemistry is peculiarly suitable for a description of this attack since it is known that most if not all of the compounds in living systems which contain molybdenum actually have the metal bound in one of two low-molecular-weight extractable units, i.e. a multimetal cluster, FeMoco, and a metal–pterin complex, but both are of unknown detailed structure and, as yet, there is no structure for any molybdenum protein. (Note immediately that non-metals

such as phosphate and sulphate are often held in such cofactors or coenzymes.) Before looking at the models we must give some biological information about molybdenum.

17.2 The availability of molybdenum

The availability of molybdenum from the sea is relatively high (see Fig. 17.1), but its distribution on land is extremely patchy. In a small country such as Great Britain there are areas where soils generate an agricultural problem due to a deficiency of molybdenum and other areas where molybdenum is in excess. Perhaps the greatest problem arises from the excess molybdenum in soils since it causes copper deficiency in animals grazing on such lands. It would appear that the difficulty arises from the metabolism of the anaerobic bacteria in the rumen of these animals. One of the reactions of molybdenum on entering the anaerobic first digestive system of the ruminant is the conversion of the molybdate into thiomolybdate by the bacteria of the stomach. Unfortunately, MoS_4^{2-} is a very good scavenging precipitation reagent for copper and so the copper content of the diet is reduced almost proportionately with the excess of molybdenum. The animals then suffer from weakened connective tissue since copper enzymes are required to cross-link collagen external to the animal cells (Chapter 15). In other geographical areas there is molybdenum deficiency which frequently relates to the difficulties of incorporation of molybdenum from molybdate of soils (see Chapter 2).

To get molybdenum chemistry in its proper biological perspective let us turn next to a brief survey of molybdenum in enzymes. Having this information in mind it will then be convenient to outline the chemistry of molybdenum relevant to biology through the study of models before turning again to a more detailed analysis of molybdenum action in the enzymes.

Fig. 17.1 The concentrations of some elements in sea water. Note that these are total concentrations and do not refer to free ion species.

17.3 The molybdenum enzymes: a first overview

Table 17.1 lists the known enzymes and their cofactors. The enzymes are functionally quite intriguing; in effect, four of them, xanthine oxidase, aldehyde oxidase, sulphate reductase, and nitrate reductase, do an oxygen atom transfer or two-electron reaction

$$X + O \longrightarrow XO.$$

(It may be that formate dehydrogenase is also in this group.) Immediately we see that this reaction avoids free radical redox steps and a good reason is apparent for the use of molybdenum. This element is one of the first metals in the periodic table to undergo relatively easy $M{=}O$ bond breakage to $M + O$, i.e. a 'two-electron leaving-group reaction'. (We know this from the exchange rates of O from M.) In fact, this is a typical reaction of a heavy element non-metal oxide and it is well known in the chemistry of selenium (SeO_2

Table 17.1 The major molybdenum enzymes

Enzymes	Reaction	Cofactors
Nitrogenase	Dinitrogen to ammonia	Fe/S, ATP
Aldehyde oxidases	Aldehyde to acid	Fe/S, flavin
Nitrate reductase	Nitrate to nitrite	Fe/S, flavin
Sulphite oxidase	Sulphite to sulphate	Fe/S, flavin
Xanthine oxidase	Purine or pyrimidine catabolism	Fe/S, flavin
Formate dehydrogenase	Formate to carbon dioxide	Fe/S

Group VA	Group VI	Group VII
	(S)	
(V)	Se	(Br)
Mo		I
W		

Fig. 17.2 Atom centres from which oxygen atoms are relatively readily transferred.

oxidations) and iodate (see Fig. 17.2 and Chapter 19) but, while selenium catalyses such reactions in the cell, molybdenum carries out these oxygen atom transfer reactions *outside* the cell cytoplasm (Fig. 17.3). (Note in passing that selenium and molybdenum belong to the same group as sulphur in the periodic table.) Thus, the outside of the cell has both a one-electron series of copper catalysts (Chapter 15) and a two-electron molybdenum series both of which (with other metals, e.g. vanadium) can be thought of as functioning in contrast with iron in the cytoplasm (see Section 17.11 on the evolution of the uses of these elements).

The second group of enzymes are the nitrogenases which are in the cell cytoplasm mostly of prokaryotes. The complexity of the enzyme is startling, but it is fairly clear that each molybdenum atom acts at a separate catalytic centre. The question is why molybdenum here? It is not to do just with the formation of $Mo{-}N_2$ complexes since this property is common to many elements and, therefore, the underlying reason has to be the ability to go between oxidation states readily, e.g. from Mo(III) to Mo(VI) (Fig. 1.7). Here lies a second property to add to the leaving group strength of molybdenum— the small change in redox potential from Mo(III)/Mo(IV) up to Mo(V)/ Mo(VI) and the low values of these potentials, see Fig. 14.1. We shall see that this is valuable when we consider all the enzyme reactions of molybdenum.

All the molybdenum-containing enzymes are very large and complex and it is notable that they have multiple units of other cofactors besides molybdenum

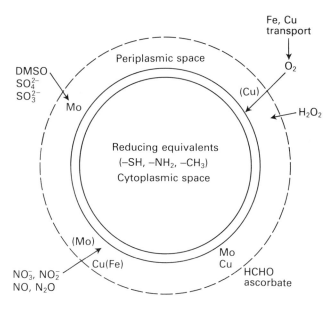

Fig. 17.3 The distribution in space of the major molybdenum (and copper) enzymes except nitrogenase. Note the types of substrate.

(Table 17.1). The reasons for some of them are obvious, e.g. a series of necessary electron transfer and storage sites, the Fe/S centres, for many electron reactions. The involvement of MgATP in an ATP-splitting reaction in nitrogenase is not so clear, but must mean that the chemical energy of $ATP \rightarrow ADP + P$, yielding 10–15 kcal mol^{-1}, is of necessity being transferred first to mechanical and then to reductive energy, i.e. the redox potential of an hydrogen atom or of an electron transfer centre. We must attempt to explain how such a unit functions and eventually relays its reductive power to the nitrogen molecule presumably associated with molybdenum.

In the cases of the O-transfer reactions that occur at molybdenum centres, coupling is also via Fe/S centres, but now includes flavin. (Once again we need to understand the reasons for the train of electron transfer centres.) The molybdenum *in these enzymes* is bound by a special organic pterin cofactor (compare iron–porphyrin), and is not held directly by the sidechains of proteins (Fig. 17.4). The pterin cofactor actually is a dithiolate complex. Another interesting feature, which we shall stress in Section 17.7, is that the molybdenum in the enzymes is not re-oxidized directly by molecular dioxygen and, clearly, the ancillary flavin and Fe/S centres have to do with the way in which dioxygen is activated; oxygen transfer by molybdenum enzymes is of oxygen atoms from water and not from dioxygen, in contrast to the reactions of most iron enzymes but parallel to those of most copper oxidases.

With this background information in mind we turn to the use of model studies. We stress that there are two very different forms of study: (1) the analysis of structure and physical properties which forms the major part of bio-inorganic chemistry and leads directly to a discussion of catalytic activity, and (2) the analysis of binding constants leading to knowledge of selective uptake, see Chapter 2. (The more biochemical reader may wish to pass directly to Section 17.6.)

Fig. 17.4 The formula of one molybdenum-containing cofactor. There are others with nucleotide phosphate extensions.

$(X = O \text{ or } S)$

17.4 Physical methods for enzyme and model studies

In order to make appropriate models for metals in enzymes of unknown structure it is conventional to use known physical as well as chemical properties of the enzyme sites as guides. Only two physical methods have been shown to be of value concerning molybdenum in enzymes. The first is electron spin resonance (ESR) which has allowed a somewhat hazardous interpretation of the co-ordination chemistry of the d^1 Mo(V) complexes in enzymes, but has led directly to firm knowledge as to when this species appears in kinetic schemes. An additional safe deduction from almost all the Mo(V) signals seen in enzymes is based on the study of model complexes. It was found that the Mo(V) ESR signal from the enzymes had g-values quite close to 2.00, in contrast with many model complexes except those involving cyanide and thiolate ligands. From this comparison it was concluded that Mo(V) must be bound by soft donors, i.e. some thiolate ligands. We return later to the more sophisticated use of the method in deciphering the presence of hydrogen atoms on the sulphur or oxygen bound to the molybdenum.

The second method is extended X-ray absorption fine structure spectroscopy (EXAFS), which is a true structural method based upon the interference patterns generated by X-rays interacting with the co-ordinated atoms (mainly) of the molybdenum. The method can be used for compounds with metals in any oxidation state and, at its best, will distinguish not only some of the ligand atoms present, for example, S from N or O, but will also give information on their number and probable geometry. However, it will not distinguish O from N or describe the further links from the ligand atom; it is unfortunate that the method is not yet definitive. Using EXAFS, a variety of workers concluded that FeMoco has Mo–S bonds and no others in the enzyme, while the other enzymes in Table 17.1 have one or possibly two Mo–O bonds, as well as three or four Mo–S bonds. The exact geometry is still in doubt and there is always the danger that more than one geometry is present in the protein sample.

In the course of some of the reactions of these enzymes, the Mo–S or Mo–O systems become bound to protons, as seen by the hyperfine structure on the Mo ESR lines. Using these Mo(V) data together with EXAFS and kinetic studies, an outline of the chemistry of molybdenum in many enzyme reactions has been delineated (see 'Further reading').

We can now see some of the central problems posed by molybdenum in biological systems which the model chemistry must face. They are: (1) the mode of uptake from potentially two major sources, MoO_4^{2-} and, possibly, MoS_4^{2-} and other thiomolybdates; (2) the modes of further transport to the site of incorporation; (3) the incorporation step into a particular organic pterin cofactor or into FeMoco (and here we must look at the thermodynamics of competition for binding sites with many other metal ions, especially iron and copper); (4) the structures of the sites and the possible advantages of their 'design'; (5) the subsequent redox reactions, both of electron and atom transfer at the site.

We shall build up some model chemistry before we turn to the biochemistry again since in the absence of structures much of our discussion of biological chemistry is based on inference. We shall start from structural knowledge and

then go to thermodynamics and kinetics, but in all cases we shall refer only to examples which are relevant to the preceding preliminary picture of molybdenum biochemistry.

17.5 Relevant model chemistry of molybdenum

17.5.1 Structures of molybdenum complexes including clusters

The structures of molybdenum complexes have been reviewed frequently and extensively in a series of conference volumes on molybdenum chemistry (see 'Further reading'). Generally, we can say that many model complexes with O- or N-donors are based on octahedral geometry with the exception of the tetrahedral oxyanion-like compounds, e.g. MoO_4^{2-}. The range of octahedral complexes extends from those of Mo(III), which resemble compounds of Cr(III), to complexes of MoO_2^{2+}. Here, the *cis* MoO_2^{2+} unit differs from the linear UO_2^{2+} and its complexes are obviously distorted from a strict octahedron (Fig. 17.5(a)). Where large ligand atoms are bound, the co-ordination number tends to fall, especially in higher oxidation states.

Now, while model structures are available for many six-co-ordinate Mo species, we must not conclude that this will be the preferred co-ordination in enzymes since experience with Cu(II) and Zn(II) chemistry and biochemistry has shown that such a direct deduction from naïve models is full of hazards. In particular, since all Mo-proteins appear to have some Mo=S as well as Mo-thiolate ligands, it may well be that five-co-ordination will occur and there are some known five-co-ordinate structures of molybdenum compounds with such ligands, illustrated here in a cluster compound (Fig. 17.5(b)).

Apart from the molybdenum complexes themselves there are also structures of a great variety of mixed metal clusters (see Fig. 17.5(c)). The relationship of all these structures to biological molecules such as FeMoco is not yet known. We can conclude that model chemistry demonstrates the great flexibility of molybdenum structural chemistry in several oxidation states. Reference to Chapter 4 will show the value of this flexibility in catalysis.

17.5.2 The elementary aqueous chemistry of molybdenum

In this section we illustrate the difficulties inherent in the discussion of the chemical equilibria of higher oxidation states of metals in aqueous solutions, and we wish to use this discussion to illustrate the problems in understanding in particular the selective uptake of molybdenum in biology. The higher oxidation states, especially, of heavy elements are often amphoteric—a property usually illustrated with aluminium in the trivalent state and associated with the change from metal to non-metal chemistry

$$Al^{3+} + 4OH^- \rightleftharpoons [Al(OH)_4]^-.$$

Here aluminium can act as either a cation or an anion, occurring as Al^{3+} in complex ions with many inorganic or organic donors, while giving salts, aluminates, with many simple cations. In the intermediate pH range, when aluminium is reacted with another amphoteric element such as silica, it

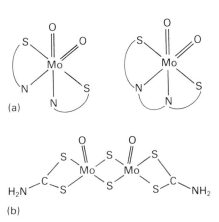

(a)

(b)

(c)

Fig. 17.5. Some model complexes of molybdenum which have been studied to assist understanding of molybdenum enzymes. (a) Mononuclear and six-co-ordinate; (b) binuclear and five-co-ordinate; (c) a mixed cluster (see 'Further reading').

produces network compounds, aluminosilicate precipitates, by reactions which are written as mutual condensations (elimination of water)

$$2Al(OH)_3 + Si(OH)_4 \longrightarrow Al_2SiO_5 + 5H_2O.$$

We have stated already that molybdenum can form aqueous compounds in oxidation states from II to VI; in the lowest oxidation states it behaves largely as a cation Mo^{2+} or Mo^{3+}, and hydrolysis is a relatively small problem, much as for Fe(III), but in all other oxidation states the amphoteric nature of the element becomes strongly apparent. We then need to look at the behaviour of each oxidation state carefully since it is obvious from the above that the central metal could have a variety of co-ordination numbers between 4 and 6 and that the oxygen attached could be water, hydroxide, or an oxene group. Moreover, in the higher oxidation states, it can also condense to form polymolybdates, or mixed polymolybdates, e.g. phosphomolybdates, with a wide range of elements. As is obvious from aluminium chemistry, polymerization occurs at intermediate pH ranges before ionization of the MOH group to an oxene. Note that a molybdenum sulphur chemistry exists parallel with that of molybdenum oxygen compounds, in which the sulphur is a terminal group or a bridging group, as S or SH, in molybdenum precipitates or complexes. This may be related to FeMoco chemistry. All of this chemistry is in competition with the binding of molybdenum to organic ligands such as proteins.

For the case of molybdenum it is found that, in oxidizing conditions and given that molybdenum is in dilute solution in the environment and in biology, the element occurs as monomeric, not polymeric, molybdate or low-molecular-weight Mo(VI) complexes and also as monomeric complexes of lower oxidation states in reducing conditions. The difficulty in understanding the formation of molybdenum complexes then lies in the competition between the formation of oxyanions, oxenes and ligand binding and from the presence of competing cations, such as Cu(II) and Fe(III), for all ligands. We turn to the relative stability of molybdenum complex ions.

17.5.3 The stability of model complexes

In the discussion of the stability of molybdenum complex ions we shall follow the discussion given in Chapter 2. Metals there were described as class 'a', hard, or class 'b', soft, and, as we consider elements lower in a group of the periodic table, we find 'b' class character developing due to the larger size of the cations, thus Cd^{2+} is more 'b' class than Zn^{2+}. We can presume that the chemistry of Mo^{3+} is somewhat more 'b' class than Cr^{3+} and it is not surprising then that quite stable Mo^{3+} model complexes are found with a variety of soft donors, e.g. RS^-, as well as with neutral nitrogen bases and phenolates. While we can then sense qualitatively how selectivity for heavy elements in a group can arise we need to extend this discussion quantitatively to the common oxidation states of molybdenum, Mo(VI) and Mo(V). Here the competition from hydrolysis is in fact so strong that the formation of complexes of such high-valent metal cations is often treated as the reactions of oxocations (oxene-cations) and we therefore should attempt this approach for molybdenum, i.e. we should start from MoO_2^{2+} and MoO^{3+} in studying complex formation. Unfortunately, there is a further difficulty in that the molybdenum oxene species themselves are not easily studied in water, due to

Table 17.2 Stability constants of several metal complexes (log K)

Ligand	VO^{2+}	UO_2^{2+}	Mn^{2+}	Cu^{2+}	Fe^{3+}
OH$^-$	8.3	8.2	3.4	6.3	11.8
Oxalate	6.5	6.4	3.2	4.8	7.5
Malonate	5.3	5.4	2.3	5.1	7.5
Acetylacetonate	8.7	7.5	4.1	8.2	9.8
8-Hydroxyquinolinate	11.0	10.0	(6.0)	12.5	12.5
Histidine	—	7.7	3.2	10.3	(10.0)
1,10-Phenanthroline	5.9	—	4.5	8.8	6.5
NTA	11.5	9.6	7.5	13.0	15.9
EDTA	18.8	7.4	13.8	18.7	25.1
Tiron	16.7	15.9	8.6	14.3	20.4

Corrections have been made to refer the data to the same ionic strength for each ligand.

MoO_2^{2+} WO_2^{2+} (UO_2^{2+})

MoO_2^+ VO_2^+ MoO^{3+}

MoO^{2+} VO^{2+} FeO^{2+}

Some oxo-cations

hydrolysis, and we shall therefore use the more stable VO^{2+} and UO_2^{2+} oxocations (see the margin) first as models to get an approximate idea of how MoO_2^{2+} and MoO^{3+} might behave (see Chapter 2). Table 17.2 shows that VO^{2+} and UO_2^{2+} behave more like trivalent ions, e.g. Fe^{3+}, than divalent ions in the sense that they bind to anions much more strongly than do M^{2+} cations of approximately the same size. By analogy, MoO_2^{2+} is then expected to behave rather like Cr^{3+} or Fe^{3+}, or better In^{3+} which has a greater 'b'-class character than Cr^{3+}. Obviously, when we look at Table 17.2 we must ask how competition from Fe^{3+} but also from Cu^{2+} is avoided since the binding strengths of all cations, except Mn^{2+}, are not too dissimilar. We mention copper in particular: (1) because of its supreme position in the Irving–Williams series; (2) because both copper and molybdenum readily form insoluble sulphides even at pH = 0; and (3) because molybdenum and copper share the biological peculiarity that they occur or are forced outside cells. Immediately there is the difficulty referred to earlier in this section concerning hydrolysis. While Cu^{2+} behaves as a divalent cation and we know its complex ion chemistry well, molybdenum is not found in solution at pH = 7 as the simple ion MoO_2^{2+}, the cation which we wish to compare with Fe^{3+} and Cu^{2+} in solution at pH = 7, but as MoO_4^{2-}. Our first problem, therefore, is to uncover the effective concentration of MoO_2^{2+}, so that competition between molybdenum and other elements for organic chelate sites can be understood. This requires a somewhat lengthy description of the hydrolysis of molybdenum species in aqueous solutions (see Chapter 2.4 and 2.6 for a full discussion of effective or conditional stability constants, all of which is very relevant to the problems of uptake and incorporation of elements).

17.5.4 Condensation and complexation

This section is given as a reminder of some of the problems of reactions of higher-valent cations in water which were introduced in Chapter 2. The problems are those of uncovering the stability of complexes of metals which can undergo either hydrolysis or ligand binding and we described the problem earlier in part under conditional stability constants (Sections 2.4, 2.6, and 2.8). Absence of understanding here removes the possibility of appreciating how biological systems can choose partners, ligands, and metal ions, so selectively. For molybdenum, which has more than one high oxidation state, there is the

further difficulty that the redox potentials are affected as are reaction pathways.

In order to estimate the concentration of the different possible complexes of molybdenum oxene species we write the reactions

$$MoO_4^{2-}+\text{H-bonding ligand} \xrightleftharpoons{K_{MoO_4L}} \text{complex (H-bonded),} \tag{17.1}$$

$$MoO_3+\text{chelating agent} \xrightleftharpoons{K_{MoO_3L}} MoO_3L, \tag{17.2}$$

$$MoO_2^{2+}+\text{chelating agent} \xrightleftharpoons{K_{MoO_2L}} MoO_2L. \tag{17.3}$$

Here it can be seen that the decision as to which species of molybdenum is preferentially complexed depends on a series of equilibria and corresponding constants. Most of these are also pH-dependent; hence we must establish their values at a fixed pH value, e.g. pH = 7, i.e. we must use 'conditional' constants.

From stability constant tables one can obtain values for the hydrolysis constants, K_1 and K_2, corresponding to the reactions

$$MoO_2^{2+} + 2H_2O \rightleftharpoons H_2MoO_4 + 2H^+ \dots\dots\log K_1 = -1.5,$$

$$H_2MoO_4 + 2H_2O \rightleftharpoons MoO_4^{2-} + 2H^+ \dots\dots\log K_2 = -7.5.$$

A way to look at the treatment of molybdenum species in solution is then to say that hydrolysis restricts the concentration of MoO_2^{2+} (and of MoO_3) to rather low levels compared with that of MoO_4^{2-}.

The hydrolysis constant of the reaction

$$MoO_2^{2+} + 4H_2O \rightleftharpoons MoO_4^{2-} + 4H^+$$

is about 10^{-9} l.moles^{-1}, which means that at pH = 7 the concentration of MoO_2^{2+} is reduced by a factor of 10^{19} relative to molybdate.

(For uranium, a similar reasoning would lead us to conclude that the equilibrium is favourable to UO_2^{2+} since neither is the simple cation U^{6+} able to form due to its huge hydrolysis constant to give UO_2^{2+} nor can the UO_4^{2-} species form since further hydrolysis of UO_2^{2+} is not more important than that of, say, Al^{3+}. Very much the same is true for vanadium in the vanadium (IV) oxidation state and at pH = 7 and the oxene species VO^{2+} is easily accessible. For vanadium (V) the situation is intermediate, the species VO_2^+ is favoured in acid solutions and VO_4^{3-} (in its protonated forms, $H_2VO_4^-$ and HVO_4^{2-}) is favoured in neutral and alkaline solutions. It is knowledge of these limiting hydrolysis reactions which allow us to build up a picture of the molybdenum constants in eqns (17.2) and (17.3) from an analogy with the UO_2^{2+} and VO^{2+} complexes. We turn to the deduced stability of MoO_2^{2+} complexes in a biological context.)

17.5.5 The competition with iron(III) and copper(II) for biological ligands

In biological systems Fe(III) is very effectively removed by iron carriers *outside*

cells. These carriers are based on phenolates and anionic nitrogen centres and on *several* co-ordination donor sites, possibly as many as five or six, which MoO_2^{2+} cannot use due to the steric hindrance from the oxene groups; good examples include transferrin and the siderophores (and compare EDTA). While iron is bound to such multidentate enolates or phenolates, Cu(II) is bound outside cells to N-donor centres of particular transfer proteins, albumins. In fact, it is the relative affinity for neutral nitrogen and even anionic nitrogen (imidazole and peptide anions) for Cu(II) and many enolate or phenolate oxygen donors for Fe(III), together no doubt with stereochemical preferences, which separate Cu and Fe carriers from one another (see phenanthroline in Table 17.2). All the data indicate that both free Cu(II) and free Fe(III) are reduced by these strong and selective carriers to very low concentrations, say, $\leq 10^{-15}$ M. This is the level of competition from these cations which MoO_2^{2+} (and VO^{2+}) face for ligands *outside* cells. From the data in Table 17.2, competition would be expected to be strong but not necessarily totally restrictive if MoO_2^{2+} behaved like UO_2^{2+} (or VO^{2+}), but the situation is made much more difficult for molybdenum since, as we have just seen, MoO_2^{2+} is reduced in concentration by a logarithmic factor of 19.0 by hydrolysis to MoO_4^{2-}, i.e. molybdenum cannot compete for any of the Cu(II) or Fe(III) types of site in Table 17.2 (but note that VO^{2+} can). How can these observations be reconciled with the fact that MoO_2^{2+} is obviously preferentially bonded by some (biological) ligands?

There is clearly an indication of a need for a special feature of molybdenum chemistry (in biology), which is reinforced by the fact that this element is indeed found with a particular ligand cofactor (mentioned in Section 17.3) in its extracellular enzymes and that this ligand is a dithiolate (Fig. 17.3). Such a ligand has not been found in association with either Cu(II) or Fe(III). We are forced to conclude that MoO_2^{2+} does have a comparatively much larger affinity for some soft anions, e.g. RS^-, than Table 17.2 might lead us to suppose. Before allowing this conclusion we return to a consideration of the chemistry of MoO_2^{2+} in proteins in order to discover if its cofactor has a peculiar protein binding for molybdenum which aids this selectivity. (The way in which we are proceeding is forced upon us by our wish to understand biological chemistry from the chemistry of inorganic systems.)

17.5.6 The special ligand for MoO_2^{2+} in biology

As described in Section 17.3, the special ligand for MoO_2^{2+} appears to be a pterin cofactor with *cis* thiolate ligands on a double-bonded carbon skeleton (Fig. 17.3). Since the metal would then be four-co-ordinated there could be some two further ligands to a protein but it would not appear that they provide the special binding feature required for selective incorporation in view of the following structural studies on these enzymes.

Although it has not been possible yet to solve any crystal structure for the enzymes, use has been made of X-ray spectroscopy, EXAFS, to uncover some of the details. The study shows that the molybdenum group bound to the pterin is a $MoOS^{2+}$ not a MoO_2^{2+} species and this still leaves the co-ordination sphere of the metal with two possible additional ligating centres from the protein. However, these binding groups must be weak since the whole

c-AMP framework

Mo(VI) dithiolate

coenzyme exchanges. In other words, the molybdenum is not retained as are copper or zinc in proteins by multidentate *protein* ligation, but is held rather in the way the central phosphorus of ATP or, even better, the phosphorus of cyclic AMP is retained (see margin), without much help from the protein. (Note that the molybdenum transfers O atoms while the central phosphate of ATP transfers phosphate; there is a need to note the close relationship of some molybdenum chemistry to that of non-metals.) Recognition by specific proteins must then be by recognition primarily of the pterin and only secondarily by molybdenum binding possibly to phenolate or thiolate groups on the enzyme. This device could well give selective uptake for molybdenum into specific proteins against competition from copper or iron, just as protoporphyrin binding ensures the selectivity of iron uptake into haemoglobin. This is to say that the selective reaction of molybdenum is first of all the result of its ability to be retained by just *two* thiolate centres of the pterin compound and then by the selective incorporation of the molybdenum pterin complex.

We have now reached the following position

1. There is no model complex ion chemistry, e.g. of UO_2^{2+} or of VO^{2+}, which leads us to suppose that MoO_2^{2+} could be bound with especially high affinity *to the usual* ligands for trivalent cations such as Fe^{3+} (Table 17.2).

2. This position is made much worse by the hydrolysis of MoO_2^{2+} to MoO_4^{2-} by a factor of some 10^{19}.

3. Biological systems are observed to have evolved a special ligand for MoO_2^{2+} or rather $MoOS^{2+}$ which is a dithiolate, i.e. a very 'b'-class donor.

4. Thus it must be the unexpectedly high 'b'-class (soft) character of molybdenum which generates the observed co-ordination chemistry.

5. The $MoOS^{2+}$ pteridin complex must be preferentially recognized by specific proteins and only two, if any, rather weak protein bonds to the metal are necessary.

Much of this chemistry leads to the somewhat surprising conclusion that molybdenum, even in its higher valence states IV to VI, is, in fact, the most 'b'-class, soft metal found in biology and that it is this chemistry that has allowed a clear-cut selection from metal ions such as Fe(III) and Cu(II). In support of this conclusion it is useful to remind ourselves of the simple chemistry revealed by the qualitative analysis tables used by all students of inorganic chemistry. Molybdenum, even as Mo(VI) in MoO_4^{2-}, is precipitated as MoS_3 in acid solution by H_2S and this precipitate redissolves as MoS_4^{2-} at alkaline pH. The explanation for this very high affinity of Mo(VI) for S^{2-} may well be the ability of the anion to donate electrons into the empty set of 4d metal orbitals forming strong σ- and π-bonds. A consequence of this chemistry is that the biochemistry of molybdenum inside the cell at low redox potential is dominated by sulpholysis to Mo/S co-ordination, e.g. in nitrogenase, while even outside the cell the competition between O^{2-} and S^{2-} is biased strongly toward S^{2-} in a manner which is unique for a metal in biology.

(We have now completed an excursion in model chemical studies. The approach illustrates the difficulties in understanding, but it also generates quite new thinking about inorganic systems themselves.)

17.6 Rates of reaction

Molybdenum is involved in three very different classes of reaction in biology. The first is *substitution* where we include addition through the expansion of the co-ordination sphere since it is generally believed that substitution for this element occurs through a higher co-ordination number intermediate. For example,

Such S_N2 mechanisms (see Chapter 4) are common amongst heavy elements. Second molybdenum may also allow relatively easy *atom transfer* in reactions such as

$$MoO_4^{2-} + SO_3^{2-} \rightleftharpoons MoO_3^{2-} + SO_4^{2-}$$

but these reactions may be solvent-assisted and not be simple reactions of MoO_4^{2-}, that is, the species are always in relatively fast equilibrium with higher co-ordination species.

The third reaction is *electron transfer* as is observed in the molybdenum blues and bronzes. The reactions are relatively fast indicating that one-electron redox reactions of molybdenum which require little rearrangement of the co-ordination sphere could be used in biology.

17.7 Biological chemistry of molybdenum: general considerations

17.7.1 Uptake of molybdate

A particular problem is posed by the *initial* uptake of molybdenum which must involve the electrostatic binding of MoO_4^{2-}. This anion resembles HPO_4^{2-} and SO_4^{2-} in geometry and is intermediate in H-bonding ability as judged by pK_a values. It is, therefore, very difficult to see how competition with SO_4^{2-}, which is highly concentrated in the sea and even in inland water, is avoided. Perhaps the two anions share the same initial capture routes (compare Ni^{2+} and Mg^{2+}). A quick look here at SO_4^{2-} and HPO_4^{2-} binding to proteins allows some insight. (This discussion will be relevant again in Chapter 18.)

The binding sites of sulphate and phosphate in proteins (Table 17.3) show that there is a diversity of sites often, but not always, based on arginine and lysine cations. There is also a multitude of H-bonds from other protein sidechains. No obvious set of selectivity factors favouring SO_4^{2-} over HPO_4^{2-} is apparent, although there is the suggestion that HPO_4^{2-} needs stronger hydrogen-bonding sites. In these conditions, it is hard to believe that much selectivity can exist for MoO_4^{2-} in its initial reactions with proteins and discrimination will probably rest on its ability to switch to octahedral $Mo(OH)_6$ type complexes at binding sites after crossing the first biological

Table 17.3 Protein–anion interactions

Protein	Anion in native structure	'Ligands' to anion	Other ions binding to site
Carbonic anhydrase	OH^- ?	Zn^{2+}, Thr 197 Lys 22, Arg 243	I^-, Br^-, Cl^- I^-, Br^-, CN^-, RSO_2NH^-
α-Chymotrypsin	SO_4^{2-} SO_4^{2-} SO_4^{2-} SO_4^{2-} SO_4^{2-} SO_4^{2-}	Ser 195, Tyr 146, His 57 Asn 95, Lys 177 Arg 154 NH_3^+, Ala 149 Asn 245 Asn 236	SeO_4^{2-} SeO_4^{2-} SeO_4^{2-} SeO_4^{2-} SeO_4^{2-} SeO_4^{2-}
Elastase (pH = 5)	SO_4^{2-} SO_4^{2-}	Ser 195, His 57, N' 193 Arg 145, Arg 230	$2H_2O$ (pH = 8.5)
Lactate dehydrogenase apoenzyme	SO_4^{2-}	Arg 173	Citrate
Myoglobin	SO_4^{2-}	His E7	
Ribonuclease	SO_4^{2-} SO_4^{2-}	His 12, His 19, N' 120, H_2O Lys 7, Lys 41	PO_4^{3-}, AsO_4^{3-}

barrier. (See Chapter 2 for a discussion of the same problem for cations.) Subsequently and as soon as more reducing conditions are met, molybdenum becomes associated with sulphide chemistry.

17.7.2 Oxidation states and redox potentials in enzymes

When discussing enzymes which are involved in catalysis of redox reactions we need to look carefully at the oxidation states of the metal in the enzyme as far as they are known. Any redox potential in the proteins is related to the relative binding constants of the appropriate two different states of the element, and to the intrinsic redox potentials of the free metal ion species in water, which we obtain from the oxidation state diagrams. The approximate oxidation state diagram of molybdenum relative to H^+/H_2 and at pH = 7 where the redox potential for the H^+/H_2 couple is -0.41 V, is shown in Figs 1.7 and 14.1. The relative stability (redox potential) between any two states is given by the slope of the line from any point to any other point. Since the stability constants of different oxidation states are not too dissimilar, we do not expect complex formation to have much influence over the redox potentials of hydrolysed ions such as MoO_2^{2+}, MoO_2^+ and MoO^{2+} (see Section 17.5.4) and, in fact, the redox potentials fall in the range 0.0 ± 0.3 V (see Table 17.4). This means that the redox potentials for molybdenum enzymes are such that this element can carry out redox changes near 0.0 V, such as mild oxidations and reductions by similar two-electron reactions. Examples are NO_3^-/NO_2^- ($+0.7$ V), SO_4^{2-}/SO_3^{2-} (-0.05 V), $-COOH/-CHO$ ($+0.05$ V). Now these are all two-electron or oxygen-atom transfer reactions and, while we have shown

Table 17.4 Redox potentials in molybdenum enzymes (pH 7 and $H^+/H_2 = -420$ mV)

Enzyme	Redox states	Potential (mV)
Aldehyde oxidase	Mo(VI)/Mo(V)	−360
	Mo(V)/Mo(IV)	−360
Sulphite oxidase	Mo(VI)/Mo(V)	+110
	Mo(V)/Mo(IV)	−110
Nitrate reductase	Mo(VI)/Mo(V)	+210
	Mo(V)/Mo(IV)	+180
Xanthine oxidase	Mo(VI)/Mo(V)	−350
	Mo(V)/Mo(IV)	−350
Formate dehydrogenase	Mo(VI)/Mo(V)	−330
	Mo(V)/Mo(IV)	−450
Flavins	$Fl^- \rightarrow Fl$	−300
	$Fl \rightarrow Fl^+$	−300
Fe/S	Cluster reaction	−300
Cytochromes b	Fe(III)→Fe(II)	−100

here that the thermodynamic potentials make molybdenum a possible catalyst, we need to see also how it functions in overcoming kinetic barriers of such reactions.

We can now turn to the molybdenum-containing enzymes listed in Table 17.1 which are in two classes:

(1) those outside cells which aid oxygen-atom transfer;

(2) nitrogenases inside cells.

17.7.3 Oxygen-atom transfer reactions

The reactions which we wish to understand are, for example,

$$MoO(VI) + NO_2^- \rightleftharpoons Mo(IV) + NO_3^-.$$

There is a parallel set of sulphur reactions, since sulphur transfer is also known in molybdenum enzymes,

$$MoS(VI) + CN^- \rightleftharpoons Mo(IV) + SCN^-.$$

The understanding we need is of the possible intermediates along the pathways since we know that bond-breaking is relatively easy for molybdenum, a heavy atom, and that its redox potential changes are close to those of the substrates.

The probable intermediates of some of the reactions are known through the EPR studies already mentioned. For example, in aldehyde oxidases we can write

$$MoOS + RCHO \longrightarrow Mo(SH)OCOR,$$
$$Mo(SH)OCOR \longrightarrow MoS + RCOOH.$$

Here there is no change in the co-ordination number of the molybdenum. The scheme indicates that the molybdenum oxene group is a good attacking group on the positive carbon atom of the carbonyl. Equally this implies that

probably the oxidation of nitrites and sulphites goes via direct binding of nitrite and sulphite to an oxygen of MoO(VI).

The reoxidation of Mo(IV) does not go via direct attack of molecular oxygen in the molybdenum oxidases. The lack of this reaction is somewhat surprising but it is essential for the proper functioning of the enzymes. They must not be used to reduce dioxygen to water. In fact, the oxidation goes by electron transfer from molybdenum to dioxygen via the ancillary flavin (Table 17.1). As in cytochrome P-450 there is, in addition, an Fe/S electron carrier site in the protein to keep the reaction centres apart. The comparison between these two very different oxidases reveals the selectivity of biology and the use of two different metals in two different parts of space for the oxidation of very different substrates but using comparable electron transfer chains. There is another parallel of unknown significance. Both the molybdenum and the iron which take part in O-atom transfer reactions are bound to thiolate.

17.7.4 Nitrogenase and its reaction centre: FeMoco

The components of nitrogenase are given in Table 17.1 and Fig. 17.6. Great effort has been put into the definition of the nitrogenase reaction centre. There is a general view that, as extracted, it has a molecular weight of around 1500, contains isocitrate, and has the probable inorganic composition Fe_6MoS_8. It is known that the iron is bound to sulphur and that some of the bonds to

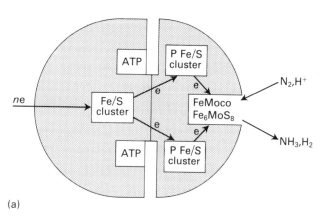

Fig. 17.6 (a) An impression of the nature of nitrogenase with its different Fe/S clusters, its ATP-reaction site, and its active site for N_2 binding and reduction. A suggested model for the active site is shown at (b).

molybdenum are also from sulphur. This biological sulphide cluster has led to many (so-called) models (see Figs 17.5 and 17.6), but most of this work is just *related* Fe/Mo/S cluster chemistry. Importantly, it is also known that FeMoco in the enzyme has the molybdenum bound to thiolate groups, some of which are replaced by oxygen on extraction. The reactions of the unit both in the enzyme and outside it with reagents such as cyanide and nitriles indicate that this binding can be to either the iron or to sulphur.

Now it is believed that the molybdenum is the site of dinitrogen reaction but that one of the necessary steps in the activation of the molybdenum is the prior formation of a hydride, Mo–H. In the reduction of nitrogen it is observed that one mole of dihydrogen is released. (It is also thought that the vanadium and pure iron nitrogenases are less effective in nitrogen reduction because they release larger quantities of hydrogen). We therefore write

$$H^+ + Mo \xrightarrow{2e} Mo\text{–}H^-$$
$$Mo\text{–}H^- + H^+ \longrightarrow Mo^* + H_2$$

where Mo* is the state to which N_2 binds

$$Mo^* + N_2 \longrightarrow MoN_2.$$

The evolution of hydrogen may be one way of keeping the enzyme free from dioxygen interference. The further reduction of N_2 is seen as a series of switches of oxidation state of molybdenum, due to the introduction of electrons , before H^+ is introduced to the N_2. Schematically,

$$Mo\text{–}N_2 \xrightarrow{2e} Mo\text{=}N\text{–}NH_2 \xrightarrow{e} Mo\text{≡}N\text{–}NH_3 \longrightarrow NH_3 + Mo\text{≡}N$$
$$Mo\text{≡}N \xrightarrow{e} Mo\text{=}NH \xrightarrow{2e} Mo\text{–}NH_3 \longrightarrow Mo + NH_3.$$

There is also the requirement for the association and hydrolysis of several molecules of ATP in this reaction which therefore uses large amounts of energy. Part of the energy is used in bringing the reactant proteins together, part in generating the special state of Mo*, i.e. a conformational change, and part, possibly, in the protonation and electron transfer reactions. It is an extremely complex reaction.

The iron sulphide clusters which form the electron transfer chain for the nitrogenase are much less unusual than those of the molybdenum site. However, we have all these deductions largely as inference and real knowledge must await a structure.

17.7.5 Molybdenum exchange and possible control functions

Molybdenum outside cells in the form of the pterin complexes exchanges from proteins quite readily but molybdenum itself does not dissociate. It is not known if the pterin–Mo complex has any control function outside cells.

In nitrogenase there is the possibility that the molybdenum is exposed in the Fe_6MoS_8 cluster and that it is associated with the isocitrate. The very presence of this organic moiety in the enzyme is reminiscent of the structure of aconitase which has an iron sulphur cluster that binds citrate. One metal atom is easily lost from the cluster and may control the citrate cycle (Chapter 12). It could be that either the molybdenum or the iron of Fe_6MoS_8 is a dissociable control

atom. The control could relate to the reaction known to be important elsewhere,

$$Fe^{3+} + (iso)citrate \rightleftharpoons [Fe^{3+} \cdot (iso)citrate].$$

In this regard the structure of the nitrogen fixation (*nif*) genes and their regulator factors, which may be iron dependent, are of interest (see Chapter 21).

17.8 Molybdenum: conclusions

We conclude Part A of this chapter with the most intriguing problem. There is no denying that molybdenum is central to *three* reactions without which biological systems would fail

$$N_2 + 3H_2 \longrightarrow 2NH_3,$$
$$NO_3^- \longrightarrow NO_2^- + (O),$$
$$SO_4^{2-} \longrightarrow SO_3^{2-} + (O).$$

These three are essential for biology's source of nitrogen and sulphur. There is here a conundrum for the chemist and the biologist or indeed for anybody interested in the nature of living systems. Is the requirement for molybdenum an absolute necessity or could the chemistry have been done in some other way and is what we observe an evolutionary accident? No matter how we answer this problem we must expect that such a need, for molybdenum, will be reflected back to the genetic apparatus (Chapter 21).

A secondary conclusion of this part of the chapter is that model chemistry is of very great value to the chemist and will help him to develop ideas as to how biology works. However, this chapter has revealed too that it is a complexity of factors, structural, thermodynamic, and kinetic, which are intertwined within biology and that the model chemistry cannot be expected to encompass them all.

17.9 Vanadium chemistry and biochemistry

Vanadium, like molybdenum, has two very different catalytic activities associated with it and two very different biological functions. Both elements have a rich oxygen chemistry associated with the oxyanion and oxycation species, VO_4^{2-} (MoO_4^{2-}), VO^{2+} and VO_2^+ (MoO^{2+}, MoO_2^+, and MoO_2^{2+}). These species are oxidizing agents which can react by one-electron steps, electron transfer, or by two electron steps, when V and Mo act as O-atom donors, but, whereas molybdate is a mild oxidant, vanadate is relatively powerful, $E \approx +0.5$ V at pH $= 7$.

The second class of chemistry of these two metals belongs to sulphur-rich reducing conditions, where vanadium, like molybdenum, can form sulphur-containing anionic centres, e.g. VS_4^{3-} (MoS_4^{2-}); sulphur-containing cationic centres, e.g. VS^{2+} and $VSSH^+$ (compare $MoOS^{2+}$); and iron–sulphur

clusters Fe_6MS_8, associated in biology with nitrogen fixation. It is quite possible that in the primitive earth, when these reducing conditions were prevalent, vanadium and molybdenum (and iron) had considerable catalytic activity (it may even have been that life became a possibility because such catalysts existed), but vanadium and iron may well have been more available since their sulphides are more soluble than that of molybdenum. (Note again the table of group separations in analytical chemistry.) Iron was then available as Fe^{2+} and Fe^{2+}/Fe^{3+} sulphide clusters and vanadium as V^{3+} and V(IV) sulphur complexes, e.g. VS^{2+}, $V(S)SH^+$, and, perhaps, $VS(S_2)_2^{2-}$.

17.9.1 The biological forms and functions of vanadium

The known forms of vanadium in biology are given in Table 17.5. It is probable that only the vanadium–iron–sulphide cluster FeVco (similar to FeMoco) is primitive and all the other forms appeared later as the level of atmospheric oxygen increased; note that only the FeVco protein is inside cells and that other vanadium compounds are probably in vesicles or outside cells. This is certainly true for the vanadium-containing haloperoxidases occurring in some brown and red algae which contain V(V) bonded to oxygen and nitrogen donors. The accumulation by tunicates and *Amanita* toadstools may have started at an earlier stage, with atmospheric oxygen still at a rather low level. Note that in the tunicates vanadium is known to be stored mostly as V(III) (or VO^{2+} in some types) in the vacuoles (vanadophores) of specialized blood cells (generally referred to as vanadocytes) and this may have been determined by the need to protect the metal ion from the oxidizing environment. Nowadays, the animals capture vanadium as vanadate from sea water and incorporate it in the vanadophores through the phosphate channel profiting from the analogy between $H_2VO_4^-$ and $H_2PO_4^-$; once inside the vacuole, vanadates are reduced to the cationic forms V^{3+} and VO^{2+} by a polyphenolic pigment, tunichrome, with which the cations may combine. In

Table 17.5 Known forms of vanadium in biology

Form	Occurrence	Function
Free as V(III) or combined with a special pigment—tunichrome	Some ascideans (tunicates), e.g. *Ascidia nigra, Phallusya mamillata, Ciona intestinalis*, etc.	Unknown; probably protection (synthesis of tunic) or, alternatively, an electron and proton sink
V(IV) bis-complex of N-hydroxyimino-di-α-propionate (amavadine)	Some *Amanita* toadstools, e.g. *Amanita muscaria, Amanita velatipes*	Unknown; probably part of protection system against injury of tissues; compare fungal tyrosinases
V(V) bonded to N,O-ligands	Some brown and red algae, e.g. *Fucus vesiculosis, Ascophyllum nodosum, Ceramium rubrum*, etc.	Bromo- and iodo-peroxidases; probably defensive
Fe, V, S cluster	Azotobacteria, e.g. *Azotobacter vinelandii, Azotobacter chroococcum*	Cofactor of nitrogenase; fixation of dinitrogen

any case the metal is then retained and kept in a reduced free or combined form.

In the toadstools, we find another curious device that biology has used, this time an apparently simple ligand—N-hydroxy-imino-di-α-propionate—which forms an extraordinary stable 2:1 complex with V(IV), (not VO^{2+}!) (see Fig. 17.7) and is therefore able to extract the metal from soils in which it is scarce relative to its potential competitors, e.g. Fe^{3+} and Cu^{2+}. Here we need to remember the stability constants of Table 17.2, which indicated that extraction of VO^{2+} would not have been easy. (We are able to discuss VO^{2+} binding directly from Table 17.2 since this ion is not hydrolysed like MoO_2^{2+}.)

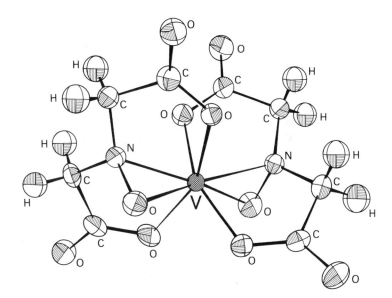

Fig. 17.7 The structure of the vanadium complex of N-hydroxy-imino-di-α-propionate, amavadine, has methyl substituents on the CH_2 carbon of this model complex.

The overall stability constant $\log \beta_2$ of this complex (named amavadine) is enough to remove the oxene group to give a complex of the naked V^{4+} ion, actually an eight-co-ordinated species in which even the N-hydroxyl groups of the ligand are ionized.

This V(IV) complex may undergo a reversible one-electron oxidation at about $E \approx 0.8$ V to the corresponding V(V) complex, but it is quite stable in aqueous solution at pH ≈ 7 and does not need to be stored where it is protected from oxygen. Indeed, it will be necessary to know where the amavadine is localized in the fungi if its function is to be correctly understood. At present it appears that it may be a kind of mediator or electron transfer catalyst at fairly high potential, very much like those found in some blue copper oxidases, having dioxygen at the terminal oxidizing end. Such copper oxidases exist in other fungi and have protective functions, e.g. fungal tyrosinases. Again, amavadine does not react directly with dioxygen, so that there must be other cofactors and, since it is involved in one-electron transfer steps, radicals are likely to be formed. We do not speculate overmuch by saying that amavadine is likely to be formed external to the cell cytoplasm in order to be kept away from DNA. Could we make drugs based on this molecule?

17.9.2 Vanadium and iron

Looking at the data in Table 17.2 and at the information summarized in Table 17.5, we find that in its location and functions vanadium does, in fact, have a chemistry related to that of iron, especially Fe^{3+}, as well as to molybdenum. VO^{2+} and Fe^{3+} are bonded in biology by similar chelating agents (catechols, oximes, hydroxamates, porphyrines, sulphur clusters, etc.; see Table 2.3). Both metals form nitrogenases, although their catalytic activities involve the liberation of much more hydrogen from water than in the case of the molybdenum enzyme which appears to have replaced the others in the present atmospheric conditions.

Another parallel is with the use of iron in peroxidases (Table 17.6), although the chemistry here appears to be different. Peroxidation using iron goes through a highly oxidizing $FeO(IV)$ species and then by direct formation of XO^- or XO^+ where X is a halogen, while the most likely route for vanadium is via known peroxo complexes $VO_2^+(O_2)$, giving XO^-, the halogenating agent, and VO_2^+, i.e. it is the peroxide, not the metal, that is involved in the reaction. Once again this kind of 'dangerous' chemistry must take place outside the cytoplasm, either in vesicles or in the external fluids of higher cells.

Table 17.6 Haloperoxidases

Protein (halide)		Source	Metal
Chloroperoxidase (I^-, Br^-, Cl^-)	Myeloperoxidase	Neutrophilic granulocytes	Iron chlorin, Iron porphyrin
	Chloroperoxidase	*Caldariomyces fumago*	Iron porphyrin
	Chloroperoxidase	*Pseudomonas pyrrocinia*	Iron porphyrin
Bromoperoxidase (I^-, Br^-)	Lactoperoxidase, Eosinophil peroxidase	Milk, saliva, tears, granulocytes	Iron porphyrin
	Bromoperoxidase	Some brown and red marine algae	Vanadium
Iodoperoxidase (I^-)	Horseradish peroxidase	Horseradish roots	Iron porphyrin
	Thyroid peroxidase	Thyroid gland	Iron porphyrin
	Iodoperoxidase	Some brown and red marine algae	Vanadium

As to the other cases in Table 17.5 it is also known that one order of ascidians which does not accumulate vanadium (the Stolidobranchia) does accumulate iron (in ferrocytes) with apparently the same function, whatever it may be—sink for electrons and protons, defence against predators, tunic synthesis or some other. As we have seen, accumulation takes place in cell vacuoles, i.e. external to the cytoplasm.

Only in the case of the *Amanita* toadstools is the analogy more with today's copper oxidases (see Section 17.9.1) than with iron enzymes, although there are iron cofactors with functions similar to that of amavadine, i.e. electron transfer catalysts, usually functional at lower redox potentials.

17.10 Tungsten biological chemistry

Tungsten as a biological metal is a most unlikely choice; there are indeed very few examples of selection for functional purposes of elements of atomic number higher than 35 (see Chapter 1). The outstanding exceptions are molybdenum and iodine and neither of them is generally required by all living species. Other rare examples are tantalum and niobium, which seem to be accumulated by certain tunicates, as well as strontium and barium used as sulphates in the skeleton of some acantharians and radiolarians (see Chapter 20). None of these last four metals are involved in any general metabolic mechanisms.

However, in the last decade, two bacterial enzymes have been purified and characterized to a fair degree and these seem to be tungsten enzymes; one is an NADP-dependent formate dehydrogenase and the other a carboxylic-acid reductase, identical to the previously identified aldehyde dehydrogenase (Table 17.7). Both have been isolated from obligate anaerobes, particularly *Clostridium thermoaceticum* and *Clostridium formicoaceticum*, and are involved in such fundamental reactions as the reduction of CO_2 to formate (the first reaction of autotrophic CO_2 fixation in anaerobic bacteria) and the oxidation of aldehydes to carboxylates. Tungsten ranks, therefore, among the primitive metals that biology seems to have used more extensively in remote times (see Fig. 21.2).

As seen in Table 17.7 the NADP-dependent formate dehydrogenase appears to be a tungsten/selenium/non-haem-iron/sulphur-enzyme operating at very low redox potential (close to or even below the redox potential for the reduction of H^+ to H_2); molybdenum can replace tungsten, but the activity decreases; other metals, e.g. vanadium, were not checked for their activation properties. The information available suggests that tungsten is incorporated in some cofactor, probably analogous to the pterin-containing molybdenum cofactor, but the possibility of a WFeS cluster cannot be excluded. Selenium

Table 17.7 Tungsten enzymes

Enzyme	Source	Function	Properties and requisites
NADP-dependent formate dehydrogenase	*Clostridium thermoaceticum* *Clostridium formicoaceticum* (possibly also *C. sticklandii*, *C. acidiurici*, *Methanococcus vanielli*, and others)	Reduction of CO_2 to formate; CO_2 fixation in anaerobic bacteria	Apparently a W–Se–Fe–S enzyme with W incorporated in a pterin cofactor: Se and Fe (non-haem) are acid-labile; W cofactor easily released. Protein type $\alpha_2\beta_2$, mol. wt. 340 kDa (α unit 96 kDa; β unit 76 kDa). Se is as selenocysteine in the α unit. Requires NADPH for CO_2 reduction at low negative redox potential. Very sensitive to O_2
Carboxylic-acid reductase/aldehyde dehydrogenase	*Clostridium thermoaceticum*	Reduction of non-activated carboxylates to aldehydes (and reverse reaction)	Apparently a W-containing enzyme, with low substrate specificity. W weakly bound. Selenite and molybdenum have little effect and may be inhibitory. Probably a 240-kDa enzyme with four 60-kDa units. Reduction at very low redox potential. Very sensitive to O_2

appears to be incorporated into the α-subunits of the $\alpha_2\beta_2$ enzyme in the form of acid-labile selenocysteine and is essential for activity.

The simplest raction catalysed by this enzyme seems to be

$$CO_2 + NADPH \rightleftharpoons HCOO^- + NADP^+, \tag{17.4}$$

but whole cells or crude extracts can reduce the carboxylates down to the corresponding alcohols (at the expense of carbon monoxide or formate). This must be due to the presence of other enzymes, namely, an aldehyde dehydrogenase (still a tungsten enzyme in the case of the bacteria studied) and an hitherto unidentified alcohol dehydrogenase/aldehyde reductase, usually a zinc enzyme. The equilibrium constant for reaction (17.4), $\approx 2 \times 10^{-2}$ at pH 2.5, is clearly towards CO_2 and NADPH, but in a gas phase of 100 per cent CO_2 the conditions are favourable to the reduction of CO_2 to formate.

The second enzyme, carboxylic acid reductase, seems to be the first to be found which is able to reduce non-activated carboxylic acids to aldehydes at the expense of some oxidizable substrate. The enzyme appears to have a mol. wt. of about 240 kDa and to be composed of four 60-kDa subunits. The highest content of W determined was 2.3 mol W/mol of enzyme. Molybdate or selenite have little effect on activity and may even be inhibitory. Tungsten is not strongly bonded and can be removed with loss of activity.

The reaction catalysed appears to be $RCOOH + 2[H] \rightarrow RCHO + H_2O$, which cannot be conducted chemically in water and is unknown without activation in biology. It requires a high negative redox potential ($E^0 \approx -560$ mV) but at pH ≈ 7 is of the order of -300 mV, hence within the range of stability of water. This reaction is similar to those catalysed by some molybdenum enzymes (see Section 17.7) and is effectively the reverse of the atom donor reaction (two-electron reduction) of a pterin centre.

Naturally, tungsten is closely similar to molybdenum, both in its redox properties and chemical behaviour, and one expects that they behave similarly in biology. This is true to some extent; both Mo and W have a high degree of 'b' character, behave more as non-metals, can be retained by only two covalent bonds, go easily from MO_4^{2-} to the $M(OH)_6$ state through five-co-ordinate intermediates, and can go through various oxidation states at redox potential close to zero. The main differences reside in their sizes, $WO_4^{2-} > MoO_4^{2-}$, and kinetics of reaction, which are slower for tungsten, particularly as a polyanion. Tungsten is, of course, less available than molybdenum. In these conditions it is difficult to explain why tungsten but not molybdenum or any other metal was selected for the above-mentioned reactions.

17.11 Molybdenum, vanadium, and tungsten in evolution

It is observed that today molybdenum is largely associated with the reactions of NO_3^-, SO_4^{2-}, aldehydes, and N_2 in biology. Of these there is good reason to suppose that the first three were rare chemicals in the period before the appearance of substantial amounts of dioxygen in the atmosphere. We can make the (risky) conclusion that molybdenum only became incorporated in biological systems in the form used to attack those substrates, i.e. all except

dinitrogen, about 10^9 years ago, life having existed at least three times as long (Fig. 1.11). The appearance of this second form of molybdenum in life would then correspond to the known change of the chemistry of the sea at that time. In the earlier long period of the evolution of life in a reducing atmosphere this metal may have been present as sparingly soluble MoS_2 and perhaps some soluble Mo/S anions (thio-complexes), generally in oxidation states lower than six. Molybdate would not have been formed and the concentration of molybdenum in the sea would be due to these soluble sulphur complexes only.

To some extent the same is probably true for vanadium and tungsten, but the sulphides of these metals are more soluble (notice that they are not precipitated together with that of molybdenum in the classical analytical group separation method using H_2S). Especially vanadium may have been present not only as solid V_3S_2 and VS_4 and soluble sulphur complexes but also as VS^{2+}, $VSSH^+$, or even soluble simple and mixed metal/sulphur clusters like those of iron. As the result of the progressive oxygenation of the atmosphere, molybdenum, vanadium, and tungsten were oxidized and appear today in sea water as soluble molybdate, vanadate, and tungstate. The concentrations of these species in the sea are shown in Fig. 17.1, together with those of several other biological and non-biological trace metals.

It is striking to see that, although molybdenum and vanadium are now even more abundant than many of the more common metals, e.g. iron, zinc, or copper, only molybdenum has a significant biological role; vanadium and tungsten are restricted to a few primitive or last remnants of primitive species and do not seem to be required by modern living organisms. Vanadate may even be objectionable since it can poison ATPases working as a phosphate mimic.

It is also interesting to note that both molybdenum and tungsten (but not vanadium), although generally available as oxygenated anions, revert to Mo–S and W–S compounds in biology, whereas the major species of vanadium today are nitrogen- and oxygen-ligated forms of V^{3+}, V(IV), and V(V). There are two reasons for this difference:

1. Higher oxidation states of vanadium are more oxidizing than those of molybdenum or tungsten so that they oxidize thiolate groups.
2. The sulphur chemistry of vanadium is less stable than that of molybdenum and tungsten, i.e., to put it differently, vanadium has a more 'a' (hard) character, whereas molybdenum and tungsten have a more 'b' (soft) character.

This does not mean that vanadium cannot form any biological sulphur compounds or complexes *inside* cells, and vanadium nitrogenase, similar to molybdenum nitrogenase, is an example.

Here, however, there is a quandary concerning evolution since we do not know whether nitrogen fixation was required early in life. Indeed, the early available forms of nitrogen are in doubt. For example, did the early atmosphere and aquatic environment have sufficient NH_3 or HCN? If the answer is affirmative then there was no need of a nitrogenase. The evidence for an early requirement of this enzyme is indirect: on one hand, nitrogenases are only found in prokaryotes and we may feel that they must have been present from a very early stage in life's history; on the other hand, both the

molybdenum and vanadium nitrogenase active centres are quite different from those in any other known molybdenum and vanadium enzymes and are inside cells, whereas the others are outside cells. Finally, the high contents of vanadium in certain oils believed to be of remote biogenic origin, e.g. some Peruvian and Venezuelan oils, show that this metal must have been widely present in the ancestors of present-day algae, fungi, and higher plants, perhaps in the giant ferns that once seemed to have populated the tropical forests. The requirement for this metal points to an essential function and nitrogen fixation or CO_2 reduction in photosynthetic processes are likely candidates.

We may add still another speculative note to the above comments on what we consider to be one of the most primitive metals used in biology. While commenting on the possible evolution of the functions of manganese (Chapter 14) we considered that, when life developed the ability to use light, a source of reducing equivalents was necessary on one side of the membrane, where light capture occurs, and hydrogen sulphide could be an ideal choice due to the reaction $H_2S \rightarrow S^0 + 2H^+ + 2e$, provided that a metal catalyst was available. We have shown that Mn(II) was a possibility, but vanadium (nickel or iron) could also be used to drive this or a similar reaction, leading, for example, to the precipitation of $V(S_2)_2$ (and FeS_2). This could be the reason for the accumulation of vanadium by ancient photosynthetic species, accounting for the unusually high contents of this metal in some oils, particularly Venezuelan and Peruvian, and also for the occurrence of this metal as the sulphide VS_4 (patronite) in the ores which were once the most important mineral source of vanadium (the Ragra's mine in Peru, curiously in the same tropical region).

A final puzzle concerning the roles of group V and VI elements of the periodic table is the absence of any firm conclusion regarding the value of chromium. This element has the peculiarities that it has virtually no available oxidation state except Cr(III) and in this state it has little affinity for sulphide and is heavily hydrolysed. It is a d^3 cation and, as such, usually has octahedral complexes which show slow ligand exchange. The verdict of biology appears to be that this element is of little interest in either redox or acid/base properties. If anything, it appears to be a poison (compare Al(III) and Ni(II)), and to be avoided by biological systems generally. Little plant life grows on Ni/Cr minerals. All these comments are made knowing that Cr(III) has now and then been reported to be essential in some function or another in higher animals. These reports have persisted for 30 years or more without definitive verification.

Further reading

Bray, R. C. (1986). The nature of the high pH–low pH transition in sulphite oxidase and nitrate reductase. *Polyhedron*, **5**, 591–5.

Bray, R. C. (1987). Molybdenum proteins. *Recueil des Travaux Chimiques des Pays-Bas*, **106**, 299–311.

Bray, R. C., Engel, P. C., and Mayhew, S. G. (ed.) (1984). *Flavins and flavoproteins*, pp. 702–22. deGruyter, Berlin.

Carrondo, M. A., Fraústo da Silva, J. J. R., Leal Duarte, M. T., Pessoa, J. C., da Silva, J. A., Vaz, M. C., Vilas-Boas, L. F. (1988). Amavadine. *Journal of the Chemical Society: Chemical Communications*, 1158–9.

Cramer, S. P., Liu, C. L., Mortensen, L. E., Spence, J. T., Liu, S. M.. Yamamoto, I., and Ljungdahl, L. G. (1985). Tungsten-containing formate dehydrogenase. *Journal of Inorganic Biochemistry*, **23**, 119–24.

De Boer, E., Boon, K., and Wever, R. (1988). Vanadium bromoperoxidases. *Biochemistry*, **27**, 1629–35.

Fraústo da Silva, J. J. R. (1989). Vanadium in toadstools. *Chemical Speciation and Bioavailability*, **1**, 139–50.

Holm, R. H. (1990). The biologically relevant oxygen atom transfer of molybdenum. *Coordination Chemistry Reviews*, **100**, 183–222.

Johnson, J. L., Bastian, N. R., and Rajagopalan, K. V. (1990). Molybdopterin guanine dinucleotide. *Proceedings of the National Academy of Science of the United States of America*, **87**, 3190–4.

Johnson, J. L. and Rajagopalan, K. V. (1982). Structural and metabolic relationship between the molybdenum cofactor and urothione. *Proceedings of the National Academy of Science of the United States of America*, **79**, 6856–60.

Karrasch, M., Börner, G., and Thaner, R. K. (1990). The molybdenum cofactor of a dehydrogenase. *Federation of European Biochemical Societies Letters*, **274**, 48–52.

Kneifel, H. and Bayer, E. (1973). Vanadium in mushrooms. *Angewandte Chemie (International Edition)*, **12**, 508–16.

Mukund, S. and Adams, M. W. W. (1990). Tungsten-containing aldehyde oxidoreductase. *Biochemistry*, **27**, 1629–35.

Michibata, H., Miyamoto, T., and Sakurai, H. (1986). Vanadium in tunicates. *Biochemical and Biophysical Research Communications*, **141**, 251–6.

Mills, C. F. and Bremmer, I. (1980). Nutritional aspects of molybdenum in animals. In *Molybdenum and molybdenum-containing enzymes* (ed. M. Coughlan), pp. 519–42. Pergamon Press, Oxford.

Newton, W. E., Schultz, F. A., Cheller, S. F., Lough, S., McDonald, J. W., Conradson, S. D., Hedman, B., and Hodgson, K. P. (1986). Iron–molybdenum cofactor of *Azobacter vinelandii* nitrogenase: oxidation–reduction properties and structural insights. *Polyhedron*, **5**, 567–82.

Orme-Johnson, W. H. (1985). Molecular basis of biological nitrogen fixation. *Annual Reviews of Biophysics and Biophysical Chemistry*, **14**, 419–59.

Polyhedron (1986). The proceedings of the Fifth Molybdenum Conference with references to the first four meetings. *Polyhedron*, **5**, 3–306.

Pope, M. T., Stille, E. R., and Williams, R. J. P. (1980). A comparison between the chemistry and biochemistry of molybdenum and related elements. In *Molybdenum and molybdenum-containing enzymes* (ed. M. Coughlan), pp. 1–40. Pergamon Press, Oxford.

Rajagopalan, K. V. (1988). Molybdenum: an essential trace element. *Annual Review of Nutrition*, **8**, 401–27.

Structure and Bonding (1983). A whole volume devoted to molybdenum and vanadium biochemistry. *Structure and Bonding*, **53**.

Wever, R. and Kustin, K. (1990). Vanadium, a biologically relevant element. *Advances in Inorganic Chemistry*, **35**, 103–37.

White, H., Strobb, G., Feicht, R., and Simon, H. (1989). Carboxylic acid reductase: a new tungsten enzyme. *European Journal of Biochemistry*, **184**, 85–96.

Yamamoto, I., Takashi, S., Shiu-Mei, L., and Ljungdahl, L. (1983). Formate dehydrogenase: a tungsten–selenium–iron protein. *Journal of Biological Chemistry*, **258**, 1826–32.

18

Phosphate, silica, and chloride: acid–base non-metals

18.1 Introduction to the non-metals 436

18.2 Phosphate biochemistry 437

18.3 The forms and energies of bound phosphate 438

18.4 Summary of phosphates 445

18.5 Anion balance: chloride, phosphate, sulphate, and carboxylates 445

18.6 Chloride channels and exchangers 445

18.7 Silicon biochemistry 446

18.8 Boron in biology 451

18.9 Non-metal trace elements 451

18.1 Introduction to the non-metals

The non-metals

It is not desirable to give in this book anything but passing reference to the non-metal chemistry of H, C, N, and O since these four elements form the central part of organic biochemistry and there are excellent volumes dealing with this subject. In fact, major features of the chemistry of hydrogen appear in Chapters 1 and 5, of that of carbon and nitrogen in Chapter 1 and toward the end of Chapter 2, and of that of oxygen in chapters on transition metal chemistry in biology (Chapters 12–17). In each case we have tried to link the biological chemistry of these non-metals to that of other inorganic elements and we have noted in particular how they are handled as elements by specific coenzymes. In this and the following chapter a somewhat more detailed account is given of the non-metal elements, P, Si, S, and Se, concentrating upon the mainly anionic chemistry corresponding to that of the cationic chemistry of metals. The anionic chemistry of chloride has been briefly described in Chapter 8 but we shall add some more detail in this chapter. Only when this anionic chemistry has been described can we go forward to the description of the way in which metals and non-metals are used together in biology: (1) in forming inorganic minerals in biology (Chapter 20); (2) in a complementary fashion to control homeostasis. In the first chapter on non-

metals we shall describe the chemistry which is associated with acid/base but not redox chemistry, that is, primarily, the chemistry of phosphorus and silicon. This division follows that of the metals.

18.2 Phosphate biochemistry

18.2.1 Introduction to the functions of phosphate in biology

Phosphate is present as the anion HPO_4^{2-} inside cells and in circulating fluids at a concentration of about 10^{-3} M. It is at considerably lower concentration in the environment and therefore it has to be pumped into biological systems (Fig. 18.1). There, in organic bound form, it has five major functions: (1) as a part of large molecules or molecular assemblies, e.g. DNA, RNA, and membranes; (2) as a carrier of substrates, e.g. in glucose phosphate and coenzymes, and as a carrier of chemical energy, e.g. ATP; (3) as a signalling device in the cytoplasm, e.g. in cAMP and IP_3; (4) as a reversible chemical modification of proteins; and (5) as a constituent of biominerals (Fig. 18.2).

18.2.2 The chemistry of phosphate

Phosphate is tribasic and two of its protonation steps are those of a weak acid ($pK_a = 13.0$ and 6.5) while the third is at $pK_a = 1.0$ which means that phosphate always carries an average charge of above one at $pH = 7$. The organic diesters and esters of phosphate also always carry a charge of one or more since they have pK_a values of 1.0, and 1.0 and 6.5, respectively. None of these anions bind appreciably to simple cations such as Mg^{2+} and Ca^{2+} but all can bind appreciably to protein surfaces and to M^{3+} ions.

Metapolyphosphates, linear P_n such as pyrophosphate, P_2, or organic esters XP_n or XP_nY behave similarly with respect to their proton affinities.

Fig. 18.1 A diagram of the uptake routes of phosphate for incorporation and silica ($Si(OH)_4$) for rejection and biomineral formation. Chloride and carboxylates are the other major anions.

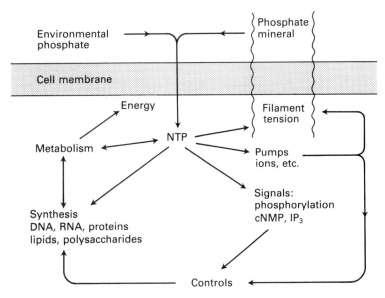

Fig. 18.2 The flow of phosphate into a cell and some of the connections it makes.

Where there is a terminal phosphate the pK_a values of it are about 1.0 and 6.5 while all the chain diester phosphates have a pK_a of 1.0. All phosphate is therefore negatively charged and all molecules with a terminal phosphate are always partially protonated. The biochemistry of phosphate is inevitably linked to that of the proton. In addition, the binding to Mg^{2+} is now of great significance since, as shown in Chapter 9, it binds all pyro- and higher metaphosphates, P_n.

The stable form of phosphate in water at $pH = 7$ is as free inorganic phosphate and all the esters and anhydrides of phosphate are unstable, i.e. they are energized. The degree of energization is small for simple phosphate esters but each pyrophosphate bond requires more than 10 kcal for its synthesis. The appearance of phosphate compounds in biology is then a sign of their kinetic stability only, which implies that they will be formed or hydrolysed at selected places and at selected times where the specific catalysts for the reactions of a given organic phosphate are to be found, e.g. glycolysis in the cytoplasm and glycosylation in the Golgi (see Chapter 5). In fact, as we have stressed, phosphorylated substrates are extremely common in biological systems.

The basis of the kinetic lability of compounds of phosphate, like those of silicate but unlike many carbon compounds, is that they react relatively quickly with attacking reagents due to the ease of going to the five-co-ordinate intermediate state

$$YO^- + X{-}O{-}\overset{\displaystyle O}{\underset{\displaystyle O^-}{P}}{-}OH \longrightarrow X{-}O{-}\overset{\displaystyle O\ \ O}{\underset{\displaystyle O}{P}}{-}O{-}Y \longrightarrow XO^- + HO{-}\overset{\displaystyle O}{\underset{\displaystyle {}_-O}{P}}{-}OY\ .$$

The intermediate state of a five-co-ordinate phosphorus is readily mimicked by stable vanadium complexes which are then able to block the phosphate transfer sites of some enzymes, e.g. Na^+/K^+ ATPases. We turn next to the directly biological chemistry of phosphate.

18.3 The forms and energies of bound phosphate

18.3.1 Phosphate in large molecules or assemblies in biology

Phosphate is a linking unit in the sugar esters of DNA(RNA) but, as we have emphasized in Chapter 7, it remains a 'deliberate' centre of mobility and bond angle variation. DNA(RNA) must be able to change conformation easily and must be open to ready hydrolysis: that is the essence of any polymer which carries rapidly reproducible instructions and which has to undergo correction of errors. It is the antithesis of a polymer which forms building blocks, e.g. proteins and polysaccharides, which are *kinetically* much more stable. We will now consider the binding power to cations.

Phosphate is also found in phospholipids both as a diester and a monoester. Ignoring at first the signalling function in release of inositol phosphates, IP_3, etc. (see Section 18.3.4), the phosphate bond is again a source of weakness

(through hydrolysis) in these lipids and a centre of mobility. Membranes are fluctuating fluid structures in shape and local composition. Phosphate, as a diester phosphate in lipids, as in DNA and RNA, has the advantage of deriving from a strong acid ($pK_a \sim 1$) and its interactions with H^+ are weak. In isolation its interactions with Mg^{2+} and Ca^{2+} are also weak but diester phosphates in membranes and DNA(RNA) are in fact large one- and two-dimensional polyelectrolytes, respectively, which, although not neutralized by direct bonding to simple cations, hold counterions in the Gouy–Chapman layer, i.e. a diffuse layer of solution near the charges. The phosphate monoesters are very different in function (see Ca^{2+} triggering of membrane fusion in Chapter 10) since they bind directly to cations and are then held in the co-ordination sphere (see Chapter 9). There is competition on all these surfaces between the inorganic cation binding, that of Mg^{2+} in cells and both Mg^{2+} and Ca^{2+} outside cells, and polyamines or positively charged proteins. The competition is of extreme importance in many associations of phosphate-carrying surfaces, e.g. DNA, RNA, membranes, polysaccharides, and even bone (see Section 18.3.7), and it may be involved in the initiation of cell division (Chapter 9).

Now, while the monoester phosphates are largely concerned with carrying small molecules in synthetic or degradative metabolic paths or in the construction of polymers and membranes, the pyrophosphates are concerned with the transfer of chemical energy and signalling related to conformational switches and only here and there to transfer of chemicals, e.g. glycosylation.

18.3.2 The energy of pyrophosphates and the resting state of cells

We have explained that all pyrophosphates, e.g. nucleotide pyrophosphates, NDP and NTP, are partially protonated at $pH = 7$. The binding constant of NDP and NTP, i.e. diphosphates and triphosphates, to magnesium ions is about $10^{3.5}$ and, since the free (Mg^{2+}) is less than 10^{-3} mM in cells, there is an equilibrium such that there is always some free Mg^{2+}, some free NTP^{4-} or NDP^{3-}, and much $(NTP \cdot Mg)^{2-}$ and $(NDPMg)^-$ in solution, the triphosphate concentration being about 10^{-3} M and the diphosphate concentration about 10^{-4} M. Thus the energy-producing reaction paths, such as glycolysis, are regulated by the H^+, Mg^{2+}, NDP^{3-}, and NTP^{4-} levels. The link between H^+, Mg^{2+}, and NTP^{4-} is seen to be very immediate and obvious direct binding and is indeed inevitable, so that, from the time in evolution when pyrophosphate became the distribution mode for chemical energy, H^+ and Mg^{2+} limited that use. Given the nature of evolution these interactions would obviously be turned to advantage and this advantage is enhanced by the way in which other cations, e.g. Ca^{2+} and adrenalin, modulate the reactions.

The maintenance of the levels of NTP in cells, e.g. all nucleotide phosphates, is through the synthesis of ATP. Even in a cell at rest there has to be constant synthesis since biological systems are not stable. ATP is slowly hydrolysed in water itself and many other compounds are more or less unstable. Again tension in the cell and hence cell shape is maintained and constantly regenerated using hydrolysis of ATP and GTP (Chapter 5). Oxidative damage has to be repaired continuously, poisons are rejected, nerves send continuous messages, and so on and so on. Each process demands use of ATP. The supply

of ATP is described in Chapter 5 and the way in which it is held constant is through feedback control of the enzymes of glycolysis or of the energy-transducing membranes. Section 9.8 describes the control in a muscle cell but the control is exerted everywhere. Phosphate is also found in other small-molecule energized forms in cells including mixed anhydrides, e.g. acyl phosphates, creatine phosphates, etc., but we have no space to include these reactions here.

18.3.3 Phosphate controls

We should perhaps have analysed the biochemistry of phosphate immediately after that of magnesium, calcium, and zinc because we consider that these four elements have a significance in the control of eukaryote cytoplasmic cellular metabolism which sets them apart (Fig. 18.3). In one sense (organic) phosphate behaves as if it were negatively charged calcium, as we shall show in this section, just as chloride behaves as negatively charged sodium or potassium in conductance. If this is a correct view then we imply that the reasons for the association of the above four elements with control systems in eukaryotic cells, and virtually only these four elements amongst all others, are that they are sufficiently abundant, bind with sufficient strength, have suitable (fast) kinetic parameters associated with their reactions, and do not undergo redox chemistry, so that together they act as basic non-oxidative homeostatic controls. The cations Mg^{2+}, Ca^{2+}, and Zn^{2+} form stable associations with organic compounds, related to their hydrated ionic reactions in chemistry, and this represents a sufficiency of binding. Their concentrations are manipulated (energized) so that, in fact, they are always in differentially balanced flow both between stable free ions and bound states as well as in gradients across membranes. Their rest and triggered states can then be governed by pumped gradients and opening of channels. The condition of phosphate compounds is different but parallel. The reference state for phosphate in *the cytoplasm of a cell* is not the same as in the above ionic chemistry since phosphate *bond* chemistry, not its cross-membrane partition, is energized to such an extent that any cross-membrane energization, though present, is not of much consequence (compare Mg^{2+}).

Because of this degree of bond energization, we shall take as reference state ATP concentration, [ATP], and not inorganic phosphate concentration. It is [ATP] which is held constant in cells and it is binding of phosphate in organic molecules relative to binding in ATP which represents the transfer potential of phosphates. (This corresponds to taking the extracellular Ca^{2+} concentration as its reference state for intracellular reactions.) The free energy ΔG of the phosphate in ATP is some 10.0 kcal (10^7 concentration units) away from that of free phosphate and ADP in cells. (Note that internal Ca^{2+} is energized by some 10^4 concentration units away from external calcium.) Thus phosphate can be said to have a 'kinetically restrained' real binding constant ($> 10^5$) to serine in proteins *relative to ATP* in regulating reactions such as

$$ATP + Protein–Serine \longrightarrow Protein\ Serine–P + ADP$$

and in reactions which give cyclic phosphate esters, e.g. cAMP, which are used as triggers, much as calcium injection is used. The protein serine phosphate is an ester (and so is cAMP), which is still somewhat energized with respect to

Fig. 18.3 Feedback loops related to phosphate controls. All connections to calcium are excluded.

free phosphate, and a final part of the trigger or control is its hydrolysis by phosphatases often using Mg, Zn, or Ca. Given that the base level from which phosphate reactions should be seen is ATP (i.e. a particular pyrophosphate), we note again that the kinetics of phosphate reactions are slow enough (they involve 'covalent' binding) so that ATP reactions are controlled in time and space by catalysts and not by reversible association/dissociation reactions as with metal ion complexes (Fig. 18.1). All the components of the phosphate trigger are in the cytoplasm and do not need to be divided by membranes because of these kinetic controls. The effect of binding phosphate to all allosteric proteins seems to be again through the adjustable nature of helices (Fig. 18.4); compare the effects in calcium and haemoglobin triggers.

Fig. 18.4 The effect of phosphorylation of Ser-14 on a helical portion of phosphorylase. The inactive state, phosphorylase b, shown in thin lines, is adjusted by phosphorylation to give the active state. NB. The effect of phosphorylation is transmitted to the active site by the adjustment of helices (see Chapters 7, 10, and 13). (After the work of L. Johnson.)

18.3.4 Phosphate messengers

Cyclic nucleotides, cNMP

Now although the primary source of *chemical* energy is adenosine triphosphate, ATP, there are other nucleotide triphosphates (NTP), e.g. of guanine, cytidine, and uridine, and also yet other pyrophosphate compounds for the distribution of saccharides and isopentyl units. All are synthesized from ATP and we have described their distribution in cellular vesicles in Chapter 3. All of these nucleotide triphosphates can be used in signalling, when they give cNMP at membrane surfaces, in controls, and in the generation of mechanical forces, for example in the construction of tubulin in the cytoplasm of eukaryote cells.

The main second messengers, i.e. cAMP and cGMP, so-called since they are generated by primary messengers such as hormones, are monovalent anions and are therefore related to chloride (Cl^-) just as the monovalent organic cations derived from the organic derivatives of ammonium, e.g. acetylcholine and adrenalin, are related to Na^+ and K^+. The skill of biological evolution has been in using the simplest inorganic ions, Na^+, K^+, and Cl^-, and relating their message-carrying capacity subsequently to the much more selective

actions of organic derivatives of ammonia and phosphate. The parallel between anion and cation chemistry is made clear in Fig. 18.5, where the role of magnesium and calcium in the whole of phosphate and message chemistry is largely put to one side (see Chapters 9 and 10). Now yet other phosphates are used in signalling through hydrolysis reactions since all phosphate esters are energized though much less so than pyrophosphates.

Fig. 18.5 Cyclic adenosine monophosphate, cAMP, and a typical set of reactions activated by it.

Inositol phosphates

The inositol phosphates are one such further group of messenger organic phosphates which can either be in a membrane-bound form or free in solution. Their release is then different in kind from the formation of cAMP from ATP which pervades the cell (Fig. 18.6). IP_3 will be released locally if through membrane composition or curvature, it is locally concentrated (in lipids), and if the stimulant is *local*. The release generates two extra signals: (1) free IP_3 which liberates calcium locally from nearby internal stores in the reticulae; and (2) free lipids locally in the cytoplasmic membrane. The subsequent stages of transmission to enzymatic activity are complex but involve GTP-dependent membrane proteins and kinases C which cause protein phosphorylations. The complexity of this set of signals in its chemistry allows a high degree of chemical selectivity and, by its locality, of physical selectivity in space and time. Again the release mechanism is through energized (but now weakly energized) chemical bonds by specific enzymes, here phospholipases which often require metal ions. Part of the selectivity of cell response to stimuli is that

Inositol 1,4,5 triphosphate (IP$_3$)

Fig. 18.6 Inositol triphosphate and some of its effects after release from phospholipid inside a cell.

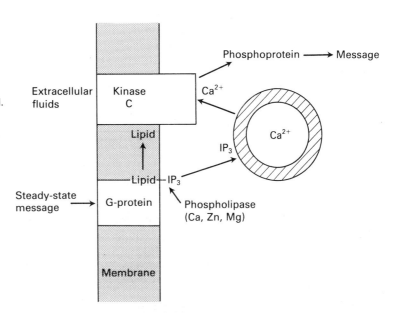

the IP$_3$ does not interact with the same calcium pools as does the cAMP response. The metal dependence of the phospholipases and the subsequent metabolism of lipids lead to complicated connections between signalling systems (Chapter 21). Here too enzymes are required to hydrolyse the IP$_3$.

18.3.5 Phosphate switches: protein phosphorylation and time hierarchy

When a signal, usually a chemical signal, causes a change in level of cNMP or of protein phosphorylation then this change must be slower and more prolonged than that due to Ca^{2+} invasion. While the calcium switch is a series of fast ionic reactions, the phosphate switch is covalent compound formation and can only be relaxed by enzymic activity. Phosphate esters are easily hydrolysed, however, so that on/off switches can be in the 1 second to 1 hour range, but not in the millisecond range. The fastest phosphate switches are those due to anhydride formation in ion, sodium and calcium pumps

$$\text{Protein–CO}_2^- + \text{ATP}^{4-} \longrightarrow \text{Protein–CO·P}^{2-} + \text{ADP}^{3-}$$

$$\text{Protein–CO·P}^{2-} + \text{H}_2\text{O} \longrightarrow \text{Protein–CO}_2^- + \text{P}^{2-} + \text{H}^+.$$

Since these reactions are slowish, the relaxation of really fast muscles uses as an intermediate an ionic buffer for calcium (Chapter 10) that is faster than the slower ATP-dependent pumps.

The reactions of phosphate which are yet slower use ester formation,

$$\text{Protein–OH} + \text{ATP}^{4-} \longrightarrow \text{Protein–OP}^{2-} + \text{ADP}^{3-},$$

$$\text{Protein–OP}^{2-} + \text{H}_2\text{O} \longrightarrow \text{Protein–OH} + \text{HP}^{2-}$$

where OH is a serine, threonine, or tyrosine centre. There is, in fact, a hierarchy of kinetic stability within protein phosphate compounds, as there is in carbon compounds, but phosphorus is always in the same oxidation state. The speed variation is due to the variation in the carbon compound to which the phosphorus is bound. This hierarchy has to be seen against the background of kinetics in the periodic table which is

$$\text{Na}^+, \text{K}^+, \text{Cl}^- \text{ diffusion: very fast}$$

$$\text{Ca}^{2+} \text{ ionic reactions: fast, close to equilibrium}$$

$$\text{IP}_3, \text{ATP highly polar covalent reactions: intermediate rates}$$

$$\text{Zn}^{2+} \text{ binding close to equilibrium: tightly bound ionic reactions,}$$
$$\text{slow rates}$$

$$\text{C–C non-polar reactions: very slow, not close to equilibrium.}$$

Phosphate in a switch device introduces, therefore, a medium-term or even a relatively long-term chemically covalent memory trace, while the calcium (electrostatic or ionic) switch is as close to instantaneous as biology can manage and leaves no trace. Carbon reactions are a permanent record or long-term memory as shown by changes of state, e.g. growth patterns.

18.3.6 Phosphates and gene transcription

We are only beginning to understand the complexities of regulation of gene transcription (DNA to RNA) and of translation (RNA to proteins) but already the two are known to be connected in certain cases. Here we refer to only the simplest control over gene transcription—direct action of a regulator gene via protein binding. The bound proteins may be in one of two conformational states but in one such state they aid transcription (positive feedback) or they prevent transcription (negative feedback). The conformational switch is needed in the control over the production of messenger RNA and, therefore, of protein product. In this chapter we are interested in phosphate. It is already known that very many important enzymes have their synthesis controlled at the gene level by proteins, the binding conformation of which is affected by phosphorylation or by association with cAMP. Much of protein synthesis is then under the control of phosphate metabolism. It is suspected in fact that there are two classes of such interactions:

(1) one which modulates protein production up and down and helps to maintain a cell in homeostatic harmony;

(2) one which causes the cell to be irretrievably sent along a new path to a particular differentiation.

While there is some difficulty with an absolute separation of these two concepts in chemical terms it is clear that, while (1) lies behind homeostasis, (2) is associated with morphogenesis. The idea of homeostasis is connected to that

of maintaining and/or controlling concentrations and we must therefore look at ways in which phosphate concentrations can be buffered.

18.3.7 Phosphate minerals

Phosphate salts are the basis of bone, where $Ca_2(OH)PO_4$ is a simple way to represent the apatite mineral. Calcium phosphates are also deposited by some lower organisms. Other phosphate nodules occur in a variety of worms. Phosphate is also found in ferritin, especially bacterial ferritin which begins to look like $Fe_4O(PO_4)_2$ rather than a derivative of ferrihydrite. It is quite possible that bacterial ferritin was an early phosphate buffer. In these cases we must not assume that the solid phosphates seen in biology are just protections, structures, or even stores of phosphate; they act as controls as well. In Chapter 21 we shall look at the controls over phosphate in body fluids. We shall not examine phosphate biological minerals here but we direct attention to Chapter 20, while noting that they are Ca^{2+} and OH^- buffers too.

18.4 Summary of phosphates

We have not attempted a wide cover of phosphate biochemistry and here we give just an outline of its enormous significance which can only be compared with that of C, H, N, O, S, and Cl among non-metals and Na, K, Mg, Fe, Zn, and Ca among metal elements. These are the elements which allow polymer formation and control of the medium, water, in which all the reactions occur.

18.5 Anion balance: chloride, phosphate, sulphate, and carboxylates

In Chapter 8 we described the basic functions of Na^+ and K^+ in osmotic and electrolytic balance. Amongst anions these roles are taken on mainly by chloride ions which, like sodium ions, were exported from cells very early in evolution. A parallel use has developed in that some cells can use chloride as a current carrier (see the next section) and in yet other cells, halobacteria, there are even light-driven chloride pumps. Among divalent ions the role of sulphate is somewhat similar to that of calcium in that it too is driven out of cells but it is not used in signalling. Of course, phosphates and carboxylates are accumulated to give the total anion pool.

18.6 Chloride channels and exchangers

The requirement for channels to allow movement of chloride is equal to that for sodium and potassium (Chapter 8). Chloride channels have been well characterized by electrical measurements and the permeability ratios for common biological anions are shown in Table 18.1. The obvious point to

Table 18.1 Chloride channels: relative permeability

Anion	Radius (Å)	Permeability (approx.)
Cl^-	1.80	1.0
Br^-	1.95	1.0
I^-	2.15	1.0
NO_3^-	1.90	0.5
ClO_4^-	2.40	0.5
Carboxylates	>2.00	Very low
Phosphates	>2.40	Very low
Sulphate	2.30	Very low
Hydroxide	1.40	Very low

From Edwards, C. (1982). *Neuroscience,* I, 1335–66. The radii are derived from thermochemical data.

Fig. 18.7 A diagram showing bicarbonate secretion coupled to chloride balance across the duodenal epithelium. Note the involvements of cAMP, Ca^{2+}, and H^+ in controls.

make is that these channels have the properties of those anion sites which give rise to the lyotropic series and therefore have somewhat hydrophobic non-hydrogen-bonding character (Section 2.15.1). It was essential that they should not accept carboxylates, sulphate, or phosphates. However, anion balances require the export of HCO_3^- and there is a chloride/bicarbonate exchanger in many cells (Fig. 18.7). This exchanger is controlled by both cAMP and Ca^{2+} ions so that anion balances are tightly connected to other cell activities, including cation balances, and also to energy (ATP) status.

18.7 Silicon biochemistry

Silicon in the form of compounds of $Si(OH)_4$ is an extremely abundant element in minerals and soils, but its role in living systems has been poorly examined since it is *required* only as a trace element in most higher animals, although there is as much silicon present in man as there is magnesium. By contrast, in plant life, especially grasses, and in many unicellular organisms silicon is a major element. The turnover of silica by these forms of life is only one order of magnitude less than that of carbon. Huge deposits of SiO_2 on the floors of the oceans and in many deserts are made from tiny silicon shells, many of which belong to the radiolaria family. Silicon as SiO_2, nH_2O, forms a major building unit of many amorphous hard structures even in lower animals. It is thought to be associated with polysaccharides whether it is found in bulk in plants or as traces in higher animals where it has a role in bone formation, in connective tissue, i.e. in glycoprotein stabilization, and perhaps elsewhere. The much wider association with plant life probably arises through the much higher concentration of saccharides there. Before looking at the chemistry it is worthwhile examining the physical forms in which silica appears since the lessons of silica in biology may be more for material scientists than for the conventional academic chemist.

Some pictures of the forms which silica takes in shells of diatoms and the baskets of certain protozoa, e.g. *Stephanoeca diplocostata ellis*, are shown in Figs 18.8 (a) and 18.9 (a). The architecture is almost beyond belief. The forms of silica in higher plants are simpler, although now and then they support

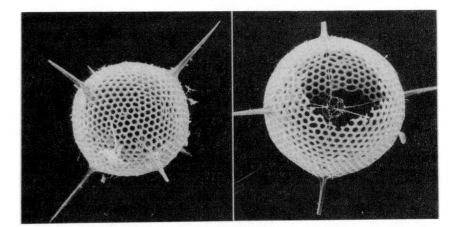

Fig. 18.8 The fine tracery of the silica shell of a diatom ($\times 4100$).

large-scale needles or hairs of well-defined structure as in nettles (see Fig. 14.8). These constructions are all found external to cells and have been made in biological moulds. The moulds are either internal vesicles (in the cells) where their forms are produced using tensions developed on the vesicle containers by the internal filaments of the cell (Fig. 18.10) or they are extracellular polysaccharide containers. The purpose of the decoration on the structures is unknown, but it is hard to believe that all these intricate patterns have a simple structural function (see Chapter 20). Here, however, we must be careful since an amateur eye glancing at fan-tracery in the ceiling of a building readily sees that it is beautiful but cannot perceive the functional significance of the whole nor of its cross-bridging. Often elaboration of struts in structures is to do with the weight effectiveness of the overall construction in relation to the load it has to bear. We do not know the stresses and strains which must be met by silica structures, remembering all the time that the structure will have to permit feeding of the organism through the shell. We return to the complexity of the forms in Chapter 20, but stress here that it is clearly related in some way to the underlying filaments belonging to cell membranes (Fig. 18.11).

The silica rods of the protozoa, *Stephanoeca diplocostata ellis* (Fig. 18.9(b)), are particularly intriguing. It appears that these protozoa are capable of making and counting roughly the number of rods produced, say up to 150 ± 5 (Fig. 18.9(a)), and that a certain small number of the rods are of a thickness different from the larger number. The rods are ejected from the cell in what looks like an oriented bundle and are sorted later so as to make a basket lying on an outer film. The rods are apparently joined together by a cement (Fig. 18.9(c)) such that the structure is relatively rigid and can be used as a kind of 'lobster pot' to catch bacteria. The rods are of hydrated SiO_2 (see Section 18.7.1) and, at first sight curiously, they dissolve from the inside (Fig. 18.9(d)). Before discussing the chemistry involved we must be aware that the loss of material from inside a rod to give a hollow tube is not disadvantageous to the mechanical properties; most scaffolding in the building industry is made from hollow tubes. It seems probable that the flow of silica from the inside of the rods to the rod joints allows rod–rod joints to be cemented by silica (Fig. 18.9(c)) without noticeable weakening of the rods themselves. The whole structure is then made in accord with well recognized principles of mechanics.

We have given this somewhat detailed picture of one or two mineral

Fig. 18.9 (a) The net of silica rods formed by *Stephanoeca diplocostata ellis*. The cone-like basket has been somewhat damaged in the process of mounting (×29 000). (b) Individual rods (×62 000). (c) The way in which individual rods are cemented together and become (d) hollow tubes (×100 000) and (e) the structure of the amorphous silica (×2 100 000).

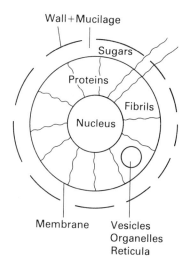

Fig. 18.10 The organization of vesicles by filaments so that eventually silica structures can be exported into the cell wall.

frameworks here in order to show where the importance of silica lies within biology. There is evidence that the building of silica frameworks is directly controlled from the DNA by silicon receptor systems in such organisms as diatoms. We know very little in depth of this possibility. Certainly silica is associated with some very early forms of life, for example in cherts found in Africa.

18.7.1 The chemistry of silica in biology

The silicon in biology is derived from $Si(OH)_4$ condensation reactions and there is no known reduction or organic chemistry involved. All the silica in biology which has been examined in detail by high-resolution electron microscopy is amorphous down to atomic levels, i.e. to less than 10 Å (Fig. 18.9(e)). This means that there is no order in the silica and it resembles a rather dense, somewhat dried silica gel (compare Fig. 18.12(a) and (b)). Nuclear magnetic resonance specroscopy shows that the silica is an ill-defined product of empirical formula $SiO_n(OH)_{4-2n}$ where n goes from 0 to 2. This agrees with its infra-red spectrum. The variable degree of hydration and the lack of structure allows the surface to redissolve readily. The surface can be stabilized by hydrogen bonding to organic matrices well-known outside biology, so that the structures of Figs 18.8 and 18.9(a) are probably based on H-bonded organic templates, see Fig. 18.11. This explains the solution of the rods of Fig. 18.9(c) from the inside which is the last part of the structure to be deposited and may well have no organic support.

While the above mechanism for the manufacture of the delicate silica forms may be reasonably clear, the detail of the movement of the silica in living

Fig. 18.11 A schematic representation of the network of crossed filaments which may well occur under many cell membranes. This network with protruding proteins through the membrane could well be the basis of patterns of silica such as that seen in Fig. 18.9(a).

Inner-sphere surface complex:
K$^+$ on vermiculite

Outer-sphere surface complex:
Ca(H$_2$O)$_6^{2+}$ on montmorillonite

(a)

Fig. 18.12 (a) A characteristic pair of silicate layer structures which are crystalline. (b) The structure of the amorphous opaline silica (hydrated) of biological structures as shown by NMR and infra-red spectroscopies (see Fig. 18.9(e)).

(b)

systems is unknown. At one time it was thought that silica was laid down in plants simply by transpiration at the leaf surface. This is clearly unsatisfactory as a general way of handling silica in organisms since some silicon constructions are known in algae which live in water. Again the concentration of silica as Si(OH)$_4$ in the sea is quite far (tenfold too dilute) from the level required to precipitate amorphous silica. (The concentration may be closer to the solubility limits of certain silicates since sea water contains much calcium and magnesium.) Since many organisms in the sea can precipitate hydrated silica it is apparent that they must accumulate Si(OH)$_4$. Again the silica content of fresh water in soils is usually low. All plants transpire, yet only some plants, especially grasses, deposit silica. It follows that many plants must have a special uptake mechanism. We can only make the suggestion that there are special hydroxy compounds, sugars or phenols, for this purpose. Silica, or rather Si(OH)$_4$, must be deliberately pumped into local regions of space, certainly vesicles. How can this be done? We do not know, but hypotheses based on the known *chemistry* of Si(OH)$_4$ can be easily developed.

Si(OH)$_4$ is known to react with cis-diols especially those of the catechols. The complexes formed are of the formula

This chemistry is reminiscent of that of B, Al, Ti, and Fe, but many other non-metals, such as As and Sb, give the same compounds. While the corresponding aliphatic complexes of B, Al, and Fe are known, made from cis-glycols, those of Si are doubtful. At present we do not know of any other way of handling $Si(OH)_4$ which would allow it to be moved around in a controlled way but it is undoubtedly associated with saccharides where it is deposited.

It is sometimes suggested that silica could replace phosphorus as the central element in condensation reactions in biology. Three factors militate against this: (1) Si–O–Si bonds are of low energy content; (2) Si–O–Si systems do not form kinetically stable organic small units in water; and (3) Si–O–Si systems readily form cross-linked solids not linear polymers. The reader is advised to look back at Fig. 6.1 and note how chemistry changes across the second short period of the periodic table. Biology knows this chemistry! One final peculiar difference between biology and geology is that there are no aluminosilicates in biology except perhaps in pathological conditions such as Alzheimer's disease. Why?

18.8 Boron in biology

Despite the known importance of boron at the trace level in plants very little is known of its value in biochemical terms, see Section 2.15.3. In this respect the status of boron in plants is exactly that of the status of silicon in animals: there is the suspicion that both elements are connected to saccharide stabilization of connective tissue. Given the absence of any detailed knowledge it is only sensible to pass on to the description of elements whose value is understood in some well defined way.

Notice that boron has one intriguing peculiarity: its nucleus has a high neutron-capture cross-section. There might be the possibility of taking advantage of this fact if boron could be directed preferentially to certain cells, e.g. tumour cells. There is just the chance that neutron therapy could then be a very valuable antitumour treatment.

18.9 Non-metal trace elements

Little is known about the distributions of other non-metal elements. We shall therefore make notes only upon them.

1. **Halides.** Here we exclude the use of the simple chloride anion (see Section 18.6). Bromine and iodine appear to be activated in vesicles to form organic compounds external to eukaryotic cells, e.g. thyroxine, or to be used external to prokaryotes as protective poisons. As this is oxidative organic halogen substitution we return to it in Chapter 19.

2. Fluoride as a simple anion is extremely low in biological systems with an odd exception: its high concentration in krill is not explained. While it is useful in biominerals, e.g. in teeth, it is a dangerous competitor for OH^- in enzymic reactions.

3. The appearance of some arsenic compounds is not yet studied in enough detail for us to appreciate the functional value. We have not included a discussion here of it or other elements which also act as poisons and pollutants. Chapter 22 will be devoted to their biochemistry, i.e. to the elements of groups III and IV, non-metals such as arsenic, and the heavy elements of the second and third transition series and beyond in the periodic table.

Further reading

No specific references to the part of the chapter on phosphorus are required since in all biochemistry textbooks the properties of all the individual compounds described here can be found and their value appreciated in detail. The purpose of this chapter is to pull together the data on phosphorus. Nevertheless, one general publication on phosphorus, Ciba Foundation Symposium, No. 57, may be useful.

Ciba Foundation Symposium (1978). *Phosphorus in the environment: its chemistry and biochemistry*, Ciba Foundation Symposium, no. 57 (ed. R. Porter and D. W. Fitzsimmons). Elsevier, Amsterdam.

Ciba Foundation Symposium (1986). *Silicon biochemistry*, Ciba Foundation Symposium, no. 121 (ed. D. Evered and M. O'Connor). Wiley, Chichester.

Bendz, G. and Lindqvist, I. (eds) (1978). *Biochemistry of silicon and related problems.* Plenum Press, New York.

Lovatt, C. J. and Dugger, W. M. (1984). Boron. In *Biochemistry of the essential ultratrace elements* (ed. E. Frieden), pp. 389–421. Plenum Press, New York.

Oesterhelt, D. and Tittor, J. (1989). Bacteriorhodopsins. *Trends in the Biochemical Sciences*, **14**, 57–61.

References to biominerals are found in the 'Further reading' for Chapter 20.

19

Sulphur, selenium, and halogens: redox non-metals

19.1	Introduction	453
19.2	Sulphur biochemistry	454
19.3	Summary of sulphur biochemistry	459
19.4	The biochemistry of selenium	460
19.5	Evolution and selenium	461
19.6	Notes on halogen redox reactions	462

19.1 Introduction

Knowledge of the early atmosphere of the planet and of the mineral elements that were present gives rise to the intuitive feeling that sulphur must have been a very important element from the very beginning of life especially at the low redox potentials which pervaded the anaerobic initial states. Hydrogen sulphide in water may have been present at above millimolar levels and anyone familiar with the properties of transition metal sulphides, such as vanadium, iron, molybdenum, nickel, and cobalt, knows that they are good semiconductors and have reactive surfaces which make excellent catalysts. Even without biological interference there must have been very considerable organic chemical activity in the first 10^9 years of the cool earth. Anyone familiar with energy-capture devices in biology also knows that the Fe_nS_n complexes are uniquely important even today in many low redox potential reactions (Chapter 12) and we have had occasion to refer to nickel, molybdenum, and even vanadium and tungsten sulphides in Chapters 16 and 17 in extremely important reactions such as those of CH_4, N_2, and H_2. In this chapter we do not wish to delve into these, origin-of-life, matters since the possible interaction of mineral sulphide chemistry and early biology can only be based on speculative discussion as yet. We do wish to summarize the involvement of especially the element sulphur in a number of important functions which have persisted throughout biological development as far as we know them (Table 19.1). However, it is not possible to give more than a glimpse of this chemistry in this book since it is so extensive. Some related biochemistry of selenium and the halogens will also be outlined.

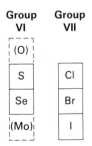

Redox non-metals in periodic table

Table 19.1 Some key functions of sulphur in biology

Function	Example
Thiolate as a base	Thiol-proteases
Thiolate as a ligand	Metal-ion binding
Sulphide as a ligand	Metal-ion binding in clusters
Sulphide as a redox group	Lipoic acid
Thiolate in group transfer	Transfer of acetyl (CoA)
–S–S– as a cross-link	Many extracellular enzymes
–S–S– in controls	Thioredoxin, glutathione
–S–CH$_3$ ethers (methyl transfer)	Coenzyme (methionyl CoA), coenzyme M
–S–CH$_3$ ether as a ligand and cross-link	Cytochrome c
Coenzyme catalysis	Biotin, thiamin, lipoic acid
Inhibitors (drugs)	Penicillin

19.2 Sulphur biochemistry

Fig. 19.1 The handling of sulphate to give both sulphated polysaccharides and the many forms of reduced sulphur requires extracellular molybdenum enzymes and a variety of intracellular processes.

19.2.1 Oxidation states and redox reactions

Today sulphur enters higher cells largely as sulphate, as opposed to the sulphide available to anaerobes, and it is processed in two very different ways (Fig. 19.1). In the first it is condensed with ATP in an energetically uphill reaction to produce APS (Fig. 19.2). This compound, formed in the Golgi apparatus, is then used to sulphate polysaccharides and some proteins and thereafter sulphate becomes a useful bound anion especially in the open matrices of connective tissue. Although we shall not describe its chemistry further, since it interacts weakly with most cations, the importance of sulphated anionic polymers, which form open networks and are not cross-linked by calcium in multicellular organisms, cannot be overstressed (Chapter 7).

The second pathway involves successive reduction to SO_3^{2-} and then to sulphur and sulphide. For this reduction biology uses a molybdenum enzyme in the first step and subsequently a haem/Fe$_n$S$_n$ protein, sulphite reductase, to take the sulphur through bound SO_2 and SO to S^{2-}. There is a parallel reaction for NO_2^- to NH_3 and both reactions require six electron reductions without release of intermediates. The initial steps of the reactions occur in the periplasm, outside cells. Some of the H_2S is then used to produce the many metal (Fe, V, Mo) sulphide units in biology, but we do not know how this is controlled. Some of the sulphide is used in energy-capture devices in anaerobes and the reversible reaction from $[-S-S-]_n$ to $2n$-SH is one of the most fascinating in bio-inorganic chemistry. Much sulphur is also incorporated into proteins as thiolate or thioether. Thiolate reactions control protein stability outside cells by the formation of –S–S– bridges but there are two much more fundamental and primitive roles for thiolate in redox reactions and in enzymes and controls of the cytoplasm, namely in glutathione (Fig. 19.2) and thioredoxin (see Fig. 21.4).

We have observed that in membranes there are many Fe$_n$S$_n$ proteins for the transfer of electrons by oxidation-state changes of the metal ions. We also noted that in the cytoplasm there was little transport of electrons by metal-ion-

Phospho-adenosine monophosphate sulphate (APS)
PAPS has an additional 3′-phosphate

Adenine

Cysteine

$$NH_3^+$$
$$CH-CH_2-SH$$
$$CO_2^-$$

Methionine

$$NH_3^+$$
$$CH-CH_2-CH_2-S-CH_3$$
$$CO_2^-$$

Fig. 19.2 The formula of sulphated adenosine phosphate, APS, which is the form in which active sulphate is transferred, together with formulae of important forms of other sulphur. See also Chapter 2.17.

α-Lipoic acid

$$CH_2-CH_2-CH-(CH_2)_4-CO_2^-$$
$$S\text{———}S$$

Glutathione

$$NH_3^+$$
$$CH-CH_2-CH_2-CO-NH-CH_2-CO-NH-CH-CO_2^-$$
$$CO_2^-\qquad\qquad\qquad\qquad\qquad\qquad CH_2$$
$$SH$$

Coenzyme M

$$HS-CH_2-CH_2-SO_3^-$$

containing enzymes, especially in eukaryotes; radicals such as Fe^{2+}, Cu^{2+}, and Mn^{2+} are too reactive and are therefore used very cautiously. Outside cells the protection from oxidative free radicals was secured by sugars such as ascorbate in higher organisms; inside the cell the exchange of redox equivalents in the cytoplasm at a potential around -0.1 V is managed in large part by glutathione. The reaction generally avoids radicals since it is a two-electron system, but even if RS˙ radicals are produced they are relatively stable. In fact, all the radicals which biology uses openly are of this stable kind—ascorbate, hydroquinone, and thiols—and they differ from the more aggressive radicals, e.g. of tyrosine, which are deeply screened in proteins. Glutathione is then a protective and homeostatic redox device in the cytoplasm, communicating the redox balance relatively slowly between metabolic pathways at around the redox potential of -0.1 V. It communicates with the low-potential NADPH on the synthesis side of metabolism only through special enzymes (Fig. 19.3). It does not equilibrate with reducing equivalents which are to be used in degradation and energy generation (NADH) but only with those to be used in the cytoplasm via NADPH.

If glutathione is to work in this way, then it should be able to communicate reductive synthesis with other reaction pathways; seemingly it does so in part through the enzyme thioredoxin (see Table 19.1 and Fig. 21.4). This enzyme is directly connected to the synthesis of DNA, since it assists the reduction of ribose, to the synthesis of proteins, and to regulation of key enzymes in

Fig. 19.3 The active site of the enzyme glutathione reductase which transfers reducing equivalents to and from NADPH via flavin. Note the close distance of separation of the active units which transfer hydride (or electrons plus protons) necessitating close approach.

photosynthetic organisms by redox reactions of –S–S– bridges. There is a further connection of the redox reactions of –S–S– bridges via α-lipoic acid (Fig. 19.2) which is a bound coenzyme of great consequence since it acts to take pyruvate to acetate in the pyruvate dehydrogenase complex. This reaction is degradative and acts with NAD, not NADP, indicating again how separated metabolic pools can be compartmentalized by kinetic restrictions (Chapter 3). The structure of this enzyme is not known in detail but both thioredoxin and glutathione reductase are based on strong β-sheet domains.

The –S–S– bridge is important in more than just redox reactions since it is used as a cross-linking unit in extracellular proteins. The way in which this oxidation occurs is quite complex and the process needs to be easily reversible to avoid folding mistakes. It is likely that this external cross-linking came after the advent of dioxygen in the atmosphere.

19.2.2 Acid–base reactions

One of the most interesting features of sulphur chemistry is its reactions as RS^-. In this state, in proteins as cysteinate, it reacts as a base binding H^+ and a range of transition metals including Fe, Zn, Mo, and Cu. The individual chapters on the metals have analysed the active sites so formed. The use of RS^- is extended to both a function as an attacking group in thiol-dependent enzymes, e.g. proteases, and in the transfer of two-carbon units, e.g. acetyl, in coenzymes (Table 19.1 and Fig. 19.4). The reason for the use of sulphur is its ability to act as a good attacking and leaving group so that it gives intermediates of low energy and fast kinetics. The redox reactions of organic sulphides are always linked to an equilibration with protons in cells.

Fig. 19.4 The steps in the synthesis of long-chain fatty acids using transfer of acetyl groups from thiolate centres.

The role of thioether in methionine is quite closely related to that of thiolate. It acts as a base forming complexes with metals such as copper and iron, e.g. in cytochrome c. It can also act as a carrier of one-carbon fragments as in methionyl coenzyme A. Coenzyme M, Fig. 19.2, acts as a simple thiolate.

19.2.3 Group transfer reactions: coupling

Looking back to Chapter 1, where we described some fundamental properties of the chemical elements, we noted that in the periodic table sulphur was the first 'inorganic' element after hydrogen that was involved in all three classes of chemistry: acid/base, redox, and covalent bond formation. Just as phosphorus as phosphate, again in the second row of the periodic table, is ideal for handling the condensation reactions of biology since it has such labile bonds, so sulphur is ideal for the handling of redox transfer reactions involving two-electron steps. In order to understand fully the appropriate nature of biological choices of functions for elements we need to see also the connection in the cytoplasm between redox reactions based on sulphur and condensation

reactions. This connection is quite clearly the thioester reactions which transfer carbonium ion fragments,

$$RS-COR \longrightarrow RS^- + {}^+COR.$$

Thus RS^- units are linked into synthesis while they also undergo formation of –S–S– units.

There is a further very important feature of thioester chemistry, RSCOR′. These esters are of 'high energy' in the sense that reaction with water is strongly downhill

$$H_2O + RSCOR' \longrightarrow RSH + RCOOH.$$

In this respect a thioester could be used to drive syntheses which rely on condensation reactions and could therefore have predated pyrophosphate (ATP). Such a pathway may not have required complex template organization since it is possible to give reaction schemes for driving the reaction toward thioester formation using transformations on the surfaces of solids or through organometallic reaction schemes. In primitive life we can picture the series of reactions leading to thioesters as follows. Assuming that both CH_4 and CO were present, then methyl and carbonyl groups can come together on the same metal atom

Such a reaction could occur on cobalt or nickel centres, Chapter 16. Today we know that the transfer of, say, acetyl groups from $RSCOCH_3$, i.e. from lipoic acid or coenzyme A, is part of the connection between non-oxidative glycolysis, the oxidative citric acid cycle, and part of the pathway for the synthesis of fats (Fig. 19.4). (Synthesis of fats for the production of membranes was a very early requirement.) The many other thiolate coenzyme-A compounds are involved in further transfers. In archaebacteria there is the very smallest compound, coenzyme M ($HS-(CH_2)_2-SO_3^-$), which operates in a similar manner in assisting transfer of one- or two-carbon fragments in metabolism (Fig. 16.3). There is here a hint that this chemistry is very primitive.

The significance of lipoic acid, like that of glutathione, derives from its redox equilibrium between the –S–S– and 2-SH forms, which is such that it connects transfer of carbon fragments to the redox reactions of the cell. However, it is a *bound* coenzyme (compare flavins), while glutathione (compare NADH) is a mobile cofactor. Thus, communication is made from oxido-reduction reactions of NADPH (and, say, glycolysis, NADH) and carbon-fragment transfer using sulphur through the balance between thiolate and disulphide chemistry of mobile glutathione.

19.2.4 The sulphur cycle

In addition to the reductive reactions of sulphur there is a reversed oxidative series from sulphide to sulphate in organisms such as phototrophic bacteria (Fig. 19.5). The corresponding chemistry does not introduce any new

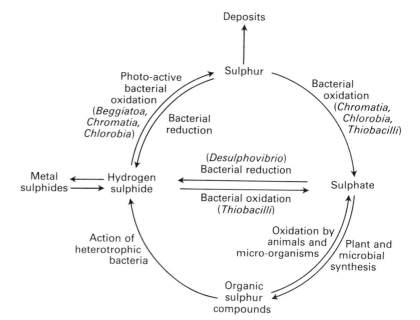

Fig. 19.5 A simple representation of the sulphur cycle.

principles differing from those above but the reduced forms of sulphur can give electrons to the energy-capture electron transfer chains which terminate in dioxygen.

19.3 Summary of sulphur biochemistry

Biotin (coenzyme R)

Closing this short section we stress that there is a huge biochemistry of sulphur compounds. Here we can only refer to some of the most important features which derive from its position in the periodic table. No other available element has such properties in combination.

1. It is a good attacking and leaving group as RS$^-$. This is of fundamental value in catalysis, e.g. in proteolysis.

2. It can polymerize to –S–S– or S$_n$ units, so it connects acid/base and redox chemistry at $E° = -0.1$ V. This is fundamental to the stability of –SH groups, to the reductive synthetic pathways in the cytoplasm of cells, to the stability of extracellular proteins, and in homeostatic control.

3. As a base, RS$^-$ or S^{2-}, it forms complexes with heavy transition metals (V, Fe, Cu, Ni, Zn, Mo) but not with Mn, Mg, and Ca, giving especially metal clusters. Metal sulphides could have supported an organo-metallic chemistry both before and in early life (see 'Further reading').

4. It forms stable organic compounds such as ethers (methionine) and cyclic thioethers, such as biotin (see margin), e.g. in pyruvate carboxylase

An element closely related in character is selenium but it is less available and its redox chemistry is rather different. Selenium has specialist not general functions in biology.

19.4 The biochemistry of selenium

Whereas it is impossible to list all the functions sulphur generates in different compounds in biology, the known activities of selenium are very limited, although it may well be that the future will reveal a much more extensive biochemistry of this trace element. Selenium occurs in some proteins as selenomethionine and has major value in oxidation reactions, but it is not known to form –Se–Se– bonds (see Table 19.2). It may well have been present in larger amounts in primitive organisms, see Chapter 17.

Table 19.2 Some biological selenium compounds and selenium enzymes

Compound or enzyme	Source
Selenocysteine	Corn and wheat seed hydrolysates
Selenocystine	Ethanol extracts of red clover and onion
Selenomethionine	Protein hydrolysates of dogs' liver; sheep's wool hydrolysates
Dimethylselenide	Astragalus plants and seeds (Se accumulators)
Formate dehydrogenase	*E. coli* and *Clostridium thermoaceticum*
Glycine reductase	*Clostridium sticklandii*
Nicotinic acid hydrolase	*Clostridium barkeri*
Xanthine dehydrogenase	Clostridium (several)
Glutathione peroxidase	Mammals
Thiolase	Mammals
Selenocysteine lyase (?)	Pig liver
Protein P	Selenium carrier in plasma

One of the great problems biology faced with the advent of dioxygen was the inevitable formation of both superoxide and hydrogen peroxide species. In Chapter 15 we have already described the various Fe, Mn, and Cu superoxide dismutases and their compartmental advantages, but a similar problem arose with the removal of peroxides. Catalase (Chapter 13) can handle H_2O_2 itself and many primary peroxides, but it is necessary to have its haem centre deeply removed from aqueous media containing organic molecules since, by single-electron transfer reactions, H_2O_2, bound to haem in catalase, could act as a dangerous radical generator just as in peroxidases; this must not happen, especially in the cytoplasm. The difficulty with a deep protected pocket is that it will not allow access to secondary or tertiary peroxides. Protection from these peroxides must use two-electron, atom-transfer reactions as far as is possible. Selenium is an ideal reagent for removal of an O atom from any peroxide and the redox potential of the bound oxygen atom is thereby greatly reduced. It is the two-electron covalent chemistry involved in the O atom transfer by selenium which allows it to be exposed to water while it reacts without radical formation, and the safe exposure makes it possible for the selenium-based peroxide reactions to take place not only with primary but also with secondary and tertiary peroxides in water. The mild oxidation catalysis in organic chemistry by SeO_2 is carried out in biology by R_2Se in proteins. The cyclic states of the selenium in the enzyme glutathione peroxidase (Fig. 19.6) require a reducing agent and this is again glutathione in the cytoplasm.

Fig. 19.6 The structure of glutathione peroxidase which has a single atom of Se and is based on a β-sheet. (After Epp, Landenstein, and Wendel, see Further reading.)

(All the reactions of selenium and haem-based enzymes with peroxides take place in aqueous media and not in membranes; to protect membranes there is a further scavenging system for peroxide based on the lipid-soluble vitamin E. This is another compound which, like ascorbic acid and glutathione, produces only innocuous free radicals.)

There are a few additional proteins which contain selenium. The variety stretches from some hydrogenases to a hydroxylase and there are also some selenium proteins of unknown function in muscle. Table 19.2 lists what is known, but we stress again that much more chemistry may be hidden as yet. The essential nature of selenium could not be better exemplified than by the human deficiency diseases (Kesham and Kashin–Beck syndrome) which cause muscular problems especially in northern central China where the soil is very low in this element. In fact, the soil is now treated with selenate with very satisfactory results.

19.5 Evolution and selenium

There is here and there the finding that selenium occurs in enzymes associated with hydrogen, i.e. in low-potential metabolism, and it is likely that some selenium was used in primitive cells. The advent of dioxygen during evolution switched the available selenide (not much would be available since many metal selenides are very insoluble) to selenate. The selenate anion could then be taken into cells through the sulphate uptake pathway, thence incorporated in enzymes, and used in the cytoplasm as an O-atom transfer centre. There is a contrast with molybdenum chemistry, although both elements are in group VI of the periodic table, in that molybdenum became used outside cells as an

oxygen-atom transfer reagent. The difference comes in the modes of their incorporations. We have described the coenzyme form of molybdenum but, quite unlike this incorporation, selenium is taken into an amino acid which is coded in the DNA, selenomethionine. While this fact correlates with use in the cytoplasm it throws up the question of the very early use of selenium in biology. Has modern biology lost the early importance of selenium and found a new use for it in detoxification from peroxides? A problem with the understanding of selenium biochemistry is that the selenium-containing amino acid is coded in DNA so that selenium incorporation is not a post-translational modification. In this feature it is only to be compared with sulphur and again the suggestion arises that its incorporation into biological systems was a very early event in evolution.

19.6 Notes on halogen redox reactions

The halide ions, Cl^-, Br^-, and I^-, appear to be activated in vesicles to form organic compounds external to eukaryotic cells, e.g. thyroxine (see margin), or to be used external to prokaryotes in particular as protective poisons. Halogen redox chemistry is a danger to DNA. Organic halogen compounds are often used by man as drugs but there is the danger of side-effects (Chapter 22).

We noted the catalysts for the production of these halogenated organic compounds in Chapters 13 and 17, i.e. haem iron and vanadium peroxidases, respectively, mentioning that these catalysts are probably in vesicles in eukaryotes. Such a system could not have developed until dioxygen entered the atmosphere and is probably a relatively recent addition to the armoury of messengers and hormones especially associated with development. At first, however, all such halogen reactions may have been used to produce poisons. A substituent iodine atom has great value in selectivity since there is no comparable unit throughout chemistry in size and shape. It can also be removed relatively easily.

We draw attention to the general contention that evolution develops from defensive strategies. The production of organic halides by biology was probably protective as was the use of halogenation. Subsequently, in a multicellular system the halogenated species can become messengers or hormones, e.g. thyroxine. A similar point arises in the case of hydroxylation generally. A problem then arises concerning the removal of iodine from organic compounds and it may be that this detoxification requires selenium, see Further reading.

Further reading

The general description of the biological chemistry of sulphur is found in standard biochemistry textbooks. This chapter is intended to stress the roles of sulphur unencumbered by the description of metabolism, but see the following references.

Ciba Foundation Symposium (1980). *Sulphur in Biology*, Ciba Foundation Symposium, no. 72 (ed. K. Elliot and J. Wheelan). Exerpta Medica, Amsterdam.

Holmgren, A. (1980). Disulphide oxidoreductases. In *Dehydrogenases requiring nicotinamide coenzymes* (ed. J. Jeffrey), pp. 149–70. Birkhauser Verlag, Basel.

Relevant selenium biochemistry is described in the following references.

Arthur, J. R., Nicol, F., and Beckett, G. J. (1990). Thyroxine, selenium and de-iodination. *Biochemical Journal*, **272**, 537–40.

Epp, O., Landenstein, R., and Wendel, A. (1983). Structure of glutathione peroxidase. *European Journal of Biochemistry*, **133**, 51.

Günzler, W. A., Steffens, G. J., Grossmann, A., Kim, S.-M. A., Otting, F., Wendel, A., and Flohé, L. (1984). The amino-acid sequence of glutathione peroxidase. *Hoppe–Sevler's Zeitschrift für Physiologische Chemie*, **365**, 195.

Huxtable, R. J. (1987). *The biochemistry of selenium*. Plenum Press, New York.

Shamberger, R. J. (1983). *Biochemistry of selenium*. Plenum Press, New York.

Stadtman, T. C. (1990). Selenium biochemistry. *Annual Reviews of Biochemistry*, **59**, 111–27.

The biochemistry of iodine is described in the following references.

Alexander, N. M. (1984). Iodine. In *Biochemistry of the essential ultratrace elements* (ed. E. Frieden). Plenum Press, New York.

Kirk, K. L. (1991). *Biochemistry of the elemental halogens and inorganic halides*. Plenum Press, New York.

PART III

Biological shape and the integration of element functions

We have seen in Part II the complicated biological chemistry of a wide variety of individual elements. Clearly prominent were the different uses of particular elements and the fact that these different elements were used in different spatial zones. At no time did we examine closely the shapes of these zones although they were described in outline in Chapters 3 and 6. It now becomes necessary to introduce the *shape* of biological systems, whether it is a whole animal, an organ, a eukaryotic cell, or even a vesicular unit, for without the description of shape it will not be possible to describe the essential flow of energy and material which was discussed in Chapter 6 where we saw, for example, that pumps and channels were positioned. In this part of the book shape will be introduced first through the description of biological minerals, hard structures (Chapter 20) before we turn to the interlinked extracellular matrix of multicellular organisms against the background of dynamic homeostasis of concentrations of elements. This, the most difficult topic, biological morphology and morphogenesis, will be described in Chapter 21. The minerals introduce all these problems since biology shows control of both the chemicals comprising the minerals, which are almost always formed from a combination of a metal and a non-metal, and of their polymorphic shape.

In Chapter 22 we look at the effect of man's activities on element availability and the use he makes of different element chemistry within biology. There is today an insistent need for man to come to terms with his chemical nature and that of both the species and the environment around him. The forms of life which exist today are in tight relationship with the environment through a common chemistry, much of which is inorganic.

20

Biological minerals and biological shapes

20.1	Introduction	467
20.2	Biominerals viewed as structures	469
20.3	Formation of inorganic mineral structures	473
20.4	Biological single crystals in relation to single-cell morphology	476
20.5	Relationships between biomineral formation and morphological structure: examples	477
20.6	Unicellular eukaryotes: summary	485
20.7	Development and growth in multicellular animals	487
20.8	Extracellular biominerals in multicellular organisms	489
20.9	Summary of skeletal crystals in multicellular organisms	491
20.10	The organic extracellular matrix for mineralization	491
20.11	Sensors	493

20.1 Introduction

The elements in biominerals

The subject of biominerals covers a wide range of inorganic salts which serve a variety of functions in biology (Table 20.1). The chemical elements involved are shown in the margin. The control over their selective precipitation, i.e. nucleation and growth patterns, the final form of their precipitates, and their locations in a living system have developed during evolution. As a consequence, the inorganic or composite (inorganic/organic) materials used by biology have very refined properties and are not only of interest in themselves but may well lead man to a rethinking of the formation and value of minerals, especially composites in industry. The 'Further reading' for this chapter suggests a few books which cover the physical properties of these composites and their biological value in detail.

The most obvious function of the minerals in biology is, in fact, to give physical strength, but biological minerals do not just provide construction materials, e.g. for shells, bones, and teeth, but are used elsewhere, for example, in many types of sensors (Table 20.2). The interaction of biological systems

Table 20.1 The main inorganic solids in biological systems

Cation	Anion	Formula	Crystal	Occurrence	Function
Calcium	Carbonate	$CaCO_3$	Calcite Aragonite Vaterite	Widespread in animals and plants	Exoskeleton Gravity sensor, Ca store Eye lens
	Phosphate	$Ca_{10}(PO_4)_6(OH)_2$	Hydroxyapatite	Shells, some bacteria, bones, and teeth	Skeletal, Ca store (piezo-electric)
	Oxalate	$Ca(COO)_2 \cdot H_2O$	Whewellite	Insect eggs, vertebrate stones,	Deterrent Cytoskeleton
		$Ca(COO)_2 \cdot 2H_2O$	Weddellite	abundant in plants	Ca store
	Sulphate	$CaSO_4 \cdot 2H_2O$	Gypsum	Coelenterate statocysts Plants (Phytoliths)	Gravity sensor S store, Ca store
	(Silicate)				
Iron	Oxide	Fe_3O_4	Magnetite	Bacteria, animals, teeth	Magnetic device
	Hydroxide	$Fe(OH)_3$	Ferritin	Widespread	Iron store
Silicon	Oxide	SiO_2	Amorphous (opaline)	Sponges, protozoa, abundant in plants	Skeletal
	Hydroxide	$SiO_n(OH)_{4-2n}$			Deterrent
Magnesium	Carbonate	$MgCO_3$	Magnesite	Reef corals	Skeletal

with gravitational and magnetic fields is usually manipulated through the use of one or other of the minerals in Table 20.1. Again, in biological solid matrices open to stress, e.g. bone, changes may be induced by stresses that produce electric fields and these fields, in turn, have consequences for the properties of the solids themselves, e.g. their (crystal) growth. We shall see that the way in which biominerals grow is linked to the big unsolved problem of biology—morphology—and it could be that the study of the growth of minerals in biology will provide an excellent base for an attack on morphogenesis (Chapter 21). The skeletons of many living creatures have peculiarities (see D'Arcy Thompson's wonderful book, *On growth and form*, which is largely about biominerals; 'Further reading') and these peculiarities are clearly a product of, and effect directly, morphogenesis since the gametes of biological systems rarely if ever have a biomineral precipitate in them.

It would be quite wrong to finish these introductory paragraphs without reference to the *chemical* significance of biological minerals. Many of these minerals are deposits which assist in the regulation of the levels of free cations

Table 20.2 Biomineral sensors

Sensor	Comment on crystallography
Calcite in otoconia (man) (gravity)	Crystal size and morphology controlled by vesicles?
Aragonite in otoliths (fish) (gravity)	Size grows with size of fish
Magnetite in bacteria (magnetic field)	Normally cubic crystal grows as a rectangular unit. Crystals are organized morphologically
Baryte in desmids (gravity)	Conventional inorganic morphology

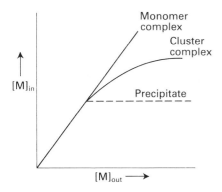

Fig. 20.1 The relationship between extracellular and intracellular concentrations of free metal ions, M, for a fixed input rate where the internal store is: (1) a soluble monomer ML complex; (2) a soluble cluster M_nL complex; and (3) a precipitate. Homeostasis is best achieved, without energy dissipation, e.g. by pumping, using a precipitate.

and anions in cellular systems. In particular, iron, calcium, and phosphate concentrations are under stringent control. A biomineral is the best 'buffer' control for such homeostasis (Fig. 20.1). It is also a simple matter for cells to exocytose mineral deposits and so eliminate unwanted excesses of elements. Indeed, there are some who believe that much of the metabolism of calcium sprang from the need to reject excess calcium and that this led to the development and temporary storage of the variety of biominerals which contain this element. Contradicting this view is the observation that many organisms build skeletons from elements which they do not need to reject, e.g. Si, Sr, and Ba. Again, deposition and redissolution of minerals, such as iron and manganese oxides, can be used as an energy source by organisms moving up and down between oxic and anoxic regions while they alternately oxidize and reduce the metal ions or they may even use the conversion of one iron sulphide to another for energy capture.

The range of biological systems which contain biominerals is vast—from man to unicellular structures. Modern molecular biology indicates that unicellular systems must be the best place to start an understanding of biological structure and this is the line of attack we shall pursue after we have remarked further upon the structures involved and the laying down of solid minerals.

20.2 Biominerals viewed as structures

20.2.1 Physicochemical character

The minerals in Table 20.1 that form structures giving internal strength, or external protection, and/or chemical 'buffering', can be either crystalline or amorphous. If they are crystalline then biological control can be exerted at various levels: (1) the chemical composition (see Table 20.3); (2) the allotropic form; (3) the crystal size and shape; (4) the spatial position of the deposit. Now, each of these choices has to be guided by an organic matrix, which respectively controls: (1) element concentration; (2) nucleation; and (3) growth, and by the cell itself which can relocate a deposit. These points will be

Table 20.3 Chemical composition of some biominerals

Biomineral	Chemical additives
Calcite ($CaCO_3$)	Inclusions of magnesium
'Apatite' ($Ca_5(OH)(PO_4)_3$)	Variable phase composition; inclusions of magnesium, carbonate, fluoride
Celestite ($SrSO_4$)	Pure phase (inclusion of some barium?)
Baryte ($BaSO_4$)	Inclusion of some strontium
Silica (SiO_2)	Widely variable phase $SiO_n(OH)_{4-2n}$ ($n<2$)
Magnetite (Fe_3O_4)	Pure phase
Iron oxides $FeO(OH)$ and $(Fe(OH)_3$)	Often ill-defined (plus phosphate)

taken in turn. If the mineral is amorphous then chemical composition is open to wide (infinite) variation, though concentration of required elements is still necessary. The typical amorphous biomineral is hydrated silica, which we write as $SiO_n(OH)_{4-2n}$ where n varies continuously from 0 to 2 (Chapter 18). Here there is no allotropic form or shape related to the chemistry or physics of the mineral and the size and shape of building blocks will be decided solely by mechanical constraints on the precipitated phase. At first sight the variety of shapes open to amorphous structures should be infinitely greater than that open to crystalline phases, but later in this chapter we may wish to revise this view. The limitation of crystal unit cell, say of $CaCO_3$, to a few local symmetries does not seem to have caused a very limited range of crystal morphologies.

The chemical character of a bio-mineral is clearly a genetically controlled feature, initially due to the deliberate movement of elements (ions) into specialized parts of biological space by ion pumps. Now the crystal which forms is limited by two considerations, the stability of the lattice (a non-biological factor) and, inside this space, the chemistry of the surface and/or of growth-inhibiting compounds will control the precise compound which is precipitated. Anybody who has worked with crystal growth in chemistry knows that the face which grows is dependent upon the added chemicals in the solution. Here again genetic control is involved. Despite the simplicity of the unit cells, the many ways in which unit cells can be put together, and hence the various types of crystallographic unit, allow a variety of crystal shapes to develop. This chemical control over structural form is well known in industry, e.g. in the growth of NaCl (common salt) crystals. It is also well illustrated by the many thousands of variations of snow-flake crystals. In biology this variation is both an advantage and a disadvantage since, on the one hand, the form is adaptable no matter what the space group of the unit cell while, on the other hand, crystal formation has to be kept under very fine control. The contrast with protein structures is very great. Here, the chemical composition, i.e. the amino sequence, is controlled and the folded form, a single-molecule 'crystal', is thereby fixed. Even the packing of protein molecules together may be very well controlled by specific surfaces due to the fold so that proteins can be built into unique forms without containers or external fields, e.g. tubulin filaments (Fig. 20.2) and also ferritin, chitin, and collagen. Unit cells of inorganic crystals do not self-assemble in a unique way. Thus, ultimately, form must be controlled by the organic matrix and genetic control is involved in part as for protein structures but indirectly.

Clearly, crystal or amorphous materials had to be made under continuous chemically and physically controlled conditions, apart from those cases with which we shall start the discussion in which irregular bits of solids were accumulated by the sticky surface of the outside of cells. This primitive form of biomineralization is not of great interest, except in evolution where, in all probability, it allowed the first steps to be taken toward cell/cell organization in colonies, not truly multicellular organisms. The interesting cases of biominerals are those under cellular control. This is managed by making physical containers. Epitaxy (chemical matching) on a surface may be involved in the nucleation step, although this is under concentration control too, but it is quite inadequate as a fully fledged control over shape in a growing system.

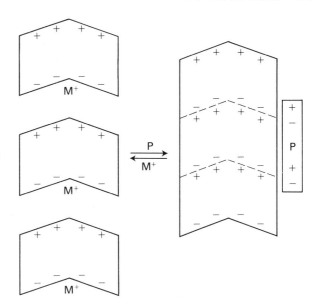

Fig. 20.2 An illustrative diagram of the reversible formation of intracellular protein filaments in linear 'crystals', e.g. tubulin. M^+ is a metal ion and P is an ancillary protein.

This description of biological minerals shows us that we shall always have to think in terms of three very different types of chemical: (1) the components of the mineral which give three-dimensional systems and fill up certain spaces; (2) deliberately added inhibitors; and (c) the organic molecules, which can form linear or layer systems as well as interact with the inorganic matrix so as to generate spaces to be filled by the minerals. At the same time we shall find that we need to notice the physical forces acting on the growing mineral and that the whole system is energized.

20.2.2 Vesicles and polymer matrices

The main types of volume control over biominerals are either: (1) separated parts of the cytoplasmic space isolated by lipid membranes (vesicles); or (2) extracellular zones (usually in multicellular species) where well defined volumes and outline shapes are developed by organic filamentous structures. Table 20.4 gives a list of some more complex examples and it is clearly seen that there are many combinations including two main types. The first, which is very common in unicellular organisms, is the simplest vesicle. Here there is

Table 20.4 Restricted vesicle or other space in biology

Type of controlled space	Control of crystals
(1) Membrane-enclosed vesicle	(a) No restraints on faces (b) Volume only limited by mechanical constraints
(2) Protein-enclosed vesicle	Volume limited and internal faces fixed
(3) Protein-enclosed continuous matrix	Volume limited and internal faces fixed; repeating units
(4) Polysaccharide-enclosed (continuous) matrix	(a) Poorly defined (b) Well defined as in (3)
(5) Vesicle enclosing fixed protein matrix	Volume and form all well defined

only an aqueous compartment surrounded by a lipid membrane which is impermeable to all ions and most organic molecules. The ions for the biomineral, typically ions such as Ca^{2+}, H^+, SO_4^{2-}, HPO_4^{2-}, and HCO_3^-, are accumulated in the vesicle by the operation of pumps. Thus much knowledge of biomineral formation rests upon an understanding of vesicles in cells and their pumps (see Chapters 3 and 9).

The second class of container is formed from proteins or polysaccharides and is permeable to ions and molecules to a lesser or greater degree depending on the system. The container may be in the cell, single and isolated, as in the case of ferritin, or outside it, multiple and continuous, e.g. collagen of bone. Here the control of precipitation rests in the maintenance of cytoplasmic or extracellular ion concentrations (Fig. 20.1). In the last case the extracellular fluids are themselves of a maintained chemical composition through the activity of control organs, such as the kidney, which are effectively macropumps.

The exact form of the protein container of ferritin is fixed, as is that for the apatite crystals of bone in the open spaces of collagen, although the whole bone shape is open to stress transmitted by extracellular fibres (see Section 20.8.2). By way of contrast, the space in a typical vesicle is not controlled directly by the organic structure since vesicles do not have internal polymer cross-links between their membranes. In fact, vesicular space is very different from cytoplasmic space which is generally criss-crossed with fibrillar stress structures, but it also differs in that: (1) ion pumps act in opposite directions (*into* vesicles as opposed to *out* of the cytoplasm); (2) the vesicle has mechanical pressures exerted on it from the cytoplasmic fibrillar network; (3) the internal face of the vesicle often contains polysaccharides which are absent in the cytoplasm, i.e. vesicle space is effectively extracellular space but without filaments; (4) vesicles can be located in special places in a cell; and (5) the vesicle, if grown by itself, would probably be of no form, i.e. it would be a sphere. (There is, of course, the rare case of an organic protein matrix in a vesicle, when the form of the vesicle would be similar to that of a simple proteinaceous container.) In evolution the creation of vesicles inside cells gave a vast potential for the development of the shapes of biominerals.

There are also other different physical as well as chemical controls over the development of mineral phases. The physical controls are limited by the physics of our world as well as by biologically produced fields, just as the chemistry of biology is limited by the properties of elements in the periodic table. The mineral and the vesicle grow together under the influence of many macrofields. We must notice that functional value as sensors depends on back-interaction with these fields through the density, the magnetic properties, the ion mobility in lattices, the deformability (elastic constants), and so on of materials, which are properties that are not under biological control (see the 'Further reading'). We refer again to sensors (Table 20.2). There is no strict separation of a structural component and a sensor—bone is always both.

We have already stated that microscopic shape is limited to some degree by principles of the symmetry of fields in crystalline but not amorphous materials. Any biomineral based on crystals has many shape limitations to which the organism has adapted. Now these limitations are of great advantage to the investigating scientist who can ask, as did D'Arcy Thompson, 'Do the observed forms of crystalline materials in biology follow expectations of

growth laws outside biology?' What are these crystal growth laws? They must have influence on the morphogenesis of biocrystals, just as they control morphology of inorganic crystals and so they can be applied to the understanding of biological morphogenesis.

20.3 Formation of inorganic mineral structures

20.3.1 Nucleation

There are two stages of formation of a crystal: nucleation and extension of a nucleus once a stable microcrystal exists under given growth conditions. Control over nucleation still allows many final shapes (and other properties); for example, calcite, aragonite, and vaterite are different allotropes, have different nucleation steps, and allow many different crystal morphologies of calcium carbonate, although generally needles grow more readily with one allotropic form, cubes with another, and spherules with the third. Concerning the control over nucleation of a compound of given formula, XY, we draw attention to Lussac's law which leads (together with modern knowledge) to the following statements: (1) thermodynamically, at the lowest degree of supersaturation, i.e. relative to the most insoluble salt, XY, only this most insoluble salt can nucleate and grow; (2) given higher degrees of supersaturation the probability of the *most soluble* form of the salt being preferentially formed increases. At very high supersaturation the most soluble form usually crystallizes first (Fig. 20.3) and obviously is amorphous in the limit (Lussac's law). This is a kinetic consequence. Nucleation can be controlled by local surface adsorption by biopolymers, i.e. 'genetic' control which will be dealt with under the title of epitaxy, but environmental factors are also important. In most cases it is the effect of organic molecules bound on surfaces upon the surface energy of the crystal faces (Fig. 20.3) that decides which phase is nucleated. This allows biology to 'choose' to form stable or unstable crystalline phases by the synthesis of matrices of organic molecules. Soluble

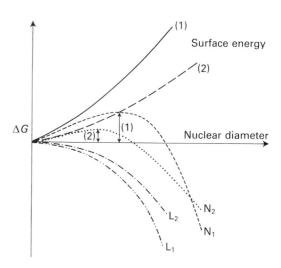

Fig. 20.3 The free energies per mole ΔG of two nucleating systems of the same compound, labelled N_1 and N_2, with change in size of nuclei during crystallization assuming two forms of crystallization from the same solution. Form (1) has the higher adverse surface energy but the more favourable lattice energy, L_1, and so meets the higher barrier to crystallization, shown by vertical arrows. N_1 is ultimately preferred by thermodynamics.

molecules are not effective nucleating agents, but very greatly affect crystal growth by inhibiting some facial development.

20.3.2 An explanation of Lussac's law

The general rule that at high supersaturation relative to the several possible forms in which a salt can grow, e.g. $CaCO_3$ that can crystallize as vaterite, aragonite, or calcite, it will be the most soluble form, here vaterite, that will crystallize first is in the opposite sense to expectation based on the thermodynamic driving energy toward crystallization (Fig. 20.3). This means that it is the *relative* surface energy of the more soluble form which has allowed its preferred nucleation and then growth. The explanation may well be as follows. The surface of a crystal remains partially hydrated, but the better the packing of ions into the lattice, i.e. the larger the lattice energy, which also applies in part to the surface of that lattice, the poorer the residual hydration must be. Thus, a well packed surface has only some of the internal lattice energy while it has lost considerably the hydration energy of the free ions. This low hydration energy of the surface is most important when the surface is large compared with the bulk and so destabilizes the crystal nucleus. The argument leads us to Lussac's law that the most soluble salt, least favoured thermodynamically, is the one which precipitates (nucleates) first, since, although it has the least stable lattice, it has the most stable surface. The law derives from the initial steps of crystallization only since the surface energy dominates the nucleating energy at this stage (Fig. 21.3). By extrapolation it leads to the rule that, under conditions of very high supersaturation, or where kinetic barriers are great, the precipitate will be amorphous, e.g. silica, since, while this is the least stable solid phase, clearly it has the most hydrated surface (compare quartz; see Fig. 18.12).

There is no need to suppose in this discussion that the initial steps in the production of a nucleus of, say, 20–50 atoms give a fixed form. It may be that as clusters of 5–20 atoms are built there are many different patterns in fluctional equilibrium. Only after, say, some 20–50 atoms come together does true form begin to arise. If this is so we must beware of looking for necessary epitaxial relationships between the surfaces of biological polymers and the *final* forms of biominerals produced in their presence. The epitaxy could well modify the nucleation by controlling the fluctional forms but, in biology, the degree of supersaturation could be kept low. Epitaxy is but one control.

20.3.3 Growth of inorganic crystals

There are two very different growth conditions observed in chemical systems: 'Equilibrium' growth and 'kinetic' growth

(i) Homogeneous, constant concentration, solutions: 'Equilibrium' growth

Constant concentration solution conditions can be maintained so long as diffusion to the growing crystal is not rate-limiting and there is a supply of material to compensate for deposition as in a chemostat. In such circumstances growth of a crystal is of constant morphology, i.e. there is so-called equilibrium growth, no matter what the crystal size. This is a consequence of a

maintained total minimum surface free energy relative to a fixed set of solution conditions. Growth of this kind, by its very nature, is slow and close to being reversible. Notice that physical (field) conditions must be constant, as well as chemical potentials. Although it is called 'equilibrium' growth, it is in fact steady-state growth. It is very probable that biological minerals grow under these equilibrium growth conditions.

If after growing under equilibrium conditions for some time a new chemical or physical field is imposed, then a new equilibrium morphology develops slowly and there is an intermediate series of rounded crystal shapes due to multinucleation and growth. A typical new 'field' is generated by, for example, the introduction of an inhibitor of growth or a constraint imposed on growth by contact with a physical barrier.

In parallel with these observations on the three-dimensional growth of inorganic crystals we can also see that the ideas of equilibrium growth apply to the linear form of a rod made from proteins (Fig. 20.2) where the terminal molecules have a steady-state energy relationship with free molecules. Here, molecules can only add in a line due to a unique single axis of recognition. Two-dimensional growth of flat membranes can also be achieved under steady-state chemostat conditions (consider Fig. 7.24). The packing of molecules can now be in a variety of forms and therefore constant shape (at the edges) of the flat membranes will require very careful control of solution conditions. Furthermore, we can consider that, if the membrane contains a variety of molecules of different shapes, it could be curved with any degree of curvature so that, if curvature is to be fixed during growth, composition will also have to be kept under very strict control. The clear implication is that if a three-dimensional cell is to grow with constant shape, its linear filaments, two-dimensional membranes, and any three-dimensional crystals it contains must grow in chemical and physical harmony. We see immediately that the requirement of a chemostat and a physicostat are those of homeostasis and in Chapter 21 we shall try to show how this can be achieved in biology.

(ii) 'Kinetic' growth

In this case local concentration is not maintained and the crystal grows where diffusion permits it (see 'Further reading'). Well known examples are in fast electrolytic deposition of metals where tree-like growths from an electrode surface are easily explained by diffusion limitations. Any tree-like feature must mean that growths of different faces are not balanced evenly at different times. Once again impurities affect the growth. This type of growth is likely to be fast and probably yields a defective form in biology. Again the same rules apply to organic filament or membrane growth. (We must remember the internal growth patterns of brain nerve cells which are, in fact, idiosyncratic and in some sense tree-like. The inside of a brain does not have a fixed shape.)

20.3.4 Ripening (ageing)

It is a well-known observation in analytical chemistry that small crystals dissolve so as to allow larger crystals to grow. Somewhat peculiarly, this 'Ostwald' ripening also occurs with amorphous silica. It is a consequence of a reduction of surface energy and a gain of 'lattice' energy. Many minerals in

biology are not in a strictly stable form, as shown by the small sizes to which particles grow. Their further growth is impeded by organic matrices or by soluble molecular inhibitors.

We shall now use this description of inorganic crystal formation in an attempt to understand the morphology of biominerals and the morphological force field in which they have grown.

20.4 Biological single crystals in relation to single-cell morphology

During equilibrium growth of a crystal the product-sum of energy over all exposed crystal planes, i.e. Σ surface area × surface energy, must be a minimum provided the medium is homogenous. The observed shape of a crystal then relates to the relative surface energies of different faces in a chemostat. On the other hand, in a cell the surface energy is an apparently unknown complex function of the solution and surface contact conditions. Therefore, using the postulate of equilibrium growth in reverse, 'If a crystal grows with constant shape then, during growth, the medium is of fixed isotropic composition or of fixed concentration gradients and the field contacts of the crystal surfaces are also fixed.' *Acantharia*, small unicellular eukaryotes found in the sea, for example, grow crystals of almost constant shape (even when the shape is complex) during long phases of their development, up to tenfold changes in size, showing that constant conditions of chemistry and physics exist in the cell over long periods of cell growth which is just expansion (Fig. 20.4). Note that there is virtually no feathering or tree-like growth typical of kinetic crystal growth. However, when cell expansion ceases the conditions change, the chemistry and physics of the cytoplasm change, and this is clearly registered in massive crystal morphology modifications in some species of *Acantharia* (see Section 20.5.1). In general, the mineral framework in a biological individual grows in size while maintaining shape, e.g. man's bones. Of course, such growth can be managed by controlled multiplication of containers of fixed sizes rather than by single crystal growth, but this implies regular organic increments added to the container system. (Note here the modular way in which plants grow.) In all cases it is the organic matrix and its contents which must therefore be under equilibrium growth conditions to ensure the constancy of shape of the crystals.

The in-filling of a *fixed size* container or vesicle cannot be described as 'equilibrium growth' since the crystal or amorphous solid just grows until it is forced by the boundaries into a fixed shape. It appears that crystals of magnetite in bacteria grow in this way. In this case, the interesting morphological point is the organization of the matrix and these systems are models for local crystals in bone development but not for bone development as a whole, which follows equilibrium growth in that the organic matrix is developed in 'equilibrium' with the mineral so as to maintain shape.

In all organisms the biominerals grow together with the membranes, the protein fibrils, and so on. They must all be in interactive balance if overall shape is to be maintained since growth of one develops stress fields for the others. This means that following the morphology in cells, here of inorganic

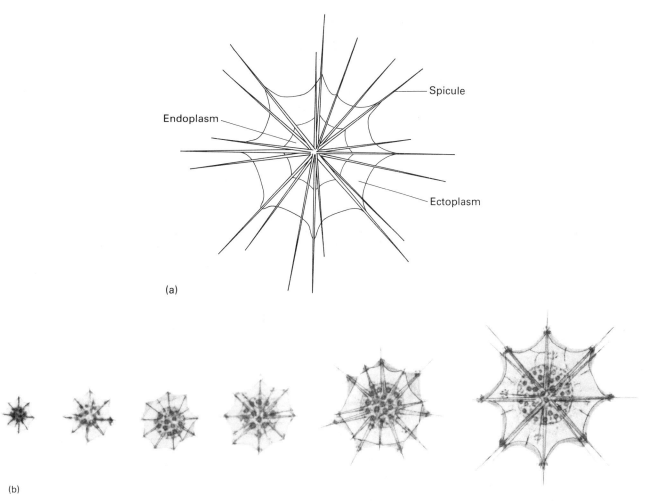

Fig. 20.4 (Top) A simple 20-spicule *Acantharia*. Note the two inner membranes. The vesicle membrane encloses the whole spicule (not shown). (Bottom) The growth of an *Acantharia* cell with constant shape.

crystals, corresponds to the study of the morphology of the whole species. The shape of the otolith of fish is fixed and, like a tree, it has age rings. The shape is maintained through vast changes in size.

Now we shall choose examples which illustrate these principles of the relationship between biomineral formation and morphological structure. Clearly, a biological system of one cell with single crystals in that cell is the easiest experimental object, since there is then a direct interaction between morphological instruction from genes and prescribed observable morphology of crystals. *Acantharia* provide the best example known to us.

(While reading the next sections, which describe a variety of mineralized shapes in biology, the reader is asked to keep in mind that these growing systems effectively 'equilibrate' with solutions under solubility product control. For example the solubility product of $CaCO_3$ controls HCO_3^-, Ca^{2+}, and H^+ concentrations to become part of the homeostasis of the whole organism, Fig. 20.1.).

20.5 Relationships between biomineral formation and morphological structure: examples

20.5.1 *Acantharia*

Acantharia skeletons are comprised of exactly 20 organized, relatively large, single crystals of strontium sulphate. In these organisms there is a 'most fortunate' relationship between a precise mathematical form of an internal *crystalline* skeleton (of strontium sulphate) which must be governed by the rules of inorganic crystal growth and biological shape simultaneously. If a framework is to evolve using strontium sulphate crystals then each piece of the framework must have the local space group (orthorhombic) symmetry of these crystals. We shall see what limitations this produces later. This symmetry is very simple, the three unequal principal axes being at right angles (Fig. 20.5).

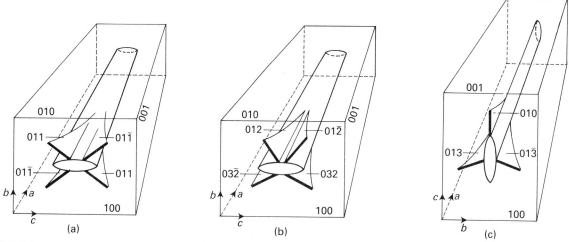

Fig. 20.5 Schematic representation of the three *Acantharia* spicule types, (a), (b), and (c), showing the relationship between spicule morphology and crystallographic orientation. The planes of all the lateral wings are labelled: *a*, *b*, and *c* refer to the orientation of the crystal axes of strontium sulphate.

There is no other stable form of $SrSO_4$ known (contrast $CaCO_3$). We must follow the axes of the crystals in relationship to cellular development and form.

Cells are never spherical; on the other hand, single cells (and often cellular systems) invariably have a rough rotational axis and unicellular prokaryotes are therefore frequently cylindrical (axial), for example *E. coli* amongst bacterial cells. There is a clear link between this type of structure and cell division which occurs at a locality associated with two separating sets of genes to give two nuclei. Examination of eukaryote cells suggests that, for many, if not all, there is also a network of bipolar stress filaments (in addition to those which must develop on chromosomal division) which stretch roughly radially with this polarity and also a series of tangential polymers under the cytoplasmic membrane. They are probably related to tubulins and spectrins, respectively (Fig. 3.5). This is apparently true of *Acantharia*. Outside the cytoplasmic membrane there is a less well organized system of glycosylated

polymers. The cell polarity can be treated as the major axis in the discussion of the relationship between the morphology of *acantharia* and the orthorhombic crystals of its internal skeleton of strontium sulphate. (In fact the simplest mature cell has apparent tetragonal symmetry, i.e. that of a square cylinder, higher than the crystals themselves (Fig. 20.5), but close examination shows it is not so.) We expect any vesicles which are involved in development and based upon a central organization with axially organized filaments to grow in two ways: (1) from the centre outwards, in general parallel to the direction of the radial filaments, controlled by these filaments, and roughly perpendicular to the outer membrane (Fig. 20.4); (2) along the membrane, i.e. parallel to it, at right angles to the filaments (see Fig. 20.8). *Acantharia* provide many examples of crystals in vesicles growing from the centre of the cell (Fig. 20.6) and only in some species is there any growth late in development parallel to the outer membrane. In fact, it is obvious that such parallel-to-the-membrane growth of crystals can only occur after cell growth has ceased since otherwise they would have to adjust curvature continuously. The crystal axes of the radial spicules (crystals) are now: *a*-axis, long axis of vesicle, radial or diametric; *b* and *c* axis, tangential to membrane. The spicule crystals have various cross-sections from species to species, e.g. roughly elliptical or inequivalently cross-girdered reflecting always their orthorhombic symmetry (Fig. 20.5) and the single crystals fit together on faces quite remarkably at the centre of the cell (Fig. 20.6). The cell morphology depends in part on genetic control over chemicals supplied to build the vesicle within the filament structure, but after that the physics and chemistry of interacting ions, molecules, and surfaces within the vesicle must be obeyed. The choice of strontium sulphate implies some limitation on preferred biological morpho-

Fig. 20.6 The fitting together of crystals of strontium sulphate at the centre of the cell. Notice how crystal planes come together. The number, 20, of spicules may be decided by crystallographic considerations! (a) Two spicules; (b) several spicules; and (c) 20 spicules coming together. A, B, and C indicate the different spicule types shown in Fig. 20.5. Crystal-face matching decides the cells morphology.

Fig. 20.7 The nature of strontium sulphate single crystals, with large and small spicules, two of which are plate-like, in some species of *Acantharia*. Note the orthorhombic habit.

logy, but the direction and relative sizes of axes of growth of $SrSO_4$ crystals are clearly decided by the biology since $SrSO_4$ in inorganic isotropic media, i.e. in a chemostat, gives flat rhombs. We now make the important statement that *Acantharia* can only grow in this way if *all* components of the cell act in a unique integrated manner. The growth of the crystals follows equilibrium growth conditions and is species-specific. Individual species must be different chemostats and physically controlled systems.

The realization that there are simultaneous inorganic crystal and biological growth controls can be seen again by looking at the morphology of different species of *Acantharia*. Figures 20.7 and 20.8 show some examples. In fact, there are species in which two, four, eight, or 12 of the 20 single crystal spicules grow preferentially and there are variations in the crystal shapes within each class making for a vast variety of species with distinct inherited morphologies all under genetic control, but each crystal can be immediately seen to be related to the orthorhombic symmetry of strontium sulphate. Moreover, *each and every cell grows with constant shape!* We are forced to conclude that all the cells grow under equilibrium growth conditions, i.e. they maintain a fixed set of physical forces and a strict chemostat. The demand is that all chemical and physical forces are in a steady state (of growth) relationship. We are forced to conclude further that the chemical potential of all the elements, not just those of strontium and sulphate, are in feedback communication so as to maintain constant relationships. The rule must apply to all organic and inorganic components alike and in turn this implies that the element concentrations connect back to the genetic information, that is, to DNA, since it is proteins, synthesized through reading of DNA instructions, which are the components of the structural control in cells. Now, apart from this homeostatic growth period, some of the *Acantharia* show a remarkable change of growth pattern later in life, that is they undergo a morphological switch of their crystals, which has to mean that the steady state of growth has switched.

Spicule development can be much more complex than in the form of simple rods in certain *Acantharia* species. When the initial spicule growth becomes limited by pressure on the outer membrane (no further cell growth) then increasing stress on the internal filaments and outer membrane is naturally generated by further attempted crystal growth. There is then the possibility that the vesicle containing the crystals, which may or may not grow further radially, can spread in directions perpendicular to the line of original growth. (The physicostat has failed.) These deviations, if unrestrained by the radial contractile apparatus, will be expected to be more closely dominated by the crystallography of $SrSO_4$ since the constraints from the polymers outside the main body of the cell are slight. This is exactly what is observed (Fig. 20.8(a)). The $SrSO_4$ crystals now grow alternately *along* and *perpendicular* to the previous direction. As they evolve they look more and more like the plate structures that would grow in a man-made chemostat. They are then only constrained in a limited way by soft tissue outside the inner membrane and by chemostat considerations, since they are still in vesicles, and not by the strong filaments of the cytoplasm. In some species the growth of the spicule vesicle extends instead *under the cytoplasmic membranes where it is still strongly constrained and now it can only grow as a sphere or ellipsoid* (Fig. 20.8(b)). The change we see here from growth in one pattern to growth in another is metamorphosis. Note how only parts of the organism are changed.

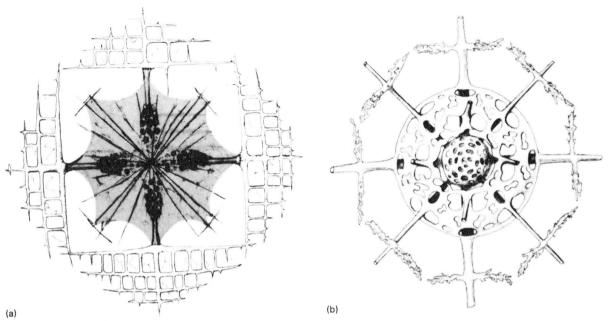

(a) (b)

Fig. 20.8 (a) A striking form of an *Acantharia*. The cell grows at first with four dominant crystals in a plane and at right angles to one another. When cell growth stops, the crystals grow further outside the plasma membrane in four rectilinear grids which remain in the plane. The growth pattern is dominated in the late stage by inorganic considerations, but also notice the inner petal-like organic growth (dark) related to this morphology and the outer membrane related to the other 16 spicules. (b) Part of the skeleton of another form of *Acantharia*. Some 12 of the 20 spicules are shown with two partly grown spherical shells of $SrSO_4$ crystals which are forced into this morphology by growth under the plasma membranes. Starting from any spicule there is, after full radial development has stopped, growth on crystal faces perpendicular to each spicule long axis which then branches at fixed angles, so that growths from different spicules meet. We believe this is forced by the transverse and radial filament structures under the cell membranes.

20.5.2 Biominerals formed outside cells

The different pattern of cells of *Acantharia* species are now seen to be due to the compromised growth of inorganic crystals and organic filaments and membranes inside and outside cells. The type of pattern depends on the supply of a variety of chemicals (genetic control), but also obeys constraints due to the physics of $SrSO_4$. It was essential that, to some degree, $SrSO_4$ morphology inside cells was biologically limited by networks of connected filaments and membranes. Outside cells similar constraints must hold in organisms where exoskeletons are biologically controlled, but their source cannot be of the same kind since the internal fibrils are controlled by internal cross-links and membrane connections. This means that external fibrils of non-cross-linked polymers which early in evolution would be organized in a random manner could only become more morphogenically effective if they were subsequently cross-linked before they were in-filled with inorganic minerals. Intriguingly, it seems likely that the development of such highly organized systems in evolution awaited the appearance of molecular dioxygen in the atmosphere and the availability of extracellular oxidases (Chapter 15). Before this stage a loose assembly of external inorganic crystals was possible, e.g. in some sponges, but a true external morphology was hardly possible. It is of some

interest to see if we can understand how external organized patterns arise around single cells. We choose to argue by reference to examples.

20.5.3 Coccolithiderae

This group of single-cell organisms produces an *organized exterior* shell (Fig. 20.9) but all the components of shell, called coccoliths, are first made in vesicles inside the cell by crystallization of $CaCO_3$. In the most primitive forms the coccoliths are composed of arrays of small crystallites. These crystals are very like inorganic rhombs which are placed in a relatively regular array outside the cell by exocytosis from a vesicle. Sophistication increases to the form seen in *Emiliania huxleyi* (Fig. 20.9(b)). In this species it has been shown recently that the curious wheel-like coccoliths are formed from units, each of which is made either from a single crystal or possibly from a combination of a single crystal plate and a series of ordered microcrystals, inside the cell in vesicles. The curvature of the surface of the crystalline coccolith is that of the cell membrane. The organism makes some 16 coccoliths for its final shell. Here we have a crystal form of enormous surface energy, which is then quite unstable, and we know therefore that its formation must be directed by physical and or chemical constraints upon its surface in the cell. We could call this a 'stress-derived' form and observe that a crystal can therefore have any form no matter what its space group or primitive unit cell (see the *Acantharia* in Section 20.5.1).

Fig. 20.9 (a) The formation of a coccolith and (b) the structure of the complete shell of a coccolithophorid, *Emiliania huxleyi* (scale bar = 1 μm). (a) The coccolith growth starts from a crystal at initiation points on a sphere in vesicles, V, attached to the nuclear membrane. The crystals grow inside the vesicle until the shaded units are built. The form is built, it is suggested, by the constraints of two types of filament, one (A) running between the nuclear (NM) and the cytoplasmic (CM) membranes and many (B and C) attaching the vesicle to these membranes. As the vesicle moves to the outer membrane B filaments disappear and C filaments (dashed) grow. The 16 or so units are completed, see b and d, when the cell is full size and they are then rejected through the outer membrane, O, to give the shell shown in (b).

Now the final positioning of the wheel-like units of the coccolithiderae after rejection outside the cells is not random and we must accept that there are positional restraints. We have met the problem before in the discussion of the silica rods which become highly organized in a network outside some cells (Chapter 18). Reference to Chapters 11 on zinc, 14 on manganese, and 15 on copper will remind the reader that it is probable that the complex external matrix of the cell could only provide an ordered matrix to receive the exocytosed minerals if these chemical elements were involved in protein hydrolysis (Zn), glycosylation (Mn), and oxidation (Cu), respectively, in the preparation of the extracellular matrix. In a sense the matrix has to be prepared for the ejection into it of the coccoliths. Once again we are forced to believe that the external matrix has evolved a potential for receiving and ordering crystals (or elsewhere amorphous solids) during growth. There is no escape from the fact that the external matrix has grown in harmony with the internal matrix and its coccoliths. This is homeostatic growth. How can the chemostat be extended outside the cell while the major growth controls are in the cell? How can all the elements required for extracellular synthesis, including of necessity cross-linking, be so controlled? (See Chapter 21.)

20.5.4 Radiolarians

Before we leave unicellular organisms we must give an example of the use of an amorphous mineral. The choice falls naturally on silica which is found very widely on the outside of higher multicellular animal species. Nearly invariably silica is laid down first *in cells*. The simplest case is that of certain *Radiolaria* (Fig. 20.10) and there is here a superficial resemblance to the *Acantharia*

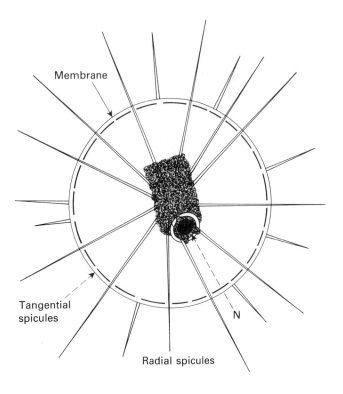

Fig. 20.10 The arrangement of spicules in the radiolarians. Notice that the numbers of straight radial spicules and their angular directions are not precisely defined (no central hub structure; compare Fig. 20.6). The arrangement of curved tangential spicules is related to the plasma membrane curvature. It is these spicules which become the fantastic shapes of radiolarian shells. It is suggested that these shapes are a precise reflection of internal stress fields imposed upon an amorphous precipitate (see Chapter 18). N is the nucleus.

(Fig. 20.4) but observe that, despite the fact that there are straight spicules, they are not centrally organized, although they are dispersed fairly evenly around the cell. These spicules could not force a well-defined internal morphology on the cell since they could not knit together through the cell centre using defined faces. In fact, we must presume that the morphology of any amorphous deposit is that imposed by field constraints of organic moieties in the cell alone when the pattern of the mineral is a strict reflection of intracellular forces of some kind. Now, while the radiolarians develop long radial spicules they also grow silica structures under the membrane which later form the outer shell. It could be that their form is dictated by the filament structure of the cytoplasm under the membrane (Fig. 18.11 and Fig. 20.10) but the external matrix must also be ready to receive the rejected silica.

The great advantage of building using *amorphous* silica is that the shape, for example, the radius of curvature, can be readily reconstructed during growth. We shall see that this advantage, which appears to be absent in the growth of minerals such as $CaCO_3$ and $SrSO_4$, re-appears in the case of bone. Be this as it may, the appearance of a fixed final shape demands a communication between events inside and outside cells as for the coccolithiderae.

The large number of pattern variations in the *Radiolaria*, many of striking beauty, like those of the *Acantharia*, are quite possibly purposeless. They represent the variety of structures which create neither genuine advantage nor disadvantage. In the exoskeleton of radiolarians they may be the product of glycoform variation as much as of isoproteins produced at the gene level. If it is the case, as we are entitled to believe at the moment, that a limited variety of sequences of a polymer will all perform a function adequately, then the ever-present errors in DNA just generate wider and wider variety within species and it is only considerable numbers of mutations which generate a break-away to a new species.

20.5.5 Sensors: barium and strontium sulphates in other organisms

In this book we wish to stress general principles and cannot give descriptions of the many species which contain these sulphate minerals in different forms. There are several species in addition to *Acantharia* which use strontium or barium sulphates, but not as frameworks. The earliest stages of growth of radiolarians use single strontium sulphate crystals as a weight to force them into anoxic layers where they are safer from predators. Barium sulphate is used in a similar manner in *Desmids*, but it is used as a gravitational sensor in *Loxedes* (Fig. 20.11). It is known that some plant cells contain barium sulphate too.

20.5.6 Magnetotactic bacteria

In magnetotactic bacteria it has been shown that the Fe_3O_4 crystals frequently grow as discontinuous rectangular bodies in vesicles of fixed size (Fig. 20.12), placed close to the membrane and in line (Fig. 20.13). The crystals are not small, so that their curvature is not noticeable. The growth pattern suggests that the shapes are governed by proteins associated with the vesicles but the

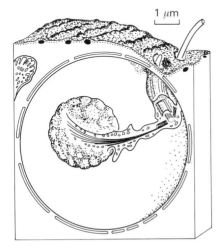

1 μm

Fig. 20.11 The balance organ of *Loxedes*. The central shaded area indicates the position of the gravity device weight of $BaSO_4$.

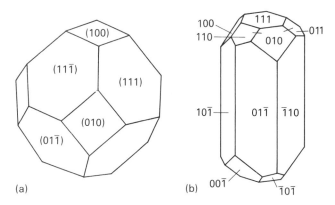

Fig. 20.12 The crystallography of (a) a conventional cubic crystal of Fe_3O_4 and (b) a crystal of Fe_3O_4 from a magnetotactic bacterium.

Fig. 20.13 A string of Fe_3O_4 crystals from a magnetotactic bacterium which follows the helical twist of the cell wall.

regular array of crystals which gives the bacteria a morphological pattern must be controlled by interaction with fibrils along the inner side of the membrane. The observed helical twist of the Fe_3O_4 series of crystals is typical of the extracellular coat structures of bacteria. The fact that the Fe_3O_4 crystals are not cubes is related then to constraints imposed by the direction of the fibrils of the external coat (Fig. 20.13). Here, the magnetic axis of the crystals is inherited from previous crystals on division and perhaps no instruction comes from DNA, except the chemistry of the units and of the outer coat. Similar magnetic sensors made from Fe_3S_4 have been found in some sulphur bacteria.

The clearest example of this type of crystal growth in a container of fixed dimensions is that of ferritin (Chapter 12). Using X-ray crystallography a detailed picture of the protein surface which controls nucleation and then growth of the crystals has emerged. These crystals have been analysed in detail by electron microscopy techniques and species-specific variations in ferritins have been revealed. It is this level of understanding that is required in the study of bone. Ferritin is not a magnetic sensor, but an iron store, and is found in cells of both unicellular and multicellular organisms (Chapter 12). We return to the discussion of its function in Chapter 21. There are other examples of the use of iron oxides, e.g. goethite in the strengthening of limpet teeth. This infilling is very similar to the structure of bone, but bone has very important differences. Still other examples, now of a different compound, are the iron sulphides, pyrite crystals, found in some anaerobic bacteria. Iron minerals may well have provided the most primitive homeostasis.

20.6 Unicellular eukaryotes: summary

In the previous sections we have shown that the eukaryotic cell learned to handle both amorphous and crystalline minerals in a delicate way. In *Acantharia* there is clear evidence that, for much of the cell growth period, the crystal morphology is partly controlled inside the cell by 'soft' structures, although the shapes of many different *Acantharia* species remain related to the inorganic crystallographic polymorphs of strontium sulphate. Clearly, this is a compromise between crystal forces and evolutionary design. Under the internal cell membranes, after the cell has stopped growing, the morphology is that of a rough sphere and shape here is that of the cell soft structures alone. At

the same time, strontium sulphate growth outside the cell is obviously very closely related to the growth of inorganic forms and the external matrix has a lesser biological control on it than the internal filaments and membranes. In the case of calcium carbonate in *Coccolitheridae* the control of the coccolith shape is dominated by internal constraints, but when cell growth stops the coccoliths are placed outside the cell. It would appear that the extracellular matrix now has kept complete control building an interlocked set of rimmed wheels (Fig. 20.9). Again, the network outside radiolarians appears to reflect internal structure and certainly the external network of rods around *Stephanoeca diplocostata* indicates that the external matrix has controlled shape, Fig. 18.9. In these last two cases the mineral itself is amorphous and hence shapeless. In fact, it is known that shape can be developed without minerals in some radiolarians, so that it has to be the case that the external matrix has become, in the course of evolution, a cross-linked and cross-strutted system, far from the fluidity associated with an extended set of random polymers. Its development has to conform to the law of equilibrium growth, however, where shape is maintained during growth.

A final example of the way an external soft matrix controls shape comes from the hair cells of plants; a typical example is that of the stinging nettle. The hair cells of many plants are quite rigid without any in-filling by minerals, but in the case of the nettle extra strength is supplied by the incorporation of amorphous silica (Chapter 18 and Fig. 20.14) which has allowed detailed study of development, Table 14.4.

Fig. 20.14 (Left) A nettle. Under the leaves are the stinging hairs. Each hair is really a break-off syringe barrel where the sphere at the end of the hair (right) is heavily mineralized with silica, see Fig. 14.8.

The problem is therefore to understand the external polymers of present-day cells and, as yet, there is rather little information about them in single-cell organisms. For this reason we turn next to the multicellular organisms where more is known about the external polymer matrix. We must introduce development patterns from a single cell to a juvenile of fixed shape before we can look at growth and, for want of space, we must do this in a very simplified manner. Just as in the case of unicellular eukaryotes, we shall deliberately search for patterns which resemble those of the development of inorganic crystals and which also have two phases: nucleation and growth. Nucleation of inorganic crystals requires some 20–50 atoms or molecules before crystal form evolves with a specific shape (see Section 20.3.2). It is not easy to find a close parallel to nucleation in the development of cell/cell interaction although there is a stage which resembles close packing of spheres, the 2–16 cell stage. Thereafter repeated changes in morphology occur until simple growth of fixed shape develops. We shall show that a parallel exists with equilibrium growth in these later stages of development. Fortunately, it is in these later stages that we can follow mineralization itself with its accompanying external polymers so we may use this as a source of information about control in whole organism growth.

20.7 Development and growth in multicellular animals

The simplest types of growth pattern of cells in organized arrays must relate to ways of packing spheres and the simplest of all cases would be modular growth of spheres in a straight line. Some plants grow in approximately this way, but have tubular cells, when it can be seen that order depends on rigidity between cells as well as of single cells. At this level minerals can be deposited along the sides of the cells in banded patterns, e.g. in Chara cells (Fig. 20.15). The next level of development in theory is organization in a plane surface but in practice the simplest observed form is the intriguing case of organization in conical forms. It is easy to see that growth of an elliptical organism from a point can be managed with maintenance of its shape and of its protective housing if, as it develops, it evolves a hard permanent external shell in the form of a simple cone. Typical approximations to this growth form are limpets which grow a conical calcium carbonate shell. If, as the cone grows, it twists either in one plane, or in two planes, a huge variety of shapes evolve (Fig. 20.16). These shells belong to some primitive multicellular animals. A further step forward is the development of a bivalve such as a mussel based on the same general mathematical form (Fig. 20.17). Now in all these cases we are discussing growth once the organism has developed its final shape as a juvenile. In all examples the organism starts life as a single cell and then goes through to a larval form before it metamorphoses to the juvenile. We are not attempting to explain the initial steps of morphogenesis but only growth of the final form. Clearly, this growth with maintained shape, that of a (spiralling) cone, has to be controlled by a matrix outside the cell. Since the synthesis of the shell and

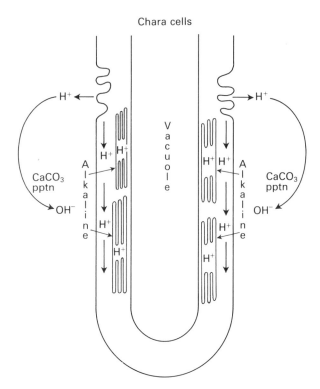

Fig. 20.15 The banded local activity of Chara cells leading to banded precipitation of calcium carbonate.

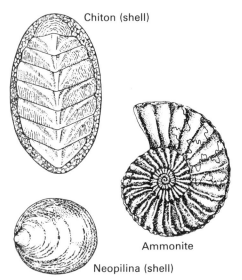

Fig. 20.16 The forms of some monovalve shellfish showing different variations on conical shape and illustrating at the right the logarithmic spiral. Here the shells are dominant in the control over shape.

the matrix is controlled from inside the cell all the system must be growing under tightly knitted integrated control, i.e. under conditions of equilibrium growth. It is this nature of control that is so intriguing (see Chapter 21) since it has to be both inorganic and organic.

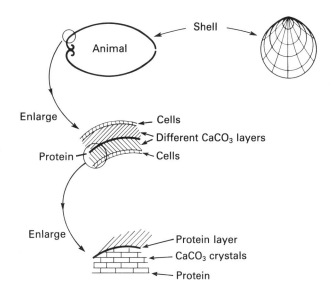

Fig. 20.17 A bivalve shell, e.g. a cockle or mussel, showing the intricate weaving of organic matrix and calcium carbonate deposits. Notice that each of the two shells illustrates logarithmic growth.

20.8 Extracellular biominerals in multicellular organisms

20.8.1 Shells: their chemistry and physics

Bivalves are based on the same mathematical theme, the logarithmic spiral, but, by developing logarithmic growth of two touching plates, shape is manipulated, so that the pattern does not require a lid. For feeding, the two halves of the shell can be open, but for protection the two halves are brought together. A mussel shell is based on this principle (Fig. 20.17). Bringing the two halves of the shell together under pressure maintains the shape of the shell.

A little thought reminds us of the principle of equilibrium growth of single crystals in a chemostat. The shape is independent of the size, but is now based on curved not flat faces for a shell. The forces acting are partly chemical and partly physical (on closing the shell), as is the case of the growth of a single crystal in *Acantharia*. The whole shell is then an example of equilibrium growth using both chemistry and physics to maintain shape and it follows that equilibrium growth with its rules must apply to the whole organism and to each cell in it. But remember the whole is self-generated.

These shells are not based on single crystals, of course, but on the building together of oriented small crystals. This is the principle of building used by man in, say, the construction of a wall (Fig. 20.17). It allows any degree of curvature of the final shape and has enormous strength since there are no fracture planes and tensions are distributed by the organic 'cement'. The only way in which such a uniform structure can be built is if each cell in the organism is in balanced communication with all others.

Finally, notice that a shell grown in this way is very different from, say, the tight cuticle of an insect, which cannot change in any dimension. A cross-linked cuticle has to be rejected and replaced, or it would act as a permanent limitation on size. The shells are also different from those of eggs, which are

deposited after growth of the cellular system and act to contain the organism. Finally, note that shell once laid down is not readjustable.

20.8.2 Bones

A bone is an internal skeleton, but it is not intracellular. We have to visualize how the principles of development of an intracellular mineral, as seen for example in *Acantharia*, can be transferred to an extracellular matrix and a multicellular system. We repeat the principles which we believe to be necessary if the bones are to grow with a fixed shape over long periods.

1. A chemostat is required to control the chemical facial energies and growth processes of all crystals, in addition to the chemical composition.
2. A physical control of the growth is also required.
3. The soft structures must grow in homeostatic harmony with the deposition of the crystallites.
4. This means that the organism grows under equilibrium growth conditions *with* its bones. The mineral is of equal importance as the proteins.

Skeleton of a lizard

There must then be some very carefully designed interaction between the growth of the cellular systems and the skeleton. The relationship is more complex than the growth of a shell since the bone has to be continuously remodelled in *three* dimensions. A shell avoids this problem and is effectively inorganic, while bone, as a growth unit, is as organic as all proteins of the soft tissue.

The nature of the protein matrix is known to be collagen-based connective tissue (Chapter 7) which is laid down *in advance* of bone crystals. It is this tissue that exerts the physical forces applied to the crystals. The collagen matrix generates flexible cavities which are in-filled in a similar fashion to the in-filling of silica in vesicles of radiolarians or of strontium sulphate in vesicles of *Acantharia*, but the only mineral known in internal skeletons is apatite, calcium hydroxyphosphate $Ca_5(OH)(PO_4)_3$, for a very good reason.

Unlike all the other minerals we have discussed, the energy of a calcium phosphate crystal is electric-field- and pressure-dependent. The so-called piezoelectric effect is found when the application of an anisotropic physical stress (pressure) induces an anisotropic charge distribution due to charge flow in the specimen. Here, stress induces proton flow in apatite giving a redistribution of surface charge on the crystal compelling growth to strictly follow the stresses applied since, under stress, one surface will dissolve while another will grow. The inorganic skeleton will mirror the stress field applied to the organism; it is this property of apatite that allows bone to be used as an internal skeleton of a multicellular organism where the skeleton is not concentric.

It follows from this discussion of bone that growth of an organism is again close to 'equilibrium' growth not just as in a chemostat but in a fixed physical force field. The implication is that surface energies due to chemical potential and physical (field) stresses (see Section 5.1) decide form. In this situation repair is achievable, so that a body may regrow after damage (see Fig. 20.18). (This is true also for crystals in a chemostat, of course.) The problem which remains is the *self-assembly of a chemostat* in the organism's body with the

Fig. 20.18 The equilibrium in a higher organism with a skeleton of bone. The bone responds to many forces and retains its shape during growth from juvenile to adult. This is due to the proton conductivity of apatite. Note again the intimate relationship with connective tissue.

increased difficulty of maintained control over internal physical forces. This is the problem of growth homeostasis (Chapter 21) and must include constant restructuring of the polymer outside the cell as well as of the mineral. Does this too depend on Ca^{2+}, HPO_4^- and $H^+(OH^-]$?

20.9 Summary of skeleton crystals in multicellular organisms

When we look at the external or the internal skeletons of complex organisms we see clearly the maintenance of macroscopic shape during growth. Additionally, at the microscopic level, the component crystals of a shell or bone are almost identical throughout growth. The building bricks of Fig. 20.17, for example, are manufactured with precision and we have to assume that equilibrium growth, i.e. the production of identical very small crystals, occurs at the *single-cell level* as well as at the level of the whole organism. Now, we also observed that the crystals grow in a matrix of proteins and, given the equilibrium growth of the crystals, it has to be the case that the protein matrix is developing under the constraints of equilibrium growth too: nucleation and growth of the inorganic solids (calcium carbonates, calcium phosphates, silica) only reflect and interact with the condition of the organic surfaces with which the crystals are in contact. Hence the proteins of the extracellular matrix need to be understood and especially the control of their synthesis and positioning. Finally, in old age, as with the *Acantharia*, there is the risk of the bone growing or failing to grow in homeostatic relationship when the nature of bone as a compatible material begins to fail.

20.10 The organic extracellular matrix of mineralization

A large number of proteins have been isolated from structures such as shells. The major feature of many is that they are acidic carrying phosphate and carboxylate residues (Fig. 20.19). The proteins themselves are not highly structured in the absence of calcium ions and it is probably this cation which helps to dictate these structures. Interestingly, in some cases they contain γ-carboxyglutamate (see Section 10.5). The case of bone is different since it is laid down as calcium hydroxyphosphate in a cross-linked collagen network. The collagen contains hydroxyproline and hydroxylysine which are products of oxidation by dioxygen under the control of iron- and copper-utilizing enzymes.

Fig. 20.19 A proposal for the way in which a β-sheet protein with alternate alanine (A) and aspartic acid (X) groups could generate an anionic sheet matching very closely the spacing of calcium (on right) in $CaCO_3$.

Now we are only beginning to understand the way in which these matrices are built and destroyed by cells. Probably the most explored topic is the synthesis of bone by osteoclasts and its destruction by osteoblasts. These cells produce the external filaments, but they also make crystal growth inhibitors and enzymes which destroy both the mineral (through acidity) and the matrix. Clearly, this constant activity of synthesis and degradation at the surface of the mineral composites is necessary for the maintenance of equilibrium growth, but it is of necessity linked to internal cell activity down to the gene level. As a foretaste of the contents of the next chapter we recall that, for example, the control of bone protein tissue is connected to vitamin D levels, which link forward to the control of calcium uptake, i.e. of bone itself, and of the bone growth inhibitor protein, osteocalcin, to connective tissue proteins.

As a final comment note that, as mentioned before, most plants do not contain organized external mineral deposits; where deposits are found they are most frequently extracellular silica in cellulose-containing structures. Is there a fundamental chemical relationship between the chosen building material of plants, cellulose, and silica which is to be contrasted with the relationship between the chosen building material of animals, protein, and calcium salts? It would be very useful for man to understand the rules underlying the choices in these composite materials. A good example for detailed study is the hair of the stinging nettle (Fig. 20.14).

The external filaments and structural proteins of bone are extremely complex since the mineral is not only supported by collagen but contains a variety of acidic proteins, many of which are highly glycosylated and phosphorylated and bind calcium. All must be synthesized in an integrated way with apatite, which itself depends on a supply of ions and a solubility product.

20.11 Sensors

A sensor is, of course, a detector of a field. In Table 20.2 we noted some biomineral sensors, the main use of which is to guide a mobile organism to or away from a target or an enemy. In the table there are the obvious sensors of gravity and of magnetic fields; however, a sensor can be used to guide growth, e.g. roots and shoots of plants, and we are well advised to look to the involvement of inorganic minerals as sensors in the development of the morphology of cells and, therefore, of other inorganic crystals during growth. A sensor crystal must be of designed size and shape, as those just mentioned, but must be connected to a sensory apparatus, e.g. a hair cell. A remarkable example is the gravity sensor in the inner ear of animals (Fig. 20.20). We see immediately from the fixed size and shape of the crystals that the principle of equilibrium growth applies to these sensor crystals, and this is true for all such crystals as well as for their matrix proteins.

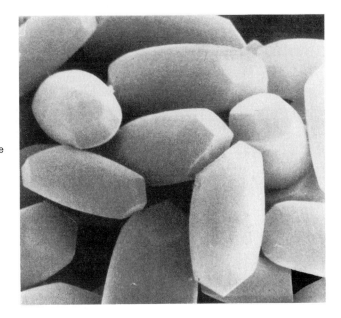

Fig. 20.20 The calcite crystals from a human balance organ. Note the very closely matched crystal shapes and sizes. Each crystal is about 1 μm.

20.11.1 Calcium carbonate in sensors

A problem with calcium carbonate is that it is difficult to remodel it as growth continues and perhaps this is why it is not used as an endoskeleton material. It is very insensitive to pressure, for example. The constancy of shape and volume once it has been formed makes calcium carbonate or other materials, such as barium sulphate, less valuable thereafter in internal structures but very valuable in gravity detectors which must not also sense pressure or electric or magnetic fields. Intriguingly, in man the only place where calcium carbonate is preferred over calcium phosphate is in the gravity sensor of the inner ear (Fig. 20.20) made from calcite crystals while in fish it is made from aragonite. Bone is a material of quite a different character from calcite.

Further reading

Crick, R. E. (ed.) (1989). *Origin, evolution and modern aspects of biomineralization in plants and animals*. Plenum Press, New York.

Frankel, R. B. and Blakemore, R. P. (eds) (1991). *Iron biominerals*. Plenum, New York.

Haeckel, E. (1966). *Reports of the scientific results of H.M.S. Challenger 1881–1898*. Reprinted by Johnson Reprint Corp, New York.

Hartman, R. (ed.) (1973). In *Crystal growth: an introduction*, pp. 367–402. North Holland, Amsterdam.

Leadbetter, B. S. C. and Riding, R. (ed.) (1986). *Biomineralization in lower plants and animals*, Systematics Association Special Volume, No. 30. Oxford University Press, Oxford.

Lowenstam, H. A. (1981). Minerals formed by organisms. *Science*, **211**, 1126–31.

Mann, S., Webb, J., and Williams, R. J. P. (ed.) (1989). *Biomineralisation*, Chapters 1, 2, 9, and 10. VCH, Weinheim, Germany.

Nancollas, G. H. (ed.) (1982). *Biological mineralisation and demineralisation*. Dahlem Konferenzen. Springer-Verlag, Berlin.

Schewiakoff, W. (1926). Die Acantharia des Golfes von Neapoli. *Fauna e Flora Neapoli Monograph*, **37**.

Suga, S. and Nakahara, H. (eds) (1991). *Mechanisms and phylogeny of mineralisation in biological systems*. Springer-Verlag, Tokyo.

Thompson, D. W. (1968). *On growth and form* (2nd edn). Cambridge University Press, Cambridge.

Wainwright, S. A., Biggs, W. D., Currey, J. D., and Gosline, J. M. (1976). *Mechanical design in organisms*. Arnold, London.

Weiner, S. (1986). Organisation of extracellularly mineralized tissues. *CRC Reviews of Biochemistry*, **20**, 365–408.

21

An integrated view: the elements in homeostasis, morphogenesis, and evolution

21.1	Introduction	495
21.2	The cytoplasmic homeostasis of basic elements	499
21.3	A detailed case: the homeostasis of iron in prokaryotes today	503
21.4	A summary of cytoplasmic homeostasis early in evolution	506
21.5	The environment of early cells	507
21.6	Eukaryotes	510
21.7	The evolutionary switch in homeostasis	512
21.8	Controls over calcium external flow and mineral formation	516
21.9	A summary of zinc controls	519
21.10	Organic signals	520
21.11	Changes in functions of other elements	525
21.12	Morphogenesis	526
21.13	Concluding remarks	527

21.1 Introduction

Throughout this book we have been insisting that many chemical elements, not just C, H, N, O, P, and S as ions or in simple compounds, e.g. coenzymes, have an absolutely essential role to play in biology. The primary observations were analytical (Chapter 1), but subsequently we have seen that each of these elements is locked into all-pervading activities, such as metabolism and its control. We saw this by examining in detail the functional value of the individual elements (Chapters 8 to 18), while Chapters 2 and 3 were concerned with the physicochemical principles which allowed specificity of element action to evolve and Chapter 7 indicated the ways in which their evolution was locked into properties of biopolymers. In this chapter an attempt will be made

to look at the ways in which the levels of all these elements, especially as *free* ions, for example, H^+, Na^+, K^+, Mg^{2+}, Ca^{2+}, Fe^{2+}, (Mn^{2+}), Zn^{2+}, Cl^-, and HPO_4^{2-}, and their dynamics are *interactively* built into states of cells. It is not possible to come to a general conclusion yet but many observations in earlier chapters demand that there is some overriding *integrated* control involving these free ion concentrations *in cells*, i.e. their homeostasis, since they are strongly linked to one another, to resting conditions, to messenger systems, to enzyme activity, and to translation and transcription, that is, effectively to all cellular activities. In Chapter 20 we extended the description in the earlier chapters to morphological patterns and observed that there was also a constancy, homeostasis, of shape during much of growth. It became obvious that there is also a very tight homeostatic control over the mineral elements, e.g. Na^+, K^+, Ca^{2+}, and Mg^{2+}, in the *extracellular* fluids of multicellular organisms, including man, which reacts back with the whole of the system of maintenance of extra cellular structures. Much of this control may well be related through inorganic precipitates to solubility products. If this description of homeostasis as a deeply integrated state of activity inside and outside cells with morphology through long periods of growth is correct, then morphogenesis must be a switch from one homeostatic metabolic and growth pattern to another. We believe that in this switch changes in the inorganic element levels are deeply involved. From the beginning we must also observe that homeostasis, in a deeply contradictory sense, is about energized dynamic conditions. Biological organizations are never static, but change continuously and the changes include growth, replacement, and death (Chapter 6).

21.1.1 Homeostasis and morphogenesis: definitions

Before we describe the theme of this chapter in detail we need to define our view of homeostasis and morphogenesis more clearly. In Fig. 21.1 we plot the changes in rate with time of two processes. Curve (1) is a synthetic activity shown as having a decreasing change of rate per unit volume which could well be due to a diminishing supply of chemicals with time. It is easy to imagine such a process and to formulate it. The second curve (2) is a degradative activity which is shown as having an increasing rate with time due to increasing concentration of certain (oxidized?) chemicals. Again it is easy to imagine such a process and to formulate it. The sum of these two processes with a suitably chosen *fixed* set of rate constants will give a growth curve for an organism up to a steady size which will persist for some time and then decline—curve (3). We shall take this whole curve to represent the growth of a biological system after an initial phase in which a shape is established. The steady, level part of curve (3) is frequently called 'homeostatic' with reference to living systems, but we shall not use the term in this way. We shall call homeostasis the total process, curve (3) in the figure, i.e. a combination of synthetic and degradative chemical reactions with fixed rate constants since the whole rate profile (3) is due only to a continuous set of changing concentrations, all of which contribute to growth, steady level, and decline. Within biological space the whole system is a pattern of flows, a dissipative system since it needs continuous input and output of material and energy and maintained organization at constant shape (see Chapter 6, and note that there will be a homeostasis in energy input too; Chapter 5).

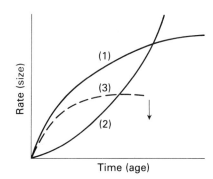

Fig. 21.1 An illustrative description of the way in which the sum of two activities, (1) a continuous synthetic activity which declines in rate with time and (2) a continuous degradative activity which increases in rate with time, leads to a rate pattern of growth (3). Initially growth, assumed to be at constant shape, is rapid, but then reaches a plateau before failing. In this article the whole of curve (3) represents growth homeostasis and in part parallels the growth with constant shape of a single crystal in a chemostat.

It is easy to see that such a simple growth pattern as (3) is consistent with growth with constant shape, and we shall take constant shape, not constant size, as the hallmark of homeostasis. The simple parallel is with equilibrium growth of crystals of fixed shape in a chemostat (Chapter 20). The suggestion is, therefore, that the shape not the size (nor yet the structure, in the sense of a body of fixed dimensions, but only the symmetry part of that structure) together with the metabolism of the organism are built into an integrated set of reactions. Many organisms grow bigger over about a quarter of their life span, remain approximately constant for, say, a further half of that span, and then decline to one degree or another over the last quarter. During the whole of this time there is little change of shape. This is a common picture for animals, but plants grow through a repeating module. The beginning of modular growth after development from a seed is morphogenesis, but the continual repeating of the pattern along the stems of some plants or the annual cycle of a tree will be treated here as growth homeostasis, using oscillating rate patterns rather than the simple pattern of Fig. 21.1. In fact, there is a constancy of species pattern (shape) during much of the growth period within many different species. At the same time, different species have different chemistries, but it was necessary for us to conclude in Chapter 20 that the chemistry of a given species during its whole growth phase is closely fixed.

In this context *morphogenesis* is defined as development of shape, but is known to be a change of metabolic pattern too. Shape and metabolic pattern are then intimately linked in any one species (or within cells, for one type of cell compared with another). For many species this change is most apparent over the first rather short period of life, e.g. during gestation, not shown in Fig. 21.1. However, some species undergo one (or more) dramatic morphogenic (shape) change, metamorphosis, which occurs at a particular time for each species. Such an effect implies that there is another set of homeostatic rate processes looking like those in Fig. 21.1 which can generate a curve (3′) which cuts curve (3). In the developmental stage of early life, i.e. embryonic development, a whole series of switches of homeostases, each following Fig. 21.1, needs to be envisaged. All developments, such as embryonic development, flowering, passage through puberty, or the emergence of a butterfly from a caterpillar, are, of course, gene-switching processes. We must try to relate rate profiles such as curve (3) with one set of gene products and pattern of rates, i.e. with one homeostatic growth condition, and morphogenesis and metamorphosis with a running change of gene products to a new stable homeostatic set. We shall suggest that by inspecting inorganic ions in biological systems we may envisage, at least in part, how this has been achieved in evolution. This implies that during a growth phase there is very tight control of all the mineral element concentrations which reacts back with the organic components. In fact, this is the conclusion we reached by a different route in Chapter 6. A simple test of such a general view is to cause deficiency (or excess) of an element to a growing organism and to observe effects on either shape or change of shape. Examples are the effects of zinc or iodine supply in the diet of animals (see Sections 21.8.2 and 21.10.2).

Initially, we shall not attempt any analysis of the origin of a shape, that is morphogenesis itself, but shall ask how the different patterns of mineral element concentrations in different homeostatic growth states are maintained. As far as morphogenesis is concerned, we shall follow the analysis summarized in the

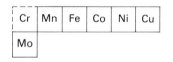

book by Peacocke (1983) (see 'Further reading'). A dissipative (kinetic) organization, i.e. a homeostatic state, is recognizable by a set of patterns of flow within steady states, and simple element concentrations, possibly all of those shown in squares in the margin, will be expected to be involved in that organization and flow. Relative ion flow changes may then be related to morphogenesis, but the search for their involvement in homeostasis is independent of whether they are morphogens responsible for morphogenesis or not.

When we attempt to describe the integrated activity in a cell we must do so under three headings concerning biological chemistry:

(1) the balanced feedback control of substrates and ionic levels;

(2) the balanced synthesis and hydrolysis of proteins (enzymes) and of mRNA which controls them, which include translation and transcription and a variety of states which are also dependent on inorganic elements;

(3) the *de novo* switch-on or switch-off of gene products which alter the balance under (1) and (2).

If we are to maintain the clear-cut distinction between homeostasis and morphogenesis, considerations under headings (1) and (2) belong to homeostasis while considerations under heading (3) belong to morphogenesis. Although we realize that the distinction between headings (2) and (3) can hardly be so clear-cut, especially perhaps in prokaryotes, we believe that it will be useful as an aid to discussion.

21.1.2 Shape in biology and its evolution

It is obvious that all biological species consist of a set of chemical reactions within a container. The simplest container is a unicellular spherical membrane and this shape is that of a bubble controlled by internal and external pressure and membrane surface forces. This very simple shape may have been that of the original form of life, but we cannot relate this isotropic shape to any partitioning of chemical element activity in the cell except between in-membrane and in aqueous solutions, inside and outside. It is not a known shape of any existing organism. The next shape to consider, that of a cylinder, has axial symmetry. This shape is apparently common amongst prokaryotes, see Section 20.5.1. How does it come about?

The bacterial prokaryote cell has no inner filaments and no inner membranes; as far as the authors know the cylindrical shape is that of the cell wall. We can only presume that this shape is a consequence of the development of the divisional process for reproduction which always would demand this symmetry at some stage of growth. Once again, given the simplicity of the cylinder, the partitioning of chemical element activity in the organism can only be in relationship to two zones in addition to in-membrane and inside against outside. The two zones are associated with the two membrane regions of different curvature—respectively at the ends and along the sides of the cell. The cell grows by elongation when shape is not strictly maintained. In fact, the interior membrane of the bacteria is just pressed against the wall by turgor (osmotic) force.

Given such simplicity of shape, do we need to consider that the spatial

construction has a homeostatic growth relationship with the metabolism? The answer to this question must be yes, to some degree, since during the life cycle the surface area must double (rather exactly), the DNA must double (exactly), and yet a lot of the concentrations of chemicals in the cell are roughly fixed in optimal growth, e.g. those of K^+, Na^+, Mg^{2+}, many phosphates, and perhaps some other substrates. In fact, the turgor pressure and the K^+ concentration are interdependent. However, bacterial metabolism is also quite adaptable, so that there must be an allowed range of the concentrations of many elements too. Whatever the relationship between shapes and homeostasis is, it cannot be very tight coupling.

When we move up to unicellular eukaryotes there is clearly a much closer definition of shape, the three axes of space grow together, a much tighter control over the metabolism is achieved, and membranes and vesicles within the cell grow with rates of their growth coupled to that of the cell despite totally divergent sizes and shapes. A new set of internal filaments must also grow in harmony with the organism as a whole.

The development of multicellular organisms brings with it even more striking control over shape and metabolism in that cells in different regions of the whole, e.g. in organs, all grow with a rather strict shape and proportionality. The requirement for such strict relationship, as we have seen for independently growing crystals in a chemostat (Chapter 20), is very good control over physical conditions, e.g. temperature, and over chemical composition (see Section 1.1). An organism in the sea has little problem with its fixed environment, but on land problems abound; hence land animals developed to become very precisely temperature-controlled and with controlled extracellular fluids (see Section 10.2). In some ways the need is self-evident if shape too is to be tightly reproduced. Curiously, the shape of a plant cycles with the temperature and even the exposure to different light intensities during the year.

Throughout all these examples the parallel with equilibrium growth of crystals is very apparent. There is one organ which is a striking exception—the brain. Here growth, which we relate to memory, is strictly connected to changes in external exposure. The brain is a plastic system and as such its growth is more related to kinetic growth of crystals which records the changes of concentration gradients with time.

It is against this background that we wish to study the nature of homeostasis associated with constant shape. We turn to the description of integrated homeostasis of many small and large organic molecules and to the role of simple elements in their control by reference, at first, to the biological chemistry of phosphate.

21.2 The cytoplasmic homeostasis of basic elements

21.2.1 Phosphate

We choose to analyse the homeostasis of phosphate first amongst all the elements because: (1) it does not change oxidation state or its basic unit (PO_4) throughout biology; (2) it is part of two major polymers, DNA and RNA, and

of many of the lipids of membranes and is associated with the metabolism of many small molecules; (3) it is at the very heart of energy distribution and of control mechanisms (Chapter 18). Phosphate must have been involved in biological systems from a very early period. We begin the discussion from the last point (3) remembering that we wish to understand growth (curve (3) of Fig. 21.1) with fixed shape. The implication is that there is a relationship between the supply of phosphate and the whole of metabolic activity and that we shall need to show later that the homeostasis of phosphate is integrated with that of many other elements.

In Chapters 6 and 18 we indicated that all major cellular processes were linked to pyrophosphate through NTPs. We shall concentrate on ATP here. The nature of ATP is such that it is in balance with ADP and inorganic phosphate. Since the energy status of the cell is not linked to a single chemical but to a steady state (not an equilibrium)

$$\text{ATP} \quad \text{ADP} + \text{P}_i \qquad (21.1)$$

the homeostasis of phosphate itself, inside the cell, is also energy homeostasis. It is known that the levels of free ATP, ADP, and P in a cell (and at this stage this could be any cell) vary only slightly during growth and are connected to a variety of feedback controls over metabolism (Fig. 21.2), including cAMP and IP_3, and to the states of protein phosphorylation, all of which we can write as being associated with the energized steady state. Short-term fluctuations are permitted to generate messages, but sustained message systems lead to altered gene expression and thence to a new steady state of growth. (In the last regard there is much greater flexibility in bacteria than, say, in man.)

The role of phosphate is then connected by feedback through internal fixed-level cytoplasmic signals using compounds of phosphate to link to metabolic pathways, polymer synthesis, etc., and therefore growth. However, the metabolism itself also supplies the energy as ATP and, in turn, this is used in part to pump in the phosphate or organic phosphates (Fig. 21.2). There is then

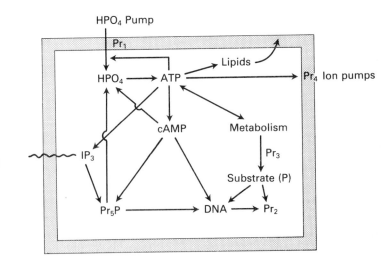

Fig. 21.2 A much simplified view of the involvements of phosphorus, as phosphate, in virtually all cellular activities, indicating their cross-relationships in mutual feedback control. Pr_n indicates different proteins. A diagram similar to this is common to all types of cell. Note that cAMP differs from other phosphates in that it is not proton dependent.

an integrated feedback homeostasis of phosphate using phosphate compounds as signals. We must show therefore that all other elements required to maintain growth are also under integrated feedback controls linked to phosphate. We look next at the relationship between phosphate, magnesium, and protons but we shall need also a supply of energy.

21.2.2 Phosphates, magnesium, and protons

The first step in recognizing the integration of the chemistry of different elements within biology can now be taken. In Fig. 21.2 we have written the phosphate species without charges. Written properly we have to include charges and counter-charges and in all cells we know that these are overwhelmingly K^+, Mg^{2+}, and H^+ and that, of these three, Mg^{2+} and H^+ interact directly with phosphate so that the homeostasis of phosphate is not possible without the homeostasis of Mg^{2+} and H^+. In Chapter 9 we have shown the role of Mg^{2+} while in Chapter 5 we have mentioned the links to the proton. A simple example of an equation found in every cell is *the equilibrium equation*

$$Mg^{2+} \cdot ATP^{4-} + H^+ \rightleftharpoons ATP^{4-} \cdot H^+ + Mg^{2+} \qquad (21.2)$$

where we note that all cells are held for much of the time at approximately $pH = 7$ and $[Mg^{2+}] = 10^{-3}$ M. Proton concentration changes frequently act as controls, but $[Mg^{2+}]$ is rarely used as a variable. At this stage we must also observe that the reactions of organic phosphate compounds do not need catalysts except Mg^{2+}, H^+ (OH^-), and protein surfaces since they are highly reactive. No other elements need be involved in these acid/base reactions.

Equation (21.2) is clearly in part a statement of the equilibrated homeostasis of Mg^{2+}, P, and H^+ (Fig. 21.3). The feedback steady-state maintenance of H^+ and Mg^{2+} involves ATP in yet other pumps but, of course, all this activity and that of phosphate itself requires an external source of energy which is either from light or unstable chemicals. We have seen that light or chemicals can generate proton gradients directly and we have described the link forward in energy transduction from H^+ gradients to ATP and then to pumps for Mg^{2+},

Fig. 21.3 A schematic figure to recall the intimate dependence of Na, K, Cl, H, Mg, and Ca movements on ATP and the feedback relationship to the states of ATP and its formation (see Chapters 8–10).

HPO_4^{2-}, and H^+ itself (Chapter 5). Now, the photo-initiated metabolism and the transformation of many chemicals for energy supply, e.g. sugar, require not only the hydrolytic reactions of phosphate but redox reactions which cannot be managed by Mg^{2+}, protons, or phosphate. We turn to this linkage next, but before doing so we must note that the requirement for stability of the $H^+/Mg^{2+}/HPO_4^{2-}$ organic chemistry needs two other controls in all cells.

21.2.3 Sodium, potassium, chloride, and calcium

It is necessary to manage both osmotic and electrolyte balance in cells, which demands the further controlled manipulation of three elements—Na^+, K^+, and Cl^- (Chapter 8). We have noted that these element concentrations are linked to ATP-dependent pumps and to exchangers (Chapter 8 and Fig. 21.3). The integration of the acid/base metals and non-metals, is completed by the rejection of Ca^{2+} from all cells by ATP pumps (Table 21.1), so that the calcium does not cross-link electrostatically the anionic complement of the cytoplasm (Chapter 10). The elements now integrated in homeostatic metabolism in the manner required, in very primitive cells and in all cells to this day, are the acid/base non-redox active Na, K, Mg, Ca, P, Cl (and H). We can link all the uptake of C, H, N, O compounds to these elements too but, as yet, no energy source has been described. It is through H that the connection is made to redox reactions and energy which are so essential for the whole of life.

We can no longer describe all cells at once and therefore start discussion with what we take to be a primitive prokaryote cell. (It is redox chemistry which has evolved most dramatically in evolution as we show in Section 21.7.)

21.2.4 Hydrogen, sulphur, and iron in primitive prokaryotes

The biochemistries of iron and hydrogen are intimately linked through the need to catalyse the following reaction in some steady state,

$$(\text{Free or bound hydrogen}) \tfrac{1}{2} H_2 \text{ (or RH)} \rightleftharpoons e + H^+ \text{ (or } R + H^+\text{)}, \quad (21.3)$$

which is central to many modes of energy supply. The catalysis involves electron transfer and may or may not need light. It was thoroughly described in Chapter 12 where we noted that, although other metals could have been used, it probably was the availability of iron in sulphides (clusters) which was decisive. These modes are mainly the membrane-bound reactions of H^+ and e for energy capture, but there must be a link also to the ongoing redox reactions of aqueous substrates in the cytoplasm. There is no doubt that through all life as we know it the redox reactions of sulphur dominate the cytoplasm through the functioning of coenzymes and cofactors such as lipoic acid, glutathione, and thioredoxin (Fig. 21.4). We have explained their central role in metabolism in Chapter 19. If we put together uptake of compounds, using ATP or ion gradients, the handling of condensation reactions of C, H, N, and O using P, and the manipulation of redox states of C, H, N, O, and S using S (see Table 21.2), we can in fact see how to make all the compounds in the cytoplasm of prokaryotes. The problem with such a controlled organic chemistry approach alone, and which is often shown on metabolic charts, lies in the energy supply and in the energy barriers and controls over the metabolic reactions, especially those involved in the making and breaking of bonds in the

Table 21.1 Elements utilized from the beginning of acid/base cytoplasmic activity

			H				
—	—	—	C	N	O	—	
Na	Mg	—	—	P	(S)	Cl	
K	Ca	—	—	—	—	—	(Zn)

☐, rejected; (), uncertain use.

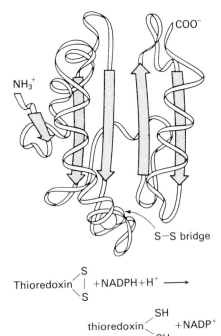

$$\text{Thioredoxin} \begin{array}{c} S \\ | \\ S \end{array} + NADPH + H^+ \longrightarrow$$

$$\text{thioredoxin} \begin{array}{c} SH \\ \\ SH \end{array} + NADP^+$$

Fig. 21.4 The structure and reaction of thioredoxin. (After A. Holmgren and C.-I. Brändén).

Table 21.2 Elements utilized from the beginnings of anaerobic redox cytoplasmic activity

				H		
				C	N	O
				—	—	S
(V)	Mo	(W)	Fe	—	—	Se

(), uncertain use.

inert feed-stock molecules from the primitive environment, e.g. H_2, CH_4, H_2O, CO_2, N_2, NO_3^-, and SO_4^{2-}. It is for these reasons that catalyst elements had to be incorporated from the beginning and perhaps the catalysed reactions existed before the beginning of life. In fact, it could well have been that reactions, not elements, were incorporated. Thus, in addition to maintaining redox poise through sulphur compounds and control over water activity through energized pyrophosphate, biology had to maintain a homeostasis of catalytic elements (see Table 21.2). Iron is the most obvious example, and in the initial stages of life, as it is now, the maintenance of iron levels was essential. However, since iron sulphide clusters were then very available, the supply of iron may have been a small problem in primitive cells.

Missing from this description of sulphur biochemical homeostasis is its uptake and rejection mechanisms. They were described in Chapter 19 using the sulphur cycle. Sulphur as H_2S was a very readily available element on the primitive earth, so that, like H, C, N, and O, there was little problem with supply but much difficulty with getting it, like all the elements, into the desirable compounds. Sulphur itself was manipulated from H_2S to S_n to give hydrogen atoms and energy supplies but, in all probability, always using iron.

21.2.5 Regulation

A problem which is not solved by linking metabolic pathways is that of the controlled synthesis of macromolecules, especially proteins. The synthesis has to be under the influence of DNA through regulation dependent on binding proteins and small molecules, i.e. on a further feedback system. In the case of phosphate the regulation depends to this day on cAMP and protein phosphorylation. (Note again the difference in proton dependence here, Fig. 21.2.) In the cases of Mg^{2+}, K^+, and H^+ we do not yet know the regulatory controls for the proteins of ATPase pumps. For sulphur there begins to emerge a link to DNA through the protein thioredoxin (Fig. 21.4). The type of feedback system which is required for all elements can be understood from a detailed study of iron metabolism in a prokaryote and we have to take a first example from an aerobic cell, *Escherichia coli*, since there is inadequate knowledge of anaerobic (primitive) systems.

21.3 A detailed case: the homeostasis of iron in prokaryotes today

While iron was readily available in the early periods of life's history it was made very difficult to obtain after the full advent of a dioxygen atmosphere, i.e. 10 per cent of atmospheric pressure, some 10^9 years ago (see Fig. 21.13). In Chapter 12 we pointed to the likelihood that levels of free iron (and of manganese in a slightly different fashion; Chapter 14) were and remain deeply involved in the control of oxidative metabolism, especially in *prokaryotes*. We noted particularly the character of the iron uptake regulatory protein, FUR. FUR is common in aerobic prokaryotes and binds to a promoter gene which, in the absence of ferrous iron, controls the synthesis of ferroxamins and other siderophores, the chelating agents which scavenge for (ferric) iron (Chapter 12), and of the proteins, TON B, of the recognition system for entry of the iron

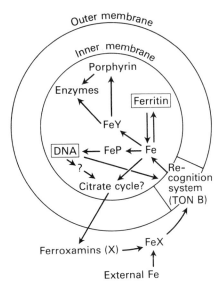

Fig. 21.5 An outlined connection between the uptake system for iron and iron utilization in *Escherichia coli*. There is intensive feedback to metabolism via regulation. The regulation of iron chelates includes that of the recognition (TON) system of membrane proteins. Most but not all of this organization may be retained in mitochondria and chloroplasts.

(III) chelates (Fig. 21.5). The siderophores themselves are secondary products of metabolism and their production of necessity diverts chemistry and energy from such pools as the citric acid cycle (Fig. 21.5). There is a feedback loop using free Fe^{2+} as the interactive messenger since FUR bound to Fe^{2+} does not promote scavenger formation or synthesis of the TON B recognition system. (The importance of the *fur* gene, the 'iron box' in DNA, is further exemplified by the finding that, through regulation of the expression of the membrane proteins, Ton B in Fig. 21.5, *fur* also controls entry of vitamin B_{12} (Co) and, separately, regulates expression of several toxins produced by bacteria, e.g. diphtheria toxin.)

It now becomes important to see the knock-on effects of changes in iron levels in the prokaryote cell. The most important and immediate influences will be upon those enzymes which require iron but from which iron dissociates freely (see Chapter 12). Several enzymes of the oxidative metabolism are involved , amongst which are enzymes in the citric acid cycle (Fig. 21.5). Dropping the level of iron switches the cell toward anaerobic metabolism, to the production of siderophores, and to a lower level of amino acids for protein synthesis and of ATP production. Immediately we have one link to the phosphate homeostasis. Of necessity, it follows that those mineral element balances which are dependent on ATP-driven pumps will change with the respiratory state. Such ions include H^+, Na^+, K^+, Ca^{2+}, and Mg^{2+}, all of which are linked directly to ATPases in membranes and several of which control a further set of metabolic processes of prokaryotes. Anion balances are also likely to be involved through ATP concentration changes. There is then the implication that *iron concentration* is a critical factor for growth within homeostasis, i.e. the position along curve (3) of Fig. 21.1, which makes it necessary to look more carefully at the maintenance of iron concentrations. Unless the iron levels are maintained at a critical level there will not be optimal growth, although homeostasis, in the sense of curve (3) of Fig. 21.1, is maintained. Here, iron levels control both a set of substrates through enzymes and the expression of proteins.

In most prokaryote cells the levels of ferric iron are apparently buffered (note that buffering is not homeostasis) by a deposit of ferritin, hydrated $Fe(OH)_3$, or ferrihydrite (Chapter 12). Unfortunately for interpretation, this is not just a simple iron deposit. The release of iron from ferritin requires acid (H^+), of course, but it also requires a redox reaction, when the buffering of iron by ferritin may still be that of the free Fe^{3+} but will also be that of the free Fe^{2+} in the cell, and both are related to pH. Ferritin is therefore a proton 'buffer' and an Fe^{2+}/Fe^{3+} 'buffer'. It is Fe^{2+} which is bound by FUR and so the metabolism requiring free iron (Table 21.2 and Fig. 21.5) will also be linked to the iron storage of ferritin and linked back to transcription. Clearly, the concentration of iron is a major dynamic link inside the prokaryote cell.

There is a further complication in the chemistry and the homeostasis of iron in ferritin in that it contains amounts of phosphate which are variable from prokaryotes, where there is a considerable amount, to that in eukaryotes, where there is little. There is no doubt that the nature of the iron in solution and the level of inorganic phosphate, as well as the pH, are all connected to the general iron homeostasis by the solubility product of $(Fe(OH)_3)_y$, $(Fe(PO_4)_x \cdot (H_2O)_z$ in prokaryotes. When we turn to the all-important Fe^{2+} concentration, via connection to FUR, iron levels now depend differently on

Table 21.3 Some primitive metabolic paths linked to non-haem iron

Metabolite	Comment
Dihydrogen	Hydrogenases need either Ni and Fe, or just Fe
Dinitrogen	Nitrogenases need Fe and Mo, or V, or only Fe
Citrate cycle	Several enzymes need Fe
Fat synthesis	Enzymes require Fe

phosphate in different organisms as well as on the redox poise of the compartment (see eqns (21.1)–(21.3)). As has been thoroughly demonstrated in Chapter 17, extensive links to regulation and control through phosphate are through cAMP and protein phosphorylation. The redox poise, phosphate, and pH controls over ferritin mean that there is iron feedback to metabolism in even more ways than described so far, which are cell and organelle dependent.

The analysis of the controlling role of iron here has been based on the observations concerning *fur* protein regulation. Now this regulation relates to the production of scavenging chelates and is not likely to have been a major control system in life early in evolution since iron was then more plentiful (Chapter 1). In fact, most if not all scavenger syntheses are based on chemistry which became possible only after the advent of dioxygen. It has become clear recently, however, that the control functions of iron may well extend to an even more primitive series of regulatory proteins where the control is over anaerobic parts of metabolism (Table 21.3). The regulatory genes concerned cover such fundamental uptake systems as those of nickel, molybdenum, vanadium, and thence the hydrogenases (Fe, Ni) and nitrogenases (Mo, V, Fe) as well as prokaryote superoxide dismutases (Mn, Fe) which were needed as soon as *any* oxygen appeared in the atmosphere. The genes are, for example, *nif*, nitrogen fixation genes, and *fnr*, aerobic/anaerobic switch genes. The possible binding unit of the DNA proteins for iron is a peptide sequence which is reminiscent in part of that found in rubredoxin, in metallothionein, and in zinc fingers (Fig. 21.6 and see Chapters 11–13 and 15 on zinc, iron, and copper

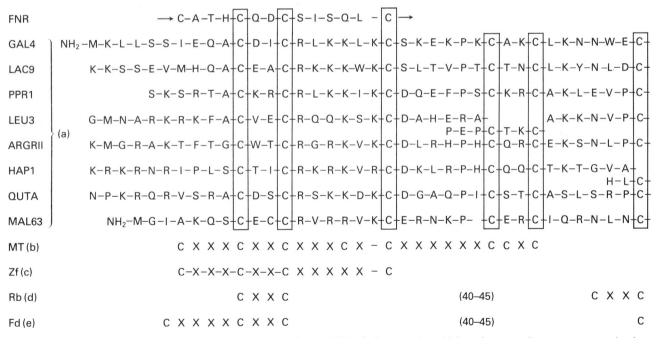

Fig. 21.6 The sequences of cysteine-containing fragments of some DNA-binding proteins which are known to be or are suggested to be iron- or zinc-binding sites and of some metalloproteins. (Other cysteine sequences are thought to be used in copper and various other heavy-metal-binding control proteins). The metallothionein (MT) sequence is typical, as is that of the zinc finger (Zf). Those of rubredoxin (Rb) and ferredoxin (Fd) are shown for comparison. These are generalized sequences from many sources: (a) yeast; (b) metallothionein; (c) bacterial metalloproteins; (d) rubredoxins; and (e) ferredoxins. Note that all metal binding is sensitive to pH, see Section 2.4, which is therefore regulatory too, and to the concentration of H_2S which may have reduced Zn^{2+} availability early in evolution, Fig. 1.12.

Fig. 21.7 The protein FUR regulates the iron-box gene so as to produce the range of proteins shown. FUR itself is produced under the control of a gene regulated by cAMP . Thus phosphate and iron are connected through the genes and in metabolism.

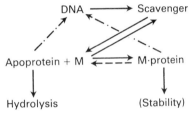

Fig. 21.8 As well as feedback regulation of DNA as a source of synthesis of proteins there is a controlled degradative system for protein hydrolysis in all cells. The ratios of protein to metal are then carefully controlled. In the figure full arrows include fast reactions, broken lines show slow reversibility, and dash-dot lines refer to regulators of expression.

proteins). The value of thiolate-containing binding centres may connect further to redox poise generally, through reversible formation of disulphide bridges, using glutathione, and of course thiolates are connected to H^+. Thus these regulatory proteins may respond to H^+, Fe^{2+}, and redox poise (compare FUR and ferritin). Finally, there is the suggestion that they may also have a cAMP binding site, e.g. in *fnr*, which will take us back to pH independent phosphate-dependent controls and regulation (Chapter 18). Even the production of FUR itself is regulated by the level of cyclic-AMP (Fig. 21.7).

While we have stressed transcriptional control here, the reader is advised to look also at translational RNA controls which are known to increase in importance for iron in higher organisms. There is a third way in which control can be exerted by free iron in prokaryotes over amounts of a protein other than through FUR or other DNA-binding proteins. In all cells proteins turn over (Chapter 4); the rate depends on the synthetic steps controlled by *fur* and *fnr*, for example, but also on protein degradative rates. The degradation rate in general depends on the strength of the protein fold (Fig. 21.8) and on proteases. Metal ions, such as Fe, Fe/S clusters, and haem Fe, cross-link and greatly stabilize proteins into which they are incorporated. This means that the amount of an iron-containing protein will greatly exceed the amount of the apoprotein inside a cell. In Fig. 21.8 we view the combination with a metal as only slowly reversible. While this set of reactions greatly increases metal ion/protein binding selectivity (see Chapter 2), both the synthesis and hydrolysis of the proteins interact with the iron uptake systems to control the steady level of activity due to Fe proteins. Note that metal ion levels now limit activity: (1) directly by equilibrating with some enzymes; (2) indirectly by controlling their protein synthesis; and (3) directly by controlling their protein destruction rates.

Concluding this description of the influence of iron we consider that there is a general connection between the levels of free Fe^{2+} and the whole of the activity of a *prokaryote* cell up to the point of the initiation of the division process. This implies a fixed set of reactions generating, in proportion, the balanced amount of proteins, lipids, polysaccharides, and nucleotides as well as of small molecules and that all of them use free iron as one in-cell messenger, which must be related to growth. There is an absolute requirement for energy and we may well think that it has always depended on iron.

21.4 A summary of cytoplasmic homeostasis early in evolution

In summary we begin to see an interesting feature of an early biological cell (Fig. 21.9). While we are familiar with discussions of buffering, often confused with homeostasis, of an ion or a compound concentration level in chemistry (thermodynamic control) and of control in chemostats involving simple feedback systems (steady-state or kinetic control) based on single substances, say, H^+ levels, we do not find such simple devices even in prokaryotes. The biological homeostasis is the homeostasis of interwoven systems of metabolism where the various reaction paths, including those of H_2, CO_2, N_2, phosphates, and H^+, are linked back to all others including polymer synthesis

Fig. 21.9 A scheme to illustrate the close connections in a prokaryote cell between a variety of elements. More than 10 elements are shown specifically but M_1 and M_2 include at least five more which are catalysts. Note how energy, \ominus, is connected in different ways to the integrated system of growth homeostasis. See the use of this figure in Chapter 6, Fig. 6.2.

and cell growth. Energy state homeostasis is a part of this system. This is, in fact, the conventional way in which today we look on links between metabolic pathways, e.g. glycolysis and fatty acid metabolism. We believe that for good reason this feedback connection will often use the simplest components common to all the cytoplasmic pools, i.e. the inorganic forms of the chemical elements of the periodic table in water. In our first example this was achieved through P, Mg^{2+}, Fe^{2+}, H^+, and bound sulphur, then through bioenergetic connections to Na^+, K^+, Ca^{2+}, and Cl^-, and finally through Ni, Co, and Mo to many redox reactions. Lying behind the homeostasis are precipitates of iron sulphides and/or iron hydroxy phosphates.

21.5 The environment of early cells

21.5.1 Prokaryote morphology

In this very simple description of the interdependence of cellular activities on many elements we have not had to be concerned with shape. Indeed, for growth of a sphere or of a simple cylinder, the necessary condition for morphology to change continuously with the internal amounts of DNA, RNA, and proteins is that lipid synthesis should keep pace with that of the other polymers while concentrations of monomers and ions remain relatively fixed. Once sufficient DNA is made, there is a critical switch in the state of DNA which is coupled to switches leading to division. We do not know the coupling between these events, although it has been suggested that changes in internal calcium are involved. Much further information is needed. Perhaps it is true that prokaryotes do not have strict homeostasis or morphological control. In the detailed case of iron homeostasis we described the particular case of an *aerobic* bacterium and showed how iron concentrations were

maintained and connected to concentrations of many other elements. We also showed that in *anaerobic* prokaryotes iron was also involved in homeostasis control. Now, some prokaryotes can switch between aerobic and anaerobic metabolism, which represents a considerable flexibility and undoubtedly involves very considerable changes in concentrations of several elements, but in both states DNA and cell-wall synthesis must increase together. The implication is that certain controls are kept under a wide variety of conditions and it is likely that these involve such simple ions as HPO_4^{2-}, Mg^{2+}, K^+, and Na^+, but the same situation cannot apply to other controls. In other words, homeostasis may be relatively loose within prokaryotes. This is seen in the relationship between element levels and shape. In prokaryotes there is no internal organization of the cell (no permanent filaments) and size increase is that of a lengthening cylinder which is not therefore of constant shape defined by constant axial ratio. However, there is little doubt that prokaryotes developed shape in their coats, since in magnetotactic bacteria we observed well-ordered Fe_3O_4 crystals (Section 20.5.6). All these situations changed with the evolution of eukaryotes. Before we look at their development we must examine the environment of the prokaryotes to understand some of the advantages eukaryotes show (noting that they illustrate true homeostatic growth) over prokaryotes in the way they interact with their environment.

21.5.2 The supply of organic chemicals: the role of zinc

The supply of organic chemicals may well have been a minor problem in early life since there would be present many highly active catalytic surfaces for the interconversion of C/H/N/O compounds. Eventually, however, there would be quantities of organic debris from living systems which would form more immediately available foodstuffs, much as they do today. A difficulty living systems faced then was the digestion of the larger rather stable organic polymers, such as proteins, polysaccharides, polynucleotides, and so on. It could well have been at this stage that lowly available zinc, see Fig. 1.12, became important as a Lewis acid in *exported* enzymes (Chapter 11). (Alternatively, it may have been utilized even earlier to bring about difficult acid/base transformations in cytoplasmic syntheses and degradations.) Homeostasis of zinc would then have come to be related to both the internal metabolism and the external food supply. While zinc controlled one set of hydrolytic/condensation reactions, iron controlled a set of redox changes. (We shall have to consider that some redox reactions outside cells may have become valuable too, requiring export of iron enzymes.) Later in evolution the new problem that developed with the coming of dioxygen was that the primary sources of elements, e.g. HCN, NH_3, H_2S, were converted to forms which are inert, e.g. N_2, NO_3^-, SO_4^{2-}, and H_2O, and at the same time iron was lost by precipitation. We shall see that this led to major alterations in the elements of enzymes outside the cytoplasm and also in the uses of Ni, Mo, and all their redox reactions (see Fig. 21.13). At this stage we introduce the information flow between the single cell and its environment since cells also learned to search for food.

21.5.3 Slow and fast controls

At the level of gene expression there is a *slow* exchange of information between

the internal systems of the cell and the outside world through such processes as the uptake of iron which we have used as our first detailed example of element regulation. It is a slow control since it involves a metal that binds quite strongly. We know that there is also gene control of the uptake of other elements such as Mg^{2+} and HPO_4^{2-}. In these cases the connection to uptake is through ion pumps rather than to chelate metabolism, but insofar as it involves protein synthesis and degradation of the pump proteins, it is again a *slow* control of homeostasis, over periods of hours if not days. By way of contrast, homeostasis of internal metabolism must have rapid signals to take advantage of external changes. In fact, the prokaryote cell reacts quickly to changes in sources of metabolites, i.e. food. It is not possible that these fast reactions depend on metal ions such as iron, nickel, or zinc since they bind too tightly (see Chapter 4). The control has to be based on faster exchange of ions, such as H^+, Na^+, K^+, Mg^{2+}, Ca^{2+}, and phosphates, apart from the use of small organic molecules themselves, e.g. the mobile coenzymes. It is difficult to be certain to what degree this sensitivity to the environment initially worked directly through ionic element feedback control over ion pumps and exchangers of, e.g. H^+, K^+, HPO_4^{2-}, and Mg^{2+}, but these pumps are the obvious communication mode with the outside environment; they can be linked to metabolite uptake and they do have feedback controls. They can also be connected to motors which allow the organism to search the environment, again using feedback systems. (When the change in the external environment is sustained, then it must also communicate to the genes to alter protein synthesis, so it was necessary to find the ways in which the signals which sense the environment become relayed to the DNA eventually. One certain connection is via the organic phosphates.)

Now, the communication with the environment could be improved by sensing devices which deliberately admitted to the cell those elements which had been rejected by pumps, e.g. Na^+, Cl^-, and Ca^{2+}. Such sensing would be useful if, for example, the levels of these ions in the environment would change where there were decaying cells, i.e. sources of food, when the organism could search *if it also developed a sense of direction, which demands a shape since it implies positioning of activities systematically.*

21.5.4 Evolution of ion currents

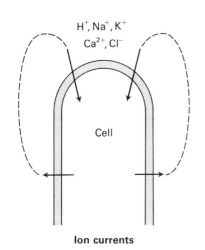

H$^+$, Na$^+$, K$^+$
Ca^{2+}, Cl$^-$

Cell

Ion currents

In Chapter 8 we referred to the evolution of ion currents as opposed to ion pumps. Pumps were required very early in cell history to remove excess Na^+, Cl^-, and probably H^+ and Ca^{2+}. We can regard excesses of these ions as poisons at this stage. Given such pumps it required nothing more than free but very selective access of K^+ to give electrostatic and osmotic balance to a cell. It is probable that primitive prokaryote systems had highly selective channels for K^+ only, and pumping out, or possibly exchanging of ions, would control Na^+, Ca^{2+}, Cl^-, and H^+. There would also be some ill-controlled electrical communication as soon as cells came together in colonies since membrane potentials were inevitable even in the most primitive cells (see Chapter 8).

Obviously, as discussed in the previous section, cell sensing of the outside environment could gain from communication and signalling networks, see margin. As far as we know, the conversion of the 'poisonous' Na^+, Cl^-, and Ca^{2+} external concentrations to useful signals came at about the same time as

eukaryotic organisms with shape appeared. It was the evolution of a variety of ion channels, controlled by external or internal circumstances, in addition to pumps for Na^+, Cl^-, Ca^{2+}, which made this possible. This utilization of an adverse factor to make a positive function during evolution will be seen again in the development of hormones from poisons and in the uses of copper, both in cell/cell systems. Now, as we have seen in several chapters, the advantages of currents based on channels and pumps were greatly enhanced by their selective positioning in the eukaryote membranes since in eukaryotes the membranes evolved with complex shapes. Before we look more deeply at the connection between the environment and the developing cell we must again note that great changes occurred inside the eukaryote cell in its evolution from the disorganized prokaryote. We start with the cytoplasmic structures which generated complex shapes in eukaryotes.

21.6 Eukaryotes

21.6.1 The value of shape: internal filaments and vesicles

It is pretentious to claim that these and further steps in evolution have already been or can be established. All we can visualize with hindsight are possibilities including advantages for organisms which were to lead on to present-day multicellular systems. One advantage is the development of complex shape since it allows the compartmental separation of functions both along the outer membrane of a cell and internally in vesicles (see Chapter 3). The development of a compartment for the nucleus heralds the beginning of eukaryotes. The advantages of a fixed position for sensing apparatus and a motor for movement in controlled directions are obvious enough. It is clear too that digestive enzymes are best held in reserve in organized internal vesicles. It takes no great imagination to envisage all such organization but, if it is to be generated and maintained, it requires a framework of internal filaments which are made from proteins. The filaments would have to be in tension to maintain shape and this leads us to the concept that the internal shape became part of the *energized homeostasis* of the organism. We may consider that the necessary filaments developed from the structures required for division processes in prokaryotes. (If we allow a coupling of the positions of channels and pumps, around a cell surface there is always a sensing current. The development of such a device leads eventually to nerve cells (Chapter 8).)

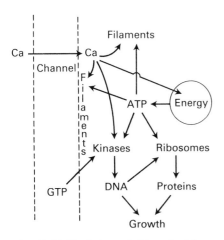

Fig. 21.10 The intimate linking of calcium (and phosphate-based compounds) to various internal cell compartments, including energy-supply, i.e. mitochondria and chloroplasts, mechanical stress, i.e. filaments, and synthesis of DNA, RNA, and proteins. Note that protons are always involved too. Figure 21.16 shows the further connections to extracellular calcium in eukaryotes and in multicellular organisms.

21.6.2 Calcium, shape, and the environment

Clearly a further step forward could be made by organizing cells so that the cell was less and less at the mercy of the environment and could be more and more sensitive to information in the environment. A little thought shows that these two developments go together and provide a major drive in evolution to develop message systems which are often based on calcium (Fig. 21.10).

The feature that calcium as a messenger using a channel introduces is that it binds on entry into a cell and therefore acts in a different capacity from sodium and chloride as messengers. Very much of the response of a cell, even to this

Fig. 21.11 A further illustration of the intimate integration of calcium and phosphate activities in higher cells. The entry 'Gene products' refers to such peptide hormones as calcitonin, which interacts with both calcium and cAMP systems.

day, is based on calcium-effected mechanical devices. We stress that it is not enough to say that calcium is a messenger since, in order to be an effective messenger, the inside of the cell has to be held at a *fixed* low level, which means that calcium had to enter into a homeostatic relationship with all the resting state components of the cell cytoplasm. This is shown in Fig. 21.10 as operating through phosphate, and it may be the phosphate system of regulation which controls the expression of the calcium-dependent trigger proteins. In a further step toward refining the calcium signals we shall see that the external fluids generated a new calcium homeostasis in multicellular organisms utilizing calcium precipitates derived from *protonated* anions.

As shapes of organisms evolved the role of calcium and its connection with the phosphate controls and regulations have grown (Fig. 21.11). The integration of the chemistry of these two elements throughout advanced biology is quite remarkable and includes the story of the development of muscle cells, for example. We are forced to consider that, in evolution, sensing

Fig. 21.12 The corresponding figure to Fig. 21.9 for a generalized eukaryote cell without external controls being shown. Notice the nucleus, vesicles and organelles. Not included are the membranous internal filaments which determine shape. The whole remains in a growth homeostatic state over long periods. Some 20 elements are involved. Symbols as in Fig. 21.9. See the use of this figure in Chapter 6, Fig. 6.4.

and growth with constant shape are established with stricter and stricter homeostasis of chemicals. Note again the use of precipitates, shell and bone, to perfect control through solubility products. (NB. In the study of morphogenesis also there is a growing body of knowledge which suggests that the use of sensing, initially based on ion currents, especially local H^+ and Ca^{2+} currents but later on organic compounds, has become the route to the development of morphology, both of single eukaryotic cells and of multicellular species.)

Before we attempt to take the next step in the description of the ever-evolving roles of the inorganic elements in life we give a summary diagram of the cellular structure of a eukaryote cell (Fig. 21.12) for comparison with that of the prokaryote (Fig. 21.9).

21.7 The evolutionary switch in homeostasis

21.7.1 The advent of dioxygen

In a complex fashion it can be seen that the regulation of prokaryote cells in early environments upon which they exerted no direct control had to be linked to elements critical for particular reaction steps. While N, C, H, and O in compounds were widely available and easy to incorporate into metabolism so long as catalysts were present and while internal controls of much of this metabolism could be based on mobile coenzymes, or even substrate feedback, elements such as Fe and P needed to be concentrated in the cell before they could be captured in chemically useful forms. Given the very central roles of these two elements perhaps we should not be too surprised that they became dominant in early *slow* internal control and regulation. The next stage of evolution introduced a more organized cell, the eukaryote, which had shape using internal systems organized within a matrix of filaments. Primitive signalling evolved, certain organic poisons could be ejected from vesicles as required, and hydrolytic enzymes could be exocytosed when food was discovered. The signalling would have used calcium, sodium, and chloride to connect to both the internal acid/base chemistry of P, Mg, H, and possibly Zn and the redox chemistry of S, Fe, and H, and both would lead on to protein activities, DNA and protein synthesis. We believe the external hydrolytic enzymes would have been based on zinc. Homeostasis would have remained relatively simple, but probably not too far from perfect. During this period of evolution there would also be a need to evolve controls and regulation in which calcium and zinc played a larger role as there was increased interaction with the environment. We know that calmodulins (Chapter 10) and zinc fingers (Chapter 11) are found in the eukaryotes (but not in prokaryotes) and that they became deeply associated with controls and regulation, e.g. in yeasts. Homeostasis of zinc and calcium in cells had to be added to that of other elements and connected to them.

Now later still in evolution there was a remarkable change, probably coincidental with the further upsurge of dioxygen in the atmosphere, which brought about increased compartmentalization and cell/cell organization beyond that of disordered cell aggregation in colonies (Fig. 21.13). It is clear

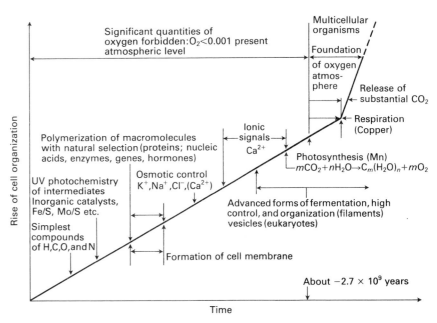

Fig. 21.13 A picture of evolutionary development where a decisive role is filled by the generation of dioxygen. The break in evolution which may well have been caused by the changed levels of the *available* chemical elements dependent on dioxygen is reflected in the development of eukaryotic compartments and multicellular organization. The very earliest cell/cell organization may have involved inorganic surfaces (not shown) as has been frequently discussed (see text). The figure is obviously related to Fig. 1.11.

that quite different control and regulatory devices were then necessary and indeed essential since elements changed in availability (Table 21.4). One additional possibility is that manganese quickly became a more involved part of these systems since it dominates dioxygen release and its availability was not much affected by the presence of dioxygen. Manganese chemistry is very like that of iron (see Chapters 12 and 14) and a similar regulatory system to *fur* can be visualized. It may still be a dominant control in plants (see Fig. 21.15). However, more fundamentally, there was and still is the need for control of extracellular space in contact with intracellular space in order that cell–cell communication and complex cell internal and external morphology could evolve and that growth conditions, which we associated with homeostasis in Fig. 21.1, could be maintained more precisely. Now, this *external* part of space is not and could not be controlled by cAMP and protein phosphorylation since NTP (ATP) cannot be made there in a steady state. Again it was not and

Table 21.4 Major element availability changes with the advent of dioxygen

Non-metals

	$H_2\downarrow$		
$CO_2\uparrow$	$N/O\uparrow$	$O_2\uparrow$	
		$S/O\uparrow$	$Cl\text{-}\uparrow$
		$Se/O\uparrow$	$Br\text{-}\uparrow$
			$I\text{-}\uparrow$

Metals

$V/O\uparrow$	—	—	$Fe/O\downarrow$	—	—	$Cu\uparrow$
—	$Mo/O\uparrow$					

\uparrow indicates increase in availability. The formulation $N/O\uparrow$ indicates an increased presence of nitrogen oxides while $Cl\text{-}\uparrow$ indicates an increase in covalently bound chlorine.

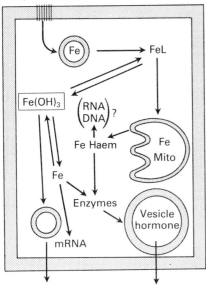

Fig. 21.14 A generalized scheme of iron uptake for multicellular animals. The differences from Fig. 21.5 arise in compartmentalization, absent in prokaryotes, in the changed nature of chemicals, e.g. ferritin, and in the pathways of control via RNA. Some controls go from haem iron to DNA, but the *fur* iron box is missing. L is a ligand and Mito refers to organelles.

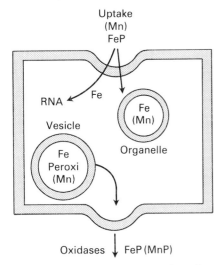

Fig. 21.15 A figure designed to show the possible development of a parallel control chemistry for manganese, compared with iron, especially in plants.

is not possible to do controlled polymer synthesis in extracellular space by condensation reactions, as for RNA, DNA, proteins, etc. Additionally, iron became extremely scarce when dioxygen increased since it was precipitated as $Fe(OH)_3$. Therefore, biological systems had to evolve with new modes of control and new filamentous external polymers. The most obvious control today is via an extension of the response to calcium ions and another is via organic transmitters and hormones. We have seen that both of these control and regulatory systems remain connected to intracellular phosphate reactions and it is phosphate controls which give the obvious continuity from early prokaryotes to modern multicellular eukaryotes. In contrast, in the developmental process the role of iron in the cell changed from a direct communication mode with DNA to one with mRNA and to a catalytic activity associated with the production of the many new organic (hormone) syntheses (see Section 21.10). In fact, as Fig. 21.14 shows, *free* iron does not even regulate eukaryotic cytoplasmic (nuclear) DNA, although it may still interact with organelle DNA, as in prokaryotes (Fig. 21.5). The 'iron-box' DNA sequences of prokaryotes have not been found in such nuclear DNA. Now, while the iron box seems to have disappeared from multicellular organisms, three series of proteins associated with control and regulation have greatly increased in sophistication in the multicellular organisms: (1) the calmodulin-related group of calcium-binding proteins (Chapter 10); (2) the calcium-binding proteins outside the cell (Chapter 10); and (3) the DNA-binding zinc fingers, which have become involved with many organic hormones (Chapter 11). (Possibly we should also include here the developed uses of Mn in plants (Fig. 21.15).) However, all of these developments demanded a truly multicellular organization and in turn this demands a new set of *cross-linked* filamentous *external* structures to control shape.

21.7.2 Cell/cell combination

Truly multicellular organisms are eukaryotes and this evolutionary development can be thought of in the following way. The study of biological mineral deposits indicates that, at first, minerals such as silica and calcium salts decorated the exterior of prokaryote cells probably in a rough and ready disorganized way. Cells also learned to attach themselves to surfaces. A combination of these two activities led to colonies of disorganized cellular systems and we may look upon the coral reefs and silica-based colonies in primitive sponges as representing early development of eukaryotic cells from these haphazard associations. D'Arcy Thomson (see 'Further reading' in Chapter 20) has shown why we might expect symmetry patterns to develop in them without any interior genetic control (see Chapter 20), but, evolution operating as it does, we can believe that shapes with advantages would also evolve through special protein production outside the cell, i.e. dependent on genetic instruction, and compatible with the calcium salt and silica surfaces. The factor that is still missing if a true multicellular organism is to be built is the control over the extracellular fluids and extracellular metabolism. As stated in Section 21.5.1, it is entirely to be expected that, initially, extracellular metabolism would be based on the export of digestive enzymes. These enzymes would be non-selective and as such would be best based on zinc and calcium (hydrolysis) and on iron (oxidation). Again all the enzymes would

develop so as to be stabilized and activated by the high external calcium. The digestive enzymes of bacteria are frequently just of this composition, i.e. they are Zn, Ca or Fe, Ca enzymes. But none of this activity could be organized until evolution generated management of extracellular space. A particular importance is given therefore to the initial development of a relatively rigid *extracellular matrix* suitable for maintaining organization.

21.7.3 Copper and zinc: the extracellular connective matrix

With the advent of dioxygen, copper and zinc increased in concentration in the environment since their sulphides were oxidized (Chapter 15), but at first copper would be a deadly poison for cells. Unlike zinc, copper is *removed* from all cells by metallothionein or a corresponding set of thiol-rich peptides in plants. Subsequently, copper became used but its enzymes were and are very largely kept in extracellular space in both (aerobic) prokaryotes and eukaryotes (Chapter 15). (This selection of copper as opposed to zinc using thiolates is possible since copper is handled as Cu(I) inside cells and Cu(I) binds very tightly to thiolates.) The intracellular zinc and the extracellular copper enzymes are therefore in communication via the buffering reactions of metallothionein discussed later in this section. We can give reasons for this complication in homeostasis by looking at the enzymic activity of copper outside cells relative to the functions of zinc with respect to extracellular matrices.

The immediate essential requirement for the evolution of multicellular structure was that there had to be an external synthetic activity making connective tissue. A glance at the external polymers used in matrix building shows, as stated in Section 21.7.1, that they are not dependent on the same polymerization reactions as those inside cells. Outside cells biology uses oxidative free-radical polymerization in the production of *cross-linked* proteins via –S–S– or other bridges. In collagens and chitins the energy for the cross-linking polymerization comes from the reaction of dioxygen and employs specific oxidases. Overwhelmingly in all this stabilization of *external polymers* it is *copper* and not iron enzymes which are involved (Table 21.5). Now, this building has to be controlled since the matrix external to a cell must

Table 21.5 Zinc and copper enzymes of the external matrix

Enzyme	Metal	Activity
		Hydrolysis of
Elastase	Zn	Elastins
Collagenase	Zn	Collagens
Stromelysin	Zn	Proteoglycans
Gelatinase	Zn	Gelatins
Protease 24.11	Zn	Peptide hormones
		Oxidation
Lysine oxidase	Cu	Cross-links collagen and elastin
Laccase	Cu	Cross-links via phenols
Phenol oxidase	Cu	Cross-links via phenols
Amine oxidase	Cu	Removes amine hormones

be in a homeostatic balance between synthesis and degradation to allow replacement of damage and to create space for systematic growth, which will require the matrix to be broken down and rebuilt. There is needed therefore a set of controlled 'digestive' enzymes, known as collagenases, stromelysins, and elastases, and a series of signals between these digestive enzymes and the external oxidative enzymes from the cellular system as a whole. Revealingly, all these digestive enzymes are zinc enzymes (Table 21.5). The two external polymerization/depolymerization systems communicate back to one another inside the cell through elementary ions, i.e. copper and zinc, probably via metallothioneins and then the genes. In turn, these genes are linked by hormones and, in particular, vitamin D becomes important for Zn and Cu through metallothionein expression and so leads back to Ca, the element for further stabilizing the external matrix (see Section 21.8.1). It would be interesting to discover how the production of connective tissue proteins is controlled at the DNA level. How much of it is connected to vitamin D (Table 21.6)?

Table 21.6 Vitamin D controlled expression

Protein	Function
Calbindin	Ca^{2+} transport
Osteocalcin	Binding to bone
Carbonic anhydrase	Carbonate/CO_2 equilibration
Alkaline phosphate	Phosphate ester hydrolysis
Collagen	Connective tissues
Metallothionein	Zinc 'buffer'?

NB. It may well be that thyroxine also links to a variety of calcium binding proteins and to connective tissue proteins and see also the role of retinoic acid.

If this description is correct we see that, through the production of a dioxygen atmosphere, copper became an essential catalyst for truly multi-cellular life. Once a cross-linked matrix could be built it had also to be digested. The connections to calcium and its minerals for stability and to extracellular zinc enzymes for break-down are clear. Homeostasis of all three elements is required. We return to the developed functions of calcium.

21.8 Controls over calcium external flow and mineral formation

21.8.1 The circulating fluids of multicellular organisms

While knowledge of *internal* calcium controls of the resting activities of cells as well as of the triggered controls of cells has increased enormously and the understanding of their relationship to phosphate controls and regulations has grown beyond recognition in the last 20 years, parallel features have emerged from the analysis of *extracellular* fluids. Figure 21.16 shows some extracellular calcium controls. Moreover, the stability of extracellular enzymes and connective tissue matrices depends on calcium binding and on the incorporation of calcium salts. In Chapters 10 and 20 we have examined the homeostasis of the calcium ion in these fluids and its relationship to the precipitates,

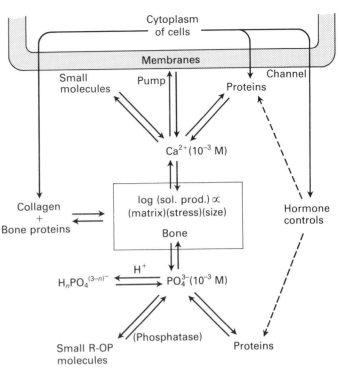

Fig. 21.16 The homeostasis of a multicellular organism containing bones can be shown to be sensitive to continuous *external* effects. Consider a local pressure stress. The stress affects the solubility product (sol. prod.) of the bone matrix. This matrix helps to stabilize the connective tissue. Thus, an extracellular pressure interaction develops, much as it does in the inside of a cell when the cell makes contact with a surface. Note that calcium effects are closely linked to those of pH and phosphate, again as in the interior of cells. The connection between inside and outside calcium is made via many means described in Chapter 10. Note that the solubility product is not absolutely fixed but varies between very narrow limits.

$CaCO_3$ (shell) and $Ca_5(OH)(PO_4)_3$ (bone). (Notice the change in the type of mineral, from domination by iron oxyphosphates, ferritins, inside cells especially prokaryotes (Fig. 21.5) to the external calcium salts of multicellular eukaryotes (Fig. 21.16) but all systems equilibriate with protons.) It is the constant exchange of internal resting cellular and external calcium levels which keeps the calcium balance seen in multicellular growth homeostasis. The fine tuning of these balances by the multiple calcium-binding sites in many of the binding proteins is discussed in Chapter 10.

Now, as we have stressed again and again in this chapter, it is not sufficient for growth homeostasis that a communication is made between compartments, here outside and inside, at the level of small molecules and ions. Macromolecular (protein) systems have to be knitted into the homeostasis so that messengers must be present which 'talk' to the DNA to generate proteins and to allow growth. Such messengers must now also go between cells in multicellular systems, much as inside prokaryotes the iron levels were related to iron uptake, to the citrate cycle, and to dinitrogen and dihydrogen metabolism via DNA/protein regulation signals. We know that in multicellular organisms many of these signals are circulating hormones (Table 21.7); how are the hormones connected to the proteins of zinc, (copper), to calcium and phosphate levels outside cells, and to the DNA inside cells? Much of the regulation appears to work through zinc itself.

21.8.2 Eukaryotic DNA and zinc: hormonal connections

Starting as for iron in prokaryotes from the regulatory interaction of elements as simple ions with DNA, a quite different set of regulatory events is found in

Table 21.7 Some types of hormones and related transmitters

Class of hormone (transmitter)	Metal association
1. Steroids	Fe in synthesis
Thyroxine	Fe in destruction
Retinoic acid	Zn fingers
2. Adrenalin	Fe, Cu in synthesis
	Cu in destruction
	Ca channel
3. Large peptides (EGF)	Ca
4. Small peptides (enkephalin)	Zn in release
	Zn in destruction
5. Indole acetate	Fe in destruction

the eukaryotes and especially in multicellular organisms. We now leave on one side the redox metabolism of mitochondria and chloroplasts, as well as the cytoplasmic immediate relationships of bioenergetics and synthesis to proton, calcium, and phosphate controls and regulation, i.e. most of primary metabolism. We turn our attention to the imperative connection between intracellular and extracellular events based on hormones. Most interesting are the set of hormones which today use the zinc fingers (Table 21.8 and Fig. 21.17) since they are the very hormones which are deeply associated with whole-organism growth homeostasis. We list in Table 21.8 the connections from the hormones to the homeostasis of other elements (M), which include the vitamin-D-dependent proteins for cellular calcium (Ca) transport and the mineral corticoid hormones which control the extracellular sodium (Na) and potassium (K) levels, Table 21.6, which in turn are linked to phosphate compounds and chloride levels. Here is a clear indication of the way in which external fluids are linked to cellular DNA. These elements have to be controlled in the extracellular space as well as intracellularly since the calcium levels, for example, are connected to almost every internal and external control system (Figs 21.10 and 21.11), and all calcium reactions are critically dependent on ionic strength as well as temperature. There is a further link from zinc fingers to the thyroid growth hormone (note the presence of the mineral element iodine and see Section 21.10) and to retinoic acid, both of which affect the extracellular matrix, and possibly to yet other growth hormones to which we return. There is probably a vast array of these regulatory systems and zinc

Table 21.8 Element connections to zinc fingers

Effector system	Element connected (M)
Glucocorticoids (sterol)	Na, K levels (kidney)
Vitamin D (sterol)	Ca transport proteins, osteocalcin and even metallothionein (Zn), alkaline phosphatase (P)
Thyroxine	Iodine metabolism (Fe), (I), Ca–ATPase (Ca), connective tissue/Ca
Retinoic acid	Zinc dehydrogenases (Zn)
Sex hormones	Not known (?)

NB. Zinc fingers are also connected to the regulation of haem synthesis enzymes.

fingers are not the only zinc-binding motifs (there are other examples in yeast). Before exploring further we must not forget the connection of zinc to a vast range of other activities of multicellular organisms which we summarize next.

21.9 A summary of zinc controls

The important features of zinc biochemistry, which have been described in Chapter 11 are (Fig. 21.17):

1. Zinc is required in a huge variety of enzymes in the cytoplasm including many involved in RNA and DNA synthesis (Table 21.8).
2. Zinc is involved in the reactions of the extracellular matrix (Table 21.5).
3. Zinc is a required element in the peptidases which release and hydrolyse many hormonal peptides.
4. Zinc is found in vesicles in nerve terminals.
5. Zinc is highly concentrated in the male reproductive tract.
6. Many of the considerations under points 4 and 5 have led to the supposition that there is a membrane pump (ATP-dependent?) for control of zinc concentrations in vesicles of some cells.
7. Zinc fingers connect to many hydrophobic hormones.

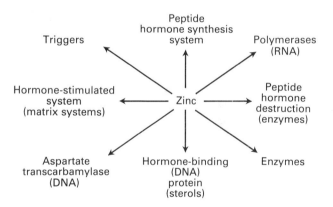

Fig. 21.17 The extensive connections of zinc to multicellular controls and regulations. Zinc links to sterol-like hormones, to peptide hormones, to the external matrix, and to a vast list of enzyme and control activities in cells.

In this list zinc has a second major function connected with control. Most of the peptide hormones or transmitters which act as signals for a vast range of multicellular activities are stored in precursor proteins and their release is due to the action of zinc proteases. A further set of zinc peptidases, peptidases 24.11, destroy the smaller peptides very locally and often close to the point of release (see Chapter 11), allowing short-range cell/cell communication unlike that of the steroids. (This control mechanism may be related to the value of stored zinc in the pancreas of some animals together with insulin and in vesicles in some nerve termini). It would appear to be likely that the peptidases are related, in evolution, to digestive enzymes both in prokaryotes and eukaryotes and to the zinc proteases used to manipulate connective tissue. It is

necessary to remember that zinc cross-links affect the stability of proteins too, compare Fig. 21.8 for iron.

There is another link with zinc which is connected to DNA synthesis. Amongst the uses of iron in prokaryotes we noted the enzymic functions leading to DNA synthesis. The parallel uses of zinc include a dependence of aspartate transcarbamylase which is necessary in the synthesis of RNA and DNA bases. Thus, the control over RNA and DNA by zinc extends from enzymes assisting *polymerization* to control over the levels of monomers (see Chapter 6 where we insisted that such connection had to exist to maintain steady growth). It is clear that zinc and calcium are so deeply involved in so many cellular and extracellular activities while they are not redox-active that they are admirably suited as general regulatory elements in organizations of eukaryotic cells. The suitability of zinc as a control might, however, be queried on the basis of its speed of exchange.

In Chapter 11 we collected some data on the zinc exchange rates. Zinc associates relatively rapidly and dissociates relatively slowly from many proteins including zinc fingers. The rates are such that zinc, like iron or manganese, could not act as a control element in the time range $1-10^{-6}$ s. Such reactions must belong to those metal and non-metal elements which are common and are bound relatively weakly, i.e. H^+, Na^+, K^+, Cl^-, Mg^{2+}, Ca^{2+}, RPO_4^{2-}, and later in evolution to the binding of a variety of designed small organic molecules such as adrenalin. However, the concentrations of Mn^{2+} and Zn^{2+} in many cell types and of Fe^{2+} in prokaryotes are not far from 10^{-10} M. Given that the on-rates of reaction for these ions can be as fast as 10^9 s^{-1}, there is no problem in achieving exchanges in the time range 10^2 to 10^4 s, i.e. from minutes to several hours (Table 11.5). These are rates commensurate with the exchange and activity rates or organic sterol hormones. There is, therefore, no reason to suppose that rate of exchange prevents zinc acting as a superhormone linking the conventional steroid-related hormones of Table 21.7. The kinetic character of the binding of the two major slow regulatory ions, i.e. iron and zinc, is seen to be closely similar. Now even the ways in which they are bound to regulatory proteins are similar and it would appear that evolution faced but a small chemical problem in switching from iron controls in prokaryotes to zinc controls in eukaryotes (see Fig. 21.6). It was then the necessary removal of iron from the cytoplasm to avoid oxidation of DNA, i.e. mutations leading to cancer, that assisted the development of zinc, not iron, regulation in higher organisms. (NB. The possibility that manganese plays an equally essential role in plants must be kept in mind (Fig. 21.15).) The next question we must ask is how are all the new organic hormones and transmitters synthesized?

21.10 Organic signals

21.10.1 Iron, dioxygen, poisons, and hormones

While the new roles for copper, calcium, and zinc evolved, the role of iron, both as free and haem iron, also changed dramatically with the advent of dioxygen. Inside cells it could not be used in direct interaction with DNA,

apparently, but it could still be used as a catalytic centre for the production of a vast range of (secondary) metabolites. The best place for this activity naturally was in vesicles. Table 21.9 lists some of the metabolites, many of which initially would have been poisons ejected from vesicles but, as cell–cell interactions increased, they, like sodium, chloride, calcium, and then copper before them, could and did become messengers, transmitters, and hormones. A more detailed look at this chemistry is given in Chapters 12 and 13. (One such compound of a different character but of extreme interest is cholesterol (Chapter 7) which completely altered the character of membranes and is absent from prokaryotes.)

Table 21.9 Secondary metabolite production catalysed by non-haem iron proteins

Reaction	Substrates/products	Location
Hydroxylation	Prostaglandins, adrenalin	Vesicle/reticulum membranes
Dioxygenation	Aromatic molecules	Extracellular (soil bacteria)
Proline hydroxylation	Proline in proteins	Golgi (?) for connective tissue synthesis
Superoxide dismutation	O_2^-	Organelle (or cytoplasm of prokaryotes)
Ribose reduction	Ribose	Cytoplasm $(Fe(III))_2O$
Lipid oxidation	Arachidonic acid	Membranes
Scavenger synthesis (Fe)	Siderophores	Prokaryote exterior
Penicillin cyclization	Penicillin	Prokaryote exterior

It may well be that plant hormone, e.g. auxin, levels are even more dependent on redox metabolism, since it is not certain that zinc and peptides have much significance for them. The extracellular matrices of plants and the cuticles of insects are usually based on lignins, which are also formed and degraded by oxidative mechanisms. Here manganese and iron act together. When a plant is damaged the peroxidases act to heal the wound, but when a higher animal is damaged it is the blood-clotting system which is the immediate protection. As shown in Chapter 10, this is a calcium-based activation of protease action to generate and moderate scar tissue. It is as if plants (and insects?) rely on the use of redox systems while animals more effectively use hydrolysis. Perhaps manganese has a more important regulatory role, both internally and externally, in plants. Naturally, both the system of hormones and the extracellular matrix are based, to some degree, on secondary metabolism.

21.10.2 Evolution of secondary metabolism

Pursuing the line of argument just given, we must seek a developmental relationship between early communication networks based on Ca^{2+} and phosphate compounds and the sophisticated systems of today based on hormones and transmitters. We shall pursue the line that protection has come before communication and that there are two modes of protection— strengthening external structures and ejecting poisons. The coming of dioxygen allowed the cross-linking and general strengthening of external matrices, at first around single cells, as described in Section 21.7.3, but it also allowed a huge wealth of new organic compounds to evolve based on dioxygen

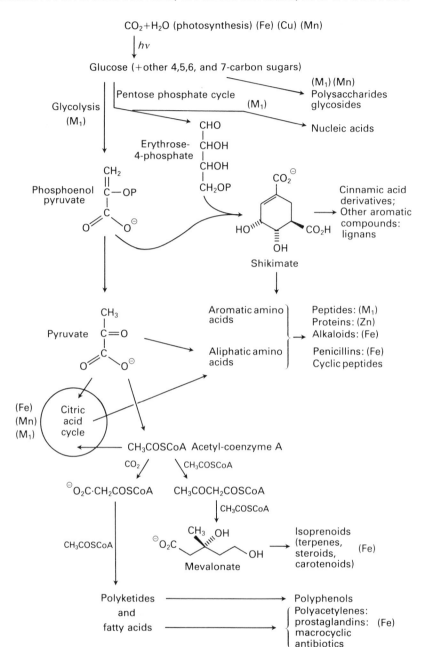

Fig. 21.18 An indication of the elaborate links between elements and metabolic paths which must be integrated into growth homeostasis. Note that both primary and secondary metabolism are involved. M_1 is Mg^{2+}, K^+, H^+. (The figure is related to Fig. 6.3.)

chemistry—much of it is called secondary metabolism (Fig. 21.18). Initially, a very valuable function of these compounds would be to poison other cells, e.g. toxins under *fur* control, including hydroxylated phenols and sterols, penicillins, halogenated hydrocarbons, etc. Later, in multicellular systems, they would become messengers, i.e. hormones (Table 21.10) and transmitters. The fact that a vast range of secondary metabolites appear to have little function, e.g. alkaloids, may be due to our wish to see functional value everywhere as opposed to trial and error selection in action. From Table 21.10 we see that much secondary metabolism is based on iron oxidases.

Table 21.10 Hormonal systems dependent on dioxygen

Hormone	Dependence on Dioxygen	Element
Sterols	Hydroxylation	Fe
Thyroxine	Hydroxylation, Iodination	Fe, I
Adrenaline	Hydroxylation	Fe, Cu
Hydroxytryptamin	Hydroxylation	Fe
Prostaglandin	Hydroxylation	Fe

All this metabolic activity must be related to protein synthesis regulated via DNA and RNA. We know little about this as yet, but there is probably some direct feedback via dioxygen-binding haem proteins. We also know that there is in higher eukaryotes a link from haem to transcriptional controls (RNA). Much of the regulation of iron proteins in eukaryote cells remains to be uncovered.

21.10.3 Overall schemes of metal linkages in controls

We can now draw up very simplified relationships between groups of inorganic elements deeply associated with connective tissue and a network of communication inside and outside cells needed for growth homeostasis (Figs 21.19–21.21). (Not shown in the almost universal link to the proton.) The intention of these figures is to stress the deep integration of these elements in growth homeostasis. All the components are in flow, i.e. are being lost and regained in a dynamic unity which is consistent with a given species of a certain shape. (The flow even extends to macrocirculation between cells by organ pumping, e.g. by the heart.) Only much further experimental work will reveal the full integration of the chemical elements in biology.

Fig. 21.19 Some of the intricate dispositions and connections in the growth homeostasis of a single cell from a multicellular eukaryote. The whole is in an energized feedback state.

Fig. 21.20 The connection in schematic form between extracellular ion levels, Na$^+$, K$^+$, Ca^{2+}, and Cl$^-$, the energy for pumps (ATP), metabolic pools, and hormones indicating especially the connected roles of iron and zinc. Protons are involved everywhere.

Fig. 21.21 An extension of Fig. 21.20 to include two cells which allows the role of iron and copper to be seen in relation to the role of zinc in the control of extracellular signals and tissues and shows also the connection to calcium and phosphate compounds from these controls. Note the further involvement of protons especially now in the precipitation reactions.

21.11 Changes in functions of other elements

21.11.1 New roles for molybdenum and vanadium

It is apparent that the internal structure of advanced eukaryotic cells protects the nuclear DNA from oxidative reactions as much as is possible. Copper is rejected to the outside of cells or is placed in vesicles or vesicle membranes, iron is placed in organelles in vesicles and in membranes, manganese is disposed of in vesicles (the Golgi) or organelles. Interestingly, molybdenum and vanadium biochemistry did not evolve further in cells from that found in nitrogen fixation. However, outside the cell the biochemistry of molybdenum has become quite extensive and associated with dioxygen reactions at low potentials (Chapter 17), while vanadium is used in a limited number of cases only (Chapter 17), but again outside cells in an oxidative capacity.

21.11.2 Nickel and cobalt in eukaryotes

The chemistry of nickel and that of cobalt have really evolved little in eukaryotes and certainly not in multicellular organisms. This may well be connected with their early uses in oxidation/reduction reactions of very elementary food molecules, e.g. H_2 and CO, in prokaryotes. Then, the photolysis of H_2O became the source of H and the source of carbon changed to CO_2, so that quite different reaction paths evolved and these metals stopped being necessary in their primitive roles.

Another element which changed roles is selenium. It became a part of the detoxification modes for removal of peroxides in cells, having been associated with species such as H_2 and formate earlier in evolution (Table 21.11).

Table 21.11 New element biochemistry after the advent of dioxygen

Element	Biochemistry
Copper	Most oxidases outside higher cells; connective tissue finalization; production of some hormones, dioxygen carrier, N/O metabolism
Molybdenum	Two-electron reactions outside cells; NO_3^-, SO_4^{2-}, aldehyde metabolism
Manganese	Higher oxidation state reactions in vesicles, organelles, and outside cells; lignin oxidation (note especially plants); O_2 production
Nickel	Virtually disappears from higher organisms
Vanadium	New haloperoxidases outside cells
Calcium	Calmodulin systems; γ-carboxyglutamate links; general value outside cells and in cell–cell links
Zinc	Zinc fingers connect to hormones produced by oxidative metabolism
Selenium	Detoxification from peroxides, de-iodination?
Halogens	New carbon–halogen chemistry, poisons, hormones
Iron	Vast range of especially membrane-bound and/or vesicular oxidases, peroxidases for the production of hydroxylated and halogenated secondary metabolites; dioxygen carrier and store

21.12 Morphogenesis

While it is apparent that homeostasis is associated with the maintenance of metabolism and shape and that obviously it allows a cell to be triggered and to relax without altering gene expression, we must also allow for moderate changes in protein levels and therefore of gene expression which can reverse with change of circumstance. For lower organisms, e.g. bacteria, the degree of change of gene expression without morphological change is considerable, while the higher and more physically complicated an organism the less is this so. Homeostasis became increasingly demanded as the sophistication increased and gave increased sensitivity.

At the same time the morphology of a complex organism has to develop from a single cell, and there is then a compulsory morphogenic stage. In fact, even within many single-cell eukaryote species, there is a developmental stage. For example, radiolarians commence life with an ill-defined shape and carrying a single crystal of strontium sulphate. This crystal dissolves and in place of the biomineral, $SrSO_4$, the adult radiolarians grow with a silica skeleton, as described in Chapters 18 and 20. Now, the appearance of a skeleton is a very useful indication of the onset of 'equilibrium' growth, i.e. growth without change of shape, which we have called homeostatic growth. More fundamentally, it is the time in development when the levels of some elements, usually silicon or calcium, in the extracellular fluids *must become fixed* by solubility product relationships (Fig. 20.1). Homeostasis is certain from then on. Before this period of mineral development it is equally clear that the same homeostatic condition does not prevail. One glance at the period of early development of animals (Fig. 21.22) shows that the structural reorganizations are not compatible with skeleton formation. A further feature is also clear: after very few divisions from the first cell, the cells are no longer a close-packed arrangement of equal spheres. Very quickly, in fact, the multiplying cells of a fertilized egg cannot replace one another. It is not likely that strongly differentiated cells have equal complements of chemical elements and certainly they have different activities. In a sense, differentiation is completed when homeostatic growth begins. At this stage organs are defined and grow. It is almost a direct conclusion that the more demanding the

Fig. 21.22 The growth pattern of animal cells from fertilization. While even cell A has polarity the distinction between different cells begins at the 8–16 cell step (D–E) and then cells become of increasingly divergent shape and function. This is the period of morphogenesis which only becomes homeostatic, it is suggested, with the appearance of a fetus with a close resemblance to the adult form. H stresses the divergence of cells as the inside becomes totally distinct from the outside of the whole organism, G, before infolding occurs. The infolding may be related to the surface currents as surfaces change during packing, and develop polarity.

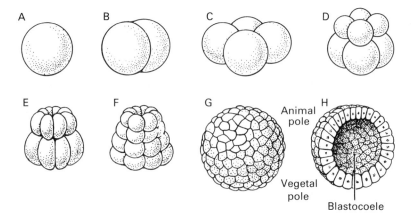

homeostasis the less the possibility of metamorphosis or of reversion to a stem cell.

While we observe the changes in the cellular structure and presumably in the extracellular matrix in the developing embryo, we need also to ask about the communication between the cells which allows parts at increasing distance from each other to grow in harmony. There is good reason to suppose that the communication is based, at least in part, on the simple ions such as calcium. In morphogenesis of simple species such as acetabularia (Fig. 21.23) the growth of the tip has been related to ionic currents, including calcium currents at the tip. Undoubtedly there are many organic compounds which are involved in complex signalling too. The exploration of the complex messenger systems in morphogenesis will surely be one of the most exciting areas for future research, since the signals must change with change of shape.

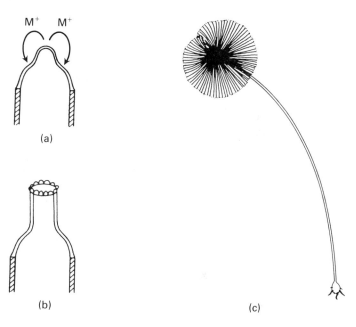

Fig. 21.23 The growth currents around the tip of an acetabularia cell which are part of the relationship between cell morphogenesis and element functions. The currents are believed to be part of the guidance system which leads to the final shape.

21.13 Concluding remarks

It is important that the last speculative paragraphs on morphogenesis do not obscure the purpose of this book. The central hypothesis is that the evolving nature of living systems (Table 21.12 and Fig. 6.2) is a product of an interaction between the properties of inorganic elements (Table 21.13) and an ever-evolving organic chemistry dependent on these properties. The possibilities for living systems to evolve are peculiarly constrained, therefore, by the basic chemistry imposed by the stability of atomic nuclei, the presence of water which also controls the availability of the elements, the irradiation from the sun, and so on, see Chapter 1, but the interacting manipulation of the elements did and does allow ever increasing but restricted sophistication. This manipulation is linked in a feed-back manner to the evolution of proteins, Chapter 7. Concentration of thinking upon linkages from DNA variations to

Table 21.12 A simplified evolution of integrated organization

Early prokaryote

Anaerobic metabolism in which developing homeostasis is based on several chemical element concentrations but note especially H, P, and Fe. Cell has no shape? Many ion pumps but only potassium channels. No morphogenic pattern. No control of environment. Internal iron precipitates in cells.

Later prokaryote (after dioxygen)

Both anaerobic and aerobic metabolism in which homeostasis is based on several chemical element concentrations. Cell has outer constructions, even walls and a periplasmic space i.e. external morphology. Many ion pumps but only potassium channels. Environment monitored and utilized via external enzymes. New systems of catalysts based on Cu, Mo, Se, and so on. Iron oxide precipitates in cells.

Eukaryote unicellular

Homeostasis now strict and based on element concentrations and morphology but external and internal shape based on organization due to filaments and vesicles. Cell has characteristic shape and phases of growth and morphogenesis. Many pumps and many channels: sensing currents around shaped exterior, e.g. Na^+, H^+, and Ca^{2+}. Vesicle rejection to give poisons and mineral protection. Connections to solubility products of calcium salts and to silica in vesicles. Iron oxide precipitates in cells.

Eukaryote multicellular

Strict homeostasis as for unicellular eukaryote but extended to external fluids and cell/cell communication. Cells and cell/cell interaction give shape, etc. Many pumps and channels connected to circulating fluids. Nervous systems of currents around and between cells. Vesicle rejection gives hormones and transmitters to circulating system. Environment of cell controlled in element composition. Development of external filaments controlling through tension external chemical potential. Long periods of morphogenesis followed by even longer periods of homeostatic growth and occasional metamorphosis.

the manipulation of organic-based structures obscures the underlying connection between living systems and basic inorganic chemistry which we have been led to believe is orthogonal to it. We wish to assist a general re-orientation so that the role of many elements is seen to be central to life. We consider that this will force man toward a deferential attitude towards his treatment of the environment. There is a oneness of the living and the dead through inorganic chemistry. Our last chapter stresses the role man has assumed in this complicated unity and some of the complications his activities introduce for himself.

Table 21.13 Functional value of the elements

Information (electrical) transfer and storage	Mechanical transmission	Acid–base catalysis	Redox catalysis	Structural role (excluding the organic polymers)	Chemical energy transmission and storage
Na^+, K^+, Cl^-, H^+, Mg^{2+}, Ca^{2+}, HPO_4^{2-}	Mg^{2+}, Ca^{2+}, HPO_4^{2-}	Non-metals and divalent and trivalent ions, e.g. H^+, N, S, Zn^{2+}, Fe^{3+}	Transition metal ions and some non-metals, e.g. Cu, Fe, Mn, Co, Mo, Se, S	Si, B, P, S, Ca, Mg, (Zn)	P, S, (C)

NB. The functional value of the elements arises partly from the chemistry and partly from the energy input to that chemistry. The energy input can be to the gradient of the free concentration of an element, to a chemical bond of given kinetic stability, or to the synthesis of a polymer in a compartment that acts as a trap (see Part I of this book).

Further reading

General

Booth, I. R. (1985). Cytoplasmic pH in bacteria. *Microbiological Reviews*, **49**, 359–78.

De Duve, C. (1991). *Blueprint for a cell*. Portland Press, Colchester, UK.

Goldbeter, A. (ed.) (1989). *Cell to cell signalling*. Academic Press, New York.

Goodwin, B. C., Briere, C., and O'Shea, P. S. (1967). In *Spatial organisation in eukaryote microbes* (ed. G. W. Gooday, D. Lloyd, and A. P. J. Trinci), pp. 1–9. IRL Press, Oxford.

Gow, N. A. R. (1989). Ionic currents in micro-organisms. *Advances in Microbiology and Physiology*, **30**, 89–123.

Harold, F. M. (1986). *The vital force. A study in bio-energetics*, Chapter 14, 'Morphogenesis'. Freeman, New York.

Hennecke, H. (1990). Regulation by metal–protein complexes *Molecular Microbiology*, **4**, 1621–8.

Mason, S. F. (1991). *Chemical evolution*, Clarendon Press, Oxford.

O'Halloran, T. V. (1989). *Metallo regulatory proteins*, Metal ions in biological systems, no. 25 (ed. H. Sigel), pp. 105–46. Dekker, New York.

Peacocke, A. R. (1983). *The physical chemistry of biological organisation*. Clarendon Press, Oxford.

Perry, C. C., Wilcock, J. R., and Williams, R. J. P. (1988). Biominerals and morphology. *Experientia*, **44**, 638–64.

Silver, S. and Misra, T. (1988). Bacterial element requirements. *Annual Reviews of Microbiology*, **42**, 717–43.

Thomson, K. S. (1988). *Morphogenesis and evolution*. Oxford University Press, New York.

Wächtershäuser, G. (1988). Evolution of life. *Microbiological Reviews*, **52**, 452–84.

Williams, R. J. P. (1990). Elements in homeostasis and morphogenesis. *Transactions of the Biochemical Society*, **18**, 689–705.

Woese, C. R. (1987). Evolution of bacteria. *Microbiological Reviews*, **51**, 221–71.

Iron proteins (see Chapters 12 and 13)

Bragg, A. and Nielands, J. B. (1987). *Microbiological Reviews*, **51**, 509–20.

Braun, V., Schäffer, S., Hantke, K., and Tröger, W. (1990). Regulation of gene expression by iron. *Colloquium Mosbach*, No 41, pp. 164–9. Springer, Berlin.

Buchman, C., Skrock, P., Welch, J., Fogel, S., and Karin, M. (1989). *Molecular and Cell biology*, **9**, 4091–5.

Fu, Y-H. and Marzluf, G. A. (1990). *Molecular and Cell Biology*, **10**, 1056–65.

Haile, D. J., Hentze, M. W., Ronault, T. A., Harford, J. B., and Klausner, R. D. (1989). *Molecular and Cell Biology*, **9**, 5055–61.

Spiro, S. and Guest, J. R. (1990). *FEMS Microbiological Reviews*, **75**, 399–428.

Trageser, M., Spiro, S., Duchêne, A., Kojvo, E., Fahrenholz, F., Guest, J. R., and Unden, G. (1990). *Molecular Microbiology*, **4**, 21–7 (and references therein).

Zinc proteins (see Chapter 11)

Berg, J. M. (1986). *Science*, **232**, 485–7.

Klug, A. and Rhodes, D. (1987). *Trends in Biochemical Science*, **12**, 464–9.

Lee, B. T. O., Brown, N. L., Rogers, S., Bergemann, A., Camakaris, J., and Rouch, (1990). In *Metal speciation in the environment* (ed. S. Gucer and J. A. C. Brockaert). Springer-Verlag, Heidelberg.

Mathisen, P. M. and Miller, L. (1989). *Molecular and Cell Biology*, **9**, 1823–31.
Miller, J., McLachlan, A. P., and Klug, A. (1985). *EMBO Journal*, **4**, 1609–14.
Vallee, B. K. L. (1983). In *Zinc enzymes* (ed. T. G. Spiro), pp. 3–24. Wiley-Interscience, New York.
Vallee, B. L. and Auld, D. S. (1990). *Biochemistry*, **29**, 5647–59.

Calcium proteins (see Chapter 10)

Jaffe, L. (1986). In *Calcium and the cell*, Ciba Foundation Symposium, no. 122, pp. 271–81. Wiley, New York.
Kater, S. B., Mattson, M. P., Cohen, C., and Connor, J. (1988). *Trends in Neurological Science*, **11**, 315–21.
Klee, C. B. and Cohen, P. (1988). *Molecular Aspects of Cell Regulation*, **5**, 225–48.
Minghetti, P. P. and Norman, A. W. (1988). *Federation of American Societies for Experimental Biology Journal*, **2**, 3043–53.
Norris, V. (1989). *Cell Calcium*, **10**, 511–17.
Williams, R. J. P. (1989). In *Calcium binding proteins in normal and transformed cells* (ed. R. Pochet, D. E. M. Lawson, and C. W. Heizman), pp. 7–14. Plenum Press, New York.

Morphogenesis

A summary of mathematical approaches to morphogenesis which does not refer to chemical species is given in *Bulletins Mathematical Biology* (1990) **52**, 119–97. Morphology of plants is similarly described by Van der Linden, F. M. J. (1990). Creating the phyllotaxis of plants. *Mathematical Biosciences* **100**, 161–200.

22

Man's use of the chemical elements in biological environments

22.1	Introduction	531
22.2	Man's agriculture	534
22.3	Industry and pollution	535
22.4	Toxicity	536
22.5	Diseases associated with abnormalities in concentrations of required elements	540
22.6	Poisons and evolution	543
22.7	Elements in medicine	544
22.8	Conclusion	550

22.1 Introduction

This chapter concerns man and his use of chemistry in the context of biology and in this sense it forms an appropriate conclusion to the book. Man's activities are clearly the most recent evolutionary development related to the chemical elements. In the last 10^5 years (contrast the 1 to 4×10^9 years of biological development) man has become the first self-conscious and powerful organism on earth and perhaps in the universe. The consequences are full simultaneously of potential benefit and risk, as is ever true of biological development. Man must, it is inevitable, much though he may wish to draw back, play the new game of self-sconscious evolution, but he cannot ignore the old rules. He cannot play with genes or fully exploit chemistry in industry in the way he has started to do today (the man-made evolution) without running into older problems concerning biological evolution. Not only his own but other forms of life are becoming at risk, both by accident and design, so he must proceed with care, not only in manipulating DNA but in handling chemical elements. Man as a species has plenty of time and only selfish exploitation is his enemy. We have understood chemistry for less than one hundred, 10^2 years; we have a fair inkling today of what biology is about, but this knowledge is much less than 10^2 years old. We need to consider that man

may survive for a very long time and it is then necessary to look at the interrelationship of chemistry, as used self-consciously by ourselves, and the nature of living systems where all the elements of the periodic table are involved. Availability, as described in Chapter 1, is now partly under the control of man and this control will become more and more powerful. It affects the atmosphere, the land surface, and the waters in and around the land masses. We do not believe it is possible for man to proceed any longer in a naïve way or to try to turn his back on this knowledge. Biological systems cannot adapt at all easily or rapidly either to a rapid change of energy input (temperature), a rapid change in DNA, or a rapid change of chemical material input. They will do their best, but the outcome of rapid change is as unpredictable as was the effect of the advent of dioxygen 2.7×10^9 years ago. Looking at what happened then, we stress that the involvement of the chemical elements in life is symbiotic with the development of genes. We must be concerned, therefore, with the way man uses those chemical elements, directly or in compounds, which are either unnatural in themselves or unnatural in the concentrations in which they are presented to biological systems. There is no hard and fast distinction between the discussion of these elements under a variety of headings: pollutants, poisons, drugs, medicines, agricultural chemicals, and even foodstuffs. Before man began to manipulate the use of chemicals there had always been present in the natural world elements and compounds which would today fall into more than one of the above classes depending solely on dose. It is only to some degree that man has altered the supply of the chemical *elements* in his surroundings (Table 22.1). We must recognize too that chemical *compounds* have an almost infinite variety and nature, living or inanimate, makes relatively few. This implies that the synthetic opportunities are very large indeed and that they too will be full of both benefit and risk simultaneously. It is an analysis of the benefit and the risk of introducing chemicals which is the subject of this chapter. The first feature to recognize is that very few chemicals are without any potential value if they are used in controlled amounts and are administered to man, animals, or plants locally. A poison may kill an 'unwanted' organism, as much as it may

Table 22.1 Some examples of element availability altered by man

Element	Change	Cause of change
Lithium	+ (Very small)	Medical applications
Boron	+ (Considerable)	Intensive agriculture
Carbon: carbon dioxide	+ (Large)	Coal and oil burning
Carbon: methane	+ (Large)	Intensive cattle farming
Nitrogen: nitrate	+ (Large)	Excess fertilizer
Oxygen	(+)(−) (Large)	Depletion of ozone layer
Fluorine	+ (Considerable)	Drinking water additive, control of tooth decay, refrigerant gases
Aluminium	+ (Considerable)	Sewage treatment, acid rain
Phosphorus	+ (Considerable)	Excess fertilizer and water softening
Sulphur	+ (Considerable)	Fossil fuel burning in the power industry (acid rain)
Chlorine	+ (Considerable)	Organic compounds in drugs, pesticides, polymers
Many metal elements	+ (Small)	Mining and industrial uses (low solubility)

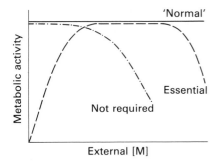

Fig. 22.1 Diagram of the effect of external (M) on metabolism. (Normal is the satisfactory optimal activity.) Essential elements can be toxic at high concentrations; 'not required' elements can be poisonous at low concentrations but may be therapeutic.

kill a human being. The effect of virtually all chemicals is also dose-related following a pattern as in Fig. 22.1. The figure shows that all compounds are deleterious if taken in very large doses, some are absolutely required in a certain range of concentration (by a given organism), while some appear to be full of danger at any dose level, but this does not mean that they are equally dangerous for all organisms or all cell types. Consider an extreme example, the effect of radioactive elements, which must be damaging to all DNA.

A radioactive element has great value in medicine since it can be used to screen for diseases (Table 22.2) and to kill tumour cells locally. On the other hand, radioactivity is a known cause of mutations and of diseases, e.g. leukaemia and other cancers. However, we cannot even state that it is desirable to prevent mutation generally since this is the mechanism which allows new organisms to evolve. The genetic pool will be exploring new possibilities for its advantage all the time in order to overcome disadvantageous circumstances which exist or develop on the planet Earth. The conclusion is simple: radioactivity can be beneficial (for man) when used carefully; otherwise it is dangerous for him, but its value and danger is unknowable for bacteria. We all know the fear that man's development of nuclear power (and bombs) could lead to the destruction of all life through radioactivity although this is extremely improbable. In a similar way some elements are generally classified as dangerous, just as radioactive elements are often so labelled. Obvious examples are lead, mercury, and thallium, no matter in what compounds they occur, but there are, or have been, recipes for the use of each of these elements by man in his battle against organisms which infect him or 'his' animals or 'his' plants. A risk is taken in using radioactive elements to shadow the interior of a patient or to kill tumour cells and a risk is taken in using any element in any compound in medicine. Such is the extreme sensitivity of living systems and the degree of risk is, in the extreme, specific to species or even to individuals (see Section 22.5.2).

Table 22.2 Some uses of nuclear diagnostics

Radionucleotide	Target organ	Carrier
^{127}Xe	Lungs	Air
^{99}Tc	Lungs	Aerosol
^{201}Tl	Cardiovascular system	Blood
^{99}Tc, ^{123}I	Central nervous system	Amines in blood
^{67}Ga, ^{111}In	Liver	Citrate, transferrin in plasma
^{99}Tc	Gastrointestinal tract and renal system	DTPA complex sulphur colloid
^{99}Tc	Vascular lymph systems	Perrhenate
^{99}Tc	Skeletal systems	Polyphosphate or phosphonates
^{123}I	Thyroid	Toluidine blue
^{67}Ga	Tumours?	Citrate

In order to illustrate the problems for living systems as well as the benefits, especially for man, we shall proceed as follows. First, we describe the introduction of elements into the biosphere by man on a massive scale in agriculture and through industry. We then ask about the toxicity of the elements so introduced and ways in which biological systems can adapt to

them. On a smaller scale man uses elements in medicines and we shall attempt to relate this activity to the account of the benefits and risks due to man's intrusive activity in the biosphere.

22.2 Man's agriculture

It is of little use to protest against the treatment of soils with chemicals. There are large areas of soil on earth which are not suitable for agriculture due to man's activities and many equally large areas which were always so poor in the relative availability of elements that they prevented plant growth or only allowed plant growth which did not sustain animal (human) life in anything like an optimal manner. Simple examples are the selenium-deficient soils of China and the molybdenum-rich soils in USSR and even in Britain which do not allow adequate intake of copper. Furthermore, the intensity of modern agriculture will not sustain long-term farming in general without the addition of chemicals. For all these reasons man has to interfere by chemical treatment of soils. Many soils need dressings of boron or manganese and, of course, potassium, phosphate (Fig. 22.2), nitrogen, and calcium have been applied for centuries. The use of cobalt licks in sheep farming has a long history too. These treatments must be continued, but there is always a risk of overtreatment and a loss of balance in availability. Too much nitrogen causes nitrate problems; too much lime reduces availability of most trace metals; too much acid phosphate gives rise to the release of aluminium; and too much sewage sludge introduces a number of unusual elements. This implies that constant monitoring is essential. Just as a patient is 'titrated' with lithium (see Section 22.7.2) so soils must be titrated with a variety of chemicals. With great care only good should be brought about, but it is essential that we come to understand the involvement of inorganic elements in plant life.

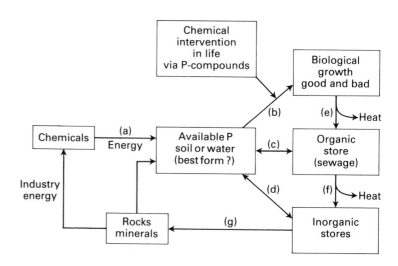

Fig. 22.2 Some of the complex relationships between use of an element, here phosphorus, by man or other living forms and the inorganic chemistry of waters and rocks.

22.3 Industry and pollution

Man has come to consider it his right to have both comfort and pleasure. These desires require great chemical activity and therefore affect the availability of elements, e.g. through mining and working of metals. We must try to summarize an acute problem which is developing from our use of elements which do have a significant role in biology, as well as of those that do not. Man, through huge exploitation of energy is adjusting the C, O, H, N chemistry, e.g. the CO_2, O_3, CH_4, N-oxide, etc. levels of gases in the atmosphere (Fig. 22.3). The energy is used for the increase of comfort and entails massive redistribution of these elements on the earth's surface. The way energy is generated using fossil fuels also generates acid rain, which mobilizes aluminium in particular. Here fish are extremely susceptible. Man also uses large quantities of a variety of heavy metals, including many 'b'-class (soft) elements in his materials, e.g. Cd, Pb, and Sn, as well as elements of intermediate character such as Ni, Cu, and Cr which are known to generate potential risks. The problem with some of these elements is that they can form

Fig. 22.3 CO_2 and CH_4 levels taken from ice core measurements showing coupling with ambient temperatures (greenhouse effect).

relatively long-lasting compounds with organic molecules and they accumulate low in the food chain. All of these elements have always been present in the environment, e.g. Pb^{2+} and CH_3Hg^+, but man, through industry, alters not only their relative availability but also the form in which they are dispersed, e.g. $Pb(C_2H_5)_4$ and $Ni(CO)_4$. Another compelling example is the use of fluorine compounds in refrigerators which compounds, when dispersed, damage the ozone layer. Once again it is a question of optimal use against the background of hazard. Man will not forego the value to himself of, say, extracted nickel (it is no good wishing for this), but he must limit to a minimum the accompanying problem of damage to living systems. This is particularly obvious in the light of the discussion of homeostasis. If we have described homeostasis correctly as *a balanced system of elements in cells*, then completely foreign compounds will always be a source of risk and even very ordinary elements taken in wrong amounts must stress the homeostatic balances (see Fig. 22.4).

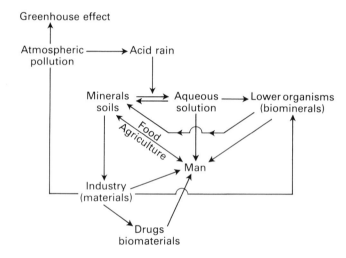

Fig. 22.4 A generalized scheme of the way in which man's industry and his environment interact with man as a living organism.

An excellent further example is provided by the problems met in using alkyl tin reagents against fouling of ship's hulls by barnacles. The compounds were extremely successful, but simultaneously they caused great damage to other shellfish which are important parts of the food chain reaching up to man since man uses the sea for agricultural purposes. We cannot redistribute any chemical anywhere without taking risks and this demands the constant monitoring of the levels of toxic elements and chemicals globally and locally.

22.4 Toxicity

22.4.1 The 'a'-class (hard) metals

Following the previous discussion we can attempt to generalize the risks referring to metal elements separately from non-metals. First, we can immediately divide metals into three groups. In the first group there are the

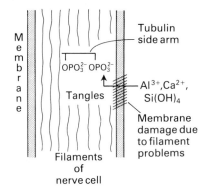

Li	Be		
Na	Mg	Al	
K	Ca	Sc	(Ti)
Rb	Sr	Y	Zr
Cs	Ba	La	Hf
Fr	Ra	Ac	Th

The 'a'-class, hard, metals

Fig. 22.5 The complexity of interactions between P, Ca, Al, Si, and neurofibrillar structures in Alzheimer's disease.

metal ions of groups IA, IIA, and IIIA of the periodic table (see margin). These metals have a single oxidation state and, with the exception of Be and Al, they bind weakly to organic ligands. Moreover, amongst them, Na^+, K^+, Mg^{2+}, and Ca^{2+} are homeostatically controlled so that biological systems are well protected against excesses of these elements, e.g. the organisms in the sea. Protection against Rb^+ and Cs^+ or Sr^{2+} and Ba^{2+} is not so secure but biological systems generally do not suffer exposure to them. They can cause problems, however, and, for example, Ba^{2+} begins to come close to K^+ in size, giving it the ability to block K^+ channels. Even so it is not usual to think of any of these elements as toxic except at very high dose. (Here we do not refer to problems associated with radioactivity such as those which developed after the Chernobyl disaster.) When we turn to Be^{2+} and Al^{3+} grave problems develop. In the case of Al^{3+} biological systems have always managed to avoid this cation, although we are not sure how this is done, and we may assume that it is only toxic in high dose, e.g. in old-fashioned dialysis machines. (Note that the association with Alzheimer's disease is more likely to be due to an interaction of aluminium salts with a pre-existing biological defect than a direct effect of a toxic feature of aluminium chemistry (Fig. 22.5).)

Unfortunately, the massive acidification of waters due to the burning of fossil fuels releases aluminium from soils and greatly damages plants and animals (fish) which suffer direct contact. There is no simple way around the problem except the avoidance of acid gases, SO_2 and NO. Too much stress on temporary benefit for man leads to long-term hazards. Beryllium is a different case; biological systems have had no opportunity to develop protection against this *rare* element and problems of beryllium binding, especially to phosphate-containing systems, are numerouss (Fig. 22.6). Beryllium is very toxic and in industry it must be used with great care or avoided altogether.

22.4.2 The intermediate metals

If we continue with the division of the metals following the periodic table we come to the second group which binds in a rather different way and more

Fig. 22.6 Ways in which Be, Al, and V can interfere with phosphate metabolism.

The first transition metal series

strongly, but with which biology is familiar. These metal ions include all those of the first transition metal series from Ti to Ga (see margin and Fig. 1.1). All these metal ions can be successfully handled by biology, except in relatively high concentration. However, two new points need to be made.

1. These metal ions can also be held in fixed chemical species without exchange. Biology itself handles cobalt virtually only as vitamin B_{12}, but it handles iron as free iron and in a variety of hidden forms in different haem complexes. Thus man can introduce these metals equally in 'hidden' states or free states when they will behave in very different ways.

2. These metals have variable valence so that biology may well be familiar with chromium as the ion, Cr^{3+}, but may have had no experience of the CrO_4^{2-} anion. The latter, as a compound, is described as *toxic* but chromium (III) ions as such are not. However, the introduction of CrO_4^{2-} into biology is probably a particular way of generating Cr^{3+} bound in a special site from which it will not exchange.

Amongst this group of elements are those used most massively in insoluble forms by industry, e.g. in steel. There is little risk here since the elements themselves are insoluble in these forms; the care that must be taken concerns not the product but the processing which can release small amounts of toxic elements, e.g. selenium and arsenic. One or two of the elements have particular risks associated with them, e.g. vanadium mimics phosphate (Fig. 22.6).

22.4.3 The 'b'-class (soft) metals

The third group of metal ions includes most of those with which biology has had little or no previous knowledge; the metals run from Mo and W to Sb and Bi and then to the transuranic series in the periodic table (Fig. 1.1). Here we come to the most 'b'-class (soft) elements, which are often placed in a triangle on their own in the table (see margin). Introducing them into an organism (Table 22.3) has its parallel with the introduction of a newly synthesized organic-molecule drug or poison. Many of these elements have the same characteristics as those of the previous section, i.e. variable valence and an ability to remain in a site but not necessarily in very slow exchange. The novel feature is that their binding may involve only two or even one organic donors so that, in effect, they begin to resemble the non-exchanging chemical combinations found for non-metals. We even saw this property emerge in molybdenum biochemistry. Thus, carbon does not exchange readily from $-S-CH_3$ binding and mercury is held strongly, though not so strongly, as $-S-Hg-CH_3$. This resemblance can be seen again in a comparison between the antitumour activity (toxicity against tumours but also against all cells) of dialkylating mustard gases and cis-$Pt(NH_3)_2Cl_2$. Both drugs cross-link DNA through two bonds which show slow exchange. Cis-platinum is effectively a dialkylating agent (see Fig. 22.7).

Biological systems are very vulnerable to poisons, toxic agents, which can react at single donor centres or even a pair of such centres since single Lewis bases, $-S^-$ or imidazole, the major attacking centres of enzymes, are usually fairly exposed and, of course, they occur commonly in most proteins. Thus, as well as the binding of Hg^{2+}, we must be aware of the risk from ions such as Tl^+

The 'b'-class, soft, metals

Table 22.3 Effects of excesses of non-biological metals on humans and animals

Metal	Effect of excess	Observations
Al	Implicated in dyalysis encephalopathies and in Alzheimer's disease	Interaction with phosphate? Cross-linking of proteins (Fig. 22.6)
Cd	Reduction of effective filtration capacity of glomerulus (renal toxicity)	Blocks sulphydryl groups in enzymes and competes with zinc for its sites. Stimulates metallothionein synthesis and interferes with Cu^{2+}, Zn^{2+} metabolism
Hg	Damage to the central nervous system; neuropsychiatric disorders	CH_3Hg^+ compounds are lipid-soluble
Pb	Saturnism; injuries to the peripheral nervous system; disturbs haem synthesis and affects the kidneys	Pb^{2+} may replace Ca^{2+} with loss of functional (and structural) integrity. Replaces Zn^{2+} in δ-aminolaevulinic acid dehydratase; reacts with sulphydryl groups
Tl^+	Poisonous to nervous systems; enters cells via K^+ channels	Tl^+ is similar to K^+ and binds more tightly, including to N and S ligands

and Pb^{2+} which are very toxic since they can bind in much the same places as K^+ and Zn^{2+}, although differently, and we must state firmly that they and their compounds must not be released into the environment no matter what their potential value in industry.

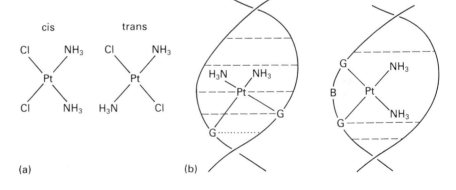

Fig. 22.7 (a) The formulae of (left) cis- and (right) trans-diamino platinum dichloride and (b) two possible modes of binding to DNA strands.

(a) (b)

22.4.4 Non-metal elements

It is easy to give an initial example of non-metal toxicity. There was a long period in evolution when there was virtually no dioxygen in the air. Its generation by some organisms must have been a major disaster for many others while some adapted. The history of the subsequent use of dioxygen in biology is not different from that of rejected Na^+, Cl^-, or Ca^{2+}. In fact, dioxygen always presented and still presents a risk for life. It generated and still generates highly poisonous elements such as free copper ions and aggressive chemicals such as OH^{\bullet}. We ourselves recognize the problems of the oxides of sulphur and nitrogen (SO_2, NO) and so on from burning oil and coal and the

The non-metals

presence of these non-metal poisons are and were to some extent an inevitable consequence of the generation of dioxygen. The organohalogen compounds, produced by oxidative reactions, are no better and the massive production by several common algae probably for protection and defensive purposes illustrates the point that problems existed before man's industry. War gases provide further examples and anaesthetics many more, since anaesthetics are mild poisons. An understanding of what may or may not be done within reasonable risk levels has to guide action by man, for example, in fluoridation of drinking water or in using NO_2^- to preserve and to improve the presentation of meat products.

22.5 Diseases associated with abnormalities in concentrations of required elements

22.5.1 General examples

We have now given some indication of general toxicity problems. Next we use diseases to illustrate what can happen if the supply of *required* elements is modified (Table 22.4). Here we exclude bacterial and viral invasions. There is a number of diseased conditions which may stem from abnormalities in the uptake of particular elements but there is a greater number of diseases where the abnormality is associated with an unusual pattern of elements generated by a metabolic fault. The two are distinguishable by the effect of adjustments to the diet since in the first class (primary deficiency) the disease should be solved by directly increasing or decreasing supply of the element. In agriculture and animal breeding many problems have been solved in this way by controlling the supplies of a wide variety of elements. It is often the case that lack of availability in soils passes up through plants to animals including man. Examples are given in Table 22.4. On the other hand, excess availability of one element may also cause problems both directly and indirectly by limiting a second element (secondary deficiency). We are not concerned here with complicated organic molecules, vitamins, and essential amino acids, although they exemplify the same principles. Occasionally, in fact, the two are seen together, e.g. where stomach bacteria of sheep lack cobalt the sheep become vitamin B_{12} deficient. When man becomes vitamin D deficient all his calcium uptake malfunctions.

We turn next to diseases caused by faults in the apparatus closely linked to the homeostasis system as we have described it. Given that homeostasis is a complex interactive system involving many elements and the supply of energy, we must expect diseased states of many kinds to be reflected in levels, even localized levels, either of particular or of several elements. Our analysis has indicated that an interactive set of elements is that of phosphate, calcium, and zinc, and hence it is likely to go astray co-operatively.

A particular case is the lack of vitamin D which causes brittle bones (rickets), but it is clear that sufferers are defective in many ways other than in their bone structure. It is all very well looking upon this as a calcium deficiency, but the connections from calcium and directly from vitamin D to many metabolic paths are becoming very clear (see Table 21.6). Now, this is an example of the effects of deficiency in the handling of non-redox elements,

Table 22.4 Examples of metabolic disturbances and diseases in humans and animals caused by deficiency or excess of trace biological elements

Element	Effect of deficiency	Effect of excess	Observations
F	Increased incidence of dental caries	Fluorosis (teeth)	Favours structural resistance of teeth
Si	Growth depression; anomalies in bone and cartilage	Silicosis (lungs)	Possible cross-linking of proteins
V	Growth depression; failure of reproduction	Unknown except at high doses	Possible effect on Na^+, K^+-ATPase; vanadate competes with phosphate
Cr	Impaired glucose tolerance; elevated serum lipids; corneal opacity	Toxic as CrO_4^{2-} (through reduction to Cr^{3+}); lung cancer; contact dermatitis mostly in men	Potentiates insulin(?); template for a component of glucose tolerance factor (?)
Mn	Skeletal and cartilage defects; depressed reproductive male function	Manganism (locura manganica)—psychiatric disorders (memory, speech, hallucinations)	Role in the synthesis of chondroitin sulphate; glycosylation
Fe	Anaemia; general weakness	Haemochromatosis	Component of haemoglobin and enzymes of respiratory chain
Co	Anaemia; anorexia; growth depression; white liver disease (in sheep)	Heart failure; hypothyroidism	Component of vitamin B_{12}; involvement in synthesis of haemoglobin
Ni	Growth depression; impaired reproduction; prenatal mortality (in animals)	Lung cancer; contact dermatitis, mostly in women	May replace other transition metals in their sites
Cu	Anaemia; ataxia; defective melanine production and keratinization	Liver necrosis, e.g. in Wilson's disease; hypertension	Component of oxidative enzymes involved in haem synthesis; cross-links elastin and collagen
Zn	Anorexia; growth depression; sexual immaturity; hypogonadism; skin lesions; hyperkeratosis; depression of immune response; acrodermatitis enteropathica; teratogenic effects	Relatively non-toxic except at high doses	Component of many hydrolytic enzymes; component of zinc fingers—affects gene expression
As	Impairment of growth and reproduction (animals)	Often poisonous	Blocks sulphydryl groups of enzymes
Se	Endemic cardiomyopathy (Kesham disease) and osteoarthropathy (Kashin's Beck disease); nutritional muscular dystrophy (ruminants); exudative diathesis (poultry)	Selenosis—hair and nails loss; blind-staggers and chronic alkali disease (grazing-cattle)	Component of glutathione peroxidase
Mo	Growth depression; defective keratinization; hyperocipurinaemia	Anaemia; persistent dysentery (grazing animals); gout-like syndrome (USSR)	Interference with Cu^{2+} absorption
I	Goiter; cretinism	Goiter; thyrotoxicosis	Component of thyroid hormones

but redox elements can also cause extensive disease. In the wrong place redox-active elements, e.g. iron compounds, or, say, chlorine covalent compounds can damage DNA. One obvious possible consequence is aberrant protein phosphorylation (due to mutation) and, in the light of our discussion, we would expect that this could cause a running series of problems in phosphate, calcium, and zinc handling, with knock-on effects such as loss of growth control, loss of morphology constraint, adjusted metabolism, incorrect assembly and disassembly of connective tissue around affected cells, and so on. This condition of a growing tissue is recognizable as concerning aberrant growth, e.g. a cancer, and following our discussion we have to say that the damage is virtually irreparable except through cell death. We may well find

that cellular levels of both zinc and calcium have been changed in such cells amongst many other adjustments. There are very many wrong features of a cancer cell.

Another disease of a related kind would be seen in the failure of appropriate growth generally. We know such stunted growth diseases can arise through zinc deficiency, but more notably zinc deficiency leads to failure to go through puberty. This is both a morphogenic and a homeostatic disease since growth pattern and growth rate are affected. Now failure of local growth could come about in other ways; indeed, unless muscles receive the correct set of nerve impulses they change their character—slow and fast muscles can be interchanged by application of incorrect pulsing and here we would say that it is the change of calcium steady states which has altered the cells' proteins. The filamentous structure is readjusted, amongst other things.

In this context let us turn to the diseases of old age where there is good reason to suppose that problems of calcium distribution and utilization abound, e.g. hardening of arteries, incorrect mineralization, and osteoporosis. (Note that if our description of homeostasis in the introduction to Chapter 21 is correct then the failure of proper element distribution in old age is inevitable. Old age is not really a disease but a stage in living.) These are all extracellular problems which are easily observed, but other features of growth could go astray if the intracellular calcium rose through increased cell membrane permeability. We would then expect changes in protein phosphorylation, aberrant tubulin constructs, some additional proteolysis, and so on. One disturbance which matches this set of changes is Alzheimer's disease (Fig. 22.5). Inability to memorize recent events is, in the terminology of this book, inability to generate *kinetically controlled cellular* growth in response to messages. A way in which this could come about is through too much calcium entering a cell on stimulation. The entrance of aluminium probably is an event after cell death, but aluminium excess will always increase problems with calcium (Fig. 22.5). Finally, there are grave problems with excess of iron, especially in patients who require repeated blood transfusions. Cases are known where iron as a form of ferritin is rejected by cells and forms a very harmful deposit, haemosiderin, all over the body (Fig. 12.4).

We have stressed here that diseases can arise through the inability of homeostasis to manage considerable excesses or deficiencies of required elements. There is also a large literature on diseases caused by the introduction of elements such as mercury into the human food chain. These are cases of simple toxicity.

22.5.2 Toxicity and the immune system

An intriguing way of seeing the differences in toxicity between the different metal ions may well be approached from their effect on the immune system. The discussion is now at the level of the individual. Here the possibilities are:

1. Stimulation of antibodies to act against metal ions or a combined form of the metal ions. This response is possible for all elements retained for reasonable lengths of time in certain proteins or polysaccharides. Sensitizations, allergic responses, are in this class and are known for many metals, including especially Ni, Cr, and Pt. Note that such responses are

even known for excess Zn, Cu, or Mn, elements which are very familiar in biology. These diseases of the immune system are quite distinct from the following possibility.

2. Auto-immune responses where the metal element in some combined form causes the immune system to attack 'self'-proteins. This reaction is thought to be due to the histocompatibility response proteins (HRP) which until perturbed are able to bind and present peptides of 'self'-proteins without initiating an antibody response. This is a necessary feature of the immune system since 'self' must be recognized without response while 'non-self' must generate a protective response. The auto-immune diseases are caused by damage to 'self'-recognition, such that a response is developed to 'self'. The major metals which cause an auto-immune reaction appear to be Au (even Au drugs) and Hg, both of which are in the third class of metals above. It could well be that they bind to the –SH groups of the HRP of the normally protective system, using either one or at most two donor groups.

A special point here is that the immune response is different for different individuals. The immune system develops as a consequence partly of genetic inheritance but partly of experience or exposure. There is then no way of generalizing risk from elements and their compounds.

Now, instead of looking at all the introductions of excesses of elements or of foreign elements into the biosphere as hazards, we wish to look at how biology might adapt by reference to how it has done so in the past.

22.6 Poisons and evolution

In the main chapters of the book we have referred to one element after another where we have reason to believe that early in the course of evolution a presently essential element was associated with hazard more than value. Even such common ions as Na^+, Cl^-, and Ca^{2+} fell into this class since they had to be excluded to some degree from cells. As organisms developed then these elements were undoubtedly used also in the development of poisons, e.g. based on ring chelates for Na^+ (Fig. 2.6), and the use of them in external protective devices also followed. An obvious example of protection is the precipitation of calcium carbonate around some single cells. It is almost a general requirement in evolution that any newly evolved general poison will be guarded against and hence poisons developed by a particular organism must be compensated by the evolution of a defence mechanism by the same organism against this very poison. Subsequently, the poison could be used in a refinement of that cell's activity, e.g. in protection or in signalling. We then looked upon the signalling and protective functions of even sodium, chlorine, and calcium as having developed from poisons. Table 22.5 gives further examples where the obvious point is the development of value within multicellular organisms. We have generalized this point to the whole of secondary metabolism in Chapter 21. It is necessary for man to see this quality of evolution and to see that adaptation is peculiarly and obviously easier for simple prokaryotes than for advanced organisms. It has been demonstrated in bacteria that protective

Table 22.5 The evolving roles of rejected elements

Rejected element	Use as a poison	Use as a signal, etc.
Na^+	Cyclic antibiotic carriers	Channel entry for messages
Ca^{2+}	?	Channel entry for messages
Halogens	Halogen organic compounds	Thyroxine
Nitric oxide	General	Muscle relaxant
Dioxygen	General	Metabolite
Copper	Antifungal	Extracellular catalyst

NB. Compare the evolution from organic chemical poisons to transmitters and hormones in secondary metabolism.

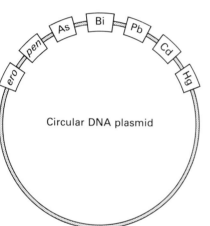

Fig. 22.8 A series of resistance genes built on to a bacterial plasmid. *ero* is erythromicin; *pen* is penicillin.

responses exist to a whole range of elements and compounds, e.g. mercury and penicillin. Such is the versatility of the mechanisms available that it has been possible to grow bacteria which carry on one DNA plasmid a range of genes the protein products of which protect against some 10 to 15 elements (Fig. 22.8). Now, the ways in which bacteria detoxify their environment sometimes prove to be lethal for other organisms. For example, metallic mercury does little harm to man, but the methylation of mercury ions to CH_3Hg^+ by some bacteria generates an extremely powerful poison. In the light of these remarks we can ask but cannot easily answer, 'Can man propel evolution in new directions by design or by accident?' We have to think on an immense timescale, not just one or two hundred years. A simple experiment would be to observe evolutionary changes at $pH = 6.0$ of an artificially doped sample of the sea and its lower organisms over, say, 100 years. Some of these detoxifying systems are also found in higher organisms, e.g. plants.

The conventional detoxifying activity of higher organisms uses removal into an organ or a vesicle where the element may just be precipitated. In plants unwanted elements such as nickel often find their way into the cell vacuoles. In animals such as snails there are found to be nodular precipitates of phosphates. Higher animals appear to detoxify to some degree through the kidney. For example, high concentrations of cadmium are often located there. We have also described the complicated handling of Zn, Cu, and Cd by metallothionein in Chapter 11. The ways by which bacteria become adapted to poisons are so different from the ways higher plants and animals can use that we must heed a general warning: increasing the number of environment chemicals increases the variety of resistant bacteria.

22.7 Elements in medicine

Against this somewhat gloomy account of the influence of mismanagement of element chemistry in the environment and noting that mismanagement in general is not necessary, being the result of haste and avarice, we introduce a more cheering approach, that of the uses of elements in medicine. The

Table 22.6 Medical uses of unusual elements in inorganic drugs

Element	Medical use
Lithium	Hyperactivity drug
Boron	Neutron capture therapy
Fluorine	Tooth protection
Aluminium	Anti-acid
Chlorine	In several antibiotics
Copper	Anti-inflammatory
Zinc	Wound healing
Platinum	Anticancer
Gold	Arthritis; anti-inflamatory
Bismuth	Drugs for gastric diseases
Arsenic	Antisyphilis (now obsolete)
Antimony	Antischistosomiasis
Barium	X-ray diagnosis (gastrointestinal tract)
Mercury	Antiseptics, diuretics (now obsolete)
Selenium	Antiseborrhoeic
Tin	Antiboil and anti-acne drugs
Gadolinium ⎫	Magnetic resonance
Manganese ⎭	imaging

extensive use of a wide range of minerals in medicine is a very ancient practice which fell into disrepute through misunderstanding and resulting poor practice. Many elements were labelled as poisonous and the medical profession trained their research staff to stay away from the heavy elements in particular. This may well be a misguided approach now that we understand to some extent the nature of interactions between inorganic and organic molecules (Chapter 2). It could be that many viral and auto-immune diseases can be tackled via the use of unconventional drugs. We give some examples in Table 22.6 following the lines of discussion used in Section 22.4 on toxicity.

22.7.1 Hard, 'a'-class metals

For a detailed discussion of elements in medicines we go back to the periodic table first and observe that many elements have virtually no essential role to play in life. Amongst the lightest elements there are only Li, Be, and Al. Earlier in this book we have mentioned that Li is well-known in medicine while Be and Al are generally thought to be poisonous. All these three elements belong to the group called 'a' or 'hard' ions which bind to O-donors better than do the hard metal ions which biology uses—Na, K, Mg, and Ca (see Fig. 1.1). We can well imagine that Li, Be, and Al find their way to block effectively O-donor centres normally found to interact with, say, Mg^{2+}. Why a particular effect is favourable in moderating a disease (Li for hyperactivity) or unfavourable (Al in Alzheimer's disease) is not open to full explanation yet.

Let us take lithium as a more detailed example. It is the 'hardest' ion, most 'a'-class, in group IA of the periodic table. The effect of lithium salts on hyperactivity appears to have been discovered by accident while a study was being made of the effect of uric acid on rodents. The lithium salt, rather than the sodium salt, was tried in order to increase the solubility of the particular drug and it was observed that this salt had quite new effects. As a consequence simple lithium salts themselves were tested on rats and were shown to reduce physical activity. They were then tested on patients suffering from hyperactivity. Although the tests were positive, lithium was deemed to be unacceptable as a useful drug at first, since several people died from its effects. Later it was found that death was due to an incorrect dose (Fig. 22.1) and today lithium is used extensively and safely by a very large number of people who are 'titrated' until safe effective lithium levels are found for the individual person. The development of this drug illustrates the importance of dose levels, even down to the treatment of individuals. The emphasis must always be, as here, on risk/benefit analysis.

The effects of lithium on a developing fertilized cell of a higher organism are much less comforting. Indeed lithium is teratogenic—in some way it interacts with the communication network which allows the correct series of messages for morphogenesis. It may well be that it is its parallel chemistry to that of magnesium (note the diagonal relationship in the periodic table, Chapter 9) and the further connection to phosphate and calcium networks of communication which lie at the root of the problems. No matter what the mechanism, we see again how alterations of simple ionic salt levels disrupt or interact with the information flow from genes.

Similar stories can be told about the good, anti-acid, and bad, renal dialysis

Li	Be		
Na	Mg	Al	
K	Ca	Sc	(Ti)
Rb	Sr	Y	Zr
Cs	Ba	La	Hf
Fr	Ra	Ac	Th

The 'a'-class, hard, metals

diseases, of aluminium salts. The aluminium ion may well compete for the calcium ion sites under conditions of low pH. As with virtually all 'a'-type elements it is the free ion, Li^+ or Al^{3+}, which is responsible for biological influence since these cations exchange relatively readily, from salts or complexes, making their ligand partners of low significance. Here lies a chemical limitation in the ways of applying one-quarter of the elements of the periodic table in medicine. The other elements have a more 'covalent' chemistry and can be used in a great variety of ways, much as carbon, sulphur, or nitrogen compounds are used in drugs.

Before passing to the elements involved in more covalent attachment, we should not fail to mention again that the non-essential elements which are deleterious, toxic, or useful as drugs from amongst the hard metal ions are the very hardest available Li^+, Be^{2+}, and Al^{3+} in each of groups I to III of the periodic table. (Boron is a non-metal.) If we now go down the groups passing over the essential elements, Na, K, Mg, and Ca, we find that, at the extreme end of each group, the elements, Rb and Cs, Sr and Ba, and lanthanides, respectively, carry less risk or benefit except at a relatively high dose. We leave aside radioactive nucleotides of all these elements. Barium can affect potassium channels and strontium can alter calcium channels, but both Sr and Ba are used by several biological systems (Chapter 20). There may be useful ways of modulating activity using these elements in a parallel fashion to the use of lithium. Within group III there is some considerable tolerance to ions in the group later than Al(III) and the use of scandium and lanthanides in medicine has been discussed, though little has yet been developed. The only other hard cations of interest at pH = 7 are oxo-ions such as VO^{2+} and UO_2^{2+}. The authors are not aware of application of or hazards from this latter cation unless we return to the problems of radioactivity, but VO^{2+} has been the object of many studies and can, of course, be oxidized to HVO_4^{2-} when it competes for HPO_4^{2-} sites. All these hard cations have a tendency to accumulate in bones which are formed from hard anions, OH^- and PO_4^{3-}. One suggested way of removing poisonous (radioactive) very hard cations from the body is to treat with solutions of very hard anions, e.g. F^-, OH^-, HPO_4^{2-}, and $SiO(OH)_3^-$, or related organic anions, e.g. enolates or phosphates.

22.7.2 Soft, 'b'-class metals

Amongst the 'b'-class (soft) metal ions we know that Pb, Cd, and Hg in many salts are extremely poisonous, yet Pt, Au, and Bi are used in medicines. These elements all bind to N, S centres of proteins and/or to DNA. Pt is used in antitumour compounds and some Au salts find application as anti-arthritic agents. Many other elements, including Sn, Sb, and As, have been used or suggested for use in one application or another. No obvious line of thought leads one to say which will be most valuable since they can be applied in so many different chemicals, at various dose levels, against so many different targets. Here lies a big difference from the hard ions since exchange from many compounds of 'b'-class elements is extremely slow; it is a specific compound which is now of interest.

Consider platinum as a different but parallel example to lithium. Until around 1970 this element and all its compounds were thought to be of no

The 'b'-class, soft, metals

biological interest except for the known problem that some people were singularly sensitive to them and developed an allergic response. In the period since 1970 a family of drugs based on 'cis-platin', or cis-[PtCl$_2$ (NH$_3$)$_2$] (Fig. 22.7) has been developed and used extremely successfully against a variety of tumours. This is a remarkable story where discovery depended, as with lithium, on a series of accidents and where acceptance by medical authority was only achieved through great persistence against great resistance. The resistance was due to known and perceived risks of heavy elements including platinum in other compounds and at other dose levels— platinum had a bad name. One genuine problem was the side-effects, especially kidney damage, and there was the danger that the drug (and almost all other drugs are similar) could do more harm than good. There is an old adage that chemicals which 'cure' cancer also 'cause' cancer and here it is true that platinum compounds attack all DNA, but cis-platinum has turned out to be a remarkably useful drug. Careful dose control is essential and much experimentation has shown how many problems of the use of these drugs can be overcome.

Now, in this case, it is not the element or its simple ions which are effective but a specific chemical, namely cis-[Pt(NH$_3$)$_2$Cl$_2$] where the group Pt(NH$_3$)$_2^{2+}$ binds to DNA (Fig. 22.7). Trans-[Pt(NH$_3$)$_2$Cl$_2$] is of no value. As with many an organic chemical drug, once the parent of a family of chemical compounds is known then variations are possible on a substitutional basis within a structural motif. There has then been and will be a lot of development of platinum drugs. The lesson here is clear. While the 'a'-group metal elements have limited potential due to fast equilibration between compounds, those elements which give rise to robust chemical combinations and especially the very 'b' (soft) class elements, almost surely must have some compounds of value in medical or agricultural practice. A sufficient number of examples is known from the chemistry of Au (anti-arthritic drugs), Re (radiological mapping), Sn (anti-fouling agents), as well as platinum, for us to know the potential and the hazards. Detailed studies are now being made so that the bio-inorganic medical chemistry of a great variety of metal elements can be developed in a valuable way. The region of the periodic table shown in the opposite margin is likely to be particularly fruitful but this is also the region of greatest toxicity!

In passing we notice that the two most effective sets of metal elements for good or evil on a dose basis are at the extreme corners of the metals of the periodic table. The reasons are based on a lack of availability of the heaviest and lightest elements to biology in evolution through one cause or another (see Chapter 1) combined with extremity of chemical behaviour. We sense that evolution will be changed if man changes significantly the availability of such elements over long periods of time. The time concerned is the time for some species to have developed a value for the element concerned and a selective scavenger system.

22.7.3 Intermediate elements

Ti	V	Cr	Mn	Fe	Co	Ni	(Cu)	Zn	Ga

The first transition metal series

Very many of the intermediate elements are in the first transition series, and of these many are required elements. Of course, doses of required elements are

used in medicine to assist homeostasis and many of us have taken iron or zinc deliberately. Vitamin B_{12} is the essential required cobalt vitamin and the cure of anaemia using it is very well known. Manganese is also regularly applied in agriculture. These facts should not hide from us the possibility of developing drugs from the required intermediate elements and also from those rather similar elements which are not required, e.g. gallium (III) resembles iron (III). There is a vast covalent organometallic chemistry of these elements, but very few have been tested in medicine. One grave problem where this approach is being tried is that of inflammation and its relationship to connective tissue damage, which is of particular concern in old age.

The desirable synthesis and degradation of connective tissue under control of zinc and copper outside cells requires also the use of manganese (glycosylation), magnesium (sulphation), and iron (oxidation) in vesicles to give the precursors of this tissue (Chapter 21). At the same time the tissue is always at risk especially through oxidative free radical attack, e.g. by OH˙ (Fig. 22.9). Inflammation and joint damage may well be due to such ill-directed reactions. One of the problems is superoxide—an inevitable product of oxidation reactions which is well-countered inside cells by a variety of superoxide dismutases. Outside cells it is a menace, yet it is part of the ammunition which higher cells, in man for example, use to fire at invading organisms. The balance between protective value of the poisonous superoxide and the protection it provides is marginal and in older people its value and its risks are very balanced. To counteract it a variety of copper and gold compounds have been prescribed with variable degrees of success but there is no really good medicine yet. This is, clearly, an open field for research.

22.7.4 Removal of excess of metal elements

A different problem is the need to remove excesses of given elements which arises with particular classes of patients who suffer from excess of, for example,

Fig. 22.9 The possible involvement of many catalytic elements in the inflammatory response. The central area of connective tissue is surrounded by cells which release and pick up metals, Fe, Cu, Zn, Mn. The connective tissue is open to damage from 'protective' response against presumed enemies, e.g. bacteria, by the generation of superoxide. The metals may cause damage, as shown, or may clear superoxide. (See Sections 12.11 and 15.5.)

iron or copper, have been poisoned by others, e.g. Pb^{2+}, or were exposed to radioactive materials. Here the use of chelation therapy is often quite successful and a whole range of regimes is now available, from the application of EDTA solutions to more elaborate complexing agents (Table 22.7). Lower forms of life have a remarkable ability to develop stratagems to detoxify themselves which can be illustrated by remarkable work on the products mainly of plasmids in *E. coli*, Fig. 22.8. It is against this background that we must view the potentials of evolution eventually to make use of any element occurring in reasonable quantity and in any available form in its environment.

Table 22.7 Chelating agents for the removal of excess of metal elements

Chelating agent	Metal to be removed	Cases when required
EDTA	Various, e.g. lead, radioactive strontium, etc.	Saturnism (Pb), radioactive contamination (Sr), etc.
DTPA	Radioactive plutonium	Radioactive contamination (Pu)
D-Penicillamine	Copper, mercury	Wilson's disease; mercury poisoning
Desferrioxamine	Iron	Haemochromatosis
BAL (British-anti-Lewisite)	Arsenic	Poisoning by war gas (Lewisite)

22.7.5 Non-metal elements in medicines

The use of compounds of C, H, N, O, S, and P will not be discussed for obvious reasons. Just as the major interest in the metals in medicine has concentrated on heavy metals so the effects of the heavy non-metals have also intrigued man. For a long time bromine and iodine have been valued as therapeutic agents and the use of arsenic, antimony, and even bismuth in many guises is found in several early pharmacopoeia (see Table 22.6 and Fig. 22.10). A very interesting feature is that the halogens in combined form, organic halides, are used in very many drugs and poisons. The history of this use once again illustrates the mixed value of the use of foreign compounds since it is known that many halogen compounds are alkylating agents and through attack on DNA cause diseases such as some cancers. As already stressed, there are no safe elements and no harmful ones. One example is the widespread use of DDT where still today man is having to attempt a risky balance of deliberate injury to different species in the hope that very unwanted creatures, such as mosquitoes, can be eliminated and so reduce malaria and other transmitted diseases without destroying too much birdlife. The use of any element in any compound is a balance of help and harm. A good notorious example is arsenic which is the culprit in many murder stories while several of its compounds have long been used to protect wood from destruction by fungi and as a curative for syphilis.

The non-metals

22.7.6 Solid-state inorganic compounds in medicine

Everybody is familiar with the mercury-containing amalgams and gold

Fig. 22.10 The periodic table showing some old remedies and some new medical potentials using elements not common in biology. Compare the first description of the periodic table in Chapter 1. (Taken from Discher (1964); see 'Further reading'.)

compounds as fillings in damaged teeth. Today there is a wide range of other possible dental fillings including colourless phosphates. Again everybody is familiar with the use of metal plates in hip joint replacements, but many new more interesting possibilities are arising from the study of ceramics. The most desirable prosthetic material is one which is compatible with connective tissue in the zone where it is placed and is such that the organic tissues actually bind to it as if it were bone. A great variety of phosphate-containing glasses are under study for exactly this reason. In principle we wish to be able to stimulate bone growth at will since there are many people who suffer during considerable periods of later life with problems which develop from bone weakness. Just as in the development of drugs the inorganic chemist can enter the field of composite materials for medical purposes and he can also learn from the composite materials of biology how to make composites of use to man in other fields of activity.

22.8 Conclusion

Man's use of a huge variety of chemical elements in medicine needs development. There is little risk here and much to be gained by careful development. Man's utilization of chemical elements in agriculture is to a degree less secure. A higher degree of risk still is associated with industrial

activity to satisfy 'economic' drives and human pleasures requiring material wealth. It does not take much imagination today to see the difficulties man faces with the burning of fossil fuels and with the release of gases into the atmosphere. We are already in the century of the depletion of the ozone layer and of the greenhouse effect. But let us suppose that all these energy-related problems could be solved by direct utilization of atomic energy. Seemingly this would allow even more extensive use of materials and inevitably of the chemical elements. Here we shall meet new problems if we are not careful for we may contaminate vast areas with 'new' 'wastes'. Man's industry is 200 years old. Man is at least 100 000 years old and life is more than 3 000 000 000 years old. How much transformation of the availability of the elements can life withstand without drastic effects? It may pay to remember that, if the thesis of this book is correct, the growth homeostasis associated with many chemical elements is at its most sophisticated and, consequently, most sensitive in the highest of animals—man.

Further reading

The literature on the topics discussed in this chapter is overwhelmingly abundant and no attempt will be made to refer the reader to more than one or two specific papers regardless of their importance.

The following list, by no means exhaustive, consists mostly of comprehensive works on the main issues involved in the interaction of living organisms with the environment (nutrition, toxicity) and on the risks for evolution that rapid changes of the environment by man's activity will eventually imply. Some reference to texts on geochemistry and environmental chemistry are also included.

Adrian, D. C. (1986). *Trace metals in the terrestrial environment*. Springer-Verlag, Berlin.

Berridge, M. J., Downes, C. P., and Hanley, M. R. (1989). Actions of lithium: an hypothesis. *Cell*, **59**, 411–19.

Birch, N. J. (ed.) (1988). *Lithium inorganic pharmacology*. IRL Press, Oxford.

Bowen, H. J. M. (1966). *Trace elements in biochemistry*. Academic Press, London.

Bowen, H. J. M. (1979). *Environmental chemistry of the elements*. Academic Press, London.

Bowie, S. H. U. and Thorton, I. (ed.) (1985). *Environmental geochemistry and health*. D. Reidel, Dordrecht.

Ferguson, J. E. (1990). *The heavy elements: chemistry, environmental impact and health effects*. Pergamon Press, Oxford.

Frantz, B. and O'Halloran, T. V. (1990). The mercury responsive gene regulatory protein. *Bochemistry*, **29**, 4747–51.

Frieden, E. (ed.) (1984). *Biochemistry of the essential ultra-trace elements*. Plenum Press, New York.

Henderson, R. (1982). *Inorganic geochemistry*. Pergamon Press, Oxford.

Hughes, M. N. (1987). Coordination compounds in biology. In *Comprehensive coordination chemistry*, Vol. 6 (ed. G. Wilkinson, R. D. Gillard, and J. A. McCleverty), pp. 541–754.

Hughes, M. N. and Poole, R. K. (1989). *Metals and micro-organisms*. Chapman-Hall, London.

Lock, H. E. H. and Lock, C. J. L. (1987). Uses of complexes in therapy. In *Comprehensive coordination chemistry*, Vol. 6 (ed. G. Wilkinson, R. D. Gillard, and J. A. McCleverty), pp. 755–78.

Marschner, H. (1986). *Mineral nutrition of higher plants*. Academic Press, London.

Martin, R. B. (1988). *Bioinorganic chemistry of aluminium*. Metal ions in biological systems, no. 24 (ed. H. Sigel). Marcel Dekker, New York.

Nriagu, J. O. (ed.) (1984). *Changing metal cycles and human health*. Springer-Verlag, Berlin.

Prasad, A. S. (1982). *Clinical, biochemical and nutritional aspects of trace elements*. Alan R. Liss, Inc, New York.

Sharp, P. F., Gemmell, H. G., and Smith, F. W. (1989). *Practical nuclear medicine*. IRL Press at Oxford University Press, Oxford.

Underwood, E. J. (1977). *Trace elements in human and animal nutrition* (4th edn). Academic Press, London.

Williams, R. J. P. and Fraústo da Silva, J. J. R. (ed.) (1987). *New trends in bioinorganic chemistry*. Academic Press, London.

Also useful are the volumes in the series 'Medical and biological effects of environmental pollution', published by the US National Academy of Sciences, Washington, DC in 1972–77, some volumes of the series 'Metal ions in biological systems' (ed. H. Sigel), published by Marcel Dekker, New York, particularly nos 5, 18, and 20, as well as:

Seiler, H. G., Sigel, H., and Sigel, A. (ed.) (1988). *Handbook on toxicity of inorganic compounds*, 3rd printing. Marcel Dekker, New York.

An unusual book on the use of inorganic compounds as medicines is

Discher, C. A. (1964). *Modern pharmaceutical chemistry*. J. Wiley & Sons, New York.

Subject Index

abundance of elements 3–10
 on earth 8
 in living systems 3–5, 15 (*table*)
 in sea 10 (*fig.*), 15 (*table*), 16, 303 (*table*), 412 (*fig.*)
accumulation of elements 4 (*table*), 10
acid rain 272, 536 (*fig.*), 537
aconitase (Fe) 330 (*fig.*)
active transport 56–7, 232–3 (*fig.*), 248, 270
alkaline-phosphatase (Zn, Mg) 310
aluminium
 amphoteric character 416, 417
 condensation reactions 417
 toxicity, Alzheimer's disease 266, 537, 539 (*table*), 542
Alzheimer's disease
 and aluminium 266, 537, 539 (*table*), 542
 neurofibrillar structure and Ca,P metabolism 85, 537 (*fig.*)
amavadine 64 (*fig.*), 429
amino acids
 modified 167 (*table*)
 natural 151, 166–7 (*table*)
 structural 167
anions, *see* non metal elements
 balance 445
 concentration in cells 223 (*table*)
 free energy change on binding 61
 hydration 60–1
 incorporation 60 (*fig.*), 64
 kinetic binding traps for 63, 65–6 (*table*)
 mechanisms of selection 60–4
 Pauling's crystal radii 60 (*table*)
 theromchemical radii 61 (*table*)
annexins (Ca) 281
archaebacteria 400, 404 (*fig.*)
aspartate transcarbamylase (Zn) 186
atom transfer reactions 109–10, 413
arsenic toxicity 452
ATP
 balance with ADP and P 500
 interrelation with ion movements 134, 232, 501 (*fig.*)
 linking to calcium and cell compartments 510
 production from gradients 126–7
ATP dependent pumps (Mg) 56, 82, 236, 250, 284
ATP synthetase (Mg) 127 (*fig.*), 233 (*fig.*)
availability of elements 8–20, 513 (*table*)
 from sea 10
 and redox states 11–14, 17 (*table*)
axonal transport 85
azurin 390 (*fig.*)

bacteriorhopsin 133
barium 545 (*table*), 546
 as sulphate in sensors 484
beryllium
 interference with phosphate 537
 toxicity 537
bio-energetics 120–38
biominerals 467–93
 amorphous silica 446, 483–4
 bone mineral 293–4, 490–2
 chemical composition of 469 (*table*)
 containers
 lipid vesicles 471–2, 479
 proteins and polysaccharides 472
 control levels 469–70, 472
 for crystals 471 (*table*)
 for mineral phases 472
 for space 471 (*table*)
 forms and occurrence 468 (*table*)
 functions 467–69
 growth of crystals
 epitaxy 470, 474
 equilibrium growth 474–5, 480
 kinetic growth 475
 Lussac's law 473–4
 nucleation 473
 in homeostasis 516–17
 Ostwald ripening 475
 outside cell patterns
 banded patterns 487–9 (*fig.*)
 coccoliths 482–3
 external shells 487–90
 sensors 468 (*table*), 484–5, 493
 single crystals
 and cell morphology 476–85
 strontium sulphate spicules 476–81
 solubility products 18, 474, 516–17
 see also individual elements
biopolymers
 backbones 195–6
 chemical modification of 218
 competition of charged species for 179
 energization 217
 excluded volume 204
 feedback control of synthesis 150 (*table*), 505
 flow of 151
 linear anionic polymers 206
 roles 163–4
 sidechains of 197
 space-filling 199
 synthesis and degradation 149–50, 191–2
 viscosity 204–5
blood-clotting (Ca) 289–92

blood-plasma
 metal ions and anions in 223
 metal-ion carriers in 49 (*table*)
biotin 459 (*fig.*)
bismuth 538, 545 (*table*)
bone
 equilibria 491 (*fig.*)
 growth with fixed shape 490
 in homeostasis 516–17
 piezoelectric effect 490
 principles of growth 490–1
 problems of old-age 491, 542
 protein matrix 292, 490, 492
 synthesis and degradation 492
boromycin 59, 63 (*fig.*)
boron 58, 63 (*fig.*), 451
bromine 430, 451, 462

cadmium
 binding by metallothionein 305, 544
 toxicity 32, 539 (*table*)
calcium 268–97
 ATPase protein pump for 286 (*fig.*)
 binding sites in proteins 274 (*table*)
 binding to small ligands 37, 271–2
 biominerals 293–4, 482–3, 487–93
 proteins of 292, 491–2
 in blood clotting 289–2
 buffering 283
 carbonate 293, 481–3, 487–90, 492–3
 carbonate in shells 487–90
 in cell morphogenesis 295–6, 526
 chemistry 271–2
 coccoliths 482–3
 competition with Mg 246, 275
 concentration 270, 276
 control of intracellular activities 157, 269 (*fig.*, *table*), 510
 in digestion 289–90
 distribution and location 82 (*fig.*), 288–9, 294
 in homeostasis 512, 516–17
 integrated role with phosphate 510–12, 516–17 (*fig.*)
 interaction with filaments 282
 internal proteases 286
 intra and extra cytoplasmic balances 294–5
 kinetics of reactions 272
 mediation by 209
 in morphogenesis 527
 movements and currents 158, 284, 509
 multiple binding sites 294

calcium (*cont.*)
 oxalate in plants 383
 phosphates in bone 490–1
 and polysaccharides 206–10
 protein ligands 272
 protein phosphorylation 283, 510 (*fig.*)
 proteins
 annexins 281
 calbindins 284
 calmodulin 272–5, 283
 calpains 273, 275, 280, 286, 294
 extracellular 289–90
 gla proteins 273–5, 292, 491
 parvalbumin 283
 S-100 269 (*table*), 278 (*table*), 280
 trigger proteins 278–80
 troponin C 169 (*fig.*), 278 (*table*), 284 (*fig.*)
 release systems 284 (*fig.*)
 release of vesicle contents 281
 resting state in cells 276
 role in biology 268–9, 296–7, 510–7
 role in cell shape 295, 510–11
 solubility of salts 293
 stability constants of complexes 26
 structure of salts and complexes 271–2 (*table*)
 transport in cells 283
 triggering processes 276–7
calmodulin (Ca)
 binding of calcium 275–80
 binding of drugs 279 (*fig.*)
 EF hands 278–9
 in evolution 512
 helix–helix motions 278 (*fig.*)
 sequence 273
carbon dioxide uptake 59–60, 308–9
carbon transport 92
carbonic anhydrase (Zn) 308–9
carboxypeptidase (Zn) 63, 316 (*fig.*)
catalases (Fe, Mn) 264–5, 370
catalysis
 entatic state 180, 182–4, 311, 327, 358, 389–90
 specific by metal ions 145, 250, 289, 308–9, 400–1, 413
 transition metal centres in 112–13, 145, 319–22, 400
cations, *see* metal ions
cells
 cell combination 514
 communication with the environment 509
 control systems 287–8
 epithelial 81 (*fig.*)
 filamentous structures 78 (*table*), 159
 integrated activities 522–4
 morphogenesis 526
 morphology and biominerals 476–85
 organelles and vesicles 72 (*fig.*), 84 (*fig.*)
 particles and membranes 73–8
 required balanced connections 152
 shape 76, 77 (*fig.*), 159, 205 (*fig.*), 282, 295

structures 72, 231
 surface packing 205–6
cellulose 196 (*fig.*)
 and silica 449
channels, *see* ion channels 53
chemical speciation 6–7, 15
chemical elements, *see* elements
chitin 196
 cross-links 515
chloride 223–40
 channels and exchangers 445–6
 pump 445
 see also sodium, potassium, *and* chlorine
chlorin ring
 insertion of magnesium 262–4
chlorophyll (Mg)
 absorption spectra 263 (*fig.*)
 binding 262
 generation of radicals 263
 in membrane proteins 262
 role of magnesium 262–4
chloroplast 73, 123, 262, 377, 380–2
cholesterol 213, 215
chondroitin-6-sulphate 207 (*fig.*)
chromaffin granules
 effect of manganese 380
 metal contents 249 (*table*)
chromium 434, 538, 541 (*table*), 542
citric-acid cycle 28–30
clusters of metal ions (Zn, Fe, Mn, Cu) 44, 312, 326, 330–1, 382, 425
 control and ion 340
 and controls 339, 381
 and homeostasis 496
coated-pits 84, 325
cobalt 400–9
 binding constants 40–4
 coenzyme B12 400–2, 406–8
 functions 400–2
 kinetics 97
 ligand-field effects 402, 403 (*table*)
 ligands 403–4
 model Lewis acids 308
 oxidation states 402
 role in evolution 525
 spin states 402
 zinc probe 318
 see also vitamin B12
coccoliths 482: *see also* biominerals
coenzymes 65–6, 148 (*table*)
 coenzyme A 66 (*fig.*), 148 (*table*)
 coenzyme B12 (Co) 401 (*fig.*), 406–8
 coenzyme F-430 (Ni) 400 (*fig.*), 401, 406 (*table*)
 coenzyme M 455 (*fig.*)
 coenzyme R 459 (*fig.*)
colicin A 229 (*fig.*)
collagen 177 (*fig.*), 336, 394, 491, 515
communication systems
 properties required 160 (*table*)
 network 317, 509–10
condensation reactions 97

control systems 103–5, 287–8, 312–15, 516–21, 523–4
coordination compounds
 binding constants 26 (*table*), 33 (*table*), 40–4, 418 (*table*)
 ligands preferred by metal ions 34 (*table*)
 preferred coordination geometry 34 (*table*)
copper 388–9
 binding constants 40–4
 binding to metallothionein 396, 515–16
 biochemistry 388–9, 525 (*table*)
 blue-centres 182–3, 389–90 (*fig.*), 392–3
 exchange rate 396
 functions in biology 397–9, 515–16
 handling in cells 398 (*fig.*)
 in homeostasis 515
 proteins and enzymes
 coenzyme PQQ 394
 cytochrome oxidase 391
 electron transfer 182–3, 389
 extra-cellular oxidases 392, 510
 galactose oxidase active site 393 (*fig.*)
 haemocyanin 391
 kinetics 103, 105
 localization and functions 389 (*table*), 414 (*fig.*), 515–16
 in melanin formation 397
 metallothioneins 305, 505, 515–16
 nitrite and nitrous oxide reduction 394 (*fig.*)
 phenol oxidase 389 (*table*), 515 (*table*)
 production of ascorbate free radicals 393
 redox potential 389 (*table*), 392
 regulatory 397
 secondary structures 393 (*table*)
 substrates 389 (*table*), 393–4
 superoxide dismutase 395
 tyrosinase 389 (*table*)
 role in evolution 397–8, 515–16
 selection 41–3
 sulphide 18, 397, 515
 transport and homeostasis 396, 515
crystal field stabilization energy 41, 319, 346 (*fig.*), 400
 and kinetics 115 (*fig.*)
cytochromes (Fe) 108 (*fig.*), 343–65
 cyt a 173–4, 345 (*table*)
 cyt b 350 (*fig.*), 352 (*table*)
 cyt c 173–4, 345 (*fig.*), 347 (*fig.*), 355–6, 360 (*fig.*)
 cyt c′ 360 (*fig.*)
 cytochrome oxidase (a, a₃) 362–3, 391
 cytochrome P-450 361–2, 363–4 (*fig.*), 365 (*table*)

deposits, inorganic 87
dermatan sulphate 207 (*fig.*)
dielectric constant 51
diffusion
 passive 100, 136, 225–7

facilitated 228
dioxygen
 atmosphere 161, 512–3 (*fig.*); *see also*
 evolution
 bio-energetics 124
 carriers 333, 358, 391
 reduction of 336–7
 uptake 59–60
DNA, *see* nucleic acids (K, Mg)
 synthesis 149
Donnan equilibrium

electrolytic messages (Na, K, Ca) 234–5
electrolytic potential 157, 216, 225–8
electron flow 105, 156–7, 182, 353–7
electron gradients 124
electronics 105, 125 (*fig.*), 235 (*fig.*), 236;
 see also electron flow
electron-rich ligands in water 408
electron transfer (Fe, Cu) 105, 125 (*fig.*),
 322, 353–7
 in mitochondria 124, 323
 in photo bacteria 106, 262
 in photosystem I 106 (*fig.*), 122–3
 in photosystem II 123 (*fig.*), 382 (*fig.*)
elements
 abundance in living systems 3–5
 abundance in sea water 11, 15 (*table*), 16,
 412
 abundance in universe 9 (*fig.*)
 accumulation by plants and animals 4
 (*table*), 10
 affinity for binding centres 25
 availability 8–11, 16, 17 (*table*), 513
 (*table*)
 binding situations in biology 28, 68
 (*table*)
 chemical properties in aqueous solution
 6, 11, 18, 53, 141
 concentration in aerobic environment
 14–16
 concentration in cell compartments 73–6
 concentration in human body 16 (*fig.*)
 concentration in marine organisms 15
 (*table*)
 concentration in the sea 11, 15 (*table*), 16
 (*fig.*), 303 (*table*), 412 (*fig.*)
 distribution in eukaryotic cells 75, 88
 (*fig.*), 153, 514 (*fig.*)
 essentiality 4–5
 evolution of use 320, 432–4, 502–25
 (*table*)
 formation of biominerals 467–9
 function in biology 7, 143 (*table*), 161
 (*fig.*), 525 (*table*), 528 (*table*)
 and homeostasis 499–528
 interactive integrated control by 496
 kinetic chemical potential 95
 links with metabolic paths 522 (*fig.*)
 local storage 134
 natural selection 3, 160
 oxidation state 14 (*fig.*), 110 (*fig.*)

periodic table (*see* Chap. 8–19) 5, 550
 radii and ionization potentials 375 (*table*)
 selective uptake at equilibrium 29 (*table*),
 30–1, 33–6
 separated functional pools 146 (*table*)
 separation and selection in biology 26–7,
 36–70, 142
 speciation 6–7, 35
 spin states 110 (*fig.*), 346 (*fig.*)
 steps in transfer 26 (*fig.*)
 traps for 26, 67–8
 uptake by biological systems 24–6, 26–31,
 33–6, 154, 514 (*fig.*)
 see also biominerals *and individual elements*
endoplasmic recticulum 84 (*fig.*), 281–2
endosome 84 (*fig.*)
energy in biological systems 8, 154–5
 chemical bonds and 121, 126, 500 (*fig.*),
 501 (*fig.*)
 comparison of aerobic and anaerobic 130
 (*table*)
 concentration storage of 120–1, 236–7
 connection of forms of 134
 early sources of 135, 458–9, 502
 electrical storage 120
 elements in capture of 134
 flow 154
 forms of storage 120–1
 and gravity fields 121, 439
 homeostasis of 499–501, 507
 intermediate mechanical storage 132, 510
 light and 121–3, 262–3
 tension storage 8, 121, 510
 transduction 107, 125 (*fig.*), 127, 354
energy generation
 aerobic 129–30, 362
 anaerobic 128–9
 evolution of compartments for 135,
 510–11 (*fig.*), 514
 feedback control of 136
 ion gradients 7, 124, 158, 240–1
 from light 122
 metabolic 130
 see also bioenergetics
enniatin 39 (*fig.*)
enolase (Mg) 253 (*fig.*)
entatic state 180, 182–4
environment 11, 512–13, 535–6
 acid rain 536 (*fig.*)
 changes in element availability 6, 11–18,
 532 (*table*)
 effect of agricultural practises 534
 effect of CO_2 and CH_4 on temperature
 535 (*fig.*)
 effect of industrial pollutants 535–6
 greenhouse effect 535 (*fig.*), 536 (*fig.*)
 man-made evolution 531–2, 536 (*fig.*)
enzymes
 active-site attacking groups 181
 entatic state 180, 182–4
 features 180
 grooves in 180
 locations for 247

required metal ions 181
required order 178
see also individual elements
eukaryotes
 compartments and reactions in 73, 153,
 514 (*fig.*)
 energy generation in 80, 129–30
 network of communications for
 homeostatic growth 523 (*fig.*), 524
 (*fig.*)
evolution
 advent of dioxygen 17, 512
 element availability changes 513 (*table*)
 new roles for elements 320, 525 (*table*)
 anaerobic cytoplasmic activity 502 (*table*)
 development
 of advantageous shapes 510–5
 of cell organization 17 (*fig.*)
 of integrated organization 528
 of life forms 6, 8, 11, 513 (*fig.*)
 of multicellular structure 515–15
 of poisons into signals 520, 543, 544
 (*table*)
 of secondary metabolism 521–3
 of uptake of Fe and Mn 514 (*fig.*)
 metabolic path linked to non haem iron
 339, 505 (*table*)
 see also elements
exchangers 237–8
extraction
 into organic solvents 52
 into proteins 50

fatty acids
 essential 217
 of lipids 213 (*fig.*)
 oxidation of 336
ferritin (Fe) 75, 340–1 (*fig.*), 366, 472, 485
 buffering of iron levels 504–5
flavin 66 (*fig.*)
flow-units of biology 148 (*table*) and
 organization 156, 526
fluoride 36, 451, 541 (*table*)
free radicals 114, 333, 393, 408, 515
fructose 195 (*fig.*)
FUR 324, 339, 503–4, 505–6; *see also* iron
 uptake regulation protein

galactose 195 (*fig.*)
 oxidase (Cu) 393 (*fig.*)
gating 107, 356, 362
gel-sol changes 207–8, 261
genes
 fnr, aerobic-anaerobic switch 505
 fur, iron uptake 504
 metal ions and 497–8, 503–6, 514
 nif, nitrogen fixation 427, 505
 resistance in bacterial plasmids 544 (*fig.*)
 switching processes 497
 transcription 444–5
glucocorticoids (Na, K) 518 (*table*)
glucose 195 (*fig.*)

glutathione 455
glutathione reductase 456 (*fig.*)
glycogen 196 (*fig.*)
glycolysis 128–9
 and magnesium 250
glycoproteins (Mn) 203–6
gold 538, 545 (*table*)
Golgi apparatus 74, 84, 380 (*table*), 454
gradients
 of ions 58 (*table*), 82
 of potential 124
 of protons 124–7
 see also elements
group transfer 111–12, 443, 457
growth
 of crystals 474
 and homeostasis 497
 whole organism 518

haem (Fe) 343–68
 high-spin and low-spin iron-porphyrins
 43, 345 (*fig.*)
 homeostasis and 366, 520–1
 incorporation in proteins 51, 344 (*table*)
 properties of isolated units 344–5
 proteins, *see* haem proteins
 receptor for NO 359
 synthesis 55, 366–7
 translation and 366
 uptake 51
haemerythrin active site (Fe) 333
haemocyanin (Cu) 391
haemoglobin (Fe) 169 (*fig.*), 358–9 (*fig.*),
 362
haem-proteins (Fe) 344, 358–9, 361
 binding constant to proteins 366
 binding groups 51
 biological function 344, 361 (*table*), 367
 classification
 by chemical binding 347–9
 by function 344
 by secondary structure 348–51
 communication with genes 366, 520
 control via RNA 366, 520
 cytochromes, *see* cytochromes
 electron-transfer/proton-transfer coupling
 353–5
 evolution 366, 520
 haloperoxidases 430 (*table*)
 in hormone synthesis or removal 365, 520
 location in cells 352–3
 monoxygenases and dioxygenases 361
 (*table*)
 multi-haem electron transfer 354
 organization in eukaryotes 352 (*table*),
 520
 oxidation of organic molecules 361
 peroxidases and catalase 364
 proton movements in cytochrome oxidase
 362–3
 reactions with dioxygen 358–9, 362
 reaction with nitrogen oxide 359–60

redox potential 51 (*table*), 346 (*fig.*)
secondary metabolism 365, 520
secondary structure 349
simple electron-transfer 353
substrates 365
surfaces 357–8
use in mechanical devices 363
halobacteria 445
halogens
 redox reactions 462
 roles in biology 451
 see also chloride, iodide
haloperoxidases (Fe, V) 430 (*table*)
hard and soft acids and bases 35 (*table*), 52,
 536, 538, 546–7
helix–helix movements 132, 278, 359 (*fig.*)
heparin 207
hinge bending, *see* proteins
histones 179
Hofmeister series, *see* lyotropic series
homeostasis 20–2, 498–528, 502
 of calcium 294, 510
 connections in eukaryote cells 511 (*fig.*)
 connections in prokaryote cells 507 (*fig.*)
 controlled synthesis of macromolecules
 149, 503
 definition 20, 496
 evolutionary switch due to dioxygen
 512–6
 evolution of integrated organization 528
 (*table*)
 and growth 496
 interdependence of elements 501 (*fig.*),
 502
 of iron in prokariotes
 connection between uptake and
 utilization 504 (*fig.*)
 FUR protein regulation 324, 339,
 503–4, 505–6
 iron binding sites in DNA-binding
 proteins 505 (*fig.*)
 iron uptake regulation 339, 503–4
 link to phosphate homeostasis 504
 recognition system for entry (TON B)
 503–4
 of levels of elements 496
 of magnesium 501
 of multicellular organisms containing
 bones 517
 overall control of homeostatic growth
 523 (*fig.*), 524 (*fig.*)
 of phosphate 499–501, 517
 of potassium 502
 of protons 502
 regulation in multicellular organisms
 517–18
 role of the proton 137–8, 502
 and shape
 of bacterial prokaryote cells 498, 507–9
 of multicellular organisms 499
 roles of calcium and phosphate 510–11
 of unicellular organisms 499, 510
 of sodium 502

whole organism growth 518
 and zinc 517–18
hop-conduction
 of electrons 105, 353, 391–2
 of protons 101, 354–5 (*fig.*)
hormones
 and calcium 510, 516
 dependence on oxygen and iron 523
 (*table*)
 and iron 337, 520
 peptides and zinc 305, 314, 319
 types and association to metals 518
 (*table*)
hyaluronate 207 (*fig.*)
hydration 38, 61
 exchange rates 103
hydride (H⁻) transfer 110
hydrogen
 biological chemistry 7
 transport 90
 see also protons and bioenergetics
hydrogenases (Ni, Fe) 405–6, 506
hydrolysis
 and cluster formation 45
 of esters, metal ion order 98 (*fig.*)
 of metal ions 45–7, 333, 376, 417–19
 by metalloproteins 98, 308, 338, 385, 409
 of phosphates 250, 338
 of proteins and other polymers 99, 289,
 300–1, 315–16
 reactions 97–8

insertion in chelating holes 55, 264, 343,
 402
insulin (Zn) 305, 314 (*table*), 519
iodine 430 (*table*), 451, 462
 deiodination (Se) 462
ion channels 53
 for calcium 225, 285–6
 for chloride 445–6
 gated 230
 permeability 102 (*table*), 230 (*table*),
 445–6 (*table*)
 for potassium 225 (*fig.*), 230–6
 for protons 53, 103
 selectivity 53, 230–1
 for sodium 225, 228 (*fig.*), 230–6
ion currents 137, 238–9, 509–10
ion flow 100–1, 158
ion gradients 82, 124–7, 157–8, 240–1
ionic messengers 238, 509
ionic potential 34, 216, 225–30
 Donnan 226
ionization potentials 40, 300
ion migration rates 101–12
ion properties 271 (*table*)
iron 322-68; *see also* haem
 binding constants 40–6
 binding sites in DNA-binding proteins
 505 (*fig.*)
 binding site in transferrin 325 (*fig.*)
 buffering by ferritin 340–1, 504–5

control of cell activities 368, 503, 522–3
distribution 324 (*fig.*)
the Fe$_2$O cluster 333
handling of dioxygen 336–7
in homeostasis 503–6, 522–3
kinetic character 103–6
ligands in non-haem proteins 338
link with citric acid cycle 340
in magnetotatic bacteria 484–5
metabolism in bacteria 324 (*fig.*)
in minerals 486
porphyrins containing 343–5
proteins and enzymes 322 (*table*)
 aconitase 330 (*fig.*)
 FUR, iron uptake regulation 324, 339,
 503–4, 505–6
 haem proteins, *see* haem proteins
 iron–sulphur proteins 335–7 (*table*); *see
 also* iron–sulphur proteins
 non-haem oxidases 335–7 (*table*)
 ribonucleotide reductase 334 (*fig.*)
 redox potential range 346 (*fig.*)
 role in biology 322, 339, 344
recognition system for entry in cells 323,
 503, 504 (*fig.*)
regulation and homeostasis 339, 522–3
role in acid-catalysis 338
role in secondary metabolism 337, 520–2
uptake 323–5, 339, 503–6, 514
iron–sulphur proteins 326–33
 coordination of iron 326
 discovery of centres 341
 as enzymes 329–30
 in evolution 331, 503
 features of centres 329 (*table*)
 ferredoxins 326, 329, 382, 505 (*fig.*)
 folds 327 (*fig.*)
 location 330
 redox potential 327–8 (*table*)
 in regulation 503
 rubredoxin 326, 328, 505 (*fig.*)
 structures 327
 synthesis of 55, 332
Irving–Williams stability order 34, 46, 98

Kashin–Beck disease (Se) 461
keratan sulphate 207 (*fig.*)
Kesham disease (Se) 461
kinases (Mg) 176 (*fig.*), 250, 499
kinetic effects and control of metal ions 20,
 54–8, 103, 496
kinetics of chemical reactions 95–118
kinetics of transport 54–5
kinetics of uptake of metal ions 54–8

laccase (Cu) 63, 389–90, 393
lactate dehydrogenase 180 (*fig.*)
lead 32, 536, 539 (*table*), 546–7, 549
 tetraethyl 536

life
 beginnings 18–21, 410, 432–4, 449, 503,
 506–7, 512–14
 chemical features 146–7
 early use of light 385, 434
 evolution of chemistry 161, 410, 512–14
 organometallic chemistry 408–9
 physical features 146
ligand-field stabilization 43–4, 319–20,
 344–6 (*fig.*), 400
 kinetic effect of 103–4
ligands
 differentiating for cations 29 (*table*), 44
 neutral O-donors 39 (*fig.*)
 selective 29 (*table*), 44
lignin oxidation 371 (*table*), 386
lipids 210–17
 in biological membranes 213 (*fig.*), 214
 (*table*)
 composition and functions 211–12 (*fig.*)
 ion-association to 215–16
 molecular characteristics 213 (*table*)
lithium
 competition with magnesium 266
 toxicity and teratogenicity 545–6
 use in medicine 545
lyotropic series 61

magnesium 243–67
 binding constants 245, 275
 binding to phosphate 250–4, 255, 266
 binding sites in enzymes 254 (*table*)
 cases of strong binding 249
 chemistry 245
 in chlorophyll 80, 262–4
 competition with calcium 275
 competition by lithium 266, 545
 competition with polyamines 189, 257–60
 in DNA and RNA 188, 256–7
 effect on cell-shape 295
 effect on gel-sol equilibria 208 (*fig.*), 261
 enzymes 247 (*table*), 250–3, 254 (*table*)
 essentiality in NTP-dependent reactions
 250–6, 501
 functions in biology 247, 250–7, 263,
 266–7
 functions in membranes 215–16, 244
 and glycolysis 253
 manganese as a probe for 247, 265
 homeostasis 501
 insertion in chlorin ring 263–4
 kinetic character 103, 246
 as a Lewis acid 246, 251
 mononucleotide complexes 250 (*table*)
 movements in kinases 171, 252 (*fig.*)
 in muscle contraction and relaxation
 255–6
 pumping 248–9
 relation to phosphate 243, 250–4, 255,
 501–2
 requirements in polynucleotides 206, 256
 (*table*), 257

requirement in transcription and
 translation 256, 501
ribozyme dependence on 257
role in synthesis of lipids 261
role in tubulin polymerization 76, 185,
 260–1
spacial distribution 80, 244
structures of salts 245 (*table*)
substrate binding 246 (*table*)
in walls and membranes 73, 123, 250
maltose 195 (*fig.*)
manganese 370–86
 abundance and availability 379
 binding constants 36–41, 373
 control systems 380–1
 clusters and cluster chemistry 437–8
 (*fig.*), 382 (*fig.*)
 deposition by bacteria 384
 dioxygen
 generation 123 (*fig.*), 124 (*fig.*), 381–2,
 512–13
 metabolism 382–3
 distribution and location in cells 72 (*fig.*),
 380, 514 (*fig.*)
 enzymes and proteins 371 (*table*)
 evolution of functions 385–6, 513, 525
 functions in biology 370–1, 385–6
 in Golgi 83, 380 (*table*)
 homeostasis 513, 523
 hydrolysis 376
 kinetics
 of Mn (II) complexes 103, 375
 of Mn (III) and Mn (IV) reactions 377
 as a Lewis acid 370
 oxidation of lignin 386
 oxidation state diagram 372 (*fig.*)
 oxidation states 372
 in photosystem II 123–4 (*fig.*), 382 (*fig.*)
 precipitates 383–4
 pumping 379–80
 as probe for magnesium 247, 265
 selection of 43–45
 stability of complexes 373–5, 376
 structure
 of Mn (II) complexes 372–3
 of Mn (III) Mn (IV) complexes 376,
 337–8
 superoxide-dismutase 383 (*fig.*)
 thermodynamics of higher oxidation states
 376
 uptake 36, 379
 in vesicles 83, 380 (*table*), 514 (*fig.*)
Marcus equation 105
mechanical coupling of helices 132–3
medical applications of elements
 anti-tumour activity of Pt complexes
 538–9, 545 (*table*), 546–7
 chelation therapy 549 (*table*)
 correction of required element deficiencies
 547–8
 effect of lithium on hyperactivity 545
 periodic table and medical applications
 550 (*fig.*)

medical applications of elements (*cont.*)
 prosthetic materials 549–50
 removal of excess elements 548–9 (*table*)
 uses of non-metals 549
 uses of radioactive diagnostics 533 (*table*)
 uses of unusual elements 545 (*table*),
 546–7, 549–50
melanin formation 393, 397
membranes 71–118
 in bio-energetics 122, 126
 clusters and patches 215 (*fig.*)
 contents of lipids 213–14 (*table*)
 curvature 205–6, 209, 215–16, 281
 internal and external particles 78–9
 lateral organization and flow 79–80, 214
 in mineralization 471–2
 potential on surfaces 216
 protein networks and organization 76
 (*fig.*), 79 (*fig.*), 80, 206 (*fig.*), 449
 (*fig.*), 510
 pumping of ions 158
 role and nature 76
 selective passage of ions 102–3
 shapes 76, 77 (*fig.*), 159, 205 (*fig.*),
 471–85, 510–12
 surface tension 205 (*fig.*), 475
 thylacoid 73 (*fig.*), 80, 123 (*fig.*)
 transmembrane organization and flow
 82–3
mercury 536, 538–9 (*table*), 544, 545 (*table*)
messengers
 organic and inorganic ions 238, 508–9,
 517–24
 phosphate messengers 441–3, 500, 503,
 511
metabolic disturbances and diseases
 allergic response 542–3
 Alzheimer's disease 85, 265, 542
 autoimmune responses 543
 effect of aberrant protein phosphorylation
 541
 effect of deficiency of required elements
 540, 541 (*table*), 542
 inflamatory responses 548 (*fig.*)
 Kashin–Beck disease 459
 Kesham disease 461
 lack of vitamin D 540
 problems of calcium distribution and
 utilization 540, 542
 problems derived of excess iron 542
 problems of old age 270, 492, 542
metabolic pathways
 aerobic in eukaryote mitochondria 130
 (*fig.*)–1, 521
 anaerobic in cell cytoplasm 128–9 (*fig.*)
 separation in compartments 131–2
metal ions
 advantages of chemistry in aqueous
 solution 141
 competition with polyamines 179, 189,
 257–60
 complexes in biological systems 33 (*table*)
 distribution and positioning 87, 507

 (*fig.*), 511 (*fig.*), 514 (*fig.*), 523 (*fig.*),
 524 (*fig.*)
 kinetics of uptake 54–8
 hydroxides 18, 45–7, 417, 418–20
 insertion in holes and rings 55–6, 262–4
 oxidation states 11–12 (*fig.*), 47–8
 rejection of 58, 75, 154
 sulphides 18–20, 46, 433
 uptake and retention 28–58
 see also individual elements
metallothionein (Zn, Cu, Cd) 305–6 (*fig.*),
 396, 505 (*fig.*), 515, 546
 binding sites for Zn, Cu 505 (*fig.*)
 buffering reactions 305–6, 515
 handling of toxic elements 546
metamorphosis 497
methane oxidase (Fe) 334
mitochondrion 73 (*fig.*), 80, 81 (*fig.*)
 oxidative energy 124–5
 respiratory chain 108 (*fig.*), 323
 shape of membranes 79
 see also cell organelles
molybdenum 411–27, 432–4
 abundance and availability in the sea 412
 'b'-class character 417, 420–1
 binding to ligands 419
 binding to proteins 421
 chemistry 12, 416–17, 421, 427
 complexation 419
 hydrolysis 45–7, 417, 418–19
 oxidation states 417
 competition from Fe and Cu 417–18,
 419–20
 complexes and clusters 416
 compounds in living systems 411
 control functions 426–7
 enzymes 413–4
 functions 427
 location 414 (*fig.*)
 oxidation states in 423
 redox potentials 423–4 (*table*)
 evolution of role 432–4, 525
 kinetic character 422
 nitrogenase 425–6
 physico-chemical model studies 415
 pterin-cofactor 414 (*fig.*), 420–2, 426
 stability of model complexes 417–8
 sulphides 19–20, 433
 O-transfer reactions 414, 424–5
 types and rates of reactions 422
 uptake 64–5, 422
 see also nitrogenase
monomer synthesis and degradation 151
morphogenesis
 changes in 156, 239, 526–7
 definition 496–7
 gene switching processes 497
 growth currents 137, 156, 527
 growth patterns of cells after fertilization
 526 (*fig.*)
 morphogenic field 81
 role of calcium 296, 527
motion of organisms 133, 232–3, 508–10

movements
 of ions, *see* ion currents
 of small chelates 83, 323, 503–4 (*fig.*)
 of vesicles in cells 83–6
muscle contraction
 role of calcium 255–6, 269, 278–80, 285
 role of magnesium 255–6
 sliding filament model 285 (*fig.*)

NADH, NADPH (H) 66 (*fig.*), 91
nerves
 action potential 235
 and Alzheimer's disease 85
 current flow in 234 (*fig.*)
neuron 234
nickel 400–10
 activation of C(1) carbon compounds 404
 binding constants 40–3
 carbonyl 536
 coenzyme F-430 400 (*fig.*), 401, 406
 (*table*)
 enzymes
 hydrogenases 405–6
 localization and function 406 (*table*)
 urease 400, 406 (*table*), 409
 functions in biology 400–2, 525
 kinetic character 103
 ligand field effects 402, 3 (*table*)
 ligands 403–4
 oxidation states 402
 reactions 404
 spin-states 402
 tendency in evolution 525
 uptake 410, 525
niobium 431; *see also* vanadium
nitrate and nitrite reduction in cells 394
 (*fig.*), 413 (*table*), 424 (*table*)
nitrogen
 NO-binding, *see* haem
 transport 92
 uptake 59–60, 426
nitrogenase (Mo, V, Fe)
 components 425
 nif genes 427, 505
 reactions 425–6
nonactin (Na) 39 (*fig.*)
non-metals
 advantages of chemistry in aqueous
 solution 141, 436–7
 in coenzymes 54, 109
 redox potentials 13, 424
 selection, *see* selection of non metals
 spacial distribution 89
 transport systems 90 (*table*)–93
 uptake and retention 26–7, 58–68
 see also elements *and individual elements*
nucleation of minerals 473
nucleic-acids 185–91
 backbones of 195
 composition and structure 185–7, 188
 (*fig.*)

nucleic acids (*cont.*)
 crucifix forms 188–9 (*fig.*)
 DNA sequences and mutation 190
 folding 187 (*fig.*)
 genetic control by metals 499–527
 grooves in 187–8
 metal enzymes related to production of
 186 (*table*)
 metal-ion binding in DNA 188–9, 239,
 260, 546
 mobility of sidechains 187
 physical properties of sidechains 197
 regulation by 503
 space-filling properties 199
 single-stranded 188
 topology of strands and helices 189 (*fig.*)
 transcription 503, 520
nucleotide triphosphates (NTP) 65–6 (*fig.*),
 437, 500; *see also* ATP *and*
 phosphorus

organelles *see individual examples*
 time-dependent organization 81
organization and tension 132–3, 159, 510
osteocalcin (Ca) 292 (*fig.*), 495
osteoporosis (Ca) 491, 542
oxidation state
 diagrams 12 (*fig.*), 13 (*fig.*), 372 (*fig.*)
 selective control of 14, 47
oxo ions 45–7, 417–19
oxygen
 bioenergetics 124
 transfer of atoms 413 (*fig.*), 414, 424–5,
 460, 461
 transport of 59, 91, 333, 358, 391
 see also dioxygen

patronite 436; *see also* vanadium
periodic table of elements 5 (*fig.*), 142 (*fig.*),
 550
 diagonal relationship 245 (*fig.*), 545
 and medical applications of elements 550
 (*fig.*)
peroxidases (Fe, Se, V) 364–5, 430 (*table*),
 459–60
peroxyzomes 72
phosphate 436–45
 binding to ions 439
 in bioenergetics 126, 154, 439
 bound phosphate
 energy of hydrolyses 63, 91, 439–40
 forms 438–9
 maintenance of levels 439
 chemistry 12–13, 437–8
 connections in cells 437 (*fig.*), 510 (*fig.*),
 511 (*fig.*)
 controls 440
 in DNA and RNA 185–90, 499
 effects of use in agriculture 534 (*fig.*)
 functions in biology 437

gene transcription and 444, 499–500
homeostasis 154, 499–501
integrated role with calcium 510–11,
 517–18 (*fig.*), 523–4
involvement in cellular activity 500 (*fig.*)
kinetic character 438
location 437
messengers
 cyclic nucleotides 441–2 (*fig.*), 500,
 503, 511
 inositol phosphate 442–3 (*fig.*), 500
minerals 445; *see also* bone
protein phosphorylation 441 (*fig.*),
 443–4, 500, 503, 511
stable forms in water 438
switches 443–4
transport 89–90
uptake routes 63, 437 (*fig.*)
phosphoglycerate kinase (Mg) 178, 252
 (*fig.*)
phospholipase A (Ca)
 calcium binding site 274
 structure 167
photosystems 122–4, 381–2 (*fig.*)
phytochelatins (Cu) 397
plastocyanin (Cu) 41, 389, 390 (*fig.*)
platinum in drugs 538–9, 546–7
polyamines
 binding to dicarboxylates 258 (*fig.*)
 competition with Mg 189, 257–60
 fitting in DNA groove 259 (*fig.*)
 structure 257 (*fig.*)
polyelectrolytes stabilization
 by Na, K, Cl 188, 239
 phase transformation 206–10
polymers, *see* biopolymers
polysaccharides 193–210
 anionic 206–7 (*fig.*)
 backbone 195–6
 chain-branching 198, 201
 chain–chain attraction or repulsion 206
 hydrophobicity 197–8
 introduction of conformational changes
 203
 information generation by 200
 interaction with cations 210
 mediation of mutual interaction 203–4
 non-interactive 193–4, 204–5
 roles 197–8, 200–1, 202–3
 sidechains 197, 201
 sol-gel changes 206–8, 261
 synthesis 149
 types 194 (*table*)
porphyrins
 selection of metal ions 56
 synthesis 366–7
 see also haem
potassium
 K-dependent enzymes 240
 K-channels 230–2, 234, 241–2, 509
 see also sodium, potassium, *and* chlorine
potential
 gradient of 7, 90, 124, 225–30

precipitates in compartments 27, 68–9; *see
 also* biominerals
prokaryote cellular activities 149, 506–7
prostaglandins 336
 and iron 336, 521 (*table*)
 synthesis 336 (*fig.*), 522 (*fig.*)
 and zinc 314, 336
proteins
 amino acid composition 173–5
 assemblies and metal ions 184
 backbones 195
 balanced synthesis of 85, 149–50, 506
 (*fig.*)
 catalysis by folded 180–1
 change of conformation 168–70
 composition and basic structure 164–5
 conformational movements 170 (*table*)
 cross-linking agents 179 (*table*)
 disulphide bridged 179 (*table*)
 folding 38, 40 (*fig.*), 51–3, 171–3
 γ-carboxy glutamate containing 276, 493
 hinge-bending 175–6
 integral and peripheral 79 (*fig.*) 209
 interaction with anions 423 (*table*)
 internal motions of sidechains 172 (*fig.*)
 matching of ions 178–9
 mechanical devices 178
 mobility in 168 (*table*)
 mobility of chains in solution 171–2
 physical properties of sidechains 197
 primary structure 165
 quaternary structure 165, 169 (*fig.*)
 ratios of hydrophobic to charged amino
 acids 174 (*table*), 175 (*fig.*)
 secondary structure 165, 169 (*fig.*)
 space-filling properties 199
 structure of redox proteins 321 (*table*)
 synthesis of 85, 150, 506
 tertiary structure 165, 169 (*fig.*), 172
 (*table*)
 see also individual proteins and elements
proton
 bioenergetics and catalysis 97–100
 channels 53, 103
 gradients 124, 125 (*fig.*), 126–7, 157
 migration 103, 126
 networks in proteins 100 (*fig.*), 354–5,
 362
 redox potential 138
 role in homeostasis 137, 502
 see also hydrogen *and* bioenergetics
pumps
 ATPases (Mg) 232–3, 236–7, 501 (*fig.*)
 for ions across membranes 56–8, 158,
 232, 248, 284
 position of 236–7
 selective 236
push–pull mechanisms 209 (*fig.*)
pyrophosphate
 energy transfer as 154–5
 production from gradients 126–7
 uses in coenzymes 155 (*table*)
 see also ATP

quinones 109, 128–30, 323 (*fig.*)

radii of ions 37, 44 (*fig.*), 60 (*table*) 61
 (*table*)
 and ligand fields 44 (*fig.*)
ras proteins 251
rate constants for substitution of inner
 sphere H_2O 102 (*fig.*), 104 (*table*),
 247
reactions
 condensation 97–8
 dissociative or associative paths 115
 electron transfer, *see* electron transfer
 free radical 114
 group transfer 111–12
 hydrolysis 97–8, 418–19
 nucleophylic substitution 115 (*fig.*)
 oxidative addition 113
 reductive elimination 113
 substitution of inner sphere H_2O 102, 104
 (*table*), 247
 two electron oxidation or reduction 110
 types 96–7
redox potential
 of complexes 47–8, 108 (*fig.*), 323 (*fig.*),
 328 (*table*), 389 (*table*), 426
 effect of σ-donors and π-acceptors 109
 in electron-transfer chains 109, 323 (*fig.*),
 328 (*table*)
 of H^+/H_2 and H_2O/OH^- systems 138
 of organic molecules 368 (*table*)
respiratory chain 108 (*fig.*), 323 (*fig.*)
retinoic acid 518 (*table*)
ribonucleotide reductase (Co, Fe, Mn) 334
 (*fig.*), 371 (*table*), 407 (*table*)–8
ribozymes (Mg) 257
RNA, *see* nucleic acids (K, Mg)

sea
 elements in 10 (*fig.*), 11 (*fig.*), 15 (*table*),
 16, 303 (*table*), 412 (*fig.*)
secondary matabolism 337, 365, 521–2
selection of metal ions
 by binding in clusters 44–5
 by charge type 36
 by control of concentration 48–50
 by insertion in chelating holes 55–6, 264,
 343, 402
 by ion size 37–40, 53
 by kinetic effects 54–8
 by liganding atoms 40–1
 by polypyrrole chelating agents 56
 by preferential coordination geometry
 41–3
 by protein folding 38
 by spin-pairing stabilization 43–4
 by transfer coefficients H_2O/proteins
 50–3
 by membrane isolation 27, 74–83
 see also individual elements

selection of non metals, *see individual*
 elements
 by charge and size of ions 60–1
 by differences in binding affinity 62–3
 by kinetic traps 63, 65–6
 by redox incorporation 27, 64–5, 425–6,
 454
selectivity 23–69
 of channels 53, 230–2
 of gates and pumps 56–7, 232–3
 tables for ions 43 (*table*)
 kinetic control 54, 263–4
 see also individual elements
selenium 460–2
 biological compounds and enzymes 460
 (*table*)
 glutatione peroxidase 460–1 (*fig.*)
 de-iodination 462, 525 (*table*)
 effects of deficiency 461
 functions 460
 reactions 460
 role in evolution 461–2, 525 (*table*)
self-organization 155; *see also* Chap. 20 and
 21
sequestration of metal ions 48
shape 496–9, 508, 510–11, 523
 of cells 77 (*fig.*), 81 (*fig.*), 295
 of organelles 80 (*fig.*), 81 (*fig.*)
 see also Chap. 20
siderophores 36, 54, 58 (*table*), 503; *see also*
 iron, uptake
signals 100–1, 209, 238, 441–3, 519, 520–4
 evolution of 509, 520–4
silica 446–51
 biochemistry 449–51, 467–9
 concentration in the sea 450
 export of 499 (*fig.*)
 forms in diatoms 446–7, 483–4, 486
 forms in plants 486
 forms in radiolarians 483–4
 occurrence in living systems 446–9, 468
 (*table*), 470, 483–4
 relation to protein filaments 449 (*fig.*)
 rods in protozoa 447–8, 486
 silicate layers 450 (*fig.*)
 see also biominerals
sodium, *see* sodium, potassium, *and* chlorine,
 Chap. 8
sodium, potassium and chlorine 223–42
 action potential 235 (*fig.*)
 active transport 56–7, 232–3
 binding constants 29, 33, 37–8, 224 (*table*)
 concentration gradients 126, 223–4, 232
 evolution of currents 509–10
 exchanges 100, 237–8, 509
 facilitated diffusion 228
 homeostasis 502, 504, 507, 509–10, 523–4
 kinetics 100–3
 ligands preferred 34 (*table*), 39 (*fig.*)
 link to electrolytic potential in cells 227
 movement in channels 225, 228 (*fig.*),
 230–6, 445–6
 movements in nerve messages 234–5

Na/K transport 56–7, 232–3
 passive diffusion 100, 225–8
 pumping 56–7, 82, 225 (*fig.*)
 rejection and accumulation 58, 224–5
 selective pumps 56–7, 236
 significance in life 240–2
 stabilization
 of polyelectrolytes 188, 239
 of membranes 214–16
soft and hard acids and bases 35 (*table*), 52,
 536, 538, 545–7; *see also* elements
sol–gel equilibria 206–8, 261
solubility of salts 18, 37–8, 473–5
solubility-product constants 18, 69 (*table*)
speciation 6–7, 15, 23–8, 319–22
spin-states for metal ions (Fe, Co, Ni) 43,
 44 (*table*), 344–6 (*fig.*), 359 (*fig.*), 360,
 402–3
stability constants
 of metal complexes 26, 29, 40 (*fig.*), 258,
 373, 418 (*table*)
 effective or conditional 31–2
 and oxidation states 47
 see also individual elements
stability orders 34, 37, 40 (*fig.*), 46, 98
stereochemistry 34, 41–4
 and ligand field 34
strontium sulphate
 in *Acantharia* and radiolarians 476–81,
 484
sulphate incorporation 454
sucrose 195 (*fig.*)
sulphides chemistry in water 18–20, 331,
 502
 precipitation as 18–20, 46–7
sulpholysis 46 (*table*), 331
sulphur
 acid–base reactions of thiolate 456
 biochemistry 459
 biological compounds 455 (*fig.*)
 glutathione 455 (*fig.*), 458
 lipoic acid 455, 458
 sulphated adenosine phosphate 455
 (*fig.*)
 thioredoxin 454–5, 504 (*fig.*), 503
 evolution and 502
 functions in biology 453–4 (*table*), 457
 handling of 454 (*fig.*)
 homeostasis and 502
 incorporation pathways 454
 oxidation states and redox reactions 454
 sulphur cycle 459 (*fig.*)
 thioster driven synthesis 457–8
 transfer of acetyl groups 457 (*fig.*)
superoxide-dismutase (Cu, Zn, Fe, Mn)
 copper–zinc 395
 iron 337
 manganese 383 (*fig.*)
 regulation 505
symbiosis 88
synthesis and degradation
 of monomers 96, 151–3
 of polymers 96, 99–100, 149–51

tantalum 431; *see also* vanadium
teichoic acid (Mg, Ca) 251 (*fig.*)
tension in biological polymers
 in eukaryotes 510
 generated by polymerization in selected
 directions 78, 130–2
 patterns and flow 157
thioesters, in bioenergetics 136, 457
thioredoxin 454–5, 502 (*fig.*), 503
thylacoid membrane 7, 80 (*fig.*), 122–3, 264
thyroxine (I) 460–2, 518 (*table*), 544 (*table*)
T0N B proteins, *see* iron, recognition system
 for entry in cells 323, 503
toxicity
 and immune system
 allergic responses 542
 autoimmune responses 543
 of 'a'-class hard metals 536–7, 546
 of 'b'-class soft metals 538–9
 of chlorine covalent compounds 539
 of excess of elements 539 (*table*)
 of first-series transition metals 536
 of non-biological metals 537 (*table*)
 of non metals 539–40
 of redox active elements 541
 of superoxide 548
 removal of unwanted elements 544, 548–9
 (*table*)
 resistance genes in bacterial plasmids 544
 (*fig.*)
transfer reactions
 mechanical devices 116–17
 required order mechanisms 116, 175–6
 (*fig.*), 363–4 (*fig.*), 406
transferrin (Fe) 325, 514 (*fig.*)
transition metals in biology 3–5, 319–21
 binding 26 (*table*), 33–45, 373, 418
 chemistry 319–21, 402–3
 ligand-field states 43–4, 319–20, 344–6
 (*fig.*), 403 (*table*)
 see also metal ions *and individual metals*
transition state 98 (*fig.*), 180
transport 54, 56–7
triggers (Ca) 104, 276, 281
tubulin (Mg, Ca) 76 (*fig.*), 78 (*table*), 159,
 185 (*fig.*), 260, 471 (*fig.*)
tungsten
 analogies with molybdenum 432
 enzymes 431 (*table*)
 functions 431 (*table*)
 properties and requirements 431
 (*table*)
 reactions 432
 occurrence in biology 431
 role in evolution 432–4

uptake, *see individual elements* 23–70
 of amino-acids 236–7
 of chemical elements 24–6, 28–31, 154,
 503–4
 of haem 51
 of metals 28–58
 of non metals 58–68
uranium, comparison with molybdenum
 417–19
urease (Ni) 400, 406 (*table*), 409, 525

valinomycin (K) 39 (*fig.*)
vanadium 417–34
 amavadine 64, 429
 analogies with iron and copper 430
 binding constants 417–18
 biological forms and functions 428
 chemistry 427–8, 433
 haloperoxidases 430 (*table*), 525 (*table*)
 interference with phosphate metabolism
 537 (*fig.*)
 occurrence and functions 427–9
 role in evolution 428, 432–4, 525 (*table*)
 uptake 64
vesicles
 acidity and redox potential 86–7
 calcium in 288–90
 digestion and 289–90, 315, 508, 519
 exocytosis 281–2
 flow in cells 83–6
 formation of precipitates in 87, 473–5
 membranes 77–82
 positioning 76, 87, 510
 redox potential in 86
 release of contents 281–2
 secondary metabolism in 520–2
 solution conditions in 86–7
 zinc in 305, 315, 519
vitamin B12 (Co)
 in archaebacteria 400–1 (*fig.*), 404
 enzymes 407 (*table*)
 reactions 406–8
 synthesis 55–6
vitamin D controlled expression 516 (*table*),
 518 (*table*)
 and calcium 516, 540
 and copper and zinc 516

zinc 299–318
 analogies with calcium 301
 binding constants 26 (*table*), 29, 40 (*fig.*),
 373 (*fig.*)
 binding by metallothionein 305–6 (*fig.*),
 505 (*fig.*), 515–6

binding sites 304 (*fig.*), 306, 505 (*fig.*)
biochemistry 308–11, 508, 515–20
catalysis 301, 308–10
chemistry 12, 299–300
concentration
 in cells 520
 in extracellular fluids 303
 in sea water 303 (*table*), 412 (*fig.*)
and connective tissue 515–16, 524 (*fig.*)
controls 301, 314, 317, 515, 517–18,
 519–20
cross-linking action 305, 313, 316–17
and DNA synthesis 520
digestive processes 315–16, 508, 519
entatic states 311
exchange rates from proteins 307 (*table*),
 520
functional advantages 300–1, 520
homeostasis 312–3, 315, 508, 512, 515–16
insulin complexes 305, 315 (*table*), 519
kinetic character 103
as a Lewis acid 97–100, 299–301, 308–10,
 311
ligand in proteins 307–8, 505
location
 in cells 302–3, 312, 314
 in pollen and sperm 302 (*fig.*), 314
as a master hormone 314–15, 517–20
multiple binding 310 (*table*)
probes for zinc sites in enzymes 311, 318
proteins and enzymes 303–6, 312 (*table*),
 315–16, 515 (*table*)
 alkaline phosphatase 310 (*fig.*)
 alcohol dehydrogenase 305, 311
 carbonic anhydrase 309 (*fig.*)
 carboxypeptidase 315–16 (*fig.*), 519
 core proteins of chromosomes 302,
 508, 515 (*table*)
 fold characteristics 304 (*table*)–6
 nucleases 301
 phosphate esterases 310 (*fig.*)
 zinc-fingers 306, 505 (*fig.*), 518
relation to peptide hormones 314 (*table*),
 517–9, 524 (*fig.*)
regulatory role 302, 312–5, 517–20
role in evolution 316, 508, 515–20
role in filamentous structures 302, 312
role in redox enzymes 311
role in transcriptase factors 313, 314
selection 41–3, 51
in solid-state devices 317
zinc fingers 306, 505 (*fig.*), 518
 binding to DNA 306, 505 (*fig.*), 520
 binding sites for 505 (*fig.*)
 connections to homeostasis 518–20